OTHER VOLUMES IN THIS SERIES

OTHER VOLUMES IN THIS SERIES

Selected Tables in Mathematical Statistics

Volume 11

**Edited by the Institute
of Mathematical Statistics**

Coeditors
R. E. Odeh
University of Victoria

J. M. Davenport
Virginia Commonwealth University

Managing Editor
N. S. Pearson
Bell Communications Research

AMERICAN MATHEMATICAL SOCIETY
PROVIDENCE, RHODE ISLAND

1980 *Mathematics Subject Classification* (1985 *Revision*)
Primary 62Q05; Secondary 62F07, 62F25, 62H10, 62J15

International Standard Serial Number 0094-8837
International Standard Book Number 0-8218-1911-9
Library of Congress Card Number 74-6283

To Joan and Connie

Contents

Preface

This volume of mathematical tables has been prepared under the aegis of the Institute of Mathematical Statistics. The Institute of Mathematical Statistics is a professional society for mathematically oriented statisticians. The purpose of the Institute is to encourage the development, dissemination, and application of mathematical statistics. The Committee on Mathematical Tables of the Institute of Mathematical Statistics is responsible for preparing and editing this series of tables. The Institute of Mathematical Statistics has entered into an agreement with the American Mathematical Society to jointly publish this series of volumes. At the time of this writing, submissions for future volumes are being solicited. No set number has been established for this series. The editors will consider publishing as many volumes as are necessary to disseminate meritorious material.

Potential authors should consider the following rules when submitting material.

1. The manuscript must be prepared by the author in a form acceptable for photo-offset. The author should assume that nothing will be set in type although the editors reserve the right to make editorial changes. This includes both the introductory material and the tables. A computer tape of the tables will be required for the checking process *and* the final printing (assuming it is accepted). The authors should contact the editors prior to submission concerning the requirements for the computer tape.

2. While there are no fixed upper and lower limits on the length of tables, authors should be aware that the purpose of this series is to provide an outlet for tables of high quality and utility which are too long to be accepted by a technical journal but too short for separate publication in book form.

3. The author must, whenever applicable, include in his introduction the following:

   (a) He should give the formula used in the calculation, and the computational procedure (or algorithm) used to generate his tables. Generally speaking FORTRAN or ALGOL programs will not be included but the description of the algorithm used should be complete enough that such programs can be easily prepared.

(b) A recommendation for interpolation in the tables should be given. The author should give the number of figures of accuracy which can be obtained with linear (and higher degree) interpolation.

(c) Adequate references must be given.

(d) The author should give the accuracy of the table and his method of rounding.

(e) In considering possible formats for his tables, the author should attempt to give as much information as possible in as little space as possible. Generally speaking, critical values of a distribution convey more information than the distribution itself, but each case must be judged on its own merits. The text portion of the tables (including column headings, titles, etc.) must be proportional to the size 5–1/4″ by 8–1/4″. Tables may be printed proportional to the size 8–1/4″ by 5–1/4″ (i.e., turned sideways on the page) when absolutely necessary; but this should be avoided and every attempt made to orient the tables in a vertical manner.

(f) The table should adequately cover the entire function. Asymptotic results should be given and tabulated if informative.

(g) Examples of the use of the tables must be included.

4. The author should submit as accurate a tabulation as he/she can. The table will be checked before publication, and any excess of errors will be considered grounds for rejection. The manuscript introduction will be subjected to refereeing. Since an inadequate introduction may lead to rejection, the author should strive for an informative manuscript, which not only establishes a need for the tables, but also explains in detail how to use the tables.

5. Authors having tables they wish to submit should send two copies to:

> Dr. Robert E. Odeh, Coeditor
> Department of Mathematics
> Univerity of Victoria
> Victoria, B. C., Canada V8W 2Y2

At the same time, a third copy should be sent to:

> James M. Davenport, Director
> Institute of Statistics
> Virginia Commonwealth University
> Richmond, Virginia 23284-2002

Additional copies may be required, as needed for the editorial process. After the editorial process is complete, a camera-ready copy must be prepared for the publisher.

Authors should check several current issues of *The Institute of Mathematical Statistics Bulletin* and *The AMSTAT News* for any up-to-date announcements about submissions to this series.

Acknowledgments

The tables included in the present volume were checked at the Universiy of Victoria. Dr. R. E. Odeh arranged for, and directed this checking with the assistance of Mr. Bruce Wilson. The editors and the Institute of Mathematical Statistics wish to express their great appreciation for this invaluable assistance. So many other people have contributed to the instigation and preparation of this volume that it would be impossible to record their names here. To all these people, who will remain anonymous, the editors and the Institute also wish to express their thanks.

Selected Tables in Mathematical Statistics
Volume 11, 1988

PERCENTAGE POINTS OF MULTIVARIATE STUDENT t DISTRIBUTIONS

Robert E. Bechhofer
Cornell University

and

Charles W. Dunnett
McMaster University

ABSTRACT

This volume presents tables of the one-sided and two-sided upper equi-coordinate percentage points of the central multivariate Student t distribution in which there is a common variance estimate in the denominators of the variates, and the numerators are equicorrelated. The parameters of the distribution are the number of variates, p; the degrees of freedom associated with the variance estimate, ν; and the common correlation coefficient, ρ. The 80, 90, 95 and 99 percentage points are tabulated to 5 decimal place accuracy for $p = 2(1)16(2)20$; $\nu = 2(1)30(5)50,60(20)120,200,\infty$; $\rho = 0.0(0.1)0.9, 1/(1+\sqrt{p})$.

In addition, two smaller tables are given in which the multivariate Student t variates have a block correlation structure with p_1 variates in the first block and p_2 in the second block; the variates within each block have $\rho = 0.5$ and the variates in different blocks have $\rho = 0$. These

---

Tables received by the editors May 1985. Text received July 1986.
AMS (MOS) Subject Classification (1980): Primary 62–Q05; Secondary 62F07, 62F25, 62H10, 62J15.

This work was partially supported at Cornell University by the U.S. Army Research Office through the Mathematical Sciences Institute of Cornell University and by U.S. Army Research Office Contract DAAL03-86-K-0046, and at McMaster University by the Natural Sciences and Engineering Research Council of Canada.

tables give one-sided upper 80, 90 and 95 percentage points tabulated to 5 decimal place accuracy for $p_1 = 1(1)4$ and $p_2 = p_1(1)6,9$; $v = 5(1)30(5)50,60(20)120,200,\infty$. The first table gives equicoordinate percentage points. The second gives percentage points of a form which is non-equicoordinate (except when $p_1 = p_2$), these being appropriate for different classes of problems.

The tables have applications in many statistical settings including selection among normal means using either the indifference-zone or subset approach, and in multiple comparisons involving contrasts among means. These and other applications are described in detail, and examples are given of the uses of the tables.

1. INTRODUCTION AND SUMMARY

The central p-variate Student t distribution in which the variates have a common variance estimate in their denominators arises in many statistical settings. These include selection problems involving p+1 normal population means and multiple comparisons problems involving p contrasts among a set of normal population means. In addition to p, the parameters of the distribution are the number of degrees of freedom (v) associated with the variance estimate, and the correlations (ρ_{ij}) between the numerators of the variates. The distribution reduces to the usual Student t distribution when p = 1, and to the standard p-variate normal distribution if $v = \infty$.

In this volume we provide one-sided and two-sided upper equicoordinate percentage points for the special case in which all of the correlation coefficients are equal to a common $\rho \geq 0$. The 80, 90, 95 and 99 percentage points are tabulated for a large number of (p, v, ρ) combinations, and are presented as Tables A (one-sided) and B (two-sided). In addition, we provide two smaller sets of tables, Tables C and D, which assume that the variates have a certain block correlation structure. These are for use in the same types of problems as those requiring Table A except that two-factor rather than single-factor experiments are involved.

The entries in these tables have been computed to 5-decimal place accuracy. Our tables are more accurate and considerably more comprehensive than previously published tables of their type. Methods of interpolation in the tables are described which extend their domain of usefulness.

The multivariate Student t distribution is defined in Section 2, and the particular correlation structures assumed in the tables are described. In

Section 3, the equicoordinate percentage points are defined and the integral
expressions needed to compute their values are given.   Section 4 contains a
review of the earlier tables of the percentage points (and the probability
integral) of this multivariate t and multivariate normal; this is done
chronologically for the percentage points in our Figure 1 (pages 64–67) with a
listing of the characteristics of the tables as well as the particular values
of the parameters on which the tables were based.   Some related tables are
also described.   Section 5 gives the methods of computation used in the
construction of the tables provided in this volume.   The accuracy of the
entries is compared with those in earlier tables.   Methods of interpolating in
the tables are given.   In Section 6 we describe many of the most important
applications for which the tables are appropriate, and indicate how the tables
can be used in these applications.   Finally, in Section 7 we give some
numerical examples of the applications of these tables.

2.   PRELIMINARIES

The multivariate Student t distribution which was introduced by Dunnett and Sobel (1954) and Cornish (1954) is defined as follows: Let $(Z_1, Z_2, \ldots, Z_p)$ have a non-singular joint p-variate standard normal distribution $(E\{Z_i\} = 0$, $\mathrm{Var}\{Z_i\} = 1)$ with $\mathrm{Corr}\{Z_i, Z_j\} = \rho_{ij}$ $(i \neq j$, $1 \leq i, j \leq p)$, and let V be distributed independently of the Z_i as a χ^2 random variable with v d.f.   If we define $T_i = Z_i / \sqrt{V/v}$ $(1 \leq i \leq p)$, then T_i is a Student t variate, and $(T_1, T_2, \ldots, T_p)$ is said to have a p-variate central Student t distribution with correlation matrix $\{\rho_{ij}\}$ and v d.f.

The joint density function of $(T_1, T_2, \ldots, T_p)$ is given by

$$f_v(t_1, t_2, \ldots, t_p; \underline{P}) = \frac{\Gamma((p+v)/2)}{(v\pi)^{p/2}\Gamma(v/2)} |\underline{P}|^{-1/2}(1 + \underline{t}'\underline{P}^{-1}\underline{t}/v)^{-(p+v)/2} \qquad (2.1)$$

where $\underline{t}' = (t_1, t_2, \ldots, t_p)$ and $\underline{P} = \{\rho_{ij}\}$. When $p = 1$ the density function (2.1) reduces to that of Student's t with v d.f. while for $v = \infty$ it becomes the p-variate standard normal density function with correlation matrix $\underline{P}$. Generalizations and specializations of (2.1), and their properties, are discussed in great detail in Chapter 37 of Johnson and Kotz (1972).

The distribution specified by the density function (2.1) underlies many statistical procedures devised to solve broad classes of multiple-decision problems, particularly those arising in the areas of "selection and ranking," "multiple comparisons," and other types of "simultaneous inference." Some of the more important of these and other problems, and the associated procedures are described in Section 6.

Each procedure requires that the correlation matrix have a specific form with particular values for the elements ρ_{ij} of $\underline{P}$. The simplest correlation structure is one in which there is a common correlation among the p variates, i.e.,

$$p_{ij} = \left\{ \begin{array}{ll} 1 & \text{if } i = j \\ \rho & \text{if } i \neq j \end{array} \right. . \qquad (2.2a)$$

Another is one in which the p variates can be divided into two sets of variates (p_1 in the first set and p_2 in the second with $p_1 + p_2 = p$) with common correlations within the three blocks generated by the two sets, i.e.,

$$p_{ij} = \left\{ \begin{array}{ll} 1 & \text{if } i = j \\ \rho_1 & \text{if } i \neq j, \ 1 \leq i, j \leq p_1 \\ \rho_2 & \text{if } i \neq j, \ p_1 + 1 \leq i, j \leq p \\ \rho_3 & \text{if } i \neq j, \ 1 \leq i \leq p_1, \ p_1 + 1 \leq j \leq p \\ & \text{or } i \neq j, \ p_1 + 1 \leq i \leq p, \ 1 \leq j \leq p_1 \ . \end{array} \right. \qquad (2.2b)$$

All of the tables in the present volume assume the structure for $\underline{P}$ given by (2.2a) or (2.2b); (2.2a) can be considered as a special case of (2.2b) with either p_1 or p_2 equal to zero.

We will be concerned with the following special cases:

Case I: This is (2.2a) with the following three particular values of ρ:

   a) $\rho = 0$ (i.e., the Z_i ($1 \leq i \leq p$) are independently distributed)

   b) $\rho = 1/2$

   c) $\rho = \rho* > 0$ where $\rho*$ is a particular value of ρ which depends on the problem under consideration.

Case II: This is (2.2b) with $\rho_3 = 0$ and $\rho_1 = \rho_2$. Corresponding to Case Ib, we are interested in $\rho_1 = \rho_2 = 1/2$.

3.   EQUICOORDINATE PERCENTAGE POINTS OF THE MULTIVARIATE t

3.1   One-sided Percentage Points

The one-sided upper equicoordinate $100(1-\alpha)$ percentage point of the multivariate t is the solution in $g = g(p,v,\underline{P},\alpha)$ of the equation

$$P\{T_i \leq g \ (1 \leq i \leq p)\} = P\{ \max_{1 \leq i \leq p} T_i \leq g\} \tag{3.1}$$

$$= \int_{-\infty}^{g} \cdots \int_{-\infty}^{g} f_v(t_1, \ldots, t_p; \underline{P}) dt_1 \cdots dt_p = 1-\alpha.$$

The probability $P\{T_i \geq -g \ (1 \leq i \leq p)\}$ which yields the one-sided lower equicoordinate $100(1-\alpha)$ percentage point is defined analogously with $(-\infty,g)$ in the limits of integration of (3.1) replaced by $(-g,\infty)$.

For Case I it has been shown (see, e.g., Dunnett and Sobel (1955) or Gupta and Sobel (1957)) that g can also be expressed as the solution of the equation

$$\int_0^\infty \left[\int_{-\infty}^\infty \left\{ \Phi\left[\frac{x\sqrt{\rho} + gy}{\sqrt{1-\rho}} \right] \right\}^p d\Phi(x) \right] q_v(y) dy = 1-\alpha. \tag{3.2}$$

Here $\Phi(\bullet)$ is the cdf of a standard normal r.v., and $q_v(y)$ is the density function of $x_v/\sqrt{v}$. The form (3.2) is much more convenient than (3.1) for computational purposes. The probability integrals of multivariate normal and multivariate t distributions are discussed in great detail in Gupta (1963). Tables A.1 to A.4 give values of g for selected values of p,v,ρ,α.

3.2   Two-sided Percentage Points

The two-sided upper equicoordinate $100(1-\alpha)$ percentage point of the multivariate t is the solution in $h = h(p,v,\underline{P},\alpha)$ of the equation

$$P\{-h \leq T_i \leq h \ \ (1 \leq i \leq p)\} = P\{|T_i| \leq h \ \ (1 \leq i \leq p)\} \tag{3.3}$$

$$= P\{ \max_{1 \leq i \leq p} |T_i| \leq h \}$$

$$= \int_{-h}^{h} \cdots \int_{-h}^{h} f_v(t_1, \ldots, t_p; \underline{P}) dt_1 \ldots dt_p = 1-\alpha.$$

For Case I, analogously to (3.2), h can be found (see equation (7.2) in Gupta and Sobel (1957) or Hahn and Hendrickson (1971)) as the solution of the equation

$$\int_0^\infty \left[\int_{-\infty}^{+\infty} \left\{ \Phi\left[\frac{x\sqrt{\rho} + hy}{\sqrt{1-\rho}} \right] - \Phi\left[\frac{x\sqrt{\rho} - hy}{\sqrt{1-\rho}} \right] \right\} d\Phi(x) \right]^p q_v(y) dy = 1-\alpha. \tag{3.4}$$

Tables B.1 to B.4 give values of h for selected values of p, v, ρ, α. Note: Throughout this volume g will always refer to one-sided percentage points as in (3.1) and h will refer to two-sided percentage points as in (3.3).

3.3   One-sided Percentage Points for Case II

For the case of two sets of variates, p_1 in the first set and p_2 in the second with $p_1 + p_2 = p$, the variates within each set having common correlation ρ and the variates in different sets having $\rho = 0$, we denote by $g_1 = g_1((p_1, p_2), v, \rho, \alpha)$ the solution of the equation

$$\int_0^\infty \prod_{i=1}^2 \left[\int_{-\infty}^{+\infty} \left\{ \Phi\left[\frac{x_i\sqrt{\rho} + g_1 y}{\sqrt{1-\rho}} \right] \right\} d\Phi(x_i) \right]^{P_i} q_v(y) dy = 1-\alpha. \tag{3.5}$$

Thus g_1 is the one-sided <u>upper</u> equicoordinate $100(1-\alpha)$ percentage point for Case II. Note that (3.5) reduces to (3.2) if $p_1 = p$ and $p_2 = 0$. Tables C.1 to C.3 give values of g_1 for selected values of $(p_1, p_2), v, \alpha$ when $\rho = 1/2$.

For the case of two sets of variates with the same correlation structure as described above we denote by $g_2 = g_2((p_1, p_2), v, \rho, \alpha)$ the solution of the equation

$$\int_0^\infty \frac{2}{\Pi} \left[\prod_{i=1}^\infty \left\{ \int_{-\infty}^{+\infty} \Phi \left[\frac{x_i\sqrt{\rho} + g_2 y\sqrt{(p_2+1)/(p_i+1)}}{\sqrt{1-\rho}} \right] \right\}^{p_i} d\Phi(x_i) \right] q_v(y) dy = 1-\alpha. \quad (3.6)$$

We point out that g_2 is <u>not</u> an equicoordinate percentage point unless $p_1 = p_2$. Tables D.1 to D.3 give values of g_2 for selected values of $(p_1, p_2), v, \alpha$ when $\rho = 1/2$. Note that when $p_1 = p_2 = a$ (say), then both (3.5) and (3.6) are given by the same expression, and $g_1 = g_2$ is a function only of (a, v, ρ, α).

The constants g_1 of equation (3.5) are necessary to implement procedures described in Sections 6.1.2 and 6.2.2 while g_2 of equation (3.6) is used for the procedure described in Section 6.3.2.

Remark 3.1: One might conceivably be interested in setting the probabilities $P\{T_i \geq g* \ (1 \leq i \leq p)\} = \alpha$ and/or $P\{|T_i| \geq h* \ (1 \leq i \leq p)\} = \alpha$; here $g*$ and $h*$ are the one-sided and two-sided <u>lower</u> equicoordinate 100α percentage points, respectively, of the multivariate t. We have elected not to tabulate these percentage points since they have at most limited practical application.

4. EARLIER TABLES OF THE PERCENTAGE POINTS AND OF THE PROBABILITY INTEGRAL
   OF THE MULTIVARIATE   t   AND MULTIVARIATE NORMAL DISTRIBUTIONS

4.1   Percentage Points

Over the years many authors have prepared tables (one-sided and/or

two-sided) of the percentage points of the multivariate Student   t   and the

multivariate normal distribution.   Most of the tables were calculated to

provide constants necessary to implement selection and ranking procedures, or

procedures for pairwise multiple comparisons among several treatments or for

comparing several test treatments with a control treatment.   In our Figure 1

(pages 64-67) we give for Case I many of the earlier articles containing such

percentage points; these are listed in the order of their publication date

along with the authors and salient features of these tables.   Although there

is much duplication, in general the tables differ in terms of the range of

arguments covered, and the number of decimal places of the entries.   In

Section 5.3 the accuracy of the entries in these earlier tables is discussed.

In addition to the tables listed in our Figure 1, several authors have

prepared tables of percentage points of multivariate Student t distributions

where the correlation structure differs from our Case I.

Freeman, Kuzmack and Maurice (1967) tabulated percentage points   g   of

(3.1) to three decimal places for   p = 2   and two decimal places for

p = 3,4,5   for the correlation structure $\rho_{ij} = -1/2$ for $|i-j| = 1$, $\rho_{ij} = 0$

for $|i-j| > 1$ $(1 \leq i,j \leq p)$,   1-α = 0.95   and $v = (p+1)s$ for

s = 9(10)99,199,499.   Freeman and Kuzmack (1972) tabulated percentage points

of the same integral with the same correlation structure for   p = 5(2)9(5)29,

1-α = 0.90,0.95,0.99,   and $v = (p+1)s$ for   s = 9,19,49,99,499;   the g-values

were given to two decimal places but since they were computed using

Monte Carlo sampling their accuracy is questionable.

Tong (1969) studied in depth a special case of the p-variate Student t density function having correlation structure given by (2.2b) with $\rho_1 = \rho_2 = 1/2$, $\rho_3 = -1/2$. In particular, he computed the one-sided equicoordinate $100(1-\alpha)$ percentage points (3.1) for the correlation matrix with $\rho = 1/2$ for $i \neq j$ and $1 \leq i,j \leq m$ or $m < i,j \leq p$, $\rho = -1/2$ for $1 \leq i \leq m$ and $m < j \leq p$ where $m = p/2$ if p is even, $m = (p+1)/2$ if p is odd. In his Table 1 he gives g-values to seven decimal places for $\upsilon = \infty$ (i.e., for the p-variate standard normal distribution), $p = 1(1)10(2)20$ and $1-\alpha = 0.50, 0.75, 0.90, 0.95, 0.975, 0.99$. In his Table 2 he gives g-values to five decimal places for $\upsilon = 5(1)10(2)20(4)60(30)120$ (i.e., for the p-variate Student t distribution), $p = 2(1)6(2)12(4)20$ and $1-\alpha = 0.50, 0.75, 0.90, 0.95, 0.975, 0.99$.

Trout and Chow (1972) calculated two-sided non-equicoordinate $100(1-\alpha)$ percentage points of the trivariate Student t-distribution (2.1) with arbitrary non-singular correlation matrix

$$\underline{P} = \left\{ \begin{matrix} 1 & \rho_{12} & \rho_{13} \\ \rho_{12} & 1 & \rho_{23} \\ \rho_{13} & \rho_{23} & 1 \end{matrix} \right\} . \tag{4.1}$$

Their two-sided percentage points are of the more general form

$$\int_{-D}^{D} \int_{-AD}^{AD} \int_{-BD}^{BD} f_\upsilon(t_1, t_2, t_3; \underline{P}) dt_1 dt_2 dt_3 = 1-\alpha. \tag{4.2}$$

They tabulated D with $1-\alpha = 0.95$ to two decimal places (three significant figures) for $\upsilon = 5(1)9(2)29$, $A = 0.5(0.1)1.5$, $B = 0.5(0.1)1.5$, for $\rho_{12} = \rho_{13} = \rho_{23} = 0$ and for all combinations of $\rho_{ij} = 0.1, 0.5, 0.9$ ($i \neq j$, $1 \leq i,j \leq 3$) except those with two of the ρ_{ij} equal to 0.9.

Also, Dutt, Mattes and Tao (1975) calculated two-sided equicoordinate $100(1-\alpha)$ percentage points of the trivariate Student t distribution where the correlations $(\rho_{12}, \rho_{13}, \rho_{23})$ for given (n_0, n_1, n_2, n_3) are of the form

$$\rho_{ij} = \{(1+n_0/n_i)(1+n_0/n_j)\}^{-1/2}$$

for $(i \neq j, 1 \leq i, j \leq 3)$ with associated d.f. $v = \Sigma_{i=0}^{3} (n_i - 1)$. Their tables were designed specifically to be used for two-sided multiple comparisons between a control and three treatments (see our Section 6.1.1) in which the numbers of observations in the four groups n_i $(0 \leq i \leq 3)$ may be unequal. The percentage points are tabulated for $3 \leq n_i \leq 12$, $0 \leq i \leq 3$, and are given to four decimal places for $1-\alpha = 0.95$ and to five decimal places for $1-\alpha = 0.99$.

4.2 Probability Integrals

4.2.1 Multivariate Normal

In addition to percentage points of multivariate normal and multivariate t distributions, there have been several extensive tabulations of multivariate normal probability integrals.

The National Bureau of Standards (1959) computed the probabilities

$$L(g_1^*, g_2^*; \rho) = \int_{g_1^*}^{\infty} \int_{g_2^*}^{\infty} f(x_1, x_2; \rho) dx_1 dx_2 \tag{4.3}$$

where $f(x_1, x_2; \rho)$ is the bivariate standard normal density function with correlation ρ. These probabilities were given to 6 decimal places for $g_1^*, g_2^* = 0(0.1)4$ and $\rho = 0(0.05)0.95(0.01)1$ and to 7 decimal places for $g_1^*, g_2^* = 0(0.1)g_{1n}^*, g_{2n}^*$; $-\rho = 0(0.05)0.95(0.01)1$ where $L(g_{1n}^*, g_{2n}^*; -\rho) \leq 10^{-7}/2$ if g_{1n}^* and g_{2n}^* are both less than 4.

Teichroew (1955) tabulated the p-variate normal probabilities

$$s \int_{-\infty}^{+\infty} \left\{ \Phi\left[\frac{x\sqrt{\rho} + g}{\sqrt{1-\rho}} \right] \right\}^{p-s+1} \{1-\Phi(x)\}^{s-1} d\Phi(x) \qquad (4.4)$$

to 6 decimal places for the following (p,s) pairs: $p = 1(1)9$, $s = 1(1)m$ where $m = p/2$ for p even and $m = (p+1)/2$ for m odd; $p = 10$, $s = 2(1)5$; $p = 11$, $s = 3(1)5$; $p = 12$, $s = 4,5$; $p = 13$, $s = 5$ for $g = 0(0.01)6.09$ and $\rho = 1/2$. These probabilities were required for equation (20) of Bechhofer (1954) where the problem was to select the s out of $p+1$ populations which have the largest population means.

Dunnett (1960) calculated (4.3) for $\rho = \pm 1/\sqrt{2}$ to six decimal places for $g_1^* = 0(0.1)4.0$ and $g_2^* = 0(0.1\sqrt{2})3.1\sqrt{2}$. In addition, Dunnett and Lamm (1960) tabulated the multivariate normal probability integral given by (4.4) with $s = 1$ for the special case $\rho = 1/3$ to 6 decimal places; for $p = 1(1)10$, their tables cover $g\sqrt{3} = 0(0.1)7.0$ and in addition, for $p = 1(1)10,13,18$ they cover $g\sqrt{3} = 1.50(0.01)2.10$.

Gupta (1963) calculated (4.4) to 5 decimal places for $s = 1$, $p = 1(1)12$, $g = -3.5(0.1)3.5$; $\rho = 0.1,0.125,0.2,0.25,0.3,1/3,0.375,0.4,0.5,0.6,0.625,2/3$, $0.7,0.75,0.8,0.875,0.9$. Milton (1963) also tabulated this same integral to 8 decimal places for $s = 1$, $p = 2(1)9(5)24$, $g = 0(0.05)5.15$ and $\rho = 0(0.05)1.00$.

Carroll and Gupta (1977) considered the problem of ordering $p+1$ normal populations in terms of their population means, a problem which had been posed earlier by Bechhofer (1954). They calculated the probability integral

$$\int_{-\infty}^{g\sqrt{2}} \int_{-\infty}^{g\sqrt{2}} \cdots \int_{-\infty}^{g\sqrt{2}} f(x_1, x_2, \ldots, x_p; \underline{P}) dx_1 dx_2 \ldots dx_p \qquad (4.5)$$

where $f(x_1, x_2, \ldots, x_p; \underline{P})$ is a p-variate standard normal density function with correlation matrix $\underline{P} = \{\rho_{ij}\}$ with

$$\rho_{ij} = \begin{cases} 1 & \text{for} \quad i = j \\ -1/2 & \text{for} \quad |i-j| = 1 \quad (1 \leq i, j \leq p) \\ 0 & \text{for} \quad |i-j| > 1. \end{cases} \tag{4.6}$$

The probabilities are given to three decimal places for $p = 1(1)9$ and $g\sqrt{2} = 0.0(0.1)4.4$.

4.2.2 Bivariate t

Dunnett and Sobel (1954) tabulated the bivariate Student t probabilities (3.1) for $\rho = 1/2$ and $\rho = -1/2$. These are given to five decimal places in their Tables 1 and 2, respectively, for $g = 0.00(0.25)2.50(0.50)10.00$ and $v = 1(1)30(3)60(15)120, 150, 300, 600, \infty$.

Carolis and Gori (1967) tabulated the bivariate Student $|t|$ probabilities (3.3) to five decimal places for $\rho = -0.75(0.25)0.75$ and $h = 0.2(0.2)6(1)50(5)100(10)200(25)300(50)500(100)900$ for $v = 1$; $h = 0.2(0.2)6(1)41$ for $v = 2$; $h = 0.2(0.2)6(1)16$ for $v = 3$; $h = 0.2(0.2)6.0(0.1)11$ for $v = 4$; $h = 0.2(0.2)6.0(0.1)8$ for $v = 5$; $h = 0.2(0.2)7.0$ for $v = 6$ and 7; $v = 0.2(0.2)5.8$ for $v = 8$; $h = 0.2(0.2)5.4$ for $v = 9$; to 5.2 for $v = 10$; to 5.0 for $v = 11$; to 4.8 for $v = 12$; to 4.6 for $v = 13, 14, 15$; to 4.4 for $v = 16, 17, 18$; to 4.2 for $v = 19(1)24$; to 4.0 for $v = 25(1)38$; to 3.8 for $v = 39(1)60$; to 3.6 for $v = \infty$.

Siotani (1964) tabulated the bivariate Student t probabilities $P\{|T_1| \geq h*, |T_2| \geq h*\}$ for $|\rho| = 0.0(0.1)0.90, 0.95,$

$v = 10(2)50(5)90, 100, 120, 150, 200, \infty$ and $h* = 2.0(0.5)4.5$. The probabilities were computed to five, six or seven decimal places for $h* = 2.0(0.5)4.5$.

Krishnaiah, Armitage and Breiter (1969a) tabulated to six decimal places the bivariate Student t probabilities (3.1) for $\rho = 0.0(0.1)0.9$ and for $-\rho = 0.0(0.1)0.9$ in their Tables I and II, respectively, for $g = 1.0(0.1)5.5$ and $v = 5(1)35$. Krishnaiah, Armitage and Breiter (1969b) also tabulated to six decimal places the bivariate Student $|t|$ probabilities (3.3) for $|\rho| = 0.0(0.1)0.9$ for $h = 1.0(0.1)5.5$ and $v = 5(1)35$.

4.2.3 Multivariate t

Dunn, Kronmal and Yee (1968) tabulated the multivariate Student t probabilities (3.3) to four decimal places for $p = 2(2)20$, $\rho = 0.0(0.1)0.9$ and $-1/(p-1)$, $h = 0.2(0.2)6.0$ and $v = 4(2)12(4)24, 30, \infty$. The probabilities were computed using Monte Carlo sampling so their accuracy is questionable.

Dutt (1975) tabulated the multivariate Student t probabilities (3.1) to six decimal places for $p = 3$ with $\rho_{12} = 0.3$, $\rho_{13} = 0.5$, $\rho_{23} = 0.7$ and $\rho_{12} = 0.1$, $\rho_{13} = 0.3$, $\rho_{23} = 0.5$, and for $p = 4$ with $\rho_{12} = 0.05$, $\rho_{13} = 0.10$, $\rho_{14} = 0.15$, $\rho_{23} = 0.25$, $\rho_{24} = 0.60$, $\rho_{34} = 0.80$ and $\rho_{12} = 0.25$, $\rho_{13} = 0.35$, $\rho_{14} = 0.50$, $\rho_{23} = 0.60$, $\rho_{24} = 0.65$, $\rho_{34} = 0.70$, all for $g = 0.0(0.5)2.0(1.0)4.0$ and $v = 8(4)40, \infty$.

5.   CONSTRUCTION OF THE PRESENT TABLES

The purpose of these tables is to provide solutions in the variables denoted by g, h, g_1 and g_2 that appear in equations (3.2), (3.4), (3.5) and (3.6), respectively. All computations were carried out on a CDC CYBER 170 computer at McMaster University. Since this computer has a 32-bit word size, it was not necessary to use double-precision arithmetic to achieve our desired degree of accuracy. Our objective in computing the tables was to achieve accuracy to the 5th decimal place in the tabulated values.

5.1 Details of Computations

In order to evaluate the integrals appearing on the left-hand side of the equations, Gaussian quadrature formulas were used (see Stroud and Secrest (1966)). Three different Gaussian formulas were employed, namely,

Gauss-Hermite:

$$\int_{-\infty}^{+\infty} \exp(-x^2)f(x)dx \cong \sum_{i=1}^{n} w_i f(x_i) \tag{5.1}$$

Gauss-Laguerre:

$$\int_{0}^{+\infty} \exp(-x)f(x)dx \cong \sum_{i=1}^{n} w_i f(x_i) \tag{5.2}$$

Gauss-Legendre:

$$\int_{-1}^{1} f(x)dx \cong \sum_{i=1}^{n} w_i f(x_i) \tag{5.3}$$

where in each case the x_i are the zeros of the relevant polynomials of degree n , and the w_i are the corresponding weight functions. The values of n employed were 120 for the Hermite, 68 for the Laguerre and 96 for the Legendre formulas. These rather large values were chosen in order to achieve a high level of accuracy in the computations. To evaluate the inner integrals in the equations, (5.1) was used. To evaluate the outer integrals, (5.2) was

used for degrees of freedom values $v \geq 10$ but was replaced by (5.3) for
$v < 10$ since the latter was found to be more accurate for the smaller values
of v. When (5.3) was used, the variable of integration x was transformed
to z = (x-1)/(x+1) in order to obtain the required limits of integration.
The numerical values of the zeros and weights of the polynomials for (5.1),
(5.2) and (5.3) were obtained from Tables 5, 6 and 1, respectively, of Stroud
and Secrest (1966); the values required for (5.3) are also available in
Table 25.4 of Abramowitz and Stegun (1964).

For these computations the special cases $\rho = 0$ and $v = \infty$ were treated
differently. For $\rho = 0$ the inner integral reduces to a product of
univariate normal cdf's which obviates the need to evaluate it by quadrature,
and for $v = \infty$ the outer integration is not needed. An algorithm to compute
the standard normal distribution function is required in the computation of
the integrands. For computers having the error function ERF(X) built in,
the cumulative normal from 0 to X can be obtained to machine accuracy as
ERF(X/SQRT(2))/2. We used the IMSL (1982) routine MDNOR. Alternatively, the
function ALNORM of Hill (1973) could be used.

The computing scheme used was to evaluate the left-hand side at three
successive values of the argument, differing by 0.05, which bracketed the
desired value of $1-\alpha$, and then to use inverse quadratic interpolation to
determine a first approximation to the required value. This was followed by
adjustments as necessary to find the solution which achieved the desired value
of $1-\alpha$. This scheme was followed for $v = 2(1)5,10,20,40$ and ∞. For other
values of v, a first approximation was obtained by inverse quadratic
interpolation on $1/v$ from the previously calculated values, and this was
followed by adjustments as before. All values were calculated to the nearest
figure in the fifth decimal place.

5.2 Accuracy of the Tables

In order to check the accuracy of the calculations based on the quadrature formulas, a computing program was developed to compute the integrals by numerical integration using Simpson's rule. To evaluate the inner integral, equal-width increments in the variable of integration were used; the increments were chosen to be proportional to $\sqrt{1-\rho}$ in order to achieve comparable levels of accuracy for all values of ρ. The numerical integration was continued until the variable of integration was large enough for the value of the integrand to become less than 1.0×10^{-12}, at which point the error from neglecting larger values of the variable was considered to be negligible.

To evaluate the outer integral by numerical integration, equal-width Simpson's rule increments were used; the widths were proportional to $v^{-1/2}$ in order to achieve comparable levels of accuracy for all values of v. The numerical integration was started at the modal value of the density function and continued in both directions until either the value of the integrand became less than 1.0×10^{-12} or the value of the variable of integration became zero.

For both the inner and outer integrals, the Simpson's rule increment was halved until no further change in the tenth decimal place of the value of the integral was observed.

Entries in the tables were checked using the numerical integration program. Some discrepancies were found which were mostly in the 5th decimal places (in a few cases the 4th decimal place was also affected) and were limited to small degrees of freedom ($v \leq 4$) and large values of ρ. Corrections were made in the entries in the tables if the errors were greater than unity in the 5th decimal place.

Another check was carried out on the entries in the tables for the special case p = 2. Here exact series expressions are available for the probability integrals (see Dunnett and Sobel (1954)). All values for p = 2 in Tables A and B were checked using computing programs employing these exact series expressions. The discrepancies found occurred only for $v \leq 4$, and corrections were made. In addition, solutions for p = 2, $v = 1$ were computed. (The case $v = 1$ had been omitted from the quadrature computations because these methods were found to be insufficiently accurate for this value of v.)

Values in these tables were independently checked by R.E. Odeh using the NAG (1983) subroutine D01DAF for accurate adaptive numerical double integration. The values returned by this routine were accurate to at least 10 decimal places.

5.3 Comparisons with Other Published Tables

Since the tables in the present volume were computed to achieve a high degree of accuracy, a comparison of their values with those of other published tables was carried out to check the accuracy achieved in the earlier tables.

The only tables computed to 5 decimal places and which are comparable to the present ones are the two-sided tables of Carolis and Gori (1967) for the case p = 2 and the one-sided tables of Gupta et al. (1985). Comparison of our values in Table B with those of Carolis and Gori (1967) revealed that the latter were inaccurate beyond the 2nd decimal place for all values that could be checked, even for $\rho = 0$ and $v = \infty$. Comparisons of our values in Table A with Gupta et al. (1985) showed agreement between the two sets of tables for $v = \infty$ and for the larger values of v, but there were discrepancies for values of $v \leq 30$ which in some cases affected even the 2nd decimal place for $v = 15$. It should be pointed out that Gupta et al. used asymptotic (in terms of $1/v$) expressions in the computation of their tables which therefore may be expected to be accurate for large values of v but are evidently insufficiently accurate to achieve the 5-decimal place accuracy to which they are tabulated for all values of v.

Comparisons for $v = \infty$ with the 6-decimal place values for the multivariate normal distribution tabulated by Milton (1963) showed agreement between our Table A and his to 5 decimal places. Similarly, agreement was obtained between our Table B and the 4-decimal place tables of Odeh (1982).

Comparisons of the entries in our Table B with those in Hahn and Hendrickson (1971) which were calculated to 3 decimal places showed agreement to within unity in the 3rd decimal place except for some larger discrepancies affecting the 3rd decimal place for $v = 3$ with their $\gamma = 1-\alpha = 0.99$ and their $k = p \geq 10$. In addition, discrepancies of 2 in their 2nd decimal place values were noted in two cases (their values for $\rho = 0.0$, $\gamma = 0.99$, $v = 5$ and 6).

Comparisons with the 2-decimal place values of Krishnaiah and Armitage (1966) showed agreement between our entries and theirs to the two decimal places to which the latter were computed.

5.4 Interpolation in the Tables

Users of the present tables may wish to interpolate in these tables for other values of v and, in the case of Tables A and B, other values of ρ.

For interpolation on v, use may be made of the fact (empirically observed) that the values are approximately linear in $1/v$. Thus we recommend linear interpolation on $1/v$ or, for greater accuracy, quadratic interpolation on $1/v$. The particular values of v included in the tables were chosen to facilitate quadratic interpolation (e.g., the values for $v = 100$, 200 and ∞ or for $v = 40$, 80 and ∞ can be used as equally-spaced interpolation values on the reciprocal d.f. scale for values of υ greater than 100 or greater than 40, respectively.)

For interpolating with respect to ρ, depending on the accuracy required, linear or quadratic interpolation on ρ can be used. For somewhat greater accuracy, interpolation with respect to $(1-\rho)^{-1}$ instead of ρ is suggested (see Dunnett (1964)).

Interpolation with respect to $1-\alpha$ should be done on the $\log \alpha$ scale and with respect to p on the $\log p$ scale.

5.5 Computer Programs

Although various authors have developed programs for the purpose of computing tables, few have been published. For computing the multivariate normal probability integral with arbitrary correlation structure, the best available programs are due to Donnelly (1973) for the bivariate normal and Schervish (1984) for the multivariate normal. Bohrer, Schervish and Sheft (1982) have published a program that computes non-central multivariate t probability integrals over arbitrary rectangular regions, but restricted to the zero correlation case. A program for multivariate t which computes the probability integral in equation (3.3) for the equal correlation coefficient case has been published by Dunlap, Marx and Agamy (1981). A program which computes the probability integral for the multivariate normal over arbitrary rectangular regions for the correlation structure $\rho_{ij} = b_i b_j$ has been developed by Dunnett (1986).

6. APPLICATIONS OF THE TABLES

6.1 Multiple Comparisons With a Control

6.1.1 Single-factor Experiments

There are available $p+1$ populations Π_i $(0 \leq i \leq p)$ where Π_0 denotes a control population and $\Pi_1, \ldots, \Pi_p$ denote $p \geq 1$ test populations. We assume the one-factor fixed effects ANOVA model for qualitative variables (i.e., no structure among the treatment effects)

$$Y_{ij} = \mu_i + \epsilon_{ij} \tag{6.1}$$

for $(0 \leq i \leq p, 1 \leq j \leq n_i)$ where Y_{ij} is the $j\underline{th}$ observation on the $i\underline{th}$ population, and the ϵ_{ij} are i.i.d. $N(0,\sigma^2)$ random variables; the μ_i $(-\infty < \mu_i < \infty)$ are assumed to be unknown while σ^2 can be known or unknown. We consider the following goal:

GOAL: To compare simultaneously the p test means $\mu_1, \mu_2, \ldots, \mu_p$
with the control mean μ_0. $\tag{6.2}$

The correct multiple comparisons solution to this problem was first given by Dunnett (1955, 1964). He proposed one-sided joint lower confidence interval estimates of the $\mu_i - \mu_0$ $(1 \leq i \leq p)$ of the form

$$\overline{y}_i - \overline{y}_0 - gs \sqrt{\frac{1}{n_i} + \frac{1}{n_0}} \leq \mu_i - \mu_0 \tag{6.3}$$

(upper confidence interval estimates being of the same form with $-g$ replaced by g), and two-sided joint confidence interval estimates of the $\mu_i - \mu_0$

$(1 \leq i \leq p)$ of the form

$$\bar{y}_i - \bar{y}_0 - hs \sqrt{\frac{1}{n_i} + \frac{1}{n_0}} \leq \mu_i - \mu_0 \leq \bar{y}_i - \bar{y}_0 + hs \sqrt{\frac{1}{n_i} + \frac{1}{n_0}} . \quad (6.4)$$

Here $\bar{y}_i = \Sigma_{j=1}^{n_i} y_{ij}/n_i$ $(0 \leq i \leq p)$, and s^2 is the usual unbiased estimate of σ^2 based on υ d.f. The constants g and h depend on $(p,\upsilon;n_0,n_1,\ldots,n_p;\alpha)$ and are chosen to achieve the specified joint confidence coefficient $1-\alpha$, i.e., they guarantee the following probability requirements which are specified by the experimenter:

PROBABILITY REQUIREMENTS:

$$P\{T_i \leq g \;\; (1 \leq i \leq p)\} = 1-\alpha \qquad (6.5a)$$

and

$$P\{-h \leq T_i \leq h \;\; (1 \leq i \leq p)\} = 1-\alpha \qquad (6.5b)$$

where

$$T_i = \frac{(\bar{Y}_i - \bar{Y}_0) - (\mu_i - \mu_0)}{\sqrt{(\frac{1}{n_i} + \frac{1}{n_0}) s^2}} \;\; (1 \leq i \leq p). \qquad (6.6)$$

The T_i $(1 \leq i \leq p)$ have a standard p-variate Student t-distribution with υ d.f. and correlation matrix $\{\rho_{ij}\}$ where $\rho_{ij} = \{(1+n_0/n_i)(1+n_0/n_j)\}^{-1/2}$ for $(i \neq j, 1 \leq i,j \leq p)$; see Dunnett (1955) or Dutt, Mattes and Tao (1975).

The tables prepared for this volume apply directly to the symmetric situation in which n_0 observations are taken from the control population and the same number n_1 (which may or may not be equal to n_0) observations are taken from each of the p test populations. In this situation

$p_{ij} = n_1/(n_0+n_1) = \rho$ (say) for $(i \neq j, 1 \leq i, j \leq p)$, and g and h depend only on $(p, v, \rho; \alpha)$. If the n_i $(1 \leq i \leq p)$ are unequal, our tables can be used to calculate conservative values of g and h; see Section 7 for an example.

Tables A.1 to A.4 and B.1 to B.4 give values of g and h, respectively, for selected values of p, v, ρ, α.

Of particular interest is the situation in which the same number of observations (n) is taken from every one of the $p+1$ populations. Then $\rho = 1/2$ and the appropriate $g-$ and $h-$values are given in the 0.5 columns of Tables A and B, respectively.

Another case of special interest involves the optimal allocation of the observations between the p test populations and the control population. This is defined as the one which for fixed total number of observations $n_T = n_0 + pn_1$ $(n_1 = \ldots = n_p)$ maximizes the joint confidence coefficient for one-sided or two-sided intervals. Based on numerical calculations Dunnett (1955) suggested that the allocation $n_0 = n_1\sqrt{p}$ is approximately optimal. This allocation was proved by Bechhofer (1969) and Bechhofer and Nocturne (1972) to be asymptotically $(n_T \to \infty)$ optimal for known σ^2 for the one-sided and two-sided intervals, respectively. For this allocation $\rho = 1/(1+\sqrt{p})$, and the corresponding values of g and h are given in the right-most columns of Tables A and B, respectively.

6.1.2 Multi-factor Experiments

In this section we extend the approach of the foregoing section to multi-factor experiments in which each factor includes a control as one of its levels. For the particular case of two-factor experiments we assume that there are available $(p_1+1)(p_2+1)$ factor-level combinations Π_{ij} $(0 \leq i \leq p_1, 0 \leq j \leq p_2)$ where Π_{i0} is the control level and the Π_{ij}

$(1 \leq j \leq p_2)$ are the test levels for the $i\underline{th}$ level $(0 \leq i \leq p_1)$ of Factor A, and Π_{0j} is the control level and the Π_{ij} $(1 \leq i \leq p_1)$ are the test levels for the jth level $(0 \leq j \leq p_2)$ of Factor B. This setup is illustrated diagramatically below.

Factor A	Control levels	Factor B			
		Level			
		0	1	...	p_2
Control levels 0		Π_{00}	Π_{01}		Π_{0p_2}
1		Π_{10}	Π_{11}		Π_{1p_2}
Level :					
p_1		Π_{p_10}	Π_{p_11}		$\Pi_{p_1p_2}$

With equal numbers of observations in each cell and provided that the effects of the two factors are additive, we can assume the balanced two-factor, no interaction, fixed-effects ANOVA model (qualitative variables)

$$Y_{ijk} = \mu + \alpha_i + \beta_j + \epsilon_{ijk} \qquad (6.7)$$

for $(0 \leq i \leq p_1,\ 0 \leq j \leq p_2,\ 1 \leq k \leq n)$ where Y_{ijk} is the $k\underline{th}$ observation on the $i\underline{th}$ level of Factor A and the $j\underline{th}$ level of Factor B, and the ϵ_{ijk} are i.i.d. $N(0,\sigma^2)$ random variables; μ, the α_i and β_j $(-\infty < \mu,\ \alpha_i,\ \beta_j < \infty)$ with $\Sigma_{i=0}^{p_1} \alpha_i = \Sigma_{j=0}^{p_2} \beta_j = 0$ are assumed to be unknown while σ^2 can be known or unknown. We consider the following goal:

GOAL: To compare <u>simultaneously</u> for Factor A the p_1 test effects

$\alpha_1, \alpha_2, \ldots, \alpha_{p_1}$ with the control effect α_0, <u>and</u> for   (6.8)

Factor B the p_2 test effects $\beta_1, \beta_2, \ldots, \beta_{p_2}$ with the

control effect β_0.

We propose <u>one-sided</u> joint lower confidence interval estimates of the $\alpha_i - \alpha_0$

$(1 \leq i \leq p_1)$ and of the $\beta_j - \beta_0$ $(1 \leq j \leq p_2)$ of the form

$$
\left\{
\begin{array}{l}
\bar{y}_{i\bullet\bullet} - \bar{y}_{0\bullet\bullet} - g_1 s \sqrt{\dfrac{2}{n(p_2+1)}} \leq \alpha_i - \alpha_0 \\[4mm]
\bar{y}_{\bullet j\bullet} - \bar{y}_{\bullet 0\bullet} - g_1 s \sqrt{\dfrac{2}{n(p_1+1)}} \leq \beta_j - \beta_0
\end{array}
\right.
\qquad (6.9)
$$

(upper confidence interval estimates being of the same form with $-g_1$

replaced by g_1). Here $\bar{y}_{i\bullet\bullet} = \Sigma_{j=0}^{p_2} \Sigma_{k=1}^{n} y_{ijk}/n(p_2+1)$ $(0 \leq i \leq p_1)$,

$\bar{y}_{\bullet j\bullet} = \Sigma_{i=0}^{p_1} \Sigma_{k=1}^{n} y_{ijk}/n(p_1+1)$ $(0 \leq j \leq p_2)$, and s^2 is an unbiased estimate

of σ^2 based on v d.f. The constant g_1 which depends on $((p_1, p_2), v; \alpha)$

is chosen to achieve the specified joint confidence coefficient $1-\alpha$, i.e.,

it guarantees the following probability requirement:

PROBABILITY REQUIREMENT:

$$
P \left\{
\begin{array}{ll}
T_{1i} \leq g_1 & (1 \leq i \leq p_1) \\[2mm]
T_{2j} \leq g_1 & (1 \leq j \leq p_2)
\end{array}
\right\} = 1-\alpha
\qquad (6.10)
$$

where

$$T_{1i} = \frac{(\bar{Y}_{i\bullet\bullet} - \bar{Y}_{0\bullet\bullet}) - (\alpha_i - \alpha_0)}{\sqrt{\frac{2}{n(p_2+1)}} \, s^2} \qquad (1 \le i \le p_1) \qquad (6.11)$$

$$T_{2j} = \frac{(\bar{Y}_{\bullet j\bullet} - \bar{Y}_{\bullet 0\bullet}) - (\beta_j - \beta_0)}{\sqrt{\frac{2}{n(p_1+1)}} \, s^2} \qquad (1 \le j \le p_2) \; . \qquad (6.12)$$

The T_{1i}, T_{2j} $(1 \le i \le p_1, \; 1 \le j \le p_2)$ have a standard joint (p_1+p_2)-variate Student t distribution with correlation matrix as defined in Case II of (2.2b) with $\rho_1 = \rho_2 = 1/2$, $\rho_3 = 0$, viz. $\{\rho_{ij}\}$ is given by $\rho_{i,j} = 1/2$ for $(i \ne j, \; 1 \le i,j \le p_1)$ and $(i \ne j, \; p_1+1 \le i,j \le p_1+p_2)$ and $\rho_{ij} = 0$ for $(1 \le i \le p_1, \; p_1+1 \le j \le p_1+p_2)$ and $(p_1+1 \le i \le p_1+p_2, \; 1 \le j \le p_1)$.

Tables C.1 to C.3 give values of g_1 for selected values of $(p_1, p_2), \nu, \alpha$. Two-sided comparisons may be appropriate in some applications but we have not prepared tables for such purposes.

Remark 6.1: The two-factor experiment described in this section is more efficient than two independent single-factor experiments (as described in Section 6.1.1) in the sense that it can guarantee the probability requirement (6.10) with a smaller total number of observations than can the two independent single-factor experiments. (This assumes that the total number of d.f. for estimating σ^2 is the same, namely $\nu = (n-1)(p_1+1)(p_2+1)$, for the two types of experiments, i.e., the sum of squares that would have been associated with interaction (with $p_1 p_2$ d.f.) for the two-factor experiment is not pooled with the error sum of squares.) See Bawa (1972) who studied this problem for $\nu = \infty$ (i.e., known σ^2) in the context of indifference-zone selection experiments as described in our Section 6.3.2.

6.2 Subset Selection

6.2.1 Single-factor Experiments

6.2.1.1 Experiments Without a Control

There are available $p+1$ test populations Π_i $(1 \leq i \leq p+1)$. We assume the model (6.1), and make the same assumptions concerning the Y_{ij}, μ_i, ϵ_{ij} $(1 \leq i \leq p+1, \ 1 \leq j \leq n)$ and σ^2 as in Section 6.1.1. We consider the following goal:

GOAL: To select a subset of the $p+1$ populations which
         contains the population associated with   (6.13)

$$\max_{1 \leq i \leq p+1} \mu_i = \mu_{[p+1]} \quad \text{(say)}.$$

This problem was first posed and solved by Gupta (1956,1965). He proposed the following subset selection procedure:

PROCEDURE: "Include in the selected subset all populations for which

$$\bar{y}_i \geq \bar{y}_{[p+1]} - g\sqrt{2}s/\sqrt{n}."$$ (6.14)

Here $\bar{y}_i = \Sigma_{j=1}^{n} y_{ij}/n$ $(1 \leq i \leq p+1)$, $\bar{y}_{[p+1]} = \max_{1 \leq i \leq p+1} \bar{y}_i$, and s^2 is the usual unbiased estimate of σ^2 based on v d.f. A correct selection (CS) is said to have been made if the selected subset contains the population associated with $\mu_{[p+1]}$. The constant g which depends on $(p,v;\alpha)$ is chosen to guarantee the following probability requirement which is specified by the experimenter.

PROBABILITY REQUIREMENT:

$$\underset{\substack{(\mu_1,\mu_2,\dots,\mu_{p+1}) \\ \sigma^2}}{\text{Inf}} \quad P\{CS\} = 1-\alpha. \tag{6.15}$$

Here $1-\alpha$ satisfies $1/(p+1) < 1-\alpha < 1$. Gupta (1965) showed that the g-value that guarantees (6.15) is the <u>same</u> as the g-value that guarantees (6.5a) when $n_i = n$ $(0 \le i \le p)$ for the latter.

If in (6.13) we replace $\underset{1 \le i \le p+1}{\max} \mu_i = \mu_{[p+1]}$ by $\underset{1 \le i \le p+1}{\min} \mu_i = \mu_{[1]}$,

then an analogous procedure can be used. See the examples in Section 7.2. Also see Section 7.3.

6.2.1.2 Experiments with a Control

The following problem with goal (6.16) combines certain aspects of the multiple comparisons with a control goal (6.2) of Section 6.1.1 with the subset selection goal (6.13) discussed in Section 6.2.1.1.

As in Section 6.1.1, there are $p+1$ populations Π_i $(0 \le i \le p)$, consisting of one <u>control</u> population and p <u>test</u> populations. We assume the model (6.1), and make the same assumptions concerning the Y_{ij}, μ_i, ϵ_{ij} $(0 \le i \le p, 1 \le j \le n_i)$ and σ^2 as in that section; n_0 observations are taken from the control population, and $n_1 = n_2 = \dots = n_p$ from each of the test populations. We consider the following <u>goal</u>:

GOAL: To select a subset of the p test populations which have
       means μ_i at least as large as the mean μ_0 of the   (6.16)
       control population.

Gupta and Sobel (1958) proposed the following subset selection procedure for this goal:

PROCEDURE: "Include in the selected subset all test populations for

$$\text{which} \quad \bar{y}_i \geq \bar{y}_0 - gs \sqrt{\frac{1}{n_0} + \frac{1}{n_1}} \text{."}$$

Here $\bar{y}_i$ $(0 \leq i \leq p)$ and s^2 are defined as in Section 6.1.1 for n_0 and $n_1 = n_2 = \ldots = n_p$, and g is the same as in that section with $p = n_1/(n_0+n_1)$. A correct selection (CS) is said to have been made if the selected subset contains all populations with $\mu_i \geq \mu_0$. This procedure with g as described above guarantees the probability requirement.

PROBABILITY REQUIREMENT:

$$\underset{\substack{(\mu_0,\mu_1,\ldots,\mu_p) \\ \sigma^2}}{\text{Inf}} \quad P\{CS\} = 1-\alpha. \tag{6.17}$$

6.2.2 Multi-factor Experiments

For two-factor experiments we assume that there are available $(p_1+1)(p_2+1)$ test populations Π_{ij} $(1 \leq i \leq p_1+1, 1 \leq j \leq p_2+1)$. We assume the model (6.7), and make the same assumptions concerning the Y_{ijk}, μ, the α_i, β_j, ϵ_{ijk} $(1 \leq i \leq p_1+1, 1 \leq j \leq p_2+1, 1 \leq k \leq n)$ and σ^2 as in Section 6.1.2. We consider the following goal:

GOAL: To select a subset of the $(p_1+1)(p_2+1)$ populations which
contains the population associated with

$$\underset{1 \leq i \leq p_1+1}{\text{max}} \alpha_i = \alpha_{[p_1+1]} \quad \text{(say)}$$

and   (6.18)

$$\underset{1 \leq j \leq p_2+1}{\text{max}} \beta_j = \beta_{[p_2+1]} \quad \text{(say)}.$$

The following subset selection procedure was proposed in Bechhofer (1977); see also Bechhofer and Dunnett (1987).

<u>PROCEDURE</u>:   "Include in the selected subset all populations for which

$$\bar{y}_{i\bullet\bullet} \geq \bar{y}_{[p_1+1]\bullet\bullet} - g_1\sqrt{2}\ s/\sqrt{n(p_2+1)}$$

and   (6.19)

$$\bar{y}_{\bullet j\bullet} \geq \bar{y}_{\bullet[p_2+1]\bullet} - g_1\sqrt{2}\ s/\sqrt{n(p_1+1)}.\text{"}$$

Here $\bar{y}_{i\bullet\bullet} = \Sigma_{j=1}^{p_2+1}\ \Sigma_{k=1}^{n}\ y_{ijk}/n(p_2+1)$ $(1 \leq i \leq p_1+1)$,

$\bar{y}_{\bullet j\bullet} = \Sigma_{i=1}^{p_1+1}\ \Sigma_{k=1}^{n}\ y_{ijk}/n(p_1+1)$ $(1 \leq j \leq p_2+1)$, $\bar{y}_{[p_1+1]\bullet\bullet} = \max\limits_{1\leq i\leq p_1+1}\ \bar{y}_{i\bullet\bullet}$,

$\bar{y}_{\bullet[p_2+1]\bullet} = \max\limits_{1\leq j\leq p_2+1}\ \bar{y}_{\bullet j\bullet}$,   and s^2 is an unbiased estimate of σ^2 based on

v d.f.   A correct selection (CS) is said to have been made if the selected

subset contains the population associated with $\alpha_{[p_1+1]}$ <u>and</u> $\beta_{[p_2+1]}$.   The

constant g_1 which depends on $(p_1,p_2,v;\alpha)$ is chosen to guarantee the

following probability requirement which is specified by the experimenter.

<u>PROBABILITY REQUIREMENT</u>:

$$\mathop{\text{Inf}}\limits_{\substack{(\alpha_1,\alpha_2,\ldots,\alpha_{p_1+1}) \\ (\beta_1,\beta_2,\ldots,\beta_{p_2+1}) \\ \mu,\sigma^2}} P\{CS\} = 1-\alpha. \qquad (6.20)$$

Here $1-\alpha$ satisfies $(1/(p_1+1)(p_2+1) < 1-\alpha < 1)$.   It can be shown that the

g_1-value that guarantees (6.20) is the same as the g_1-value that guarantees

(6.10).

6.3   <u>Two-stage Indifference-zone Selection</u>

6.3.1   <u>Single-factor Experiments</u>

There are available p+1   test populations Π_i $(1 \leq i \leq p+1)$.   We assume

the model (6.1), and make the same assumptions concerning the Y_{ij}, μ_i, ϵ_{ij}

$(1 \leq i \leq p+1, \quad 1 \leq j \leq n)$ as in Section 6.1.1 except that now we assume that σ^2 is unknown. (If σ^2 were known, a single-stage procedure that guarantees the probability requirement (6.22), below, could be used. See Bechhofer (1954).) We consider the following goal:

GOAL: To select the population associated with

$$\max_{1 \leq i \leq p_1+1} \mu_i = \mu_{[p+1]} \quad \text{(say)}. \qquad (6.21)$$

A correct selection (CS) is said to have been made if the population selected by the procedure is the one associated with $\mu_{[p+1]}$. Consideration is restricted to selection procedures which guarantee the following indifference-zone probability requirement:

PROBABILITY REQUIREMENT:

$$P\{CS\} \geq 1-\alpha \quad \text{whenever} \quad \mu_{[p+1]} - \mu_{[p]} \geq \delta*. \qquad (6.22)$$

Here $\mu_{[1]} \leq \cdots \leq \mu_{[p+1]}$ are the ordered values of $\mu_1, \mu_2, \ldots, \mu_{p+1}$, and $\{\delta*, 1-\alpha\}$ with $0 < \delta* < \infty$, $1/(p+1) < 1-\alpha < 1$ is specified prior to the start of experimentation.

The two-stage procedure described below was proposed by Bechhofer, Dunnett and Sobel (1954). (Note: It was formally proved by Dudewicz (1971) that the probability requirement (6.22) cannot be guaranteed using any single-stage procedure if the common variance σ^2 is unknown.)

TWO-STAGE PROCEDURE:

"a) In the first stage take an arbitrary common number $n_0 > 1$

of observations from each of the p+1 populations Π_i

$(1 \leq i \leq p+1)$.

b) Calculate

$$s^2 = \sum_{i=1}^{p+1} \sum_{j=1}^{n_0} (y_{ij} - \bar{y}_{i\bullet})^2 / v$$

where $v = (p+1)(n_0-1)$ d.f.; s^2 is an unbiased estimate of σ^2

based on v d.f. (Note: This assumes that a completely randomized

design was used. If a different design were used for the selection

experiment, the formula for s^2 and the value of v would be

changed accordingly.)

c) Enter Table A in the column headed $\rho = 0.5$ with the appropriate

$(p, v; 1-\alpha)$ and obtain the associated constant g.

d) In the second stage, take the same number $n-n_0$ of additional

observations from each of the p+1 populations where

$$n = n_0 \quad \text{if} \quad M \leq n_0, \quad n = [M] \quad \text{if} \quad M > n_0.$$

Here $M = 2(gs/\delta^*)^2$, [M] denotes the smallest integer equal to or

greater than M, and $g = g(p, v, \rho; \alpha)$ is the solution of equation

(3.2) for $\rho = 1/2$ (which is the same as the g-value required by

(6.15)).

e) Calculate the p+1 overall sample means $\bar{y}_i = \sum_{j=1}^{n} y_{ij} / n$

$(1 \leq i \leq p+1)$ based on first- and second-stage observations.

f) Select the population that yielded $\max\limits_{1\leq i\leq p+1} \bar{y}_i = \bar{y}_{[p+1]}$ as the one

associated with $\mu_{[p+1]}$."

<u>Remark 6.2</u>: Other two-stage selection procedures have been proposed which are different from the one that we have just described. Dudewicz and Dalal (1975) and Rinott (1978) have proposed procedures which, unlike ours, are applicable when the p+1 variances are <u>completely unknown and possibly unequal</u>; tables of constants necessary to implement the Dudewicz-Dalal procedure are included in their article while tables for the Rinott procedure are given by Wilcox (1984). Tamhane and Bechhofer (1977,1979) proposed a minimax procedure with screening which is closed (i.e., has a predetermined upper limit on the total sample size) for the case of <u>common known</u> variance; tables of constants necessary to implement that procedure are given in their articles.

6.3.2. <u>Multi-factor Experiments</u>

For <u>two-factor</u> experiments we assume that there are available $(p_1+1)(p_2+1)$ test populations π_{ij} $(1 \leq i \leq p_1+1, 1 \leq j \leq p_2+1)$. We assume the model (6.7), and make the same assumptions concerning the Y_{ijk}, μ, α_i, β_j, ϵ_{ijk} $(1 \leq i \leq p_1+1, 1 \leq j \leq p_2+1, 1 \leq k \leq n)$ as in Section 6.1.2, except that we now assume that σ^2 is <u>unknown</u>. We consider the following <u>goal</u>:

GOAL: To select the population associated with

$$\max\limits_{1\leq i\leq p_1+1} \alpha_i = \alpha_{[p_1+1]} \quad \text{(say)}$$

and   (6.23)

$$\max\limits_{1\leq j\leq p_2+1} \beta_j = \beta_{[p_2+1]} \quad \text{(say)}.$$

A correct selection (CS) is said to have been made if the population selected is the one associated with $\alpha_{[p_1+1]}$ and $\beta_{[p_2+1]}$. Consideration is restricted to procedures which guarantee the following indifference-zone probability requirement:

PROBABILITY REQUIREMENT

$$P\{CS\} \geq 1-\alpha \quad \text{whenever} \quad \begin{cases} \alpha_{[p_1+1]} - \alpha_{[p_1]} \geq \delta* \\ \text{and} \\ \beta_{[p_2+1]} - \beta_{[p_2]} \geq \delta* \end{cases} \quad (6.24)$$

Here $\alpha_{[1]} \leq \alpha_{[2]} \leq \cdots \leq \alpha_{[p_1+1]}$ and $\beta_{[1]} \leq \beta_{[2]} \leq \cdots \leq \beta_{[p_2+1]}$ are the ordered values of $\alpha_1, \alpha_2, \ldots, \alpha_{p_1+1}$ and $\beta_1, \beta_2, \ldots, \beta_{p_2+1}$, respectively, and $\{\delta*, 1-\alpha\}$ with $0 < \delta* < \infty$, $1/(p_1+1)(p_2+1) < 1-\alpha < 1$ are specified prior to the start of experimentation. The following two-stage procedure was proposed in Bechhofer (1977):

TWO-STAGE PROCEDURE:

"a)   In the <u>first</u> stage take an arbitrary common number $n_0 > 1$ of observations from each of the $(p_1+1)(p_2+1)$ populations Π_{ij} $(1 \leq i \leq p_1+1, \ 1 \leq j \leq p_2+1)$.

b)   Calculate

$$s^2 = \sum_{i=1}^{p_1+1} \sum_{j=1}^{p_2+1} \sum_{k=1}^{n_0} (y_{ijk} - \sum_{k=1}^{n_0} y_{ijk}/n_0)^2/v$$

which is an unbiased estimate of σ^2 based on $v = (p_1+1)(p_2+1)(n_0-1)$ d.f.

c)   Enter Table D with the appropriate $((p_1, p_2), v; 1-\alpha)$ and obtain the associated constant g_2.

d)   In the <u>second</u> stage, take the same number $n-n_0$ of additional observations from each of the $(p_1+1)(p_2+1)$ populations where

$$n = n_0 \quad \text{if} \quad M \leq n_0, \quad n = [M] \quad \text{if} \quad M > n_0.$$

Here $M = 2(g_2 s/\delta*)^2/(p_1+1)$, $[M]$ denotes the smallest integer equal to or greater than M, and $g_2 = g_2((p_1, p_2), v; \alpha)$ is the solution of equation (3.6) for $\rho = 1/2$.

e) Calculate the $(p_1+1) + (p_2+1)$ overall row and column means

$$\bar{y}_{i\bullet\bullet} = \sum_{j=1}^{p_2+1} \sum_{k=1}^{n} y_{ijk}/(p_2+1)n \quad (1 \leq i \leq p_1+1)$$

$$\bar{y}_{\bullet j\bullet} = \sum_{i=1}^{p_1+1} \sum_{k=1}^{n} y_{ijk}/(p_1+1)n \quad (1 \leq j \leq p_2+1)$$

based on the combined first- and second-stage observations.

f) Select the population that yielded

$$\max_{1 \leq i \leq p_1+1} \bar{y}_{i\bullet\bullet} \quad \underline{\text{and}} \quad \max_{1 \leq j \leq p_2+1} \bar{y}_{\bullet j\bullet}$$

as the one associated with $\alpha_{[p_1+1]}$ and $\beta_{[p_2+1]}$."

Bechhofer and Dunnett (1986) proved that the two-stage procedure described above guarantees (6.24). In their article they showed how the indifference-zone selection procedure of this section can be generalized in an obvious way to multi-factor experiments (including fractional factorial experiments) with no interaction between the factor levels.

6.4 Multiple Comparisons for Orthogonal Contrasts

6.4.1. Single-factor Experiments

There are available k test populations Π_i $(1 \leq i \leq k)$. We assume the model (6.1), and make the same assumptions concerning the Y_{ij}, μ_i, ϵ_{ij}

$(1 \leq i \leq k, 1 \leq j \leq n_i)$ and σ^2 as in Section 6.1.1. Let $\theta_m = \Sigma_{i=1}^{k} c_{mi}\mu_i$

where the c_{mi} are specified constants such that $\Sigma_{i=1}^{k} c_{mi} = 0$

$(1 \leq m \leq p < k)$ and the $p \times k$ matrix of the c_{mi} has rank p. The θ_m

represent a family of <u>contrasts</u> among the μ_i. We consider the following

<u>goal</u>:

<u>GOAL</u>: To estimate simultaneously the p contrasts $\theta_1, \theta_2, \ldots, \theta_p$.   (6.25)

<u>One-sided</u> joint lower confidence interval estimates of the θ_m

$(1 \leq m \leq p)$ are of the form

$$\hat{\theta}_m - gs \sqrt{\sum_{i=1}^{k} c_{mi}^2/n_i} \leq \theta_m \qquad (6.26)$$

while <u>two-sided</u> joint confidence interval estimates of the θ_m $(1 \leq m \leq p)$

are of the form

$$\hat{\theta}_m - hs \sqrt{\sum_{i=1}^{k} c_{mi}^2/n_i} \leq \theta_m \leq \hat{\theta}_m + hs \sqrt{\sum_{i=1}^{k} c_{mi}^2/n_i}. \qquad (6.27)$$

(See Bechhofer and Dunnett (1982).) Here $\hat{\theta}_m = \Sigma_{i=1}^{k} c_{mi}\bar{y}_i$, $\bar{y}_i = \Sigma_{j=1}^{n_i} y_{ij}/n_i$

$(1 \leq i \leq k)$, and s^2 is the usual unbiased estimate of σ^2 based on υ

d.f. The constants g and h which depend on $(p, \upsilon,$ the n_i and c_{im}

$(1 \leq i \leq k, 1 \leq m \leq p)$; $\alpha)$ are chosen to guarantee the probability

requirements (6.5a) and (6.5b), respectively, where now

$$T_m = \frac{\hat{\theta}_m - \theta_m}{\sqrt{\sum_{i=1}^{k} c_{mi}^2 s^2/n_i}} \quad (1 \leq m \leq p). \qquad (6.28)$$

The T_m $(1 \leq m \leq p)$ have a standard p-variate Student t-distribution with correlation matrix $\{\rho_{m_1,m_2}\}$ given by

$$\rho_{m_1,m_2} = \frac{\sum\limits_{i=1}^{k} c_{m_1 i} c_{m_2 i}/n_i}{[(\sum\limits_{i=1}^{k} c_{m_1 i}^2/n_i)(\sum\limits_{j=1}^{k} c_{m_2 j}^2/n_j)]^{1/2}} \qquad (1 \leq m_1,m_2 \leq p). \qquad (6.29)$$

If $\sum\limits_{i=1}^{k} c_{m_1 i} c_{m_2 i} = 0$ $(m_1 \neq m_2, 1 \leq m_1,m_2 \leq p)$, i.e., the p contrasts are mutually orthogonal, and $n_1 = n_2 = \ldots = n_k$, then $\rho_{m_1,m_2} = 0$ $(m_1 \neq m_2,$ $1 \leq m_1,m_2 \leq p)$. The g- and h-values necessary to guarantee (6.5a) and (6.5b), respectively, then depend only on $(p,\nu;\alpha)$ with $\rho = 0$, and are given in the $\rho = 0.0$ column of Tables A and B, respectively.

Note: When $\rho = 0$ the statistic $\max\limits_{1 \leq m \leq p} |T_m|$ is known as the Studentized maximum modulus and the statistic $\max\limits_{1 \leq m \leq p} T_m$ is known as the Studentized maximum.

6.4.2. Multi-factor Experiments

For m-factor $(m \geq 2)$ 2-level experiments we assume that there are available 2^m test populations $\Pi_{i_1 i_2 \ldots i_m}$ $(i_t = 1,2; 1 \leq t \leq m)$. We consider the model

$$Y_{i_1 i_2 \ldots i_m, j} = \mu_{i_1 i_2 \ldots i_m} + \epsilon_{i_1 i_2 \ldots i_m, j} \qquad (6.30)$$

for $(i_t = 1,2; 1 \leq t \leq m, 1 \leq j \leq n)$ where $Y_{i_1 i_2 \ldots i_m, j}$ is the jth observation on the i_1-th level of Factor $1, \ldots, i_m$-th level of Factor m, and

the $\epsilon_{i_1 i_2 \ldots i_m, j}$ are i.i.d. $N(0, \sigma^2)$ random variables; the $\mu_{i_1 i_2 \ldots i_m}$

$(i_t = 1, 2; \; 1 \leq t \leq m)$ and σ^2 are assumed to be <u>unknown</u>. Let

$\sum_{i_1=1}^{2} \cdots \sum_{i_m=1}^{2} c_{i_1 \ldots i_m} \mu_{i_1 \ldots i_m}$ denote a family of $2^m - 1$ <u>orthogonal</u>

<u>contrasts</u> among the $\mu_{i_1 \ldots i_m}$.

    <u>GOAL</u>: To estimate simultaneously any subset of $p \leq 2^m - 1$ of

           these contrasts.   (6.31)

One-sided joint <u>lower</u>, and two-sided confidence interval estimates of these

orthogonal contrasts are found by replacing $\hat{\theta}_m$ in (6.26) and (6.27),

respectively, by $\sum_{i_1=1}^{2} \cdots \sum_{i_m=1}^{2} c_{i_1 \ldots i_m} \bar{y}_{i_1 \ldots i_m}$ where $\bar{y}_{i_1 \ldots i_m} =$

$\sum_{j=1}^{n} y_{i_1 \ldots i_m, j} / n$, and replacing $\sqrt{\sum_{i=1}^{k} c_{mi}^2 / n_i}$ by

$\sqrt{\sum_{i_1=1}^{2} \cdots \sum_{i_m=1}^{2} c_{i_1 \ldots i_m}^2 / n}$; the g- and h-values again depend only on (p, v, α)

with $\rho = 0$. (See, e.g., Davies (1978), Section 7.7.)

<u>Remark 6.8</u>: An analogous development can be employed for m-factor $(m \geq 2)$

3-level experiments. (See, e.g., Davies (1978), Appendix 8G.)

6.5   <u>Simultaneous Confidence Intervals to Contain p Regression Coefficients</u>

      <u>of Orthogonal Polynomials</u>

      There are n populations Π_x $(x = 1, 2, \ldots, n)$. We assume the one-factor

fixed effects ANOVA regression model

$$Y_x = \alpha_0 + \alpha_1 \xi_1'(x) + \alpha_2 \xi_2'(x) + \cdots + \alpha_k \xi_k'(x) + \epsilon_x \qquad (6.32)$$

for x = 1,2,...,n where Y_x is the observation at level x

$(x = 1,2,...,n)$, $\xi_i'(x)$ $(0 \leq i \leq k)$ is a polynomial of degree i in x

where

$$\sum_{x=1}^{n} \xi_i'(x)\xi_j'(x) = 0 \quad \text{for} \quad i \neq j, \quad 0 \leq i,j \leq k; \quad \xi_0'(x) \equiv 1, \qquad (6.33)$$

i.e., equation (6.32) is the <u>orthogonal</u> representation of a k<u>th</u> degree

polynomial in n equally-spaced x-values. (See Fisher and Yates (1938–

1963).) The ϵ_x are i.i.d. $N(0,\sigma^2)$ random variables. The α_i $(-\infty < \alpha_i < \infty)$

and σ^2 are assumed to be <u>unknown</u>. We consider the following <u>goal</u>:

<u>GOAL</u>: To estimate simultaneously any $p \leq k$ of the α_i

$\quad\quad (0 \leq i \leq k)$. $\hfill (6.34)$

For convenience of notation label the α's of interest $\alpha_1, \alpha_2, ..., \alpha_p$. Then

<u>one-sided</u> joint lower confidence interval estimates of $\alpha_1, \alpha_2, ..., \alpha_p$ are of

the form

$$\hat{\alpha}_i - gs \sqrt{1/\sum_{x=1}^{n} [\xi_i'(x)]^2} \leq \alpha_i \quad\quad (1 \leq i \leq p), \qquad (6.35)$$

(upper confidence interval estimates being of the same form with $-g$ replaced

by g), and <u>two-sided</u> joint confidence intervals of the α_i are of the form

$$\hat{\alpha}_i - hs \sqrt{1/\sum_{x=1}^{n} [\xi_i'(x)]^2} \leq \alpha_i \leq \hat{\alpha}_i + hs \sqrt{1/\sum_{x=1}^{n} [\xi_i'(x)]^2} \quad (1 \leq i \leq p). \quad (6.36)$$

Here

$$\hat{\alpha}_i = \sum_{x=1}^{n} y_x \xi_i'(x) / \sum_{x=1}^{n} [\xi_i'(x)]^2 \quad (1 \leq i \leq p) \tag{6.37}$$

where y_x is the observed valued of Y_x at x, and

$$s^2 = \left[\sum_{x=1}^{n} y_x^2 - \frac{[\sum_{x=1}^{n} y_x \xi_i'(x)]^2}{\sum_{x=1}^{n} [\xi_i'(x)]^2} \right] / (n-k-1). \tag{6.38}$$

is an unbiased estimate of σ^2 based on $v = n-k-1$ d.f. The constants g and h which depend only on $(p, v; \alpha)$ are chosen to guarantee the probability requirements (6.5a) and (6.5b), respectively where now

$$T_i = \frac{\hat{\alpha}_i - \alpha_i}{\sqrt{\frac{1}{\sum_{x=1}^{n} [\xi_i'(x)]^2} s^2}} \quad (1 \leq i \leq p). \tag{6.39}$$

The T_i $(1 \leq i \leq p)$ have a standard p-variate Student t-distribution with $\{\rho_{ij}\}$ given by

$$\rho_{ij} = \begin{cases} 1 & \text{if } i = j \\ 0 & \text{if } i \neq j \end{cases} \quad (1 \leq i, j \leq p). \tag{6.40}$$

The required g- and h-values are given in the $\rho = 0.0$ column of Tables A and B, respectively.

6.6 Additional Applications

Additional applications requiring equicoordinate percentage points of the multivariate Student t distribution are described in Hahn and Hendrickson (1971). Included among these are "prediction intervals to contain all of k future means and when the estimate of σ^2 is pooled from several samples."

Bohrer (1979) reports a need for the upper $1-\alpha$ point of the distribution of the <u>Studentized maximum</u> (i.e., p-variate t-distribution with $\rho_{ij} = 0$ ($i \neq j$)) when deciding the signs of $p \geq 2$ parameters $\theta_1, \theta_2, \ldots, \theta_p$, simultaneously, the estimators of which are normally and independently distributed.

Hochberg (1974) proposed a procedure (GT2) for pairwise comparisons of treatment means in <u>unbalanced</u> designs which employs the percentage points of the Studentized maximum modulus distribution. Hayter (1984) proved the truth of the so-called Tukey conjecture concerning the conservativeness of the procedure based on the Studentized range for the <u>unbalanced</u> one-way ANOVA setup, and hence the latter procedure is preferable to GT2 for this case. However, in more complex settings such as the one-way ANCOVA, the Tukey conjecture is known to be true only for $p = 2$ (Brown (1984)) and hence the GT2 procedure may be more appropriate. The reader is referred to Hochberg and Tamhane (1987) for a comprehensive treatment of multiple comparisons problems, some of which have been discussed in the present volume.

Hsu (1984) proposed an interesting procedure concerning multiple comparisons with the best. For equal sample sizes the constants required to implement his procedure are given in our Table A with $\rho = 0.5$. For unequal sample sizes approximate constants can be computed using our Table A and a method analogous to that described in our Section 7.1 for Case (c).

7. EXAMPLES ILLUSTRATING APPLICATIONS OF THE TABLES

7.1 Multiple Comparisons With a Control

To illustrate the application of Tables A and B in problems involving multiple comparisons between two or more test treatments and a control treatment, we consider the data given by Snedecor and Cochran (1980) in their Table 12.2.1. These data came from an experiment to compare the fat absorption of four oils used in the making of doughnuts. There were six batches made with each of the oils, and the amount of fat absorbed was measured for each batch. The means for the four oils were 72, 85, 76 and 62. Assuming that the batch to batch variability is the same for each oil, the estimate of the common variance is $s^2 = 100.9$, based on 20 degrees of freedom. For illustrative purposes, we suppose that the fourth oil was the control oil and that the purpose of the experiment was to compare the first three oils with the control oil. Thus, we are interested in the differences between the means for the first 3 oils and the control; the observed values of these differences are 10, 23 and 14 in this example.

Case (a):   Equal sample sizes, n.

For the equal sample size case, the correlation coefficients ρ_{ij} given in Section 6.1.1 have a common value $\rho = 0.5$. Also, the estimated standard error for each treatment difference is $\sqrt{2s^2/n}$, which has the value $\sqrt{2(100.9)/6} = 5.80$ in this example. Suppose that we require 2-sided joint 95% confidence intervals for the 3 treatment differences. Referring to Table B.3 with $p = 3$, $\nu = 20$ and $\rho = 0.5$, we obtain the value $h = 2.540$ (rounding from the 5-decimal place value of 2.54035 given in the table). Then the joint confidence intervals are

$$
\left\{
\begin{array}{l}
10 \pm (2.540)(5.80) = -4.7 \text{ to } 24.7 \\
23 \pm (2.540)(5.80) = 8.3 \text{ to } 37.7 \\
14 \pm (2.540)(5.80) = -0.7 \text{ to } 28.7
\end{array}
\right.
$$

for the three differences. Table A.3 would have been used in the same way for one-sided joint 95% confidence intervals.

Case (b): Equal sample size n_1 for the treatments and a different sample size n_0 for the control.

For this case, there is a common value ρ for the correlation coefficients given by the formula in Section 6.1.1 with the value of ρ depending on n_0/n_1. Suppose that the sample sizes for the four oils had been $n_1 = 5$ for each of the first three oils and $n_0 = 9$ for the control oil. Substituting into the formula for the common value of ρ yields $\rho = 1/(1+9/5) = 0.357$. To obtain the corresponding value of h, it is necessary to interpolate in the tables. For 2-sided joint 95% confidence intervals, we obtain for $p = 3$, $v = 20$ the values $h = 2.57586$ for $\rho = 0.3$ and $h = 2.56090$ for $\rho = 0.4$. Linear interpolation gives the value $h = (0.43)(2.57586) + (0.57)(2.56090) = 2.567$ for $\rho = 0.357$. The estimated common standard error for each treatment difference is $\sqrt{s^2(1/n_1 + 1/n_0)} = \sqrt{(100.9)(1/5 + 1/9)} = 5.603$. Then, the joint confidence intervals are

$$
\left\{
\begin{array}{l}
10 \pm (2.567)(5.603) = -4.4 \text{ to } 24.4 \\
23 \pm (2.567)(5.603) = 8.6 \text{ to } 37.4 \\
14 \pm (2.567)(5.603) = -0.4 \text{ to } 28.4
\end{array}
\right.
$$

for the three differences.

For somewhat greater accuracy, linear interpolation on $1/(1-\rho)$ instead of linear interpolation on ρ is recommended. This gives the value $h = 2.568$ in the present example. The exact value, computed by numerical integration is $h = 2.5679$.

It should be noted that the confidence intervals are slightly narrower with this allocation of the sample sizes than they were previously, for the same total sample size. The allocation that approximately minimizes the confidence interval width for fixed total sample size when $1-\alpha$ exceeds 0.95 (approximately) is given by $n_0/n_1 = \sqrt{p}$, which here has the value $\sqrt{3} = 1.7$. The value of h for this allocation can be obtained directly from the right-most column of Tables A and B without using interpolation.

Case (c):   Unequal sample sizes

If the sample sizes for the treatment groups are unequal, the values of the correlation coefficients given by the formula for ρ_{ij} in Section 6.1.1 are unequal and, strictly speaking, the tables are no longer applicable. However, they still can be used to obtain an approximate value for h.

For example, suppose that the sample sizes in the doughnut problem are 3, 8, 7 and 6, respectively, for the four oils. First, we calculate $b_i = 1/\sqrt{1+n_0/n_i}$ $(1 \le i \le 3)$. With $n_0 = 6$ and $n_1 = 3$, $n_2 = 8$, and $n_3 = 7$, we obtain the values $b_1 = 0.577$, $b_2 = 0.756$ and $b_3 = 0.734$, respectively. Then we calculate the correlation coefficients $\rho_{ij} = b_i b_j$ $(i \ne j, 1 \le i,j \le 3)$. Following the recommendation of Dunnett (1985), we find the arithmetic average of the ρ_{ij} values and use this value for ρ to enter the tables. In this example, the correlation coefficients are $\rho_{12} = 0.436$, $\rho_{13} = 0.424$ and $\rho_{23} = 0.555$ with a mean of 0.472. Reading from Table B.3 for 2-sided joint 95% confidence intervals, we obtain for $p = 3$, $v = 20$ the values h = 2.56090 for ρ = 0.4 and h = 2.54035 for ρ = 0.5; by linear interpolation we obtain h = 2.546 for ρ = 0.472.   The joint confidence intervals for the three differences are

$$\left\{ \begin{array}{l} 10 \pm (2.546) \ \sqrt{100.9(1/3 + 1/6)} \ = -8.1 \ \text{to} \ 28.1 \\[2mm] 23 \pm (2.546) \ \sqrt{100.9(1/8 + 1/6)} \ = \ \ 9.2 \ \text{to} \ 36.8 \\[2mm] 14 \pm (2.546) \ \sqrt{100.9(1/7 + 1/6)} \ = -0.2 \ \text{to} \ 28.2. \end{array} \right.$$

The exact value of h can be calculated by numerical integration using the computer algorithm given by Dunnett (1986).   For this example, the value h = 2.5455 is obtained.   It can also be obtained from the tables for p = 3 of Dutt, Mattes and Tao (1975).

7.2 Subset Selection

We consider the same example from Table 12.2.1 of Snedecor and Cochran (1980) to illustrate the application of Table A in subset selection.   Now we are interested in selecting the best of the four oils (i.e., the one with the smallest population mean), and we wish to do this by selecting a subset of the four oils, the selected subset being such that the best oil will be included in it with a probability of at least the specified 1-α.

Case (a):   Equal sample sizes, no control treatment

The p+1 = 4 sample means ranked in ascending order are 62, 72, 76 and 85.   Suppose that we require a probability of 0.95 of including the best oil in the selected subset.   Referring to Table A.3 with p = 3, v = 20 and ρ = 0.5 we obtain the value g = 2.19228.   Applying the modification of (6.14) when the best one is the one with the smallest population mean, we include in the selected subset all oils whose sample means are less than or equal to

$$62 + 2.192\sqrt{100.9/6} = 70.99.$$

Hence, in this example, the selection rule has resulted in the selection of the single oil having the sample mean 62 as the best one of the four.

Case (b): Equal sample sizes, a specified one of the treatments being the control treatment

Here, one of the treatments is specified in advance as being a control treatment and the goal is to select a subset containing all treatments at least as good as the control treatment. Suppose that we require a probability of at least $1-\alpha = 0.95$ that the selected subset will contain all oils that have population means which are less than or equal to the population mean of the control. Then $g = 2.19228$, the same value as for Case (a). If the oil which is the control treatment yielded the sample mean of 62, then the selected subset consists of all oils having sample means less than or equal to

$$62 + 2.192\sqrt{100.9(2/6)} = 74.7.$$

Here the oil that gave the sample value of 72, as well as the standard oil itself, are included in the selected subset.

Case (c): Sample size n_0 for the control treatment and n_1 for each of the other treatments

Suppose that the ordered sample means are the same as before, i.e. 62, 72, 76 and 85 with the value 62 being the sample mean for the control treatment, but with sample sizes of $n_0 = 9$ for the control oil and $n_1 = 5$ for each of the other three oils. The value of the correlation coefficient is then $\rho = 5/(9+5) = 0.357$. For $1-\alpha = 0.95$, we refer to Table A.3 with

$p = 3$, $v = 20$, and must interpolate between $g = 2.23607$ for $\rho = 0.3$ and $g = 2.21687$ for $\rho = 0.4$. Using linear interpolation, we obtain $g = 2.225$. Then the selected subset consists of all oils having sample means less than or equal to

$$62 + 2.225\sqrt{100.9(1/9 + 1/5)} = 74.5.$$

The selected subset thus contains the control oil and the oil which yielded the sample mean 72.

7.3   Indifference-zone Selection

To illustrate the application of Table A in indifference zone selection, we again use the data from Table 12.2.1 of Snedecor and Cochran (1980), and suppose that they constitute a _first_ sample of size $n_0 = 6$ from each of $p+1 = 4$ oils. If we use the two-stage selection procedure of Section 6.3.1, the problem is to determine how many additional observations are needed on each oil so that we can guarantee a probability of at least $1-\alpha$ that the best oil will be selected whenever the smallest population mean differs from the second smallest by at least $\delta*$.

Suppose that we specify $\delta* = 10.0$ and $1-\alpha = 0.95$. Consulting Table A.3 with $p = 3$, $v = 20$ and $\rho = 0.5$, we obtain the value $g = 2.19228$. The formula for the total size M in Section 6.3.1 is $M = 2(gs/\delta*)^2$. Here $M = 2(100.9)(2.192/10.0)^2 = 9.7$. Therefore a total sample size of $n = 10$ is required, and an additional sample of size $10 - 6 = 4$ must be taken from each oil.

7.4 Two-factor Selection

To illustrate the application of Tables C and D in two-factor selection,
we consider the experiment described in Smith (1969). The experiment was a
factorial with 4 reagents (Factor A) and 3 catalysts (Factor B), and two
replicates of each of the 12 factor-level combinations. The experimenter was
interested in determining which combination of reagent and catalyst yielded
the highest production rate. The coded sample mean yields for the four
reagents (averaged over catalysts and replicates) were 7, 9, 13 and 11, and
the sample mean yields for the three catalysts (averaged over reagents and
replicates) were 9, 12 and 9. In his initial analysis, Smith assumed a linear
model without interaction, and tested the two main effects against the
residual variance $(s^2 = 7.33$ based on 18 d.f.). He found that reagents
differed significantly at the one percent level, but catalysts were not
significantly different at the five percent level. Then he posed a linear
model with interaction, and tested interaction against the new residual
$(s^2 = 4.00$ based on 12 d.f.) and found the interaction to be significant at
the five percent level; both reagents and catalysts were now significant at
the one percent and the five percent level, respectively, when tested against
the new residual.

Remark 7.1: A crucial issue in factorial experiments is whether or not there
are sizeable interactions between the factors. (In almost all experiments
there is at least some interaction since the response differences between two
levels of one factor are unlikely to be identical for all levels of the other
factor.) This information is important because the presence or absence of
interaction will dictate the statistical procedures to be employed, and the

final inferences that can legitimately be made. Sometimes the experimenter
may be prepared to make the assumption of "no" interaction based on previous
experience or knowledge of the factors being studied. At other times he may
feel that he has little basis a priori for making this assumption. We propose
to indicate here how the experimenter might proceed in each of these two
situations. For illustrative purposes we shall first show how to use Smith's
data under the assumption that there is no interaction. Then we will indicate
how one might proceed if there were little or no basis a priori for assuming
no interaction. We emphasize that the procedure that we will propose for this
latter situation is an ad hoc one; to the best of our knowledge this problem
has not yet been studied analytically.

Case (a): Subset selection

    i) Interaction assumed to be zero or negligible

In subset selection, we wish to select a subset of the $p_1+1 = 3$
catalysts and a subset of the $p_2+1 = 4$ reagents. The procedure is to have
the property that with a probability equal to or greater than a specified
value of $1-\alpha = 0.95$ (say) we will simultaneously include the best catalyst
and the best reagent in the selected subset. Entering Table C.3 with
$(p_1, p_2) = (2,3)$ and $v = 18$, we obtain the tabulated value $g_1 = 2.475$
(rounded). Following the procedure described in equation (6.19), we note that
the largest means observed for the three catalysts and for the four reagents
were 12 and 13, respectively, and therefore we include in the selected subset
all catalysts having sample means that equal or exceed

$$12 - \frac{2.475\sqrt{2(7.33)}}{\sqrt{2(4)}} = 8.6,$$

and all reagents having sample means that equal or exceed

$$13 - \frac{2.475\sqrt{2(7.33)}}{\sqrt{2(3)}} = 9.1.$$

Hence we select all three catalysts , and the two reagents that yielded means of 11 and 13.

ii) Presence or absence of interaction to be assessed using a
     preliminary test of significance

Our goal and probability requirement are the same as in i), above. However, because we are unprepared a priori to assume zero or negligible interaction, we propose to assess its existence using a preliminary test of significance. The outcome of this test will suggest the procedure that we use and the inference that we make.

One might proceed as follows: Start by testing for interaction at some prespecified level of significance: if a non-significant F is obtained, continue as in i), above; if a significant F is obtained, employ the procedure for single-factor experiments as described in Section 7.2, Case (a).

Note: The procedure described immediately above is an ad hoc one, and although it appears to be reasonable, its performance characteristics are not known. Because it is a composite procedure, we cannot assert that it will guarantee the probability requirement (6.20). This problem is being studied by the authors.

If we adopted this ad hoc procedure with Smith's data, and tested for interaction at the one percent level (or less), a non-significant result would be obtained; then we would continue as in i), above. However, if the test were conducted at the five percent level (or greater), then (as noted above) a significant result would be obtained, and we would continue as in Section 7.2, Case (a), with p+1 = 12. For $1-\alpha = 0.95$ we enter Table A.3 wih p = 11, $v = 12$, $\rho = 0.5$, and obtain the tabulated value g = 2.812 (rounded). The 12 reagent-catalyst sample means can be computed from Smith's Table 1; in increasing magnitude they are 5,5,7,8,8,9,12,12,13,13,14,14. Using the procedure (6.14) we would then include in the selected subset all reagent-catalyst combinations with sample means at least as large as

$$14 - 2.812\sqrt{4.0/2} = 10.02.$$

Thus the 6 combinations with sample means 12, 13 and 14 would be included in the selected subset. (Interestingly, these are the same combinations that were selected by Smith.)

Case (b): <u>Indifference zone selection</u> (Interaction assumed to be zero or
             negligible)

For the two-stage indifference-zone selection procedure described in Section 6.3.2, we again consider the data in Smith (1969), and assume that they constitute the first stage of $n_0 = 2$ observations from each of the $3 \times 4 = 12$ factor-level combinations. We seek to determine how many additional observations are needed in the second stage so that a specified probability $1-\alpha$ or greater is achieved for simultaneously selecting the best catalyst and the best reagent for a specified $\delta*$. For $1-\alpha = 0.95$ we enter Table D.3 with $(p_1, p_2) = (2,3)$ and $v = 18$, and obtain the value $g_2 = 2.352$ (rounded). Now we must specify a value for $\delta*$, which denotes the width of the indifference zone (assumed to be the same for <u>both</u> factors).

If we specify $\delta* = 2.0$, then the <u>total</u> number of observations needed on <u>each</u> factor-level combination is obtained by calculating the value of M from the formula in Section 6.3.2, namely,

$$M = \frac{2(2.352)^2(7.33)}{(2.0)^2 3} = 6.8$$

Thus, a <u>total</u> of 7 observations is needed with each factor-level combination. Since $n_0 = 2$ observations have already been taken, 5 additional observations are needed with each of the 12 factor-level combinations.

<u>Note</u>: If the experimenter is unprepared a priori to assume zero or negligible interaction he could use the analogue of the ad hoc procedure described for subset selection, i.e., use a preliminary test of significance for interaction. If non-significance is obtained, proceed as above; if significance is obtained, proceed as in Section 7.3.

8. ACKNOWLEDGMENTS

The writers are pleased to acknowledge the assistance of Professor Ajit Tamhane who read an early draft of this paper and made many helpful comments. We are particularly indebted to Ms. Kathy King for a superb job of typing the manuscript. The research was partially supported at Cornell University by the U.S. Army Research Office through the Mathematical Sciences Institute of Cornell University and by U.S. Army Research Office Contract DAAL03-86-K-0046, and at McMaster University by the Natural Sciences and Engineering Research Council of Canada.

REFERENCES

Abramowitz, M. and Stegun, I. (1964). Handbook of Mathematical Functions.
National Bureau of Standards Applied Mathematics Ser. 55, Washington,
D.C.: U.S. Govt. Printing Office.

Bawa, V.S. (1972). Asymptotic efficiency of one R-factor experiment relative
to R one-factor experiments for selecting the best normal population.
Journal of the American Statistical Association 67, 660-661.

Bechhofer, R.E. (1954). A single-sample multiple decision procedure for
ranking means of normal populations with known variances. Annals of
Mathematical Statistics 25, 16-39.

Bechhofer, R.E. (1969). Optimal allocation of observations when comparing
several treatments with a control. Multivariate Analysis-II (ed. by P.R.
Krishnaiah). New York, Academic Press, 463-473.

Bechhofer, R.E. (1977). Selection in factorial experiments. Proceedings of
the 1977 Winter Simulation Conference (Ed. H.J. Highland, R.G. Sargent
and J.W. Schmidt) held at the National Bureau of Standards, Gaithersburg,
Maryland, pp. 65-70.

Bechhofer, R.E. and Dunnett, C.W. (1982). Multiple comparisons for
orthogonal contrasts: examples and tables. Technometrics 24, 213-222.

Bechhofer, R.E. and Dunnett, C.W. (1986). Two-stage selection of the best
factor-level combination in multi-factor experiments: common unknown
variance. Statistical Design - Theory and Practice. Proceedings of a
conference in honor of W.T. Federer, Biometrics Unit, Cornell University.

Bechhofer, R.E. and Dunnett, C.W. (1987). Subset selection for
normal means in multi-factor experiments. Communications in
Statistics--Theory and Methods A16, No. 8.

Bechhofer, R.E., Dunnett, C.W. and Sobel, M. (1954). A two-sample multiple-
    decision procedure for ranking means of normal populations with a common
    unknown variance. Biometrika 41, 170-176.

Bechhofer, R.E. and Nocturne, D.J. (1972). Optimal allocation of observations
    when comparing several treatments with a control, II: 2-sided
    comparisons. Technometrics 14, 423-436.

Bohrer, R. (1979). Multiple three-decision rules for parametric signs.
    Journal of the American Statistical Association 74, 432-437.

Bohrer, R., Schervish, M. and Sheft, J. (1982). Non-central Studentized
    maximum and related multiple-t probabilities. Algorithm AS 184, Applied
    Statistics 31, 309-317.

Brown, L.D. (1984). A note on the Tukey-Kramer procedure for pairwise compar-
    isons of correlated means. Design of Experiments: Ranking and Selection
    (Ed. T.J. Santner and A.C. Tamhane), New York, Marcel Dekker, 1-6.

Carolis, L.V. de and Gori, F. (1967). Tavole numeriche della t doppia,
    Pubblicazioni dell'Instituto de Calcolo della Probabilita dell'
    Universita di Roma, Serie 2, 64, 1-36.

Carroll, R.J. and Gupta, S.S. (1977). On the probabilities of rankings of k
    populations with applications. Journal of Statistical Computation and
    Simulation 5, 145-157.

Cornish, E.A. (1954). The multivariate t-distribution associated with a set
    of normal sample deviates. Australian Journal of Physics 7, 531-542.

Davies, O.L. (ed.) (1978). The Design and Analysis of Industrial Experiments
    (2nd ed.), London and New York: Longman Group Limited.

Donnelly, T.G. (1973). Bivariate normal distribution. Algorithm 462.
    Communications of the Association of Computing Machines 16, 638.

Dudewicz, E.J. (1971). Non-existence of a single-sample selection procedure whose P(CS) is independent of the variances. South African Statistical Journal 5, 37–39.

Dudewicz, E.J. and Dalal, S.R. (1975). Allocation of observations in ranking and selection with unequal variances. Sankhyā Ser. B 37, 28–78.

Dunlap, W.P., Marx, M.S. and Agamy, G.J. (1981). Fortran IV functions for calculating probabilities associated with Dunnett's test. Behavior Research Methods and Instruction 13(3), 363–366.

Dunn, O.J., Kronmal, R.A. and Yee, W.J. (1968). Tables of the multivariate t-distribution. School of Public Health, University of California at Los Angeles.

Dunn, O.J. and Massey, F.J. (1965). Estimation of multiple contrasts using t-distributions. Journal of the American Statistical Association 60, 573–583.

Dunnett, C.W. (1955). A multiple comparison procedure for comparing several treatments with a control. Journal of the American Statistical Association 50, 1096–1121.

Dunnett, C.W. (1960). Tables of the bivariate normal distribution with correlation $1/\sqrt{2}$. Mathematics of Computation 14, 79.

Dunnett, C.W. (1964). New tables for multiple comparisons with a control. Biometrics 20, 482–491.

Dunnett, C.W. (1985). Multiple comparisons between several treatments and a specified treatment. Lecture Notes in Statistics, 35: Linear Statistical Inference. Berlin: Springer-Verlag, 39–47.

Dunnett, C.W. (1986). Multivariate normal probability integrals with product correlation structure. To appear in Applied Statistics.

Dunnett, C.W. and Lamm, R.A. (1960). Some tables of the multivariate normal probability integral with correlation coefficients 1/3. Mathematics of Computation 14, 290.

Dunnett, C.W. and Sobel, M. (1954). A bivariate generalization of Student's t distribution, with tables for certain special cases. Biometrika 41, 153–169.

Dunnett, C.W. and Sobel, M. (1955). Approximations to the probability integral and certain percentage points of a multivariate analogue of Student's t-distribution. Biometrika 42, 258–260.

Dutt, J.E. (1975). On computing the probability integral of a general multivariate t. Biometrika 62, 201–205.

Dutt, J.E., Mattes, K.D. and Tao, L.C. (1975). Tables of the trivariate t for comparing three treatments to a control with unequal sample sizes. Tech. Report No. 3, Mathematical/Statistical Services, G.D. Searle & Company.

Fisher, R.A. and Yates, F. (1938–1963). Statistical Tables for Biological, Agricultural and Medical Research (6th ed.), New York: Hafner.

Freeman, H. and Kuzmack, A.M. (1972). Tables of multivariate t in six and more dimensions. Biometrika 59, 217–219.

Freeman, H., Kuzmack, A.M., and Maurice, R.J. (1967). Multivariate t and the ranking problem. Biometrika 54, 305–308.

Gupta, S.S. (1956). On a decision rule for a problem in ranking means. Ph.D. thesis (Mimeo Ser. No. 150). Institute of Statistics, University of North Carolina, Chapel Hill.

Gupta, S.S. (1963). Probability integrals of multivariate normal and multivariate t. Annals of Mathematical Statistics 34, 792–828.

Gupta, S.S. (1965). On some multiple decision (selection and ranking) rules. Technometrics 7, 225-245.

Gupta, S.S., Nagel, K. and Panchapakesan, S. (1973). On the order statistics from equally correlated normal random variables. Biometrika 60, 403-413.

Gupta, S.S., Panchapakesan, S. and Sohn, J.K. (1985). On the distribution of the Studentized maximum of equally correlated normal random variables. Communications in Statistics--Simulation and Computation 14(1), 103-135.

Gupta, S.S. and Sobel, M. (1957). On a statistic which arises in selection and ranking problems. Annals of Mathematical Statistics 28, 957-967.

Gupta, S.S. and Sobel, M. (1958). On selecting a subset which contains all populations better than a standard. Annals of Mathematical Statistics 29, 235-244.

Hahn, G.J. and Hendrickson, R.W. (1971). A table of percentage points of the distribution of the largest absolute value of k Student t variates and its applications. Biometrika 58, 323-332.

Halperin, M., Greenhouse, S.W., Cornfield, J. and Zalokar, J. (1955). Tables of percentage points for the studentized maximum absolute deviate in normal samples. Journal of the American Statistical Association 50, 185-195.

Hayter, A.J. (1984). A proof of the conjecture that the Tukey-Kramer multiple comparisons procedure is conservative. Annals of Statistics 12, 61-75.

Hill, I.D. (1973). The normal integral. Algorithm AS66, Applied Statistics 22, 424-427.

Hochberg, Y. (1974). Some conservative generalizations of the T-method in simultaneous inference. Journal of Multivariate Analysis 4, 224-234.

Hochberg, Y. and Tamhane, A.C. (1987). Multiple Comparisons Procedures, New York: John Wiley (to be published).

Hsu, J.C. (1984). Ranking and selection and multiple comparisons with the best. Design of Experiments: Ranking and Selection (Ed. T.J. Santner and A.C. Tamhane), New York, Marcel Dekker, 23-33.

IMSL Library Reference Manual, Edition 9 (1982), Vol. 3, Houston: International Mathematical and Statistical Libraries, Inc.

Johnson, N.L. and Kotz, S. (1972). Distributions in Statistics: Continuous Multivariate Distributions, New York: Wiley.

Krishnaiah, P.R. and Armitage, J.V. (1965). Percentage points of the multivariate t distribution. ARL65-199. Aerospace Research Laboratories, Wright-Patterson Air Force Base, Ohio.

Krishnaiah, P.R. and Armitage, J.V. (1966). Tables for multivariate t-distribution. Sankhyā Ser. B 28, 31-56.

Krishnaish, P.R. and Armitage, J.V. (1970). On a multivariate F distribution, in Essays in Probability and Statistics (S.N. Roy Memorial Volume), Chapel Hill: University of North Carolina Press, pp. 439-468.

Krishnaiah, P.R., Armitage, J.V. and Breiter, M.C. (1969a). Tables for the probability integrals of the bivariate t distribution, ARL-69-0060. Aerospace Research Laboratories, Wright-Patterson Air Force Base, Ohio.

Krishnaiah, P.R., Armitage, J.V. and Breiter, M.C. (1969b). Tables for the bivariate |t| distribution, ARL-69-0210. Aerospace Research Laboratories, Wright-Patterson Air Force Base, Ohio.

Milton, R.C. (1963). Tables of the equally correlated multivariate normal probability integral. Tech. Report No. 27, Department of Statistics, University of Minnesota, Minneapolis.

National Bureau of Standards (1959). Tables of the Bivariate Normal
    Distibution Function and Related Functions. Applied Mathematics Series,
    50, U.S. Department of Commerce, Washington, D.C.

Numerical Algorithms Group (1983). NAG Fortran Library Manual, Mark 10.
    Oxford, U.K.

Odeh, R.E. (1982). Tables of percentage points of the distribution of the
    maximum absolute value of equally correlated normal random variables.
    Communications in Statistics--Simulation and Computation 11(1), 65-87.

Pillai, K.C.S. and Ramachandran, K.V. (1954). On the distribution of the
    ratio of the ith observation in an ordered sample from a normal
    population to an independent estimate of the standard deviation. Annals
    of Mathematical Statistics 25, 565-572.

Rinott, Y. (1978). On two-stage procedures and related probability inequali-
    ties. Communications in Statistics--Theory and Methods A8, 799-811.

Schervish, M.J. (1984). Multivariate normal probabilites with error bound.
    Algorithm AS 195, Applied Statistics 33, 81-94. Corrections in Applied
    Statistics 34 (1985), 103-104.

Siotani, M. (1964). Interval estimation for linear combinations of means.
    Journal of the American Statistical Association 59, 1141-1164.

Smith, H. (1969). The analysis of data from a designed experiment. Journal
    of Quality Technology 1, 259-263.

Snedecor, G.W. and Cochran, W.G. (1980). Statistical Methods, 7th Edition.
    The Iowa State University Press.

Steffens, F.E. (1969). Critical values for bivariate Student t-tests.
    Journal of the American Statistical Association 64, 637-646.

Stoline, M.R. and Ury, H.K. (1979). Tables of the Studentized maximum modulus
    and an application to multiple comparisons among means. Technometrics
    21, 87-93.

Stroud, A.H. and Secrest, D. (1966). <u>Gaussian Quadrature Formulas</u>. Englewood Cliffs, New Jersey: Prentice-Hall.

Tamhane, A.C. and Bechhofer, R.E. (1977). A two-stage minimax procedure with screening for selecting the largest normal mean. <u>Communications in Statistics--Theory and Methods</u> A6, 1003-1033.

Tamhane, A.C. and Bechhofer, R.E. (1979). A two-stage minimax procedure with screening for selecting the largest normal mean (II): an improved PCS lower bound and associated tables. <u>Communications in Statistics--Theory and Methods</u> A8, 337-358.

Teichroew, D. (1955). <u>Probabilities Associated with Order Statistics in Samples from Two Normal Populations with Equal Variances</u>. Chemical Corps Engineering Agency, Army Chemical Center, Maryland.

Tong, Y.L. (1969). On partitioning a set of normal populations by their locations with respect to a control. <u>Annals of Mathematical Statistics</u> 40, 1300-1324.

Trout, R. and Chow, B. (1972). Table of the percentage points of the trivariate t-distribution with an application to uniform confidence bands. <u>Technometrics</u> 14, 855-879.

Ury, H.K., Stoline, M.R., and Mitchell, B.T. (1980). Further tables of the Studentized maximum modulus distribution. <u>Communications in Statistics--Simulation and Computation</u> B9(2), 167-178.

Wilcox, R.R. (1984). A table for Rinott's selection procedure. <u>Journal of Quality Technology</u> 16, 97-100.

EARLIER TABLES

FIGURE 1

Earlier Tables of Percentage Points of Equicorrelated (Case I)
Multivariate t and Multivariate Normal

Reference	Quantity Tabulated	Number of variates(p)	$(1-\alpha)$-values
Pillai, K.C.S. and Ramachandran, K.V. (1954)	g	1(1)8	0.95
	h	1(1)8	0.95
Bechhofer, R.E. (1954)	$g\sqrt{2}$	1(1)9	0.50(0.05)0.80(0.02) 0.90(0.01)0.99, 0.995,0.999,0.9995
	$g\sqrt{2}$	2	0.20(0.05)0.80(0.02) 0.90(0.01)0.99
Dunnett, C.W. and Sobel, M. (1954)	g	2	0.50,0.75,0.90, 0.95,0.99
Dunnett, C.W. (1955)	g and h	1(1)9	0.95,0.99
Gupta, S.S. and Sobel, M. (1957)	$g\sqrt{2}$	1,4,9(1)15 (2)19(5)39, 49	0.75,0.90,0.95 0.975,0.99
Gupta, S.S. (1963)	g	1(1)50	0.75,0.90,0.95 0.975,0.99
Halperin, M., Greenhouse, S.W., Cornfield, J. and Zalokar, J. (1955)	h	2(1)9(5)19 (10)39,59	0.95,0.99
Milton, R.C. (1963)	g	2(1)9(5)24	0.5,0.75,0.9,0.95,0.975, 0.99,0.995,0.999,0.9995, 0.9999 (Table IA)
	g	2(1)9(5)24	0.3(0.05)0.95,0.975,0.99, 0.995,0.999,0.9995,0.9999 (Table IB)
Dunnett, C.W. (1964)	h	1(1)12,15, 20	0.95,0.99
Dunn, O.J. and Massey, F.J., Jr. (1965)	h	2,6,10,20	0.5(0.1)0.9,0.95, 0.975,0.999
Krishnaiah, P.R. and Armitage, J.V. (1965)	g	1(1)10	0.9,0.95,0.975,0.99
Krishnaiah, P.R. and Armitage, J.V. (1966)	g	1(1)10	0.95,0.99
Carolis, L.V. de and Gori, F. (1967)	h	2	0.95,0.99,0.999

ρ-values	d.f.	Number of decimal places
0	3(1)10,12,14(1)16(2)20, 24,30,40,60,120,∞	2
0	5(5)20,24,30,40,60,120,∞	2
0.5	∞	4
-0.5	∞	4
-0.5,0.5	1(1)30(3)60(15)120, 150,300,600,∞	3
0.5	5(1)20,24,30,40,60,120,∞	2
0.5	12(1)20,24,30,36,40,48, 60(20)120,360,∞	2
0.5	∞	3
-1/p	3(1)10(5)20(10)40,60,120,∞	2
0(0.05)1.00,21/41, 11/21,5/9,2/3	∞	6
0.5	∞	6
0.5	5(1)20,24,30,40,60,120,∞	2
0(0.1)1.0	4,10,30,∞	2
0(0.1)0.9	5(1)35	2
0(0.1)0.9	5(1)35	2
-0.75(0.25) +0.75	1(1)40(5)50(10)100,∞	5

Reference	Quantity Tabulated	Number of variates(p)	$(1-\alpha)$-values
Steffens, F.E. (1969)	g	2	0.90,0.95,0.99
	h	2	0.90,0.95,0.99
Krishnaiah, P.R. and Armitage, J.V. (1970)	h^2	1(1)10	0.95,0.99
Hahn, G.J. and Hendrickson, R.W. (1971)	h	1(1)6(2)12, 15,20	0.90,0.95,0.99
Gupta, S.S., Nagel, K. and Panchapakesan, S. (1973)	g	1(1)10(2)50	0.75,0.90,0.95, 0.975,0.99
Dutt, J.E., Mattes, K.D. and Tao, L.C. (1975)	h	3	0.95
			0.99
Stoline, M.R. and Ury, H.K. (1979)	h	p=t(t-1)/2 for t=3(1)20	0.80,0.90,0.95,0.99
Ury, H.K., Stoline, M.R. and Mitchell, B.T. (1980)	h	p=t(t-1)/2 for t=20(2)50 (5)80,90,100	0.80,0.90,0.95,0.99
Bechhofer, R.E. and Dunnett, C.W. (1982)	g	1(1)31	0.90,0.95,0.99
Odeh, R.E. (1982)	h	2(1)40(2)50	0.75,0.90,0.95,0.975 0.99,0.995,0.999
Gupta, S.S., Panchapakesan, S. and Sohn, J.K. (1985)	g	1(1)9(2)19	0.75,0.90,0.95,0.99
	g	1(1)9(2)19	0.90,0.95
Bechhofer, R.E. and Dunnett, C.W.	g	2	0.80,0.90,0.95,0.99
(PRESENT VOLUME)	g	3(1)16,18,20	0.80,0.90,0.95,0.99
	h	2	0.80,0.90,0.95,0.99
	h	3(1)16,18,20	0.80,0.90,0.95,0.99

p-values	d.f.	Number of decimal places
0	$1(1)20(2)30,36,48,60,120,240,600,\infty$	4
0	$1(1)20(2)30,36,48,60,120,240,600,\infty$	4
0.1(0.1)0.9	5(1)35	2
0,0.2,0.4,0.5	$3(1)12,15(5)30,40,60$	3
0.1,0.125,0.2,0.25, 0.3,1/3,0.375,0.4, 0.5,0.6,0.625,2/3, 0.7,0.75,0.8,0.875,0.9	∞	4
selected $(\rho_{12},\rho_{13},\rho_{23})$	selected	4
selected $(\rho_{12},\rho_{13},\rho_{23})$	selected	5
0	$5,7,10,12(4)24,30,40,60,120,\infty$	3
0	$20(1)40(2)60(5)120,240,480,\infty$	3
0	$2(1)12(2)20,24,30,40,60,\infty$	2
0.1,0.125,0.2,0.25,0.3, 1/3,0.375,0.4,0.5,0.6, 0.625,2/3,0.7,0.75,0.8, 0.875,0.9,$1/(1+\sqrt{p})$	∞	4
0.1(0.1)0.6	$15(1)20,24,30,36,48,60,120,\infty$	5
0.7(0.1)0.9	$15,17,20,24,36,60,120,\infty$	5
0.0(0.1)0.9,$1/(1+\sqrt{2})$	$1(1)30(5)50,60,80(20)120,200,\infty$	5
0.0(0.1)0.9,$1/(1+\sqrt{p})$	$2(1)30(5)50,60,80(20)120,200,\infty$	5
0.0(0.1)0.9,$1/(1+\sqrt{2})$	$1(1)30(5)50,60,80(20)120,200,\infty$	5
0.0(0.1)0.9,$1/(1+\sqrt{p})$	$2(1)30(5)50,60,80(20)120,200,\infty$	5

TABLES A.1 to A.4

These tables give the one-sided upper equicoordinate $100(1-\alpha)$ percentage point $g = g(p,v,\rho;\alpha)$ of a central p-variate Student t-distribution based on v d.f. with a correlation matrix $\underline{P} = \{\rho_{ij}\}$ where $\rho_{ij} = \begin{cases} 1 & \text{if } i = j \\ \rho & \text{if } i \ne j \end{cases}$

$(1 \le i,j \le p)$. That is, it is the solution of the equation

$$P\{ \max_{1 \le i \le p} T_i \le g\} = \int_{-\infty}^{g} \cdots \int_{-\infty}^{g} f_v(t_1,\ldots,t_p;\underline{P})dt_1 \ldots dt_p$$

$$= \int_0^\infty \left[\int_{-\infty}^{+\infty} \left\{ \Phi\left[\frac{x\sqrt{\rho} + gy}{\sqrt{1-\rho}} \right] \right\}^p d\Phi(x) \right] q_v(y)dy = 1-\alpha$$

where $f_v(t_1,\ldots,t_p;\underline{P})$ is the p-variate Student t density function, $\Phi(\bullet)$ is the cdf of a standard normal r.v., and $q_v(\bullet)$ is the density function of $x_v/\sqrt{v}$, i.e.,

$$q_v(y) = \frac{2}{\Gamma(v/2)} (v/2)^{v/2} y^{v-1} \exp(-vy^2/2).$$

The constant g is tabulated to 5 decimal places for

$$p = 2(1)16(2)20$$

$$\rho = 0.0(0.1)0.9, \ 1/(1+\sqrt{p})$$

$$1-\alpha = 0.80 \text{ (Table A.1)}, \ 0.90 \text{ (Table A.2)},$$

$$0.95 \text{ (Table A.3)}, \ 0.99 \text{ (Table A.4)}.$$

For $p = 2$: $\quad v = 1(1)30(5)50,60(20)120,200,\infty$

$p > 2$: $\quad v = 2(1)30(5)50,60(20)120,200,\infty$

BECHHOFER and DUNNETT

TABLE A.1

One-sided Percentage Point (g) for Equal Correlations $\begin{pmatrix} Case\ I\ with\ Correlation\ \rho \end{pmatrix}$

$p = 2$, $1-\alpha = 0.80$

	ρ					
ν ↓	0.0	0.1	0.2	0.3	0.4	0.5
1	2.57938	2.51765	2.45238	2.38289	2.30822	2.22703
2	1.73337	1.70636	1.67661	1.64369	1.60702	1.56571
3	1.54173	1.52139	1.49852	1.47277	1.44360	1.41024
4	1.45847	1.44085	1.42081	1.39799	1.37191	1.34183
5	1.41210	1.39596	1.37744	1.35621	1.33179	1.30349
6	1.38259	1.36737	1.34980	1.32956	1.30619	1.27900
7	1.36217	1.34758	1.33066	1.31110	1.28844	1.26201
8	1.34721	1.33308	1.31663	1.29757	1.27542	1.24954
9	1.33578	1.32199	1.30591	1.28721	1.26546	1.24000
10	1.32676	1.31325	1.29744	1.27904	1.25760	1.23246
11	1.31946	1.30617	1.29059	1.27243	1.25123	1.22636
12	1.31344	1.30033	1.28494	1.26697	1.24598	1.22132
13	1.30838	1.29542	1.28019	1.26238	1.24156	1.21709
14	1.30407	1.29124	1.27614	1.25847	1.23780	1.21348
15	1.30036	1.28764	1.27266	1.25511	1.23456	1.21037
16	1.29713	1.28451	1.26962	1.25217	1.23173	1.20767
17	1.29429	1.28175	1.26696	1.24960	1.22925	1.20529
18	1.29178	1.27931	1.26459	1.24732	1.22706	1.20318
19	1.28954	1.27714	1.26249	1.24528	1.22510	1.20130
20	1.28753	1.27519	1.26060	1.24346	1.22334	1.19961
21	1.28571	1.27343	1.25890	1.24181	1.22175	1.19809
22	1.28407	1.27183	1.25735	1.24032	1.22031	1.19671
23	1.28257	1.27038	1.25594	1.23896	1.21900	1.19545
24	1.28120	1.26905	1.25465	1.23771	1.21780	1.19430
25	1.27994	1.26783	1.25347	1.23657	1.21670	1.19325
26	1.27878	1.26670	1.25238	1.23552	1.21569	1.19227
27	1.27771	1.26566	1.25137	1.23454	1.21475	1.19137
28	1.27671	1.26470	1.25044	1.23364	1.21388	1.19054
29	1.27579	1.26380	1.24957	1.23280	1.21307	1.18976
30	1.27493	1.26296	1.24876	1.23202	1.21232	1.18904
35	1.27137	1.25951	1.24542	1.22879	1.20920	1.18605
40	1.26872	1.25694	1.24292	1.22637	1.20688	1.18382
45	1.26666	1.25494	1.24099	1.22450	1.20508	1.18209
50	1.26502	1.25335	1.23944	1.22301	1.20364	1.18071
60	1.26256	1.25096	1.23713	1.22078	1.20149	1.17865
80	1.25950	1.24800	1.23426	1.21800	1.19881	1.17608
100	1.25768	1.24622	1.23254	1.21634	1.19721	1.17454
120	1.25646	1.24504	1.23140	1.21524	1.19615	1.17352
200	1.25404	1.24269	1.22912	1.21303	1.19402	1.17148
∞	1.25042	1.23918	1.22571	1.20974	1.19085	1.16843

TABLE A.1

One-sided Percentage Point (g) for Equal Correlations $\left(Case\ I\ with\ Correlation\ \rho\right)$

$p = 2,\ 1-\alpha = 0.80$

$\nu \downarrow$	0.6	0.7	0.8	0.9	$1/(1+\sqrt{2})$
1	2.13723	2.03529	1.91438	1.75680	2.29712
2	1.51840	1.46278	1.39432	1.30134	1.60146
3	1.37150	1.32533	1.26775	1.18842	1.43913
4	1.30663	1.26440	1.21136	1.13777	1.36790
5	1.27022	1.23011	1.17955	1.10910	1.32803
6	1.24693	1.20816	1.15915	1.09068	1.30258
7	1.23076	1.19291	1.14496	1.07785	1.28493
8	1.21889	1.18170	1.13453	1.06840	1.27199
9	1.20980	1.17312	1.12653	1.06116	1.26209
10	1.20262	1.16634	1.12021	1.05543	1.25427
11	1.19681	1.16084	1.11509	1.05078	1.24794
12	1.19200	1.15630	1.11086	1.04694	1.24271
13	1.18797	1.15249	1.10730	1.04371	1.23832
14	1.18453	1.14923	1.10426	1.04096	1.23458
15	1.18156	1.14643	1.10165	1.03858	1.23136
16	1.17898	1.14399	1.09937	1.03651	1.22855
17	1.17671	1.14184	1.09736	1.03469	1.22609
18	1.17470	1.13994	1.09559	1.03308	1.22390
19	1.17291	1.13824	1.09400	1.03164	1.22195
20	1.17130	1.13672	1.09258	1.03035	1.22020
21	1.16985	1.13535	1.09130	1.02918	1.21863
22	1.16853	1.13410	1.09014	1.02812	1.21720
23	1.16733	1.13296	1.08908	1.02716	1.21589
24	1.16623	1.13193	1.08811	1.02628	1.21470
25	1.16522	1.13097	1.08721	1.02547	1.21360
26	1.16430	1.13009	1.08639	1.02472	1.21260
27	1.16344	1.12928	1.08563	1.02403	1.21166
28	1.16264	1.12852	1.08493	1.02339	1.21080
29	1.16190	1.12782	1.08427	1.02279	1.20999
30	1.16121	1.12717	1.08366	1.02224	1.20924
35	1.15836	1.12447	1.08114	1.01994	1.20615
40	1.15623	1.12245	1.07926	1.01823	1.20384
45	1.15458	1.12089	1.07780	1.01690	1.20204
50	1.15326	1.11964	1.07663	1.01584	1.20061
60	1.15129	1.11778	1.07489	1.01425	1.19848
80	1.14883	1.11545	1.07271	1.01227	1.19581
100	1.14737	1.11406	1.07141	1.01109	1.19422
120	1.14639	1.11314	1.07055	1.01030	1.19316
200	1.14444	1.11129	1.06882	1.00873	1.19105
∞	1.14154	1.10853	1.06624	1.00638	1.18789

BECHHOFER and DUNNETT

TABLE A.1

One-sided Percentage Point (g) for Equal Correlations (*Case I with Correlation* ρ)

p = 3, 1-α = 0.80

$\nu \downarrow$	ρ					
	0.0	0.1	0.2	0.3	0.4	0.5
2	2.14689	2.09757	2.04442	1.98680	1.92377	1.85402
3	1.87068	1.83522	1.79599	1.75246	1.70385	1.64902
4	1.75271	1.72289	1.68933	1.65155	1.60883	1.56011
5	1.68756	1.66079	1.63030	1.59562	1.55608	1.51065
6	1.64630	1.62145	1.59287	1.56013	1.52257	1.47920
7	1.61785	1.59431	1.56704	1.53563	1.49942	1.45745
8	1.59705	1.57446	1.54815	1.51770	1.48248	1.44153
9	1.58118	1.55931	1.53373	1.50401	1.46954	1.42936
10	1.56868	1.54738	1.52237	1.49322	1.45934	1.41977
11	1.55858	1.53774	1.51318	1.48450	1.45109	1.41200
12	1.55024	1.52978	1.50561	1.47731	1.44428	1.40560
13	1.54325	1.52310	1.49925	1.47127	1.43857	1.40022
14	1.53729	1.51742	1.49384	1.46613	1.43370	1.39564
15	1.53217	1.51253	1.48917	1.46170	1.42951	1.39170
16	1.52771	1.50827	1.48512	1.45785	1.42587	1.38826
17	1.52379	1.50453	1.48155	1.45446	1.42266	1.38525
18	1.52032	1.50122	1.47840	1.45147	1.41983	1.38258
19	1.51723	1.49827	1.47559	1.44879	1.41730	1.38020
20	1.51445	1.49562	1.47307	1.44640	1.41503	1.37806
21	1.51195	1.49323	1.47079	1.44424	1.41298	1.37613
22	1.50968	1.49107	1.46873	1.44228	1.41113	1.37439
23	1.50762	1.48910	1.46685	1.44049	1.40944	1.37279
24	1.50573	1.48729	1.46513	1.43886	1.40789	1.37134
25	1.50399	1.48564	1.46356	1.43736	1.40647	1.37000
26	1.50239	1.48411	1.46210	1.43598	1.40517	1.36877
27	1.50092	1.48270	1.46076	1.43470	1.40396	1.36763
28	1.49955	1.48139	1.45951	1.43352	1.40284	1.36658
29	1.49827	1.48018	1.45836	1.43242	1.40180	1.36560
30	1.49709	1.47905	1.45728	1.43140	1.40082	1.36468
35	1.49219	1.47437	1.45282	1.42716	1.39682	1.36090
40	1.48854	1.47088	1.44950	1.42400	1.39383	1.35809
45	1.48570	1.46818	1.44692	1.42156	1.39151	1.35590
50	1.48344	1.46602	1.44487	1.41960	1.38966	1.35416
60	1.48007	1.46280	1.44180	1.41668	1.38689	1.35155
80	1.47586	1.45879	1.43798	1.41305	1.38345	1.34831
100	1.47335	1.45639	1.43569	1.41088	1.38140	1.34637
120	1.47168	1.45479	1.43417	1.40944	1.38003	1.34508
200	1.46835	1.45162	1.43114	1.40656	1.37730	1.34251
∞	1.46338	1.44687	1.42662	1.40226	1.37323	1.33867

TABLE A.1

One-sided Percentage Point (g) for Equal Correlations $\left(Case\ I\ with\ Correlation\ \rho\right)$

p = 3, 1-α = 0.80

ν ↓	ρ				
	0.6	0.7	0.8	0.9	$1/(1+\sqrt{3})$
2	1.77548	1.68469	1.57485	1.42842	1.94586
3	1.58617	1.51226	1.42130	1.29780	1.72100
4	1.50370	1.43674	1.35360	1.23967	1.62396
5	1.45772	1.39452	1.31562	1.20691	1.57012
6	1.42844	1.36759	1.29134	1.18590	1.53593
7	1.40817	1.34893	1.27450	1.17129	1.51231
8	1.39332	1.33524	1.26213	1.16056	1.49503
9	1.38197	1.32478	1.25266	1.15233	1.48183
10	1.37302	1.31652	1.24519	1.14582	1.47143
11	1.36577	1.30983	1.23913	1.14055	1.46302
12	1.35979	1.30431	1.23413	1.13620	1.45608
13	1.35477	1.29967	1.22993	1.13253	1.45025
14	1.35049	1.29572	1.22635	1.12941	1.44530
15	1.34680	1.29231	1.22326	1.12672	1.44103
16	1.34359	1.28935	1.22057	1.12438	1.43731
17	1.34077	1.28674	1.21821	1.12232	1.43404
18	1.33828	1.28444	1.21612	1.12049	1.43115
19	1.33605	1.28238	1.21425	1.11886	1.42857
20	1.33406	1.28053	1.21258	1.11740	1.42626
21	1.33225	1.27887	1.21106	1.11608	1.42418
22	1.33062	1.27736	1.20969	1.11488	1.42228
23	1.32913	1.27598	1.20844	1.11379	1.42056
24	1.32777	1.27472	1.20730	1.11279	1.41899
25	1.32652	1.27357	1.20625	1.11188	1.41754
26	1.32537	1.27250	1.20529	1.11103	1.41621
27	1.32430	1.27152	1.20439	1.11025	1.41498
28	1.32332	1.27060	1.20356	1.10953	1.41383
29	1.32240	1.26975	1.20279	1.10885	1.41277
30	1.32154	1.26896	1.20207	1.10823	1.41178
35	1.31801	1.26569	1.19911	1.10563	1.40770
40	1.31537	1.26325	1.19689	1.10369	1.40465
45	1.31333	1.26136	1.19517	1.10219	1.40229
50	1.31170	1.25985	1.19380	1.10099	1.40040
60	1.30926	1.25760	1.19175	1.09920	1.39759
80	1.30623	1.25479	1.18920	1.09697	1.39408
100	1.30441	1.25311	1.18767	1.09563	1.39198
120	1.30321	1.25199	1.18666	1.09474	1.39059
200	1.30080	1.24976	1.18463	1.09297	1.38781
∞	1.29721	1.24644	1.18160	1.09032	1.38366

BECHHOFER and DUNNETT

TABLE A.1

One-sided Percentage Point (g) for Equal
Correlations (*Case I with Correlation* ρ)

$p = 4, \ 1-\alpha = 0.80$

$\nu \downarrow$	ρ					
	0.0	0.1	0.2	0.3	0.4	0.5
2	2.44549	2.37770	2.30570	2.22862	2.14530	2.05408
3	2.10355	2.05577	2.00359	1.94633	1.88305	1.81234
4	1.95850	1.91893	1.87486	1.82571	1.77062	1.70830
5	1.87865	1.84354	1.80387	1.75911	1.70845	1.65065
6	1.82818	1.79587	1.75896	1.71695	1.66905	1.61407
7	1.79340	1.76302	1.72801	1.68788	1.64187	1.58883
8	1.76799	1.73903	1.70540	1.66664	1.62200	1.57036
9	1.74862	1.72073	1.68815	1.65043	1.60685	1.55627
10	1.73336	1.70632	1.67457	1.63767	1.59490	1.54516
11	1.72103	1.69467	1.66359	1.62735	1.58525	1.53618
12	1.71086	1.68506	1.65454	1.61885	1.57729	1.52877
13	1.70232	1.67701	1.64695	1.61171	1.57061	1.52255
14	1.69506	1.67015	1.64049	1.60564	1.56492	1.51726
15	1.68881	1.66425	1.63492	1.60041	1.56003	1.51270
16	1.68336	1.65911	1.63008	1.59586	1.55576	1.50873
17	1.67858	1.65460	1.62583	1.59186	1.55202	1.50525
18	1.67435	1.65060	1.62207	1.58832	1.54871	1.50217
19	1.67058	1.64704	1.61871	1.58517	1.54576	1.49942
20	1.66720	1.64385	1.61570	1.58234	1.54311	1.49695
21	1.66415	1.64097	1.61299	1.57979	1.54072	1.49473
22	1.66138	1.63836	1.61053	1.57748	1.53856	1.49271
23	1.65886	1.63598	1.60829	1.57537	1.53659	1.49087
24	1.65656	1.63381	1.60624	1.57345	1.53478	1.48920
25	1.65444	1.63181	1.60436	1.57168	1.53313	1.48765
26	1.65249	1.62997	1.60262	1.57005	1.53160	1.48623
27	1.65069	1.62827	1.60102	1.56855	1.53019	1.48492
28	1.64902	1.62669	1.59954	1.56715	1.52888	1.48370
29	1.64747	1.62523	1.59816	1.56585	1.52767	1.48257
30	1.64602	1.62386	1.59687	1.56464	1.52654	1.48151
35	1.64005	1.61823	1.59156	1.55965	1.52186	1.47716
40	1.63559	1.61402	1.58760	1.55593	1.51837	1.47391
45	1.63213	1.61076	1.58453	1.55304	1.51567	1.47139
50	1.62938	1.60816	1.58208	1.55074	1.51352	1.46938
60	1.62526	1.60428	1.57842	1.54730	1.51029	1.46638
80	1.62013	1.59944	1.57387	1.54302	1.50628	1.46264
100	1.61706	1.59655	1.57114	1.54046	1.50389	1.46041
120	1.61503	1.59463	1.56934	1.53876	1.50229	1.45893
200	1.61096	1.59080	1.56573	1.53537	1.49912	1.45597
∞	1.60490	1.58508	1.56034	1.53031	1.49438	1.45155

TABLE A.1

One-sided Percentage Point (g) for Equal Correlations $\left(Case\ I\ with\ Correlation\ \rho\right)$

p = 4, 1-α = 0.80

ν ↓	ρ				
	0.6	0.7	0.8	0.9	$1/(1+\sqrt{4})$
2	1.95243	1.83610	1.69682	1.51324	2.20161
3	1.73201	1.63835	1.52406	1.37030	1.92597
4	1.63670	1.55233	1.44835	1.30700	1.80806
5	1.58375	1.50442	1.40602	1.27140	1.74293
6	1.55012	1.47392	1.37901	1.24861	1.70168
7	1.52688	1.45282	1.36029	1.23279	1.67324
8	1.50987	1.43736	1.34656	1.22116	1.65246
9	1.49688	1.42555	1.33606	1.21226	1.63660
10	1.48664	1.41623	1.32778	1.20522	1.62411
11	1.47835	1.40869	1.32107	1.19952	1.61402
12	1.47152	1.40247	1.31553	1.19481	1.60569
13	1.46578	1.39725	1.31087	1.19085	1.59870
14	1.46090	1.39280	1.30691	1.18748	1.59276
15	1.45669	1.38897	1.30349	1.18457	1.58764
16	1.45303	1.38563	1.30052	1.18204	1.58319
17	1.44981	1.38270	1.29791	1.17981	1.57928
18	1.44697	1.38010	1.29559	1.17784	1.57581
19	1.44443	1.37779	1.29353	1.17608	1.57273
20	1.44215	1.37571	1.29167	1.17451	1.56996
21	1.44010	1.37384	1.29000	1.17308	1.56746
22	1.43823	1.37214	1.28849	1.17179	1.56520
23	1.43654	1.37059	1.28711	1.17061	1.56314
24	1.43499	1.36918	1.28584	1.16953	1.56125
25	1.43356	1.36788	1.28468	1.16854	1.55952
26	1.43225	1.36668	1.28362	1.16763	1.55793
27	1.43104	1.36557	1.28263	1.16679	1.55646
28	1.42991	1.36455	1.28171	1.16601	1.55509
29	1.42886	1.36359	1.28086	1.16528	1.55382
30	1.42789	1.36270	1.28007	1.16460	1.55264
35	1.42387	1.35903	1.27679	1.16180	1.54775
40	1.42086	1.35629	1.27434	1.15971	1.54411
45	1.41854	1.35417	1.27244	1.15809	1.54128
50	1.41668	1.35247	1.27093	1.15680	1.53903
60	1.41390	1.34994	1.26866	1.15487	1.53566
80	1.41045	1.34678	1.26584	1.15246	1.53147
100	1.40838	1.34490	1.26416	1.15102	1.52897
120	1.40701	1.34365	1.26304	1.15006	1.52730
200	1.40427	1.34114	1.26080	1.14815	1.52398
∞	1.40019	1.33741	1.25746	1.14529	1.51903

BECHHOFER and DUNNETT

TABLE A.1

One-sided Percentage Point (g) for Equal
Correlations (*Case I with Correlation* ρ)

p = 5, 1-α = 0.80

$\nu\ \downarrow$	0.0	0.1	0.2	0.3	0.4	0.5
			ρ			
2	2.67744	2.59438	2.50695	2.41412	2.31451	2.20622
3	2.28311	2.22502	2.16218	2.09381	2.01881	1.93556
4	2.11627	2.06854	2.01584	1.95753	1.89260	1.81960
5	2.02452	1.98245	1.93529	1.88243	1.82297	1.75550
6	1.96654	1.92806	1.88438	1.83495	1.77891	1.71490
7	1.92659	1.89059	1.84932	1.80224	1.74854	1.68690
8	1.89741	1.86323	1.82371	1.77835	1.72635	1.66643
9	1.87515	1.84236	1.80418	1.76014	1.70943	1.65081
10	1.85762	1.82593	1.78881	1.74579	1.69610	1.63851
11	1.84345	1.81265	1.77638	1.73420	1.68534	1.62857
12	1.83176	1.80170	1.76614	1.72464	1.67646	1.62037
13	1.82195	1.79251	1.75755	1.71663	1.66901	1.61349
14	1.81360	1.78469	1.75024	1.70981	1.66267	1.60764
15	1.80641	1.77796	1.74394	1.70394	1.65722	1.60260
16	1.80016	1.77210	1.73846	1.69882	1.65247	1.59821
17	1.79466	1.76695	1.73365	1.69434	1.64830	1.59436
18	1.78980	1.76240	1.72939	1.69037	1.64461	1.59095
19	1.78546	1.75834	1.72560	1.68683	1.64132	1.58791
20	1.78157	1.75470	1.72220	1.68365	1.63837	1.58519
21	1.77806	1.75141	1.71913	1.68079	1.63571	1.58273
22	1.77488	1.74844	1.71634	1.67819	1.63330	1.58050
23	1.77198	1.74572	1.71381	1.67583	1.63110	1.57847
24	1.76933	1.74324	1.71149	1.67367	1.62909	1.57662
25	1.76689	1.74097	1.70936	1.67168	1.62725	1.57491
26	1.76465	1.73887	1.70740	1.66986	1.62555	1.57334
27	1.76258	1.73693	1.70559	1.66817	1.62398	1.57189
28	1.76065	1.73513	1.70391	1.66660	1.62252	1.57055
29	1.75887	1.73346	1.70235	1.66514	1.62117	1.56929
30	1.75720	1.73190	1.70089	1.66379	1.61991	1.56813
35	1.75033	1.72547	1.69489	1.65819	1.61470	1.56332
40	1.74519	1.72067	1.69041	1.65401	1.61082	1.55973
45	1.74122	1.71696	1.68693	1.65077	1.60781	1.55695
50	1.73804	1.71399	1.68416	1.64819	1.60541	1.55473
60	1.73330	1.70956	1.68002	1.64433	1.60183	1.55141
80	1.72739	1.70404	1.67487	1.63953	1.59736	1.54729
100	1.72386	1.70074	1.67179	1.63666	1.59470	1.54482
120	1.72152	1.69855	1.66974	1.63475	1.59292	1.54318
200	1.71683	1.69417	1.66566	1.63094	1.58939	1.53992
∞	1.70984	1.68764	1.65957	1.62527	1.58411	1.53504

TABLE A.1

One-sided Percentage Point (g) for Equal Correlations $\left(Case\ I\ with\ Correlation\ \rho\right)$

$$p = 5, \quad 1-\alpha = 0.80$$

$\nu \downarrow$	ρ				
	0.6	0.7	0.8	0.9	$1/(1+\sqrt{5})$
2	2.08636	1.95012	1.78813	1.57626	2.40544
3	1.84157	1.73264	1.60052	1.42393	2.08734
4	1.73620	1.63847	1.51865	1.35669	1.95196
5	1.67782	1.58613	1.47297	1.31894	1.87736
6	1.64078	1.55287	1.44386	1.29481	1.83019
7	1.61522	1.52988	1.42371	1.27805	1.79769
8	1.59652	1.51305	1.40894	1.26575	1.77395
9	1.58225	1.50019	1.39764	1.25634	1.75585
10	1.57100	1.49006	1.38873	1.24890	1.74160
11	1.56191	1.48186	1.38152	1.24288	1.73008
12	1.55441	1.47510	1.37557	1.23790	1.72059
13	1.54811	1.46942	1.37057	1.23372	1.71262
14	1.54275	1.46459	1.36632	1.23016	1.70585
15	1.53814	1.46042	1.36265	1.22708	1.70001
16	1.53412	1.45680	1.35945	1.22441	1.69493
17	1.53060	1.45361	1.35665	1.22206	1.69048
18	1.52748	1.45079	1.35416	1.21998	1.68653
19	1.52469	1.44828	1.35195	1.21812	1.68301
20	1.52220	1.44603	1.34996	1.21645	1.67986
21	1.51995	1.44399	1.34816	1.21495	1.67701
22	1.51790	1.44215	1.34654	1.21358	1.67444
23	1.51604	1.44047	1.34506	1.21234	1.67209
24	1.51434	1.43893	1.34370	1.21120	1.66994
25	1.51278	1.43752	1.34246	1.21016	1.66797
26	1.51134	1.43622	1.34131	1.20920	1.66615
27	1.51001	1.43502	1.34025	1.20831	1.66447
28	1.50878	1.43391	1.33927	1.20748	1.66292
29	1.50764	1.43287	1.33835	1.20671	1.66147
30	1.50657	1.43191	1.33750	1.20600	1.66012
35	1.50216	1.42792	1.33398	1.20304	1.65456
40	1.49887	1.42495	1.33136	1.20084	1.65040
45	1.49632	1.42264	1.32932	1.19913	1.64719
50	1.49429	1.42081	1.32770	1.19776	1.64462
60	1.49125	1.41806	1.32527	1.19572	1.64079
80	1.48747	1.41464	1.32225	1.19318	1.63602
100	1.48521	1.41259	1.32044	1.19166	1.63317
120	1.48371	1.41123	1.31924	1.19065	1.63127
200	1.48071	1.40852	1.31684	1.18864	1.62749
∞	1.47624	1.40447	1.31327	1.18562	1.62185

TABLE A.1

*One-sided Percentage Point (g) for Equal
Correlations* $\begin{bmatrix} Case\ I\ with\ Correlation\ \rho \end{bmatrix}$

p = 6, 1-α = 0.80

ν ↓	0.0	0.1	0.2	0.3	0.4	0.5
2	2.86580	2.77001	2.66974	2.56383	2.45074	2.32838
3	2.42863	2.36177	2.28995	2.21230	2.12757	2.03400
4	2.24380	2.18907	2.12909	2.06310	1.99003	1.90826
5	2.14214	2.09411	2.04061	1.98099	1.91425	1.83887
6	2.07788	2.03411	1.98472	1.92911	1.86635	1.79495
7	2.03360	1.99279	1.94624	1.89339	1.83335	1.76469
8	2.00123	1.96260	1.91813	1.86730	1.80924	1.74257
9	1.97654	1.93958	1.89670	1.84741	1.79087	1.72571
10	1.95708	1.92144	1.87983	1.83175	1.77640	1.71243
11	1.94135	1.90679	1.86619	1.81910	1.76471	1.70170
12	1.92837	1.89470	1.85495	1.80867	1.75507	1.69285
13	1.91747	1.88456	1.84552	1.79992	1.74699	1.68543
14	1.90820	1.87593	1.83750	1.79248	1.74011	1.67911
15	1.90021	1.86850	1.83059	1.78607	1.73419	1.67367
16	1.89326	1.86203	1.82458	1.78050	1.72904	1.66894
17	1.88715	1.85635	1.81930	1.77560	1.72451	1.66478
18	1.88174	1.85132	1.81463	1.77127	1.72051	1.66111
19	1.87692	1.84684	1.81046	1.76740	1.71694	1.65783
20	1.87260	1.84281	1.80672	1.76394	1.71374	1.65489
21	1.86869	1.83919	1.80336	1.76082	1.71085	1.65224
22	1.86515	1.83590	1.80030	1.75799	1.70824	1.64983
23	1.86193	1.83290	1.79752	1.75541	1.70585	1.64765
24	1.85898	1.83016	1.79497	1.75305	1.70368	1.64564
25	1.85627	1.82765	1.79264	1.75088	1.70167	1.64381
26	1.85378	1.82533	1.79049	1.74889	1.69983	1.64211
27	1.85147	1.82319	1.78850	1.74704	1.69813	1.64055
28	1.84933	1.82120	1.78665	1.74534	1.69655	1.63910
29	1.84735	1.81935	1.78494	1.74375	1.69508	1.63775
30	1.84549	1.81763	1.78334	1.74227	1.69372	1.63649
35	1.83784	1.81053	1.77675	1.73616	1.68807	1.63131
40	1.83213	1.80522	1.77183	1.73160	1.68386	1.62744
45	1.82770	1.80111	1.76802	1.72807	1.68060	1.62444
50	1.82417	1.79784	1.76498	1.72525	1.67799	1.62205
60	1.81888	1.79293	1.76043	1.72104	1.67411	1.61848
80	1.81230	1.78683	1.75477	1.71580	1.66927	1.61403
100	1.80837	1.78319	1.75139	1.71267	1.66637	1.61137
120	1.80575	1.78076	1.74914	1.71059	1.66445	1.60961
200	1.80053	1.77592	1.74466	1.70644	1.66062	1.60609
∞	1.79274	1.76870	1.73797	1.70024	1.65490	1.60083

TABLE A.1

One-sided Percentage Point (g) for Equal
Correlations $\left(\text{Case I with Correlation } \rho\right)$

$p = 6, \ 1-\alpha = 0.80$

$\nu \downarrow$	ρ				
	0.6	0.7	0.8	0.9	$1/(1+\sqrt{6})$
2	2.19359	2.04110	1.86070	1.62608	2.57482
3	1.92885	1.80752	1.66104	1.46619	2.22044
4	1.81525	1.70670	1.57417	1.39578	2.07007
5	1.75241	1.65077	1.52577	1.35631	1.98732
6	1.71260	1.61525	1.49497	1.33109	1.93504
7	1.68513	1.59073	1.47366	1.31360	1.89904
8	1.66505	1.57278	1.45804	1.30076	1.87274
9	1.64973	1.55907	1.44611	1.29093	1.85270
10	1.63766	1.54827	1.43669	1.28317	1.83692
11	1.62790	1.53954	1.42908	1.27689	1.82417
12	1.61986	1.53233	1.42279	1.27170	1.81366
13	1.61311	1.52629	1.41751	1.26734	1.80485
14	1.60736	1.52114	1.41302	1.26362	1.79735
15	1.60241	1.51670	1.40915	1.26042	1.79089
16	1.59811	1.51285	1.40577	1.25763	1.78527
17	1.59433	1.50946	1.40281	1.25518	1.78033
18	1.59098	1.50646	1.40019	1.25301	1.77597
19	1.58800	1.50378	1.39785	1.25107	1.77208
20	1.58532	1.50138	1.39576	1.24934	1.76858
21	1.58291	1.49922	1.39386	1.24777	1.76544
22	1.58072	1.49725	1.39215	1.24635	1.76258
23	1.57873	1.49547	1.39058	1.24505	1.75998
24	1.57691	1.49383	1.38915	1.24387	1.75761
25	1.57523	1.49233	1.38784	1.24278	1.75542
26	1.57369	1.49095	1.38663	1.24178	1.75341
27	1.57227	1.48967	1.38551	1.24085	1.75156
28	1.57095	1.48848	1.38448	1.23999	1.74983
29	1.56972	1.48738	1.38351	1.23919	1.74823
30	1.56858	1.48636	1.38261	1.23844	1.74674
35	1.56385	1.48212	1.37890	1.23537	1.74058
40	1.56033	1.47895	1.37613	1.23307	1.73599
45	1.55760	1.47650	1.37399	1.23129	1.73243
50	1.55542	1.47455	1.37228	1.22987	1.72959
60	1.55217	1.47163	1.36972	1.22774	1.72535
80	1.54812	1.46799	1.36653	1.22509	1.72006
100	1.54570	1.46581	1.36463	1.22351	1.71691
120	1.54409	1.46437	1.36336	1.22246	1.71481
200	1.54088	1.46149	1.36083	1.22036	1.71063
∞	1.53609	1.45718	1.35706	1.21722	1.70439

TABLE A.1

One-sided Percentage Point (g) for Equal Correlations $\bigl($ *Case I with Correlation* $\rho\,\bigr)$

$p = 7,\ 1-\alpha = 0.80$

$\nu\ \downarrow$	0.0	0.1	0.2	0.3	0.4	0.5
			ρ			
2	3.02350	2.91698	2.80585	2.68885	2.56435	2.43009
3	2.55055	2.47611	2.39658	2.31099	2.21799	2.11567
4	2.35053	2.28970	2.22342	2.15087	2.07087	1.98167
5	2.24046	2.18720	2.12823	2.06281	1.98988	1.90780
6	2.17083	2.12243	2.06810	2.00719	1.93871	1.86108
7	2.12283	2.07780	2.02669	1.96890	1.90347	1.82890
8	2.08772	2.04519	1.99645	1.94094	1.87774	1.80539
9	2.06092	2.02032	1.97339	1.91962	1.85813	1.78747
10	2.03980	2.00073	1.95523	1.90284	1.84269	1.77336
11	2.02272	1.98489	1.94057	1.88929	1.83022	1.76196
12	2.00862	1.97182	1.92847	1.87811	1.81993	1.75255
13	1.99679	1.96086	1.91832	1.86874	1.81131	1.74467
14	1.98671	1.95153	1.90968	1.86076	1.80397	1.73796
15	1.97803	1.94349	1.90225	1.85389	1.79765	1.73219
16	1.97046	1.93649	1.89577	1.84792	1.79216	1.72716
17	1.96382	1.93035	1.89009	1.84267	1.78733	1.72275
18	1.95794	1.92491	1.88506	1.83803	1.78306	1.71884
19	1.95269	1.92006	1.88058	1.83389	1.77925	1.71536
20	1.94799	1.91571	1.87656	1.83018	1.77584	1.71224
21	1.94374	1.91178	1.87293	1.82683	1.77276	1.70943
22	1.93989	1.90822	1.86964	1.82380	1.76997	1.70687
23	1.93638	1.90498	1.86664	1.82103	1.76743	1.70455
24	1.93317	1.90201	1.86390	1.81851	1.76510	1.70242
25	1.93022	1.89929	1.86139	1.81619	1.76297	1.70047
26	1.92751	1.89678	1.85907	1.81405	1.76100	1.69868
27	1.92500	1.89446	1.85693	1.81207	1.75919	1.69702
28	1.92267	1.89231	1.85494	1.81024	1.75750	1.69548
29	1.92050	1.89031	1.85310	1.80854	1.75594	1.69404
30	1.91848	1.88845	1.85138	1.80695	1.75448	1.69271
35	1.91015	1.88076	1.84428	1.80041	1.74846	1.68721
40	1.90392	1.87501	1.83898	1.79552	1.74397	1.68310
45	1.89910	1.87056	1.83487	1.79173	1.74049	1.67992
50	1.89524	1.86701	1.83160	1.78872	1.73771	1.67738
60	1.88948	1.86170	1.82670	1.78420	1.73357	1.67359
80	1.88231	1.85509	1.82061	1.77859	1.72840	1.66887
100	1.87802	1.85114	1.81696	1.77523	1.72532	1.66605
120	1.87516	1.84851	1.81454	1.77300	1.72327	1.66417
200	1.86947	1.84326	1.80971	1.76855	1.71918	1.66044
∞	1.86096	1.83543	1.80250	1.76192	1.71308	1.65486

TABLE A.1

One-sided Percentage Point (g) for Equal
Correlations   (*Case I with Correlation* ρ)

p = 7, 1-α = 0.80

ν ↓	ρ				
	0.6	0.7	0.8	0.9	$1/(1+\sqrt{7})$
2	2.28270	2.11653	1.92069	1.66711	2.71957
3	2.00111	1.86939	1.71091	1.50090	2.33364
4	1.88057	1.76297	1.61985	1.42785	2.17018
5	1.81398	1.70400	1.56917	1.38695	2.08030
6	1.77181	1.66659	1.53695	1.36083	2.02353
7	1.74274	1.64077	1.51466	1.34272	1.98444
8	1.72149	1.62188	1.49833	1.32943	1.95590
9	1.70528	1.60746	1.48587	1.31926	1.93414
10	1.69252	1.59610	1.47603	1.31123	1.91701
11	1.68220	1.58692	1.46808	1.30473	1.90317
12	1.67369	1.57934	1.46151	1.29936	1.89176
13	1.66656	1.57298	1.45600	1.29485	1.88219
14	1.66049	1.56757	1.45131	1.29101	1.87405
15	1.65526	1.56291	1.44726	1.28770	1.86704
16	1.65071	1.55886	1.44374	1.28482	1.86094
17	1.64671	1.55529	1.44065	1.28228	1.85558
18	1.64317	1.55214	1.43791	1.28004	1.85084
19	1.64002	1.54933	1.43547	1.27804	1.84661
20	1.63719	1.54681	1.43328	1.27624	1.84282
21	1.63464	1.54453	1.43131	1.27462	1.83941
22	1.63233	1.54247	1.42952	1.27315	1.83631
23	1.63023	1.54059	1.42789	1.27181	1.83349
24	1.62830	1.53887	1.42639	1.27059	1.83090
25	1.62654	1.53730	1.42502	1.26946	1.82854
26	1.62491	1.53585	1.42376	1.26842	1.82635
27	1.62340	1.53450	1.42259	1.26747	1.82434
28	1.62201	1.53326	1.42151	1.26658	1.82247
29	1.62071	1.53210	1.42051	1.26575	1.82073
30	1.61950	1.53102	1.41957	1.26498	1.81911
35	1.61451	1.52657	1.41570	1.26180	1.81242
40	1.61079	1.52325	1.41281	1.25943	1.80743
45	1.60791	1.52067	1.41057	1.25758	1.80356
50	1.60561	1.51862	1.40878	1.25612	1.80048
60	1.60217	1.51555	1.40611	1.25392	1.79587
80	1.59789	1.51173	1.40279	1.25119	1.79014
100	1.59534	1.50945	1.40080	1.24955	1.78671
120	1.59364	1.50793	1.39948	1.24847	1.78443
200	1.59025	1.50490	1.39685	1.24630	1.77989
∞	1.58520	1.50038	1.39291	1.24305	1.77311

BECHHOFER and DUNNETT

TABLE A.1

One-sided Percentage Point (g) for Equal Correlations (*Case I with Correlation* ρ)

p = 8, 1-α = 0.80

ν ↓	ρ					
	0.0	0.1	0.2	0.3	0.4	0.5
2	3.15857	3.04289	2.92243	2.79587	2.66152	2.51699
3	2.65515	2.57409	2.48785	2.39535	2.29517	2.18529
4	2.44209	2.37588	2.30408	2.22580	2.13977	2.04415
5	2.32474	2.26686	2.20306	2.13259	2.05428	1.96640
6	2.25046	2.19794	2.13925	2.07372	2.00028	1.91726
7	2.19921	2.15044	2.09530	2.03320	1.96310	1.88342
8	2.16171	2.11572	2.06320	2.00361	1.93596	1.85870
9	2.13307	2.08923	2.03873	1.98105	1.91527	1.83986
10	2.11049	2.06835	2.01945	1.96330	1.89898	1.82503
11	2.09222	2.05148	2.00388	1.94895	1.88583	1.81305
12	2.07713	2.03755	1.99103	1.93712	1.87498	1.80317
13	2.06447	2.02587	1.98025	1.92720	1.86589	1.79488
14	2.05368	2.01592	1.97108	1.91876	1.85815	1.78783
15	2.04438	2.00735	1.96318	1.91149	1.85149	1.78176
16	2.03628	1.99989	1.95631	1.90517	1.84569	1.77648
17	2.02917	1.99333	1.95027	1.89962	1.84060	1.77185
18	2.02286	1.98753	1.94493	1.89470	1.83610	1.76775
19	2.01724	1.98236	1.94016	1.89033	1.83208	1.76409
20	2.01220	1.97771	1.93589	1.88640	1.82848	1.76081
21	2.00764	1.97352	1.93204	1.88285	1.82524	1.75785
22	2.00351	1.96973	1.92854	1.87964	1.82229	1.75517
23	1.99975	1.96627	1.92536	1.87672	1.81961	1.75273
24	1.99631	1.96310	1.92245	1.87404	1.81716	1.75050
25	1.99315	1.96020	1.91978	1.87159	1.81491	1.74845
26	1.99023	1.95752	1.91731	1.86932	1.81284	1.74656
27	1.98754	1.95504	1.91504	1.86723	1.81093	1.74482
28	1.98504	1.95275	1.91293	1.86529	1.80915	1.74320
29	1.98272	1.95061	1.91097	1.86349	1.80750	1.74170
30	1.98055	1.94862	1.90914	1.86181	1.80596	1.74029
35	1.97161	1.94041	1.90159	1.85488	1.79961	1.73451
40	1.96493	1.93428	1.89596	1.84971	1.79488	1.73020
45	1.95974	1.92952	1.89159	1.84570	1.79121	1.72686
50	1.95561	1.92573	1.88811	1.84251	1.78828	1.72419
60	1.94942	1.92006	1.88290	1.83773	1.78391	1.72021
80	1.94171	1.91299	1.87642	1.83179	1.77847	1.71525
100	1.93710	1.90877	1.87255	1.82824	1.77522	1.71229
120	1.93403	1.90596	1.86998	1.82587	1.77306	1.71032
200	1.92791	1.90035	1.86484	1.82117	1.76875	1.70640
∞	1.91876	1.89198	1.85717	1.81414	1.76232	1.70054

TABLE A.1

One-sided Percentage Point (g) for Equal
Correlations $\left(\text{Case I with Correlation } \rho\right)$

$p = 8, \quad 1-\alpha = 0.80$

$\nu \downarrow$	ρ				
	0.6	0.7	0.8	0.9	$1/(1+\sqrt{8})$
2	2.35872	2.18078	1.97168	1.70188	2.84579
3	2.06261	1.92195	1.75319	1.53025	2.43206
4	1.93608	1.81070	1.65852	1.45495	2.25701
5	1.86624	1.74912	1.60589	1.41282	2.16078
6	1.82204	1.71008	1.57244	1.38592	2.10001
7	1.79158	1.68314	1.54932	1.36729	2.05817
8	1.76932	1.66344	1.53239	1.35361	2.02762
9	1.75235	1.64841	1.51946	1.34315	2.00433
10	1.73898	1.63656	1.50926	1.33489	1.98599
11	1.72818	1.62699	1.50102	1.32821	1.97117
12	1.71928	1.61909	1.49421	1.32269	1.95896
13	1.71181	1.61247	1.48850	1.31805	1.94871
14	1.70545	1.60683	1.48363	1.31410	1.93999
15	1.69998	1.60197	1.47944	1.31070	1.93248
16	1.69521	1.59775	1.47580	1.30773	1.92595
17	1.69103	1.59403	1.47259	1.30513	1.92021
18	1.68733	1.59075	1.46976	1.30282	1.91513
19	1.68403	1.58782	1.46723	1.30076	1.91061
20	1.68108	1.58519	1.46496	1.29892	1.90655
21	1.67841	1.58282	1.46291	1.29725	1.90288
22	1.67599	1.58068	1.46106	1.29574	1.89956
23	1.67379	1.57872	1.45937	1.29437	1.89654
24	1.67177	1.57693	1.45782	1.29311	1.89378
25	1.66992	1.57529	1.45640	1.29195	1.89124
26	1.66822	1.57378	1.45510	1.29088	1.88890
27	1.66665	1.57238	1.45389	1.28990	1.88674
28	1.66519	1.57108	1.45277	1.28899	1.88474
29	1.66383	1.56988	1.45172	1.28814	1.88287
30	1.66257	1.56875	1.45075	1.28735	1.88114
35	1.65735	1.56412	1.44674	1.28408	1.87397
40	1.65345	1.56066	1.44375	1.28164	1.86863
45	1.65044	1.55797	1.44143	1.27975	1.86448
50	1.64803	1.55584	1.43958	1.27824	1.86118
60	1.64444	1.55264	1.43682	1.27598	1.85624
80	1.63996	1.54866	1.43338	1.27317	1.85009
100	1.63729	1.54629	1.43132	1.27149	1.84642
120	1.63551	1.54471	1.42995	1.27038	1.84397
200	1.63197	1.54155	1.42722	1.26816	1.83910
∞	1.62668	1.53685	1.42315	1.26481	1.83183

TABLE A.1

One-sided Percentage Point (g) for Equal
Correlations $\left(\text{Case I with Correlation } \rho\right)$

$$p = 9, \quad 1-\alpha = 0.80$$

$\nu \downarrow$	ρ 0.0	0.1	0.2	0.3	0.4	0.5
2	3.27633	3.15272	3.02413	2.88922	2.74624	2.59270
3	2.74654	2.65964	2.56745	2.46886	2.36236	2.24583
4	2.52212	2.45110	2.37439	2.29103	2.19969	2.09842
5	2.39839	2.33635	2.26826	2.19329	2.11023	2.01727
6	2.32001	2.26377	2.20120	2.13157	2.05374	1.96599
7	2.26589	2.21374	2.15502	2.08908	2.01486	1.93068
8	2.22627	2.17715	2.12127	2.05805	1.98646	1.90490
9	2.19600	2.14923	2.09554	2.03440	1.96483	1.88525
10	2.17211	2.12722	2.07527	2.01578	1.94780	1.86979
11	2.15279	2.10942	2.05889	2.00073	1.93404	1.85729
12	2.13682	2.09474	2.04538	1.98833	1.92270	1.84699
13	2.12341	2.08241	2.03405	1.97793	1.91319	1.83835
14	2.11199	2.07191	2.02440	1.96907	1.90509	1.83100
15	2.10214	2.06286	2.01609	1.96145	1.89813	1.82468
16	2.09356	2.05499	2.00886	1.95482	1.89207	1.81917
17	2.08602	2.04807	2.00251	1.94900	1.88674	1.81434
18	2.07933	2.04194	1.99688	1.94384	1.88204	1.81006
19	2.07338	2.03648	1.99187	1.93925	1.87784	1.80625
20	2.06803	2.03157	1.98737	1.93513	1.87407	1.80283
21	2.06320	2.02715	1.98332	1.93141	1.87068	1.79975
22	2.05882	2.02314	1.97964	1.92804	1.86760	1.79695
23	2.05483	2.01948	1.97629	1.92497	1.86480	1.79441
24	2.05118	2.01614	1.97323	1.92217	1.86224	1.79208
25	2.04782	2.01307	1.97041	1.91959	1.85988	1.78994
26	2.04473	2.01024	1.96782	1.91722	1.85772	1.78798
27	2.04187	2.00762	1.96543	1.91502	1.85571	1.78616
28	2.03922	2.00520	1.96320	1.91299	1.85386	1.78447
29	2.03676	2.00294	1.96114	1.91110	1.85213	1.78291
30	2.03446	2.00084	1.95922	1.90934	1.85052	1.78144
35	2.02496	1.99216	1.95127	1.90207	1.84388	1.77542
40	2.01787	1.98568	1.94534	1.89664	1.83893	1.77092
45	2.01236	1.98065	1.94074	1.89244	1.83510	1.76744
50	2.00797	1.97664	1.93708	1.88908	1.83204	1.76466
60	2.00139	1.97064	1.93159	1.88407	1.82746	1.76051
80	1.99320	1.96317	1.92477	1.87783	1.82177	1.75534
100	1.98830	1.95870	1.92069	1.87411	1.81837	1.75226
120	1.98503	1.95573	1.91798	1.87163	1.81611	1.75021
200	1.97852	1.94980	1.91256	1.86669	1.81161	1.74612
∞	1.96879	1.94093	1.90448	1.85931	1.80488	1.74001

TABLE A.1

One-sided Percentage Point (g) for Equal
Correlations (*Case I with Correlation* ρ)

$$p = 9, \; 1-\alpha = 0.80$$

$\nu \downarrow$	ρ				
	0.6	0.7	0.8	0.9	$1/(1+\sqrt{9})$
2.	2.42490	2.23663	2.01593	1.73198	2.95756
3	2.11603	1.96754	1.78981	1.55562	2.51904
4	1.98423	1.85206	1.69198	1.47835	2.33362
5	1.91154	1.78819	1.63764	1.43514	2.23170
6	1.86556	1.74771	1.60312	1.40758	2.16733
7	1.83388	1.71979	1.57926	1.38848	2.12300
8	1.81073	1.69938	1.56180	1.37447	2.09063
9	1.79308	1.68380	1.54847	1.36376	2.06595
10	1.77919	1.67154	1.53795	1.35530	2.04652
11	1.76796	1.66162	1.52945	1.34845	2.03081
12	1.75870	1.65344	1.52244	1.34280	2.01786
13	1.75094	1.64658	1.51655	1.33805	2.00700
14	1.74434	1.64074	1.51153	1.33401	1.99775
15	1.73865	1.63572	1.50722	1.33052	1.98979
16	1.73370	1.63134	1.50346	1.32749	1.98287
17	1.72935	1.62750	1.50016	1.32482	1.97678
18	1.72551	1.62410	1.49723	1.32246	1.97140
19	1.72208	1.62107	1.49463	1.32035	1.96660
20	1.71901	1.61835	1.49229	1.31846	1.96229
21	1.71624	1.61589	1.49018	1.31676	1.95841
22	1.71372	1.61367	1.48827	1.31521	1.95489
23	1.71143	1.61165	1.48653	1.31380	1.95168
24	1.70934	1.60979	1.48494	1.31251	1.94875
25	1.70742	1.60809	1.48347	1.31133	1.94606
26	1.70565	1.60653	1.48213	1.31024	1.94357
27	1.70402	1.60508	1.48088	1.30923	1.94128
28	1.70250	1.60374	1.47973	1.30830	1.93916
29	1.70109	1.60249	1.47865	1.30743	1.93718
30	1.69978	1.60133	1.47765	1.30662	1.93534
35	1.69436	1.59653	1.47352	1.30327	1.92774
40	1.69031	1.59295	1.47044	1.30078	1.92206
45	1.68718	1.59017	1.46805	1.29884	1.91767
50	1.68468	1.58796	1.46615	1.29730	1.91416
60	1.68095	1.58465	1.46330	1.29499	1.90892
80	1.67630	1.58054	1.45975	1.29211	1.90239
100	1.67352	1.57808	1.45763	1.29040	1.89849
120	1.67168	1.57644	1.45622	1.28926	1.89590
200	1.66800	1.57318	1.45341	1.28699	1.89073
∞	1.66251	1.56832	1.44922	1.28356	1.88301

BECHHOFER and DUNNETT

TABLE A.1

One-sided Percentage Point (g) for Equal
Correlations ⎡*Case I with Correlation* ρ⎤

p = 10, 1-α = 0.80

ν ↓	ρ					
	0.0	0.1	0.2	0.3	0.4	0.5
2	3.38046	3.24991	3.11415	2.97185	2.82121	2.65967
3	2.82754	2.73542	2.63794	2.53391	2.42177	2.29931
4	2.59307	2.51774	2.43663	2.34872	2.25262	2.14631
5	2.46369	2.39789	2.32593	2.24694	2.15963	2.06213
6	2.38166	2.32206	2.25599	2.18266	2.10092	2.00894
7	2.32499	2.26976	2.20780	2.13841	2.06051	1.97233
8	2.28347	2.23150	2.17258	2.10610	2.03100	1.94560
9	2.25173	2.20230	2.14573	2.08147	2.00852	1.92523
10	2.22668	2.17927	2.12457	2.06208	1.99082	1.90919
11	2.20640	2.16065	2.10747	2.04641	1.97652	1.89624
12	2.18964	2.14527	2.09336	2.03349	1.96474	1.88556
13	2.17556	2.13236	2.08153	2.02265	1.95485	1.87661
14	2.16356	2.12137	2.07145	2.01343	1.94644	1.86899
15	2.15321	2.11189	2.06277	2.00549	1.93920	1.86243
16	2.14419	2.10364	2.05522	1.99858	1.93290	1.85673
17	2.13627	2.09640	2.04858	1.99251	1.92737	1.85172
18	2.12925	2.08998	2.04271	1.98714	1.92248	1.84729
19	2.12298	2.08425	2.03747	1.98235	1.91812	1.84334
20	2.11736	2.07911	2.03277	1.97806	1.91421	1.83979
21	2.11228	2.07448	2.02854	1.97419	1.91068	1.83660
22	2.10767	2.07027	2.02469	1.97068	1.90748	1.83370
23	2.10347	2.06644	2.02119	1.96748	1.90457	1.83106
24	2.09963	2.06294	2.01799	1.96455	1.90190	1.82865
25	2.09610	2.05972	2.01505	1.96187	1.89946	1.82644
26	2.09285	2.05675	2.01234	1.95940	1.89721	1.82440
27	2.08984	2.05401	2.00984	1.95711	1.89513	1.82251
28	2.08705	2.05147	2.00752	1.95499	1.89320	1.82077
29	2.08445	2.04910	2.00536	1.95302	1.89140	1.81914
30	2.08204	2.04690	2.00335	1.95119	1.88973	1.81763
35	2.07204	2.03779	1.99504	1.94361	1.88283	1.81138
40	2.06456	2.03099	1.98884	1.93795	1.87769	1.80672
45	2.05876	2.02572	1.98404	1.93357	1.87370	1.80311
50	2.05413	2.02151	1.98020	1.93008	1.87052	1.80023
60	2.04720	2.01521	1.97447	1.92485	1.86577	1.79593
80	2.03857	2.00737	1.96733	1.91835	1.85985	1.79058
100	2.03340	2.00268	1.96307	1.91447	1.85632	1.78738
120	2.02996	1.99956	1.96023	1.91188	1.85397	1.78526
200	2.02309	1.99333	1.95457	1.90673	1.84929	1.78102
∞	2.01281	1.98402	1.94612	1.89904	1.84230	1.77469

TABLE A.1

One-sided Percentage Point (g) for Equal
Correlations   (*Case I with Correlation* ρ)

p = 10, 1-α = 0.80

ν ↓	ρ				
	0.6	0.7	0.8	0.9	1/(1+√10)
2	2.48339	2.28596	2.05496	1.75847	3.05774
3	2.16317	2.00773	1.82204	1.57792	2.59693
4	2.02669	1.88848	1.72141	1.49890	2.40212
5	1.95145	1.82257	1.66555	1.45474	2.29504
6	1.90388	1.78081	1.63008	1.42658	2.22740
7	1.87111	1.75202	1.60557	1.40707	2.18081
8	1.84717	1.73097	1.58763	1.39277	2.14678
9	1.82892	1.71492	1.57394	1.38183	2.12083
10	1.81456	1.70227	1.56315	1.37320	2.10039
11	1.80295	1.69206	1.55442	1.36621	2.08388
12	1.79338	1.68363	1.54722	1.36044	2.07026
13	1.78535	1.67656	1.54117	1.35559	2.05883
14	1.77853	1.67054	1.53603	1.35146	2.04910
15	1.77265	1.66536	1.53159	1.34791	2.04073
16	1.76753	1.66085	1.52774	1.34481	2.03344
17	1.76304	1.65689	1.52435	1.34209	2.02704
18	1.75907	1.65339	1.52135	1.33968	2.02137
19	1.75552	1.65026	1.51867	1.33753	2.01632
20	1.75235	1.64746	1.51628	1.33560	2.01179
21	1.74948	1.64494	1.51411	1.33386	2.00770
22	1.74689	1.64265	1.51215	1.33228	2.00399
23	1.74452	1.64056	1.51036	1.33084	2.00062
24	1.74236	1.63865	1.50873	1.32953	1.99753
25	1.74037	1.63690	1.50723	1.32832	1.99470
26	1.73854	1.63529	1.50585	1.32721	1.99208
27	1.73685	1.63380	1.50457	1.32618	1.98967
28	1.73529	1.63241	1.50338	1.32523	1.98743
29	1.73383	1.63113	1.50228	1.32434	1.98535
30	1.73247	1.62993	1.50126	1.32351	1.98341
35	1.72687	1.62499	1.49702	1.32010	1.97541
40	1.72269	1.62130	1.49386	1.31755	1.96943
45	1.71946	1.61844	1.49140	1.31558	1.96480
50	1.71687	1.61616	1.48945	1.31400	1.96111
60	1.71301	1.61276	1.48653	1.31165	1.95559
80	1.70821	1.60852	1.48289	1.30872	1.94871
100	1.70535	1.60599	1.48072	1.30697	1.94460
120	1.70344	1.60430	1.47927	1.30581	1.94187
200	1.69964	1.60094	1.47639	1.30350	1.93642
∞	1.69396	1.59593	1.47209	1.29999	1.92829

BECHHOFER and DUNNETT

TABLE A.1

One-sided Percentage Point (g) for Equal
Correlations   (Case I with Correlation ρ)

p = 11, 1-α = 0.80

ν ↓	ρ					
	0.0	0.1	0.2	0.3	0.4	0.5
2	3.47361	3.33690	3.19476	3.04586	2.88836	2.71963
3	2.90016	2.80335	2.70109	2.59216	2.47494	2.34714
4	2.65673	2.57748	2.49237	2.40035	2.29996	2.18912
5	2.52228	2.45306	2.37758	2.29494	2.20380	2.10220
6	2.43698	2.37430	2.30503	2.22837	2.14308	2.04730
7	2.37800	2.31995	2.25504	2.18253	2.10129	2.00951
8	2.33476	2.28019	2.21850	2.14906	2.07078	1.98192
9	2.30170	2.24982	2.19062	2.12354	2.04753	1.96090
10	2.27559	2.22587	2.16866	2.10344	2.02923	1.94435
11	2.25444	2.20649	2.15091	2.08721	2.01444	1.93098
12	2.23696	2.19049	2.13626	2.07382	2.00225	1.91996
13	2.22227	2.17706	2.12397	2.06259	1.99203	1.91072
14	2.20975	2.16562	2.11350	2.05303	1.98334	1.90286
15	2.19894	2.15575	2.10449	2.04480	1.97585	1.89610
16	2.18953	2.14716	2.09664	2.03764	1.96934	1.89021
17	2.18125	2.13962	2.08975	2.03135	1.96362	1.88504
18	2.17392	2.13293	2.08365	2.02579	1.95856	1.88047
19	2.16737	2.12697	2.07821	2.02083	1.95405	1.87639
20	2.16150	2.12161	2.07333	2.01638	1.95000	1.87274
21	2.15619	2.11678	2.06892	2.01236	1.94635	1.86944
22	2.15138	2.11240	2.06493	2.00872	1.94304	1.86645
23	2.14699	2.10841	2.06129	2.00541	1.94003	1.86373
24	2.14297	2.10476	2.05797	2.00238	1.93728	1.86124
25	2.13928	2.10141	2.05491	1.99960	1.93475	1.85896
26	2.13588	2.09831	2.05210	1.99703	1.93242	1.85685
27	2.13273	2.09546	2.04949	1.99466	1.93027	1.85491
28	2.12981	2.09281	2.04708	1.99247	1.92827	1.85310
29	2.12710	2.09034	2.04484	1.99043	1.92642	1.85143
30	2.12457	2.08804	2.04275	1.98852	1.92469	1.84987
35	2.11411	2.07855	2.03411	1.98067	1.91756	1.84342
40	2.10629	2.07146	2.02767	1.97480	1.91223	1.83862
45	2.10022	2.06596	2.02267	1.97026	1.90811	1.83489
50	2.09537	2.06157	2.01869	1.96664	1.90482	1.83192
60	2.08811	2.05500	2.01273	1.96122	1.89991	1.82748
80	2.07907	2.04682	2.00530	1.95448	1.89379	1.82196
100	2.07365	2.04192	2.00087	1.95045	1.89014	1.81866
120	2.07004	2.03866	1.99792	1.94777	1.88771	1.81647
200	2.06284	2.03216	1.99203	1.94243	1.88286	1.81209
∞	2.05207	2.02244	1.98323	1.93445	1.87563	1.80557

TABLE A.1

One-sided Percentage Point (g) for Equal
Correlations (*Case I with Correlation* ρ)

p = 11, 1-α = 0.80

ν ↓	ρ				
	0.6	0.7	0.8	0.9	$1/(1+\sqrt{11})$
2	2.53574	2.33006	2.08982	1.78210	3.14842
3	2.20530	2.04361	1.85080	1.59777	2.66739
4	2.06460	1.92098	1.74764	1.51719	2.46404
5	1.98708	1.85322	1.69041	1.47218	2.35225
6	1.93807	1.81032	1.65408	1.44349	2.28160
7	1.90432	1.78074	1.62899	1.42361	2.23294
8	1.87967	1.75913	1.61063	1.40904	2.19739
9	1.86088	1.74264	1.59662	1.39790	2.17027
10	1.84608	1.72965	1.58557	1.38911	2.14891
11	1.83413	1.71916	1.57664	1.38199	2.13165
12	1.82428	1.71051	1.56927	1.37612	2.11741
13	1.81602	1.70325	1.56308	1.37118	2.10546
14	1.80899	1.69707	1.55782	1.36698	2.09529
15	1.80294	1.69175	1.55328	1.36336	2.08653
16	1.79767	1.68712	1.54934	1.36020	2.07891
17	1.79305	1.68306	1.54587	1.35743	2.07221
18	1.78896	1.67946	1.54280	1.35498	2.06628
19	1.78531	1.67626	1.54006	1.35279	2.06100
20	1.78204	1.67338	1.53761	1.35083	2.05626
21	1.77910	1.67079	1.53540	1.34906	2.05198
22	1.77642	1.66844	1.53339	1.34745	2.04810
23	1.77399	1.66629	1.53156	1.34599	2.04457
24	1.77176	1.66434	1.52989	1.34465	2.04134
25	1.76972	1.66254	1.52836	1.34342	2.03837
26	1.76784	1.66088	1.52694	1.34229	2.03564
27	1.76610	1.65935	1.52563	1.34124	2.03311
28	1.76448	1.65793	1.52442	1.34027	2.03077
29	1.76299	1.65662	1.52330	1.33937	2.02859
30	1.76159	1.65539	1.52225	1.33853	2.02656
35	1.75582	1.65031	1.51791	1.33505	2.01819
40	1.75152	1.64653	1.51468	1.33246	2.01193
45	1.74819	1.64360	1.51217	1.33045	2.00708
50	1.74553	1.64126	1.51017	1.32885	2.00321
60	1.74156	1.63776	1.50718	1.32645	1.99743
80	1.73662	1.63341	1.50346	1.32347	1.99023
100	1.73367	1.63081	1.50124	1.32169	1.98593
120	1.73171	1.62908	1.49976	1.32051	1.98307
200	1.72780	1.62564	1.49682	1.31816	1.97736
∞	1.72196	1.62050	1.49242	1.31459	1.96883

TABLE A.1

One-sided Percentage Point (g) for Equal
Correlations $\left(\text{Case I with Correlation } \rho\right)$

p = 12, 1-α = 0.80

ν ↓	ρ					
	0.0	0.1	0.2	0.3	0.4	0.5
2	3.55775	3.41554	3.26767	3.11280	2.94909	2.77386
3	2.96588	2.86482	2.75823	2.64485	2.52301	2.39037
4	2.71439	2.63156	2.54281	2.44704	2.34275	2.22778
5	2.57537	2.50300	2.42431	2.33833	2.24369	2.13837
6	2.48709	2.42158	2.34938	2.26966	2.18115	2.08191
7	2.42602	2.36538	2.29775	2.22238	2.13810	2.04305
8	2.38122	2.32423	2.26000	2.18785	2.10668	2.01468
9	2.34695	2.29281	2.23120	2.16153	2.08273	1.99306
10	2.31987	2.26802	2.20850	2.14079	2.06388	1.97604
11	2.29793	2.24795	2.19015	2.12404	2.04865	1.96230
12	2.27979	2.23138	2.17501	2.11023	2.03609	1.95097
13	2.26454	2.21747	2.16230	2.09864	2.02556	1.94147
14	2.25154	2.20561	2.15148	2.08878	2.01661	1.93339
15	2.24032	2.19539	2.14216	2.08028	2.00890	1.92643
16	2.23054	2.18649	2.13405	2.07289	2.00219	1.92038
17	2.22194	2.17867	2.12692	2.06640	1.99629	1.91507
18	2.21432	2.17174	2.12061	2.06066	1.99108	1.91036
19	2.20752	2.16556	2.11498	2.05554	1.98643	1.90617
20	2.20141	2.16001	2.10993	2.05094	1.98227	1.90241
21	2.19589	2.15500	2.10538	2.04680	1.97851	1.89902
22	2.19088	2.15046	2.10125	2.04304	1.97510	1.89595
23	2.18632	2.14632	2.09748	2.03962	1.97200	1.89315
24	2.18214	2.14253	2.09404	2.03649	1.96916	1.89060
25	2.17830	2.13905	2.09088	2.03362	1.96656	1.88825
26	2.17476	2.13585	2.08797	2.03097	1.96416	1.88608
27	2.17149	2.13288	2.08528	2.02853	1.96194	1.88409
28	2.16845	2.13013	2.08278	2.02626	1.95988	1.88223
29	2.16563	2.12757	2.08046	2.02415	1.95797	1.88051
30	2.16300	2.12519	2.07830	2.02219	1.95619	1.87890
35	2.15211	2.11534	2.06936	2.01408	1.94884	1.87228
40	2.14396	2.10798	2.06269	2.00802	1.94336	1.86734
45	2.13764	2.10227	2.05752	2.00333	1.93911	1.86351
50	2.13259	2.09772	2.05339	1.99959	1.93572	1.86045
60	2.12503	2.09090	2.04722	1.99400	1.93066	1.85589
80	2.11561	2.08240	2.03954	1.98703	1.92436	1.85021
100	2.10996	2.07732	2.03494	1.98287	1.92059	1.84682
120	2.10620	2.07393	2.03189	1.98011	1.91809	1.84457
200	2.09869	2.06718	2.02579	1.97459	1.91310	1.84007
∞	2.08746	2.05708	2.01668	1.96635	1.90565	1.83336

TABLE A.1

One-sided Percentage Point (g) for Equal
Correlations (Case I with Correlation ρ)

p = 12, 1-α = 0.80

ν ↓	ρ				
	0.6	0.7	0.8	0.9	1/(1+√12)
2	2.58306	2.36991	2.12129	1.80339	3.23118
3	2.24335	2.07600	1.87672	1.61565	2.73168
4	2.09882	1.95028	1.77127	1.53365	2.52051
5	2.01921	1.88085	1.71280	1.48786	2.40437
6	1.96890	1.83691	1.67569	1.45869	2.33097
7	1.93425	1.80662	1.65007	1.43848	2.28039
8	1.90895	1.78448	1.63133	1.42367	2.24343
9	1.88967	1.76760	1.61702	1.41235	2.21524
10	1.87449	1.75430	1.60574	1.40341	2.19302
11	1.86223	1.74356	1.59662	1.39618	2.17506
12	1.85212	1.73470	1.58910	1.39021	2.16025
13	1.84364	1.72727	1.58279	1.38520	2.14781
14	1.83643	1.72095	1.57742	1.38093	2.13723
15	1.83022	1.71550	1.57279	1.37725	2.12812
16	1.82482	1.71077	1.56876	1.37404	2.12018
17	1.82007	1.70661	1.56522	1.37123	2.11321
18	1.81588	1.70293	1.56209	1.36873	2.10704
19	1.81214	1.69964	1.55930	1.36651	2.10154
20	1.80878	1.69670	1.55680	1.36452	2.09660
21	1.80576	1.69405	1.55454	1.36272	2.09215
22	1.80301	1.69164	1.55249	1.36109	2.08811
23	1.80052	1.68945	1.55063	1.35960	2.08443
24	1.79823	1.68745	1.54892	1.35824	2.08107
25	1.79614	1.68561	1.54735	1.35699	2.07798
26	1.79421	1.68391	1.54591	1.35584	2.07513
27	1.79242	1.68235	1.54458	1.35478	2.07250
28	1.79077	1.68089	1.54334	1.35379	2.07006
29	1.78923	1.67955	1.54219	1.35287	2.06779
30	1.78780	1.67829	1.54112	1.35202	2.06568
35	1.78188	1.67310	1.53670	1.34849	2.05695
40	1.77747	1.66922	1.53340	1.34586	2.05043
45	1.77405	1.66622	1.53084	1.34382	2.04538
50	1.77133	1.66383	1.52880	1.34219	2.04135
60	1.76725	1.66025	1.52575	1.33975	2.03533
80	1.76219	1.65580	1.52196	1.33673	2.02782
100	1.75916	1.65314	1.51969	1.33492	2.02334
120	1.75715	1.65137	1.51818	1.33372	2.02035
200	1.75313	1.64785	1.51518	1.33134	2.01440
∞	1.74714	1.64259	1.51069	1.32770	2.00551

TABLE A.1

One-sided Percentage Point (g) for Equal
Correlations (*Case I with Correlation ρ*)

p = 13, 1-α = 0.80

ν ↓	ρ					
	0.0	0.1	0.2	0.3	0.4	0.5
2	3.63437	3.48720	3.33413	3.17385	3.00448	2.82332
3	3.02585	2.92091	2.81036	2.69290	2.56684	2.42976
4	2.76704	2.68092	2.58883	2.48962	2.38174	2.26299
5	2.62385	2.54858	2.46693	2.37788	2.28003	2.17130
6	2.53286	2.46473	2.38983	2.30730	2.21582	2.11341
7	2.46988	2.40682	2.33669	2.25870	2.17163	2.07357
8	2.42365	2.36442	2.29783	2.22320	2.13936	2.04448
9	2.38827	2.33202	2.26818	2.19613	2.11477	2.02232
10	2.36030	2.30646	2.24481	2.17481	2.09541	2.00487
11	2.33764	2.28577	2.22591	2.15758	2.07978	1.99078
12	2.31889	2.26867	2.21031	2.14338	2.06689	1.97917
13	2.30312	2.25431	2.19722	2.13146	2.05607	1.96943
14	2.28968	2.24208	2.18608	2.12131	2.04688	1.96115
15	2.27807	2.23153	2.17648	2.11258	2.03896	1.95401
16	2.26796	2.22234	2.16811	2.10497	2.03207	1.94781
17	2.25906	2.21426	2.16077	2.09830	2.02602	1.94236
18	2.25117	2.20711	2.15427	2.09239	2.02066	1.93754
19	2.24413	2.20073	2.14847	2.08712	2.01589	1.93324
20	2.23780	2.19500	2.14326	2.08240	2.01161	1.92939
21	2.23209	2.18982	2.13857	2.07813	2.00775	1.92592
22	2.22690	2.18513	2.13431	2.07427	2.00425	1.92277
23	2.22218	2.18085	2.13043	2.07075	2.00107	1.91990
24	2.21785	2.17694	2.12688	2.06753	1.99815	1.91728
25	2.21388	2.17335	2.12363	2.06457	1.99548	1.91487
26	2.21021	2.17003	2.12062	2.06185	1.99301	1.91265
27	2.20682	2.16697	2.11785	2.05933	1.99074	1.91060
28	2.20367	2.16413	2.11527	2.05700	1.98862	1.90870
29	2.20075	2.16149	2.11288	2.05483	1.98666	1.90694
30	2.19802	2.15902	2.11065	2.05281	1.98483	1.90529
35	2.18673	2.14884	2.10144	2.04446	1.97728	1.89850
40	2.17829	2.14123	2.09456	2.03823	1.97165	1.89343
45	2.17173	2.13533	2.08923	2.03341	1.96729	1.88951
50	2.16649	2.13062	2.08497	2.02956	1.96381	1.88638
60	2.15865	2.12357	2.07861	2.02380	1.95861	1.88170
80	2.14887	2.11478	2.07068	2.01663	1.95213	1.87588
100	2.14301	2.10952	2.06594	2.01235	1.94827	1.87240
120	2.13910	2.10602	2.06279	2.00950	1.94570	1.87009
200	2.13131	2.09903	2.05650	2.00382	1.94057	1.86548
∞	2.11964	2.08858	2.04710	1.99534	1.93292	1.85861

TABLE A.1

One-sided Percentage Point (g) for Equal
Correlations (Case I with Correlation ρ)

p = 13, 1-α = 0.80

ν ↓	ρ				
	0.6	0.7	0.8	0.9	1/(1+√13)
2	2.62620	2.40622	2.14994	1.82275	3.30723
3	2.27801	2.10547	1.90030	1.63190	2.79077
4	2.12996	1.97693	1.79275	1.54859	2.57238
5	2.04845	1.90598	1.73314	1.50210	2.45223
6	1.99694	1.86108	1.69533	1.47249	2.37627
7	1.96148	1.83013	1.66922	1.45198	2.32392
8	1.93558	1.80752	1.65012	1.43695	2.28564
9	1.91585	1.79028	1.63554	1.42546	2.25644
10	1.90031	1.77670	1.62406	1.41639	2.23343
11	1.88776	1.76573	1.61477	1.40905	2.21483
12	1.87742	1.75668	1.60711	1.40300	2.19948
13	1.86874	1.74909	1.60068	1.39791	2.18659
14	1.86136	1.74264	1.59521	1.39358	2.17562
15	1.85501	1.73708	1.59050	1.38984	2.16617
16	1.84948	1.73224	1.58639	1.38659	2.15795
17	1.84463	1.72799	1.58279	1.38374	2.15072
18	1.84033	1.72423	1.57960	1.38121	2.14432
19	1.83650	1.72088	1.57676	1.37896	2.13862
20	1.83307	1.71788	1.57421	1.37693	2.13350
21	1.82998	1.71517	1.57191	1.37511	2.12888
22	1.82717	1.71271	1.56983	1.37345	2.12469
23	1.82462	1.71048	1.56793	1.37194	2.12088
24	1.82228	1.70843	1.56619	1.37057	2.11739
25	1.82014	1.70655	1.56460	1.36930	2.11418
26	1.81816	1.70482	1.56313	1.36813	2.11123
27	1.81633	1.70322	1.56177	1.36705	2.10850
28	1.81464	1.70174	1.56051	1.36605	2.10597
29	1.81307	1.70036	1.55934	1.36512	2.10362
30	1.81160	1.69908	1.55825	1.36426	2.10143
35	1.80555	1.69378	1.55375	1.36068	2.09237
40	1.80104	1.68983	1.55039	1.35801	2.08560
45	1.79754	1.68676	1.54778	1.35594	2.08036
50	1.79475	1.68432	1.54571	1.35428	2.07618
60	1.79059	1.68067	1.54260	1.35182	2.06992
80	1.78540	1.67613	1.53874	1.34875	2.06213
100	1.78230	1.67341	1.53643	1.34692	2.05748
120	1.78024	1.67161	1.53490	1.34570	2.05438
200	1.77614	1.66801	1.53184	1.34330	2.04820
∞	1.77001	1.66264	1.52727	1.33959	2.03896

TABLE A.1

One-sided Percentage Point (g) for Equal Correlations $\left(Case\ I\ with\ Correlation\ \rho\right)$

$p = 14,\ 1-\alpha = 0.80$

ν ↓	0.0	0.1	0.2	0.3	0.4	0.5
2	3.70464	3.55296	3.39515	3.22992	3.05536	2.86873
3	3.08094	2.97244	2.85825	2.73704	2.60709	2.46592
4	2.81544	2.72629	2.63110	2.52872	2.41754	2.29530
5	2.66844	2.59048	2.50608	2.41420	2.31339	2.20151
6	2.57496	2.50439	2.42699	2.34186	2.24763	2.14229
7	2.51021	2.44492	2.37245	2.29203	2.20238	2.10154
8	2.46267	2.40135	2.33257	2.25563	2.16933	2.07180
9	2.42627	2.36806	2.30213	2.22788	2.14415	2.04913
10	2.39748	2.34177	2.27814	2.20602	2.12433	2.03129
11	2.37414	2.32050	2.25873	2.18835	2.10832	2.01688
12	2.35483	2.30292	2.24272	2.17378	2.09511	2.00500
13	2.33858	2.28815	2.22927	2.16155	2.08404	1.99504
14	2.32473	2.27556	2.21783	2.15115	2.07462	1.98657
15	2.31277	2.26471	2.20796	2.14219	2.06651	1.97928
16	2.30234	2.25526	2.19937	2.13439	2.05945	1.97293
17	2.29316	2.24694	2.19183	2.12754	2.05325	1.96736
18	2.28502	2.23958	2.18514	2.12148	2.04777	1.96243
19	2.27776	2.23301	2.17918	2.11608	2.04288	1.95804
20	2.27123	2.22711	2.17384	2.11123	2.03850	1.95410
21	2.26534	2.22178	2.16901	2.10686	2.03454	1.95054
22	2.25999	2.21695	2.16464	2.10289	2.03096	1.94732
23	2.25511	2.21255	2.16065	2.09928	2.02770	1.94439
24	2.25064	2.20852	2.15700	2.09598	2.02471	1.94171
25	2.24654	2.20482	2.15366	2.09295	2.02197	1.93925
26	2.24275	2.20141	2.15057	2.09015	2.01945	1.93698
27	2.23925	2.19825	2.14772	2.08757	2.01711	1.93488
28	2.23600	2.19532	2.14507	2.08517	2.01495	1.93294
29	2.23298	2.19260	2.14261	2.08295	2.01294	1.93113
30	2.23016	2.19006	2.14032	2.08087	2.01107	1.92945
35	2.21850	2.17958	2.13085	2.07231	2.00334	1.92250
40	2.20978	2.17174	2.12377	2.06592	1.99757	1.91732
45	2.20301	2.16565	2.11829	2.06096	1.99310	1.91331
50	2.19759	2.16080	2.11392	2.05701	1.98953	1.91011
60	2.18949	2.15353	2.10737	2.05110	1.98420	1.90532
80	2.17937	2.14446	2.09922	2.04375	1.97757	1.89937
100	2.17331	2.13904	2.09434	2.03935	1.97361	1.89582
120	2.16927	2.13543	2.09110	2.03643	1.97098	1.89345
200	2.16120	2.12822	2.08463	2.03060	1.96572	1.88874
∞	2.14912	2.11744	2.07496	2.02190	1.95788	1.88171

TABLE A.1

One-sided Percentage Point (g) for Equal
Correlations   (Case I with Correlation ρ)

p = 14, 1-α = 0.80

ν ↓	ρ				
	0.6	0.7	0.8	0.9	1/(1+√14)
2	2.66582	2.43955	2.17622	1.84050	3.37754
3	2.30980	2.13250	1.92190	1.64676	2.84541
4	2.15852	2.00136	1.81242	1.56227	2.62032
5	2.07525	1.92900	1.75177	1.51513	2.49645
6	2.02264	1.88321	1.71330	1.48511	2.41811
7	1.98642	1.85166	1.68674	1.46432	2.36410
8	1.95998	1.82861	1.66732	1.44909	2.32460
9	1.93982	1.81104	1.65249	1.43744	2.29446
10	1.92396	1.79720	1.64081	1.42826	2.27071
11	1.91115	1.78602	1.63137	1.42082	2.25150
12	1.90058	1.77680	1.62358	1.41469	2.23564
13	1.89172	1.76906	1.61704	1.40953	2.22234
14	1.88419	1.76248	1.61148	1.40514	2.21101
15	1.87770	1.75682	1.60669	1.40136	2.20124
16	1.87206	1.75189	1.60252	1.39807	2.19274
17	1.86710	1.74756	1.59886	1.39518	2.18528
18	1.86272	1.74373	1.59562	1.39261	2.17866
19	1.85881	1.74032	1.59273	1.39033	2.17277
20	1.85531	1.73726	1.59014	1.38828	2.16748
21	1.85215	1.73450	1.58780	1.38643	2.16270
22	1.84928	1.73199	1.58568	1.38476	2.15838
23	1.84668	1.72971	1.58375	1.38323	2.15443
24	1.84429	1.72763	1.58198	1.38183	2.15082
25	1.84210	1.72571	1.58036	1.38055	2.14751
26	1.84008	1.72395	1.57887	1.37937	2.14446
27	1.83822	1.72232	1.57749	1.37827	2.14164
28	1.83649	1.72081	1.57621	1.37726	2.13902
29	1.83489	1.71941	1.57502	1.37632	2.13659
30	1.83339	1.71810	1.57391	1.37544	2.13432
35	1.82721	1.71270	1.56934	1.37182	2.12495
40	1.82260	1.70867	1.56592	1.36911	2.11796
45	1.81903	1.70555	1.56327	1.36701	2.11253
50	1.81619	1.70306	1.56116	1.36534	2.10821
60	1.81193	1.69934	1.55801	1.36284	2.10173
80	1.80664	1.69471	1.55408	1.35973	2.09367
100	1.80348	1.69195	1.55174	1.35788	2.08885
120	1.80138	1.69011	1.55018	1.35665	2.08565
200	1.79719	1.68644	1.54707	1.35422	2.07925
∞	1.79093	1.68098	1.54242	1.35046	2.06969

TABLE A.1

One-sided Percentage Point (g) for Equal
Correlations (*Case I with Correlation* ρ)

p = 15, 1-α = 0.80

ν ↓	ρ					
	0.0	0.1	0.2	0.3	0.4	0.5
2	3.76947	3.61367	3.45151	3.28171	3.10237	2.91070
3	3.13185	3.02007	2.90250	2.77783	2.64427	2.49931
4	2.86020	2.76823	2.67018	2.56484	2.45060	2.32513
5	2.70968	2.62922	2.54226	2.44775	2.34419	2.22939
6	2.61391	2.54106	2.46132	2.37377	2.27700	2.16895
7	2.54753	2.48013	2.40550	2.32281	2.23076	2.12736
8	2.49877	2.43549	2.36467	2.28558	2.19700	2.09700
9	2.46142	2.40136	2.33350	2.25719	2.17127	2.07386
10	2.43187	2.37442	2.30893	2.23482	2.15101	2.05565
11	2.40790	2.35260	2.28905	2.21675	2.13464	2.04095
12	2.38807	2.33457	2.27264	2.20184	2.12115	2.02883
13	2.37138	2.31942	2.25887	2.18933	2.10984	2.01866
14	2.35714	2.30650	2.24714	2.17869	2.10021	2.01001
15	2.34484	2.29537	2.23703	2.16952	2.09192	2.00257
16	2.33412	2.28566	2.22823	2.16154	2.08471	1.99609
17	2.32468	2.27713	2.22050	2.15453	2.07837	1.99041
18	2.31632	2.26957	2.21365	2.14832	2.07277	1.98538
19	2.30884	2.26282	2.20754	2.14279	2.06777	1.98089
20	2.30213	2.25677	2.20206	2.13783	2.06329	1.97687
21	2.29607	2.25130	2.19711	2.13336	2.05925	1.97324
22	2.29056	2.24634	2.19263	2.12930	2.05559	1.96996
23	2.28554	2.24182	2.18854	2.12560	2.05225	1.96696
24	2.28095	2.23768	2.18480	2.12222	2.04920	1.96423
25	2.27672	2.23388	2.18137	2.11912	2.04640	1.96171
26	2.27282	2.23037	2.17821	2.11626	2.04382	1.95940
27	2.26922	2.22713	2.17528	2.11361	2.04144	1.95726
28	2.26588	2.22412	2.17257	2.11116	2.03923	1.95528
29	2.26276	2.22133	2.17005	2.10888	2.03717	1.95343
30	2.25986	2.21872	2.16769	2.10676	2.03526	1.95171
35	2.24785	2.20794	2.15798	2.09799	2.02735	1.94463
40	2.23887	2.19989	2.15073	2.09145	2.02145	1.93934
45	2.23189	2.19364	2.14511	2.08638	2.01688	1.93524
50	2.22631	2.18864	2.14062	2.08233	2.01324	1.93197
60	2.21795	2.18117	2.13390	2.07628	2.00779	1.92709
80	2.20752	2.17185	2.12554	2.06875	2.00102	1.92102
100	2.20127	2.16628	2.12054	2.06425	1.99697	1.91739
120	2.19710	2.16256	2.11721	2.06125	1.99427	1.91498
200	2.18878	2.15515	2.11057	2.05528	1.98890	1.91017
∞	2.17631	2.14405	2.10064	2.04637	1.98089	1.90299

TABLE A.1

One-sided Percentage Point (g) for Equal Correlations (*Case I with Correlation* ρ)

$$p = 15, \quad 1-\alpha = 0.80$$

$\nu \downarrow$	ρ				
	0.6	0.7	0.8	0.9	$1/(1+\sqrt{15})$
2	2.70242	2.47034	2.20048	1.85686	3.44286
3	2.33916	2.15745	1.94183	1.66046	2.89619
4	2.18488	2.02389	1.83056	1.57486	2.66488
5	2.09998	1.95022	1.76893	1.52713	2.53753
6	2.04634	1.90362	1.72985	1.49673	2.45696
7	2.00942	1.87151	1.70288	1.47569	2.40140
8	1.98247	1.84805	1.68316	1.46026	2.36076
9	1.96193	1.83017	1.66811	1.44848	2.32974
10	1.94576	1.81609	1.65625	1.43918	2.30528
11	1.93270	1.80471	1.64666	1.43165	2.28550
12	1.92193	1.79533	1.63875	1.42544	2.26918
13	1.91291	1.78746	1.63211	1.42023	2.25547
14	1.90523	1.78077	1.62647	1.41579	2.24380
15	1.89862	1.77500	1.62160	1.41196	2.23374
16	1.89287	1.76999	1.61737	1.40863	2.22499
17	1.88782	1.76559	1.61365	1.40570	2.21729
18	1.88335	1.76169	1.61036	1.40311	2.21048
19	1.87937	1.75822	1.60743	1.40080	2.20440
20	1.87580	1.75510	1.60480	1.39872	2.19895
21	1.87258	1.75229	1.60242	1.39685	2.19403
22	1.86966	1.74975	1.60027	1.39516	2.18957
23	1.86700	1.74743	1.59831	1.39361	2.18550
24	1.86457	1.74531	1.59652	1.39220	2.18178
25	1.86234	1.74336	1.59488	1.39090	2.17837
26	1.86028	1.74157	1.59336	1.38970	2.17522
27	1.85838	1.73991	1.59196	1.38860	2.17231
28	1.85662	1.73838	1.59066	1.38757	2.16961
29	1.85498	1.73695	1.58945	1.38662	2.16710
30	1.85346	1.73562	1.58833	1.38573	2.16477
35	1.84717	1.73012	1.58368	1.38206	2.15511
40	1.84247	1.72603	1.58022	1.37932	2.14789
45	1.83883	1.72285	1.57753	1.37720	2.14230
50	1.83593	1.72032	1.57539	1.37551	2.13783
60	1.83160	1.71654	1.57219	1.37298	2.13116
80	1.82620	1.71183	1.56820	1.36984	2.12284
100	1.82298	1.70902	1.56582	1.36797	2.11786
120	1.82084	1.70715	1.56424	1.36673	2.11455
200	1.81657	1.70342	1.56108	1.36427	2.10795
∞	1.81020	1.69785	1.55637	1.36046	2.09808

BECHHOFER and DUNNETT

One-sided Percentage Point (g) for Equal
Correlations [*Case I with Correlation* ρ]

p = ˙6, 1-α = 0.80

ν ↓	ρ					
	0.0	0.1	0.2	0.3	0.4	0.5
2	3.82960	3.67000	3.50383	3.32981	3.14603	2.94968
3	3.17914	3.06431	2.94362	2.81572	2.67880	2.53032
4	2.90181	2.80721	2.70648	2.59840	2.48130	2.35281
5	2.74803	2.66522	2.57588	2.47891	2.37278	2.25526
6	2.65013	2.57514	2.49322	2.40340	2.30426	2.19368
7	2.58224	2.51286	2.43620	2.35139	2.25710	2.15130
8	2.53235	2.46722	2.39448	2.31338	2.22267	2.12037
9	2.49411	2.43231	2.36263	2.28440	2.19642	2.09680
10	2.46385	2.40475	2.33752	2.26156	2.17576	2.07825
11	2.43930	2.38242	2.31720	2.24311	2.15907	2.06326
12	2.41897	2.36397	2.30043	2.22788	2.14531	2.05091
13	2.40187	2.34846	2.28635	2.21511	2.13376	2.04056
14	2.38727	2.33524	2.27435	2.20424	2.12394	2.03175
15	2.37466	2.32384	2.26402	2.19487	2.11549	2.02416
16	2.36366	2.31391	2.25502	2.18672	2.10813	2.01756
17	2.35398	2.30517	2.24711	2.17956	2.10167	2.01177
18	2.34540	2.29742	2.24010	2.17323	2.09595	2.00664
19	2.33773	2.29051	2.23386	2.16758	2.09085	2.00207
20	2.33084	2.28431	2.22825	2.16251	2.08628	1.99798
21	2.32462	2.27871	2.22319	2.15794	2.08216	1.99428
22	2.31897	2.27363	2.21860	2.15379	2.07842	1.99093
23	2.31382	2.26899	2.21442	2.15001	2.07502	1.98788
24	2.30910	2.26475	2.21060	2.14656	2.07191	1.98509
25	2.30476	2.26086	2.20709	2.14339	2.06905	1.98253
26	2.30076	2.25727	2.20385	2.14047	2.06642	1.98018
27	2.29706	2.25394	2.20086	2.13777	2.06398	1.97800
28	2.29362	2.25086	2.19808	2.13526	2.06173	1.97598
29	2.29042	2.24800	2.19550	2.13294	2.05963	1.97410
30	2.28744	2.24532	2.19309	2.13077	2.05768	1.97235
35	2.27511	2.23428	2.18316	2.12181	2.04961	1.96512
40	2.26587	2.22602	2.17573	2.11512	2.04360	1.95974
45	2.25870	2.21961	2.16998	2.10994	2.03893	1.95556
50	2.25296	2.21449	2.16538	2.10580	2.03522	1.95223
60	2.24437	2.20682	2.15851	2.09962	2.02966	1.94726
80	2.23364	2.19726	2.14995	2.09193	2.02274	1.94107
100	2.22721	2.19154	2.14483	2.08733	2.01861	1.93737
120	2.22292	2.18773	2.14142	2.08427	2.01586	1.93491
200	2.21436	2.18012	2.13462	2.07816	2.01038	1.93001
∞	2.20152	2.16873	2.12446	2.06905	2.00220	1.92270

TABLE A.1

One-sided Percentage Point (g) for Equal
Correlations $\left(\text{Case I with Correlation } \rho\right)$

$$p = 16, \quad 1-\alpha = 0.80$$

ν ↓	ρ 0.6	0.7	0.8	0.9	$1/(1+\sqrt{16})$
2	2.73641	2.49892	2.22299	1.87203	3.50383
3	2.36641	2.18059	1.96030	1.67316	2.94362
4	2.20933	2.04479	1.84736	1.58653	2.70648
5	2.12291	1.96990	1.78483	1.53823	2.57588
6	2.06832	1.92254	1.74519	1.50748	2.49322
7	2.03075	1.88990	1.71784	1.48620	2.43620
8	2.00332	1.86607	1.69783	1.47061	2.39448
9	1.98242	1.84790	1.68256	1.45869	2.36263
10	1.96597	1.83359	1.67054	1.44929	2.33752
11	1.95268	1.82203	1.66082	1.44168	2.31720
12	1.94172	1.81250	1.65280	1.43540	2.30043
13	1.93254	1.80450	1.64607	1.43013	2.28635
14	1.92473	1.79770	1.64034	1.42564	2.27435
15	1.91800	1.79185	1.63541	1.42177	2.26402
16	1.91215	1.78675	1.63112	1.41840	2.25502
17	1.90701	1.78228	1.62735	1.41544	2.24711
18	1.90246	1.77832	1.62401	1.41282	2.24010
19	1.89841	1.77480	1.62104	1.41048	2.23386
20	1.89478	1.77163	1.61837	1.40839	2.22825
21	1.89150	1.76878	1.61597	1.40649	2.22319
22	1.88853	1.76619	1.61378	1.40478	2.21860
23	1.88583	1.76384	1.61180	1.40322	2.21442
24	1.88335	1.76168	1.60998	1.40179	2.21060
25	1.88108	1.75970	1.60831	1.40048	2.20709
26	1.87899	1.75788	1.60678	1.39927	2.20385
27	1.87706	1.75620	1.60536	1.39815	2.20086
28	1.87527	1.75464	1.60404	1.39711	2.19808
29	1.87360	1.75319	1.60282	1.39615	2.19550
30	1.87205	1.75184	1.60167	1.39525	2.19309
35	1.86565	1.74626	1.59697	1.39155	2.18316
40	1.86087	1.74210	1.59345	1.38878	2.17573
45	1.85717	1.73887	1.59073	1.38663	2.16998
50	1.85422	1.73630	1.58856	1.38492	2.16538
60	1.84981	1.73246	1.58531	1.38237	2.15851
80	1.84432	1.72768	1.58127	1.37919	2.14995
100	1.84105	1.72482	1.57886	1.37730	2.14483
120	1.83887	1.72292	1.57726	1.37605	2.14142
200	1.83452	1.71913	1.57406	1.37357	2.13462
∞	1.82804	1.71348	1.56928	1.36971	2.12446

TABLE A.1

One-sided Percentage Point (g) for Equal
Correlations   (Case I with Correlation ρ)

p = 18, 1-α = 0.80

	ρ					
ν ↓	0.0	0.1	0.2	0.3	0.4	0.5
2	3.93807	3.77168	3.59832	3.41671	3.22494	3.02014
3	3.26459	3.14428	3.01794	2.88420	2.74121	2.58635
4	2.97706	2.87770	2.77213	2.65905	2.53676	2.40282
5	2.81743	2.73035	2.63667	2.53522	2.42443	2.30197
6	2.71569	2.63680	2.55089	2.45695	2.35349	2.23831
7	2.64507	2.57207	2.49170	2.40302	2.30466	2.19451
8	2.59312	2.52461	2.44837	2.36361	2.26901	2.16254
9	2.55328	2.48830	2.41528	2.33354	2.24183	2.13817
10	2.52173	2.45960	2.38918	2.30985	2.22043	2.11900
11	2.49612	2.43636	2.36807	2.29070	2.20314	2.10351
12	2.47491	2.41714	2.35063	2.27490	2.18889	2.09074
13	2.45704	2.40098	2.33599	2.26165	2.17693	2.08004
14	2.44179	2.38720	2.32352	2.25036	2.16676	2.07093
15	2.42862	2.37532	2.31277	2.24064	2.15800	2.06309
16	2.41712	2.36496	2.30340	2.23218	2.15038	2.05627
17	2.40699	2.35584	2.29517	2.22475	2.14368	2.05028
18	2.39801	2.34777	2.28788	2.21817	2.13776	2.04498
19	2.38999	2.34056	2.28138	2.21230	2.13248	2.04026
20	2.38278	2.33408	2.27555	2.20704	2.12774	2.03602
21	2.37626	2.32824	2.27028	2.20229	2.12347	2.03220
22	2.37034	2.32293	2.26551	2.19799	2.11960	2.02874
23	2.36494	2.31809	2.26115	2.19406	2.11607	2.02559
24	2.36000	2.31367	2.25717	2.19048	2.11285	2.02271
25	2.35545	2.30960	2.25351	2.18718	2.10989	2.02006
26	2.35126	2.30585	2.25014	2.18415	2.10716	2.01762
27	2.34738	2.30238	2.24703	2.18134	2.10464	2.01537
28	2.34378	2.29916	2.24414	2.17874	2.10230	2.01328
29	2.34043	2.29617	2.24145	2.17632	2.10013	2.01134
30	2.33730	2.29337	2.23894	2.17407	2.09811	2.00953
35	2.32436	2.28183	2.22859	2.16476	2.08975	2.00206
40	2.31466	2.27320	2.22086	2.15782	2.08351	1.99649
45	2.30713	2.26649	2.21486	2.15243	2.07868	1.99218
50	2.30111	2.26114	2.21007	2.14814	2.07483	1.98874
60	2.29208	2.25312	2.20291	2.14171	2.06907	1.98359
80	2.28080	2.24312	2.19398	2.13371	2.06190	1.97720
100	2.27403	2.23713	2.18864	2.12893	2.05762	1.97337
120	2.26952	2.23314	2.18509	2.12575	2.05477	1.97083
200	2.26051	2.22517	2.17800	2.11941	2.04909	1.96576
∞	2.24699	2.21324	2.16740	2.10993	2.04061	1.95820

TABLE A.1

One-sided Percentage Point (g) for Equal
Correlations (*Case I with Correlation* ρ)

p = 18, 1-α = 0.80

ν ↓	ρ				
	0.6	0.7	0.8	0.9	1/(1+√18)
2	2.79784	2.55056	2.26364	1.89939	3.61469
3	2.41562	2.22237	1.99363	1.69603	3.02992
4	2.25348	2.08249	1.87766	1.60754	2.78219
5	2.16429	2.00539	1.81349	1.55823	2.64564
6	2.10798	1.95664	1.77283	1.52685	2.55916
7	2.06922	1.92306	1.74477	1.50514	2.49947
8	2.04092	1.89853	1.72425	1.48923	2.45577
9	2.01936	1.87984	1.70860	1.47707	2.42239
10	2.00239	1.86512	1.69627	1.46748	2.39606
11	1.98869	1.85323	1.68630	1.45972	2.37476
12	1.97739	1.84343	1.67808	1.45331	2.35716
13	1.96792	1.83520	1.67119	1.44794	2.34239
14	1.95986	1.82821	1.66532	1.44336	2.32980
15	1.95292	1.82219	1.66026	1.43941	2.31895
16	1.94689	1.81695	1.65586	1.43598	2.30950
17	1.94159	1.81235	1.65200	1.43296	2.30119
18	1.93690	1.80828	1.64858	1.43029	2.29383
19	1.93272	1.80465	1.64554	1.42791	2.28727
20	1.92898	1.80140	1.64280	1.42577	2.28138
21	1.92560	1.79846	1.64034	1.42384	2.27606
22	1.92254	1.79580	1.63810	1.42209	2.27124
23	1.91975	1.79338	1.63607	1.42050	2.26685
24	1.91720	1.79117	1.63420	1.41904	2.26282
25	1.91486	1.78913	1.63250	1.41771	2.25913
26	1.91270	1.78726	1.63092	1.41647	2.25573
27	1.91071	1.78553	1.62947	1.41533	2.25258
28	1.90886	1.78392	1.62812	1.41428	2.24966
29	1.90714	1.78243	1.62686	1.41330	2.24695
30	1.90554	1.78104	1.62569	1.41238	2.24442
35	1.89894	1.77531	1.62087	1.40860	2.23396
40	1.89401	1.77103	1.61727	1.40578	2.22615
45	1.89020	1.76771	1.61448	1.40359	2.22009
50	1.88716	1.76507	1.61226	1.40185	2.21525
60	1.88261	1.76112	1.60893	1.39925	2.20802
80	1.87695	1.75620	1.60479	1.39601	2.19900
100	1.87357	1.75326	1.60232	1.39409	2.19360
120	1.87133	1.75131	1.60068	1.39281	2.19001
200	1.86685	1.74742	1.59740	1.39030	2.18285
∞	1.86016	1.74161	1.59250	1.38634	2.17213

TABLE A.1

One-sided Percentage Point (g) for Equal Correlations (*Case I with Correlation* ρ)

p = 20, 1-α = 0.80

ν ↓	ρ					
	0.0	0.1	0.2	0.3	0.4	0.5
2	4.03371	3.86140	3.68175	3.49349	3.29469	3.08242
3	3.34011	3.21497	3.08364	2.94474	2.79638	2.63586
4	3.04364	2.94006	2.83018	2.71268	2.58578	2.44698
5	2.87886	2.78798	2.69043	2.58500	2.47005	2.34321
6	2.77374	2.69135	2.60189	2.50428	2.39698	2.27771
7	2.70071	2.62447	2.54077	2.44864	2.34666	2.23264
8	2.64695	2.57539	2.49602	2.40798	2.30992	2.19974
9	2.60569	2.53783	2.46183	2.37695	2.28191	2.17467
10	2.57300	2.50814	2.43485	2.35250	2.25986	2.15494
11	2.54644	2.48407	2.41303	2.33273	2.24204	2.13900
12	2.52444	2.46417	2.39500	2.31642	2.22734	2.12587
13	2.50590	2.44743	2.37985	2.30273	2.21502	2.11485
14	2.49007	2.43316	2.36695	2.29108	2.20453	2.10548
15	2.47638	2.42084	2.35583	2.28104	2.19550	2.09741
16	2.46443	2.41010	2.34614	2.27230	2.18764	2.09039
17	2.45391	2.40065	2.33763	2.26463	2.18074	2.08423
18	2.44458	2.39227	2.33008	2.25783	2.17463	2.07877
19	2.43623	2.38479	2.32336	2.25177	2.16919	2.07391
20	2.42873	2.37808	2.31732	2.24633	2.16431	2.06956
21	2.42195	2.37201	2.31187	2.24143	2.15990	2.06562
22	2.41579	2.36651	2.30692	2.23698	2.15591	2.06206
23	2.41018	2.36149	2.30242	2.23293	2.15228	2.05882
24	2.40503	2.35689	2.29829	2.22922	2.14895	2.05585
25	2.40030	2.35267	2.29451	2.22582	2.14590	2.05313
26	2.39593	2.34878	2.29102	2.22269	2.14308	2.05062
27	2.39189	2.34518	2.28779	2.21979	2.14048	2.04830
28	2.38814	2.34183	2.28479	2.21710	2.13807	2.04615
29	2.38465	2.33873	2.28201	2.21460	2.13583	2.04415
30	2.38139	2.33583	2.27942	2.21227	2.13375	2.04229
35	2.36790	2.32384	2.26869	2.20265	2.12513	2.03461
40	2.35779	2.31486	2.26068	2.19547	2.11870	2.02887
45	2.34993	2.30790	2.25447	2.18991	2.11371	2.02443
50	2.34365	2.30233	2.24950	2.18547	2.10974	2.02089
60	2.33422	2.29400	2.24208	2.17882	2.10380	2.01560
80	2.32244	2.28360	2.23283	2.17056	2.09640	2.00901
100	2.31537	2.27736	2.22729	2.16561	2.09198	2.00508
120	2.31066	2.27321	2.22361	2.16232	2.08904	2.00246
200	2.30124	2.26492	2.21625	2.15576	2.08318	1.99725
∞	2.28709	2.25249	2.20525	2.14596	2.07444	1.98946

TABLE A.1

One-sided Percentage Point (g) for Equal
Correlations ⎡*Case I with Correlation* ρ⎤

p = 20, 1-α = 0.80

ν ↓	ρ				
	0.6	0.7	0.8	0.9	1/(1+√20)
2	2.85215	2.59619	2.29954	1.92352	3.71332
3	2.45909	2.25924	2.02302	1.71618	3.10680
4	2.29244	2.11574	1.90437	1.62604	2.84964
5	2.20081	2.03668	1.83874	1.57583	2.70778
6	2.14295	1.98670	1.79716	1.54389	2.61788
7	2.10314	1.95227	1.76849	1.52179	2.55578
8	2.07407	1.92714	1.74752	1.50560	2.51030
9	2.05193	1.90798	1.73152	1.49323	2.47555
10	2.03450	1.89289	1.71892	1.48348	2.44812
11	2.02042	1.88071	1.70874	1.47558	2.42592
12	2.00882	1.87066	1.70034	1.46907	2.40758
13	1.99909	1.86224	1.69329	1.46360	2.39218
14	1.99082	1.85507	1.68730	1.45894	2.37905
15	1.98369	1.84890	1.68213	1.45493	2.36773
16	1.97750	1.84353	1.67764	1.45143	2.35787
17	1.97205	1.83882	1.67369	1.44837	2.34920
18	1.96724	1.83465	1.67020	1.44565	2.34152
19	1.96295	1.83093	1.66709	1.44323	2.33467
20	1.95910	1.82760	1.66430	1.44105	2.32852
21	1.95563	1.82459	1.66178	1.43909	2.32297
22	1.95249	1.82187	1.65950	1.43731	2.31793
23	1.94962	1.81939	1.65742	1.43569	2.31334
24	1.94700	1.81712	1.65551	1.43421	2.30914
25	1.94460	1.81503	1.65377	1.43285	2.30529
26	1.94238	1.81312	1.65216	1.43160	2.30173
27	1.94034	1.81134	1.65067	1.43044	2.29844
28	1.93844	1.80970	1.64930	1.42937	2.29539
29	1.93668	1.80817	1.64802	1.42837	2.29255
30	1.93504	1.80675	1.64682	1.42744	2.28991
35	1.92825	1.80087	1.64189	1.42359	2.27898
40	1.92320	1.79649	1.63822	1.42072	2.27081
45	1.91928	1.79309	1.63537	1.41850	2.26448
50	1.91615	1.79038	1.63310	1.41673	2.25942
60	1.91148	1.78633	1.62970	1.41408	2.25185
80	1.90567	1.78130	1.62548	1.41080	2.24241
100	1.90220	1.77829	1.62295	1.40884	2.23676
120	1.89990	1.77629	1.62127	1.40755	2.23300
200	1.89530	1.77230	1.61792	1.40500	2.22550
∞	1.88843	1.76635	1.61292	1.40096	2.21427

TABLE A.2

One-sided Percentage Point (g) for Equal
Correlations $\left(\text{Case I with Correlation } \rho\right)$

$$p = 2, \quad 1-\alpha = 0.90$$

ν ↓	\(\rho\) 0.0	0.1	0.2	0.3	0.4	0.5
1	5.36593	5.24850	5.12435	4.99217	4.85015	4.69572
2	2.74328	2.71079	2.67474	2.63455	2.58942	2.53820
3	2.26497	2.24477	2.22170	2.19529	2.16491	2.12963
4	2.07220	2.05653	2.03827	2.01700	1.99215	1.96289
5	1.96893	1.95557	1.93977	1.92115	1.89916	1.87301
6	1.90475	1.89278	1.87848	1.86146	1.84120	1.81694
7	1.86105	1.85002	1.83671	1.82077	1.80166	1.77867
8	1.82940	1.81903	1.80644	1.79126	1.77299	1.75090
9	1.80543	1.79556	1.78350	1.76889	1.75125	1.72984
10	1.78664	1.77716	1.76552	1.75136	1.73420	1.71332
11	1.77153	1.76235	1.75105	1.73725	1.72047	1.70002
12	1.75911	1.75019	1.73915	1.72564	1.70919	1.68908
13	1.74872	1.74001	1.72920	1.71594	1.69974	1.67993
14	1.73990	1.73136	1.72075	1.70769	1.69172	1.67216
15	1.73232	1.72394	1.71349	1.70061	1.68483	1.66548
16	1.72574	1.71749	1.70718	1.69445	1.67884	1.65967
17	1.71997	1.71183	1.70165	1.68906	1.67359	1.65458
18	1.71487	1.70684	1.69676	1.68429	1.66895	1.65008
19	1.71033	1.70239	1.69241	1.68004	1.66482	1.64607
20	1.70626	1.69840	1.68851	1.67624	1.66112	1.64248
21	1.70260	1.69481	1.68499	1.67281	1.65778	1.63924
22	1.69928	1.69155	1.68181	1.66970	1.65476	1.63631
23	1.69626	1.68859	1.67892	1.66688	1.65201	1.63364
24	1.69350	1.68589	1.67627	1.66429	1.64949	1.63120
25	1.69096	1.68341	1.67384	1.66192	1.64719	1.62897
26	1.68863	1.68112	1.67161	1.65974	1.64506	1.62691
27	1.68648	1.67901	1.66954	1.65773	1.64310	1.62500
28	1.68448	1.67705	1.66763	1.65586	1.64128	1.62324
29	1.68263	1.67524	1.66585	1.65412	1.63959	1.62160
30	1.68090	1.67354	1.66419	1.65251	1.63802	1.62007
35	1.67379	1.66657	1.65737	1.64584	1.63153	1.61378
40	1.66848	1.66137	1.65228	1.64088	1.62670	1.60909
45	1.66438	1.65735	1.64835	1.63704	1.62296	1.60546
50	1.66111	1.65414	1.64521	1.63397	1.61998	1.60257
60	1.65623	1.64935	1.64053	1.62940	1.61553	1.59825
80	1.65017	1.64341	1.63471	1.62372	1.61000	1.59288
100	1.64655	1.63986	1.63124	1.62033	1.60669	1.58968
120	1.64414	1.63750	1.62893	1.61808	1.60450	1.58755
200	1.63935	1.63281	1.62433	1.61359	1.60013	1.58331
∞	1.63221	1.62580	1.61749	1.60690	1.59362	1.57698

TABLE A.2

One-sided Percentage Point (g) for Equal
Correlations (*Case I with Correlation* ρ)

$p = 2, \ 1-\alpha = 0.90$

$\nu \downarrow$	ρ				
	0.6	0.7	0.8	0.9	$1/(1+\sqrt{2})$
1	4.52490	4.33101	4.10102	3.80129	4.82903
2	2.47905	2.40891	2.32177	2.20205	2.58255
3	2.08802	2.03763	1.97369	1.88385	2.16022
4	1.92793	1.88506	1.83003	1.75173	1.98829
5	1.84148	1.80253	1.75213	1.67987	1.89571
6	1.78751	1.75094	1.70337	1.63481	1.83802
7	1.75065	1.71568	1.67002	1.60394	1.79865
8	1.72389	1.69007	1.64577	1.58148	1.77010
9	1.70359	1.67063	1.62736	1.56442	1.74845
10	1.68766	1.65538	1.61291	1.55101	1.73148
11	1.67483	1.64308	1.60126	1.54020	1.71781
12	1.66428	1.63297	1.59167	1.53131	1.70657
13	1.65545	1.62451	1.58364	1.52385	1.69717
14	1.64794	1.61732	1.57682	1.51752	1.68918
15	1.64150	1.61114	1.57096	1.51207	1.68232
16	1.63589	1.60576	1.56586	1.50734	1.67635
17	1.63098	1.60105	1.56139	1.50318	1.67113
18	1.62663	1.59688	1.55744	1.49951	1.66650
19	1.62276	1.59317	1.55392	1.49624	1.66239
20	1.61930	1.58985	1.55076	1.49330	1.65870
21	1.61617	1.58685	1.54792	1.49066	1.65538
22	1.61334	1.58414	1.54534	1.48826	1.65237
23	1.61077	1.58167	1.54299	1.48608	1.64963
24	1.60841	1.57941	1.54085	1.48409	1.64712
25	1.60625	1.57734	1.53888	1.48226	1.64483
26	1.60426	1.57543	1.53707	1.48057	1.64271
27	1.60242	1.57366	1.53540	1.47902	1.64076
28	1.60072	1.57203	1.53384	1.47757	1.63895
29	1.59914	1.57051	1.53240	1.47623	1.63727
30	1.59767	1.56910	1.53106	1.47498	1.63570
35	1.59159	1.56327	1.52552	1.46983	1.62924
40	1.58706	1.55892	1.52139	1.46599	1.62442
45	1.58355	1.55555	1.51820	1.46302	1.62070
50	1.58076	1.55287	1.51565	1.46064	1.61773
60	1.57659	1.54887	1.51184	1.45710	1.61330
80	1.57140	1.54389	1.50712	1.45270	1.60779
100	1.56831	1.54092	1.50429	1.45007	1.60450
120	1.56625	1.53894	1.50241	1.44832	1.60232
200	1.56215	1.53501	1.49868	1.44484	1.59796
∞	1.55604	1.52914	1.49310	1.43965	1.59147

BECHHOFER and DUNNETT

TABLE A.2

One-sided Percentage Point (g) for Equal Correlations $\left(\text{Case I with Correlation } \rho \right)$

$p = 3, \; 1-\alpha = 0.90$

$\nu \downarrow$	0.0	0.1	0.2	0.3	0.4	0.5
			ρ			
2	3.29491	3.23297	3.16594	3.09295	3.01277	2.92363
3	2.64806	2.61107	2.56966	2.52318	2.47068	2.41080
4	2.39184	2.36396	2.33198	2.29533	2.25316	2.20425
5	2.25576	2.23252	2.20537	2.17376	2.13690	2.09364
6	2.17163	2.15119	2.12695	2.09839	2.06476	2.02493
7	2.11456	2.09597	2.07369	2.04717	2.01569	1.97817
8	2.07331	2.05606	2.03517	2.01011	1.98018	1.94430
9	2.04213	2.02587	2.00602	1.98207	1.95329	1.91865
10	2.01773	2.00225	1.98321	1.96011	1.93224	1.89856
11	1.99812	1.98325	1.96487	1.94245	1.91530	1.88240
12	1.98202	1.96766	1.94980	1.92795	1.90139	1.86911
13	1.96856	1.95462	1.93721	1.91582	1.88975	1.85800
14	1.95715	1.94356	1.92652	1.90553	1.87988	1.84858
15	1.94734	1.93406	1.91734	1.89669	1.87139	1.84047
16	1.93883	1.92581	1.90937	1.88901	1.86403	1.83344
17	1.93137	1.91858	1.90239	1.88228	1.85757	1.82727
18	1.92478	1.91220	1.89622	1.87633	1.85186	1.82182
19	1.91892	1.90651	1.89072	1.87104	1.84678	1.81697
20	1.91366	1.90142	1.88580	1.86630	1.84223	1.81262
21	1.90893	1.89683	1.88137	1.86203	1.83813	1.80870
22	1.90465	1.89268	1.87736	1.85816	1.83442	1.80515
23	1.90075	1.88890	1.87370	1.85464	1.83104	1.80193
24	1.89719	1.88545	1.87037	1.85143	1.82796	1.79898
25	1.89392	1.88228	1.86731	1.84848	1.82512	1.79627
26	1.89091	1.87937	1.86449	1.84576	1.82252	1.79378
27	1.88814	1.87667	1.86188	1.84326	1.82011	1.79148
28	1.88556	1.87418	1.85947	1.84093	1.81788	1.78935
29	1.88317	1.87186	1.85723	1.83877	1.81581	1.78737
30	1.88095	1.86970	1.85515	1.83676	1.81388	1.78552
35	1.87177	1.86081	1.84655	1.82848	1.80592	1.77792
40	1.86494	1.85418	1.84014	1.82230	1.79999	1.77225
45	1.85966	1.84906	1.83519	1.81753	1.79541	1.76787
50	1.85545	1.84498	1.83124	1.81373	1.79175	1.76438
60	1.84916	1.83888	1.82535	1.80805	1.78630	1.75916
80	1.84136	1.83132	1.81803	1.80099	1.77953	1.75269
100	1.83670	1.82680	1.81367	1.79678	1.77548	1.74882
120	1.83361	1.82380	1.81077	1.79399	1.77280	1.74625
200	1.82745	1.81783	1.80499	1.78842	1.76745	1.74114
∞	1.81828	1.80893	1.79638	1.78012	1.75948	1.73352

TABLE A.2

One-sided Percentage Point (g) for Equal
Correlations   (Case I with Correlation ρ)

p = 3, 1-α = 0.90

ν ↓	ρ				
	0.6	0.7	0.8	0.9	1/(1+√3)
2	2.82279	2.70564	2.56313	2.37188	3.04091
3	2.34139	2.25879	2.15584	2.01402	2.48927
4	2.14667	2.07716	1.98931	1.86648	2.26817
5	2.04218	1.97946	1.89944	1.78653	2.15007
6	1.97719	1.91859	1.84336	1.73650	2.07682
7	1.93292	1.87709	1.80506	1.70228	2.02700
8	1.90084	1.84700	1.77727	1.67741	1.99095
9	1.87654	1.82418	1.75619	1.65853	1.96367
10	1.85749	1.80630	1.73965	1.64371	1.94230
11	1.84216	1.79190	1.72633	1.63176	1.92511
12	1.82956	1.78006	1.71538	1.62193	1.91099
13	1.81902	1.77016	1.70621	1.61370	1.89919
14	1.81008	1.76175	1.69843	1.60671	1.88917
15	1.80239	1.75452	1.69173	1.60070	1.88056
16	1.79571	1.74824	1.68592	1.59548	1.87309
17	1.78986	1.74274	1.68082	1.59089	1.86653
18	1.78468	1.73787	1.67631	1.58684	1.86074
19	1.78008	1.73354	1.67230	1.58323	1.85559
20	1.77595	1.72965	1.66870	1.58000	1.85097
21	1.77223	1.72616	1.66546	1.57708	1.84682
22	1.76886	1.72299	1.66252	1.57444	1.84305
23	1.76580	1.72010	1.65985	1.57204	1.83962
24	1.76300	1.71747	1.65740	1.56984	1.83649
25	1.76043	1.71505	1.65516	1.56782	1.83362
26	1.75806	1.71283	1.65310	1.56597	1.83097
27	1.75588	1.71077	1.65119	1.56425	1.82853
28	1.75385	1.70886	1.64942	1.56266	1.82627
29	1.75197	1.70709	1.64778	1.56118	1.82417
30	1.75022	1.70545	1.64625	1.55981	1.82221
35	1.74300	1.69865	1.63995	1.55413	1.81414
40	1.73761	1.69358	1.63525	1.54990	1.80813
45	1.73345	1.68966	1.63161	1.54662	1.80348
50	1.73013	1.68654	1.62871	1.54401	1.79977
60	1.72518	1.68187	1.62439	1.54012	1.79424
80	1.71903	1.67608	1.61901	1.53527	1.78737
100	1.71535	1.67262	1.61580	1.53238	1.78327
120	1.71291	1.67032	1.61366	1.53045	1.78054
200	1.70805	1.66574	1.60941	1.52662	1.77512
∞	1.70081	1.65892	1.60308	1.52091	1.76704

TABLE A.2

*One-sided Percentage Point (g) for Equal
Correlations   (Case 1 with Correlation ρ)*

p = 4, 1-α = 0.90

ν ↓	ρ					
	0.0	0.1	0.2	0.3	0.4	0.5
2	3.70087	3.61380	3.52106	3.42150	3.31358	3.19508
3	2.92504	2.87376	2.81724	2.75465	2.68486	2.60618
4	2.62013	2.58200	2.53885	2.48999	2.43442	2.37065
5	2.45882	2.42740	2.39111	2.34931	2.30107	2.24499
6	2.35932	2.33196	2.29982	2.26230	2.21850	2.16710
7	2.29191	2.26725	2.23790	2.20324	2.16243	2.11416
8	2.24325	2.22053	2.19316	2.16056	2.12188	2.07586
9	2.20649	2.18521	2.15934	2.12829	2.09121	2.04688
10	2.17774	2.15759	2.13288	2.10303	2.06721	2.02419
11	2.15464	2.13539	2.11162	2.08273	2.04791	2.00594
12	2.13568	2.11717	2.09416	2.06606	2.03206	1.99096
13	2.11984	2.10194	2.07957	2.05213	2.01881	1.97842
14	2.10641	2.08903	2.06719	2.04031	2.00757	1.96779
15	2.09487	2.07794	2.05656	2.03015	1.99791	1.95866
16	2.08486	2.06831	2.04733	2.02134	1.98953	1.95073
17	2.07608	2.05987	2.03925	2.01362	1.98218	1.94378
18	2.06833	2.05242	2.03210	2.00679	1.97569	1.93763
19	2.06144	2.04578	2.02574	2.00072	1.96991	1.93217
20	2.05526	2.03984	2.02005	1.99528	1.96474	1.92727
21	2.04970	2.03449	2.01492	1.99038	1.96008	1.92286
22	2.04466	2.02965	2.01027	1.98594	1.95586	1.91886
23	2.04008	2.02524	2.00605	1.98190	1.95202	1.91523
24	2.03589	2.02121	2.00219	1.97822	1.94851	1.91191
25	2.03205	2.01752	1.99865	1.97483	1.94529	1.90886
26	2.02852	2.01412	1.99539	1.97172	1.94232	1.90606
27	2.02525	2.01098	1.99238	1.96884	1.93959	1.90347
28	2.02223	2.00807	1.98959	1.96618	1.93705	1.90107
29	2.01942	2.00537	1.98700	1.96370	1.93470	1.89884
30	2.01680	2.00285	1.98458	1.96140	1.93250	1.89676
35	2.00602	1.99248	1.97464	1.95190	1.92346	1.88820
40	1.99800	1.98476	1.96723	1.94482	1.91673	1.88183
45	1.99179	1.97878	1.96151	1.93935	1.91152	1.87689
50	1.98684	1.97402	1.95694	1.93499	1.90737	1.87297
60	1.97946	1.96692	1.95013	1.92848	1.90118	1.86710
80	1.97030	1.95810	1.94167	1.92040	1.89348	1.85982
100	1.96483	1.95284	1.93662	1.91557	1.88889	1.85547
120	1.96120	1.94934	1.93327	1.91237	1.88584	1.85258
200	1.95397	1.94238	1.92660	1.90599	1.87977	1.84683
∞	1.94320	1.93201	1.91665	1.89649	1.87073	1.83827

TABLE A.2

One-sided Percentage Point (g) for Equal Correlations (*Case I with Correlation* ρ)

p = 4, 1-α = 0.90

ν ↓	ρ				
	0.6	0.7	0.8	0.9	$1/(1+\sqrt{4})$
2	3.06263	2.91058	2.72789	2.48611	3.38655
3	2.51599	2.40983	2.27901	2.10095	2.73226
4	2.29636	2.20756	2.09646	1.94283	2.47227
5	2.17891	2.09909	1.99823	1.85733	2.33400
6	2.10601	2.03166	1.93703	1.80390	2.24845
7	2.05642	1.98574	1.89530	1.76739	2.19037
8	2.02052	1.95247	1.86503	1.74088	2.14840
9	1.99334	1.92727	1.84209	1.72075	2.11665
10	1.97206	1.90752	1.82410	1.70496	2.09180
11	1.95493	1.89163	1.80962	1.69224	2.07183
12	1.94087	1.87857	1.79771	1.68178	2.05543
13	1.92910	1.86765	1.78775	1.67302	2.04172
14	1.91912	1.85838	1.77929	1.66558	2.03009
15	1.91054	1.85041	1.77202	1.65918	2.02010
16	1.90310	1.84349	1.76571	1.65362	2.01143
17	1.89657	1.83742	1.76017	1.64875	2.00383
18	1.89080	1.83206	1.75527	1.64444	1.99712
19	1.88566	1.82729	1.75092	1.64060	1.99114
20	1.88106	1.82301	1.74701	1.63716	1.98579
21	1.87692	1.81916	1.74349	1.63406	1.98097
22	1.87316	1.81567	1.74030	1.63125	1.97660
23	1.86975	1.81249	1.73740	1.62869	1.97263
24	1.86663	1.80959	1.73475	1.62636	1.96900
25	1.86376	1.80693	1.73232	1.62421	1.96567
26	1.86113	1.80448	1.73008	1.62224	1.96260
27	1.85869	1.80222	1.72801	1.62041	1.95977
28	1.85644	1.80012	1.72610	1.61872	1.95715
29	1.85434	1.79817	1.72432	1.61715	1.95472
30	1.85239	1.79636	1.72266	1.61569	1.95245
35	1.84435	1.78887	1.71582	1.60966	1.94310
40	1.83835	1.78330	1.71072	1.60516	1.93613
45	1.83372	1.77899	1.70678	1.60168	1.93075
50	1.83002	1.77555	1.70364	1.59891	1.92646
60	1.82451	1.77042	1.69894	1.59477	1.92005
80	1.81766	1.76405	1.69311	1.58962	1.91210
100	1.81357	1.76024	1.68963	1.58654	1.90735
120	1.81086	1.75772	1.68732	1.58450	1.90420
200	1.80545	1.75268	1.68272	1.58043	1.89792
∞	1.79739	1.74518	1.67585	1.57437	1.88857

BECHHOFER and DUNNETT

TABLE A.2

One-sided Percentage Point (g) for Equal
Correlations (*Case I with Correlation* ρ)

p = 5, 1-α = 0.90

ν ↓	ρ					
	0.0	0.1	0.2	0.3	0.4	0.5
2	4.01972	3.91151	3.79740	3.67599	3.54547	3.40330
3	3.14140	3.07788	3.00863	2.93271	2.84881	2.75500
4	2.79750	2.75055	2.69798	2.63902	2.57254	2.49686
5	2.61588	2.57743	2.53345	2.48324	2.42576	2.35944
6	2.50395	2.47066	2.43190	2.38701	2.33499	2.27437
7	2.42816	2.39833	2.36307	2.32175	2.27340	2.21660
8	2.37347	2.34611	2.31337	2.27461	2.22890	2.17483
9	2.33216	2.30666	2.27581	2.23897	2.19524	2.14325
10	2.29986	2.27581	2.24643	2.21110	2.16891	2.11852
11	2.27392	2.25102	2.22282	2.18870	2.14775	2.09865
12	2.25262	2.23067	2.20344	2.17030	2.13037	2.08232
13	2.23483	2.21367	2.18724	2.15493	2.11585	2.06868
14	2.21974	2.19925	2.17351	2.14189	2.10353	2.05710
15	2.20679	2.18687	2.16171	2.13070	2.09295	2.04716
16	2.19554	2.17612	2.15147	2.12098	2.08376	2.03853
17	2.18569	2.16670	2.14250	2.11246	2.07571	2.03096
18	2.17698	2.15838	2.13457	2.10494	2.06860	2.02428
19	2.16924	2.15097	2.12752	2.09824	2.06227	2.01833
20	2.16230	2.14434	2.12120	2.09224	2.05660	2.01300
21	2.15605	2.13837	2.11551	2.08684	2.05150	2.00820
22	2.15040	2.13296	2.11036	2.08195	2.04687	2.00386
23	2.14525	2.12804	2.10567	2.07750	2.04266	1.99990
24	2.14055	2.12354	2.10138	2.07343	2.03882	1.99629
25	2.13624	2.11942	2.09746	2.06970	2.03530	1.99298
26	2.13227	2.11563	2.09384	2.06627	2.03205	1.98993
27	2.12861	2.11212	2.09050	2.06310	2.02905	1.98711
28	2.12521	2.10887	2.08741	2.06016	2.02628	1.98450
29	2.12206	2.10586	2.08453	2.05743	2.02370	1.98207
30	2.11912	2.10305	2.08186	2.05489	2.02130	1.97981
35	2.10701	2.09147	2.07082	2.04442	2.01140	1.97051
40	2.09800	2.08285	2.06261	2.03662	2.00402	1.96358
45	2.09103	2.07618	2.05626	2.03059	1.99832	1.95821
50	2.08548	2.07087	2.05120	2.02578	1.99378	1.95394
60	2.07719	2.06294	2.04364	2.01861	1.98700	1.94757
80	2.06690	2.05309	2.03426	2.00970	1.97858	1.93965
100	2.06076	2.04722	2.02866	2.00439	1.97355	1.93492
120	2.05668	2.04332	2.02494	2.00086	1.97022	1.93178
200	2.04856	2.03555	2.01754	1.99383	1.96357	1.92554
∞	2.03647	2.02397	2.00651	1.98336	1.95367	1.91623

TABLE A.2

One-sided Percentage Point (g) for Equal
Correlations $\left(\text{Case I with Correlation } \rho\right)$

p = 5, 1-α = 0.90

ν ↓	ρ 0.6	0.7	0.8	0.9	1/(1+√5)
2	3.24564	3.06604	2.85202	2.57140	3.66463
3	2.64829	2.52367	2.37125	2.16555	2.92550
4	2.40933	2.30547	2.17645	1.99943	2.63336
5	2.28182	2.18871	2.07184	1.90974	2.47838
6	2.20278	2.11621	2.00673	1.85375	2.38263
7	2.14906	2.06688	1.96237	1.81551	2.31769
8	2.11020	2.03117	1.93022	1.78775	2.27079
9	2.08079	2.00413	1.90585	1.76669	2.23533
10	2.05777	1.98295	1.88675	1.75017	2.20759
11	2.03925	1.96591	1.87139	1.73687	2.18530
12	2.02404	1.95191	1.85875	1.72593	2.16700
13	2.01133	1.94020	1.84818	1.71677	2.15170
14	2.00054	1.93027	1.83921	1.70899	2.13873
15	1.99127	1.92173	1.83150	1.70230	2.12759
16	1.98322	1.91431	1.82481	1.69649	2.11791
17	1.97617	1.90782	1.81893	1.69139	2.10944
18	1.96994	1.90207	1.81374	1.68689	2.10195
19	1.96439	1.89696	1.80912	1.68288	2.09529
20	1.95942	1.89238	1.80498	1.67928	2.08932
21	1.95495	1.88825	1.80125	1.67604	2.08394
22	1.95089	1.88452	1.79788	1.67311	2.07907
23	1.94720	1.88112	1.79480	1.67044	2.07464
24	1.94383	1.87801	1.79199	1.66800	2.07060
25	1.94074	1.87516	1.78941	1.66576	2.06689
26	1.93790	1.87254	1.78704	1.66369	2.06347
27	1.93527	1.87011	1.78485	1.66179	2.06032
28	1.93284	1.86787	1.78282	1.66002	2.05739
29	1.93057	1.86578	1.78093	1.65838	2.05468
30	1.92847	1.86384	1.77918	1.65685	2.05215
35	1.91978	1.85583	1.77193	1.65055	2.04173
40	1.91332	1.84987	1.76653	1.64586	2.03397
45	1.90831	1.84525	1.76235	1.64222	2.02796
50	1.90433	1.84157	1.75902	1.63932	2.02318
60	1.89838	1.83608	1.75406	1.63500	2.01605
80	1.89099	1.82926	1.74788	1.62962	2.00718
100	1.88658	1.82519	1.74420	1.62641	2.00189
120	1.88365	1.82249	1.74175	1.62428	1.99838
200	1.87781	1.81711	1.73687	1.62003	1.99139
∞	1.86912	1.80909	1.72960	1.61370	1.98097

TABLE A.2

One-sided Percentage Point (g) for Equal
Correlations   (Case I with Correlation ρ)

p = 6, 1-α = 0.90

ν ↓	ρ					
	0.0	0.1	0.2	0.3	0.4	0.5
2	4.28053	4.15451	4.02243	3.88270	3.73332	3.57148
3	3.31835	3.24423	3.16409	3.07687	2.98111	2.87469
4	2.94225	2.88758	2.82690	2.75935	2.68370	2.59808
5	2.74374	2.69913	2.64852	2.59115	2.52589	2.45105
6	2.62144	2.58295	2.53848	2.48732	2.42840	2.36011
7	2.53863	2.50426	2.46392	2.41693	2.36228	2.29839
8	2.47888	2.44746	2.41009	2.36611	2.31451	2.25379
9	2.43374	2.40455	2.36942	2.32770	2.27840	2.22006
10	2.39845	2.37099	2.33761	2.29765	2.25016	2.19367
11	2.37010	2.34403	2.31205	2.27351	2.22746	2.17246
12	2.34683	2.32190	2.29107	2.25369	2.20882	2.15504
13	2.32738	2.30341	2.27354	2.23713	2.19325	2.14049
14	2.31089	2.28772	2.25867	2.22308	2.18004	2.12814
15	2.29673	2.27425	2.24590	2.21102	2.16869	2.11753
16	2.28444	2.26256	2.23481	2.20054	2.15884	2.10832
17	2.27367	2.25232	2.22510	2.19137	2.15021	2.10026
18	2.26416	2.24326	2.21652	2.18326	2.14259	2.09313
19	2.25569	2.23521	2.20888	2.17605	2.13580	2.08678
20	2.24811	2.22800	2.20204	2.16959	2.12973	2.08110
21	2.24128	2.22150	2.19588	2.16377	2.12425	2.07598
22	2.23510	2.21561	2.19030	2.15850	2.11930	2.07135
23	2.22947	2.21026	2.18522	2.15370	2.11479	2.06713
24	2.22434	2.20537	2.18059	2.14932	2.11067	2.06328
25	2.21962	2.20089	2.17633	2.14530	2.10689	2.05974
26	2.21528	2.19676	2.17242	2.14161	2.10341	2.05649
27	2.21128	2.19294	2.16880	2.13819	2.10020	2.05349
28	2.20756	2.18941	2.16546	2.13503	2.09722	2.05071
29	2.20412	2.18613	2.16234	2.13209	2.09446	2.04812
30	2.20091	2.18307	2.15945	2.12935	2.09189	2.04571
35	2.18767	2.17047	2.14750	2.11807	2.08127	2.03579
40	2.17782	2.16109	2.13861	2.10967	2.07337	2.02840
45	2.17019	2.15384	2.13173	2.10317	2.06726	2.02268
50	2.16412	2.14806	2.12625	2.09800	2.06239	2.01813
60	2.15506	2.13943	2.11807	2.09027	2.05513	2.01134
80	2.14380	2.12871	2.10791	2.08067	2.04610	2.00289
100	2.13709	2.12232	2.10185	2.07495	2.04072	1.99786
120	2.13263	2.11807	2.09782	2.07115	2.03714	1.99451
200	2.12375	2.10961	2.08981	2.06358	2.03002	1.98785
∞	2.11052	2.09702	2.07786	2.05230	2.01942	1.97793

TABLE A.2

*One-sided Percentage Point (g) for Equal
Correlations* (*Case I with Correlation* ρ)

p = 6, 1−α = 0.90

ν ↓	ρ				
	0.6	0.7	0.8	0.9	1/(1+√6)
2	3.39295	3.19070	2.95111	2.63906	3.89721
3	2.75434	2.61456	2.44459	2.21663	3.08604
4	2.49964	2.38345	2.23991	2.04410	2.76652
5	2.36393	2.25997	2.13014	1.95106	2.59728
6	2.27989	2.18337	2.06189	1.89302	2.49282
7	2.22281	2.13128	2.01541	1.85340	2.42201
8	2.18153	2.09358	1.98173	1.82465	2.37088
9	2.15030	2.06505	1.95622	1.80285	2.33224
10	2.12586	2.04270	1.93623	1.78574	2.30201
11	2.10621	2.02473	1.92015	1.77197	2.27773
12	2.09007	2.00997	1.90693	1.76065	2.25779
13	2.07658	1.99762	1.89587	1.75117	2.24113
14	2.06514	1.98715	1.88649	1.74312	2.22700
15	2.05530	1.97815	1.87843	1.73620	2.21486
16	2.04677	1.97034	1.87142	1.73019	2.20433
17	2.03929	1.96349	1.86528	1.72491	2.19509
18	2.03268	1.95743	1.85985	1.72025	2.18694
19	2.02679	1.95205	1.85502	1.71610	2.17968
20	2.02152	1.94722	1.85069	1.71239	2.17318
21	2.01678	1.94287	1.84679	1.70903	2.16733
22	2.01248	1.93894	1.84326	1.70600	2.16203
23	2.00857	1.93535	1.84005	1.70324	2.15720
24	2.00499	1.93208	1.83711	1.70071	2.15280
25	2.00172	1.92908	1.83441	1.69839	2.14876
26	1.99870	1.92631	1.83193	1.69626	2.14504
27	1.99591	1.92376	1.82964	1.69429	2.14160
28	1.99333	1.92139	1.82752	1.69247	2.13842
29	1.99093	1.91920	1.82555	1.69077	2.13546
30	1.98870	1.91715	1.82371	1.68919	2.13271
35	1.97949	1.90871	1.81613	1.68267	2.12136
40	1.97264	1.90243	1.81049	1.67782	2.11291
45	1.96734	1.89757	1.80613	1.67406	2.10637
50	1.96311	1.89370	1.80265	1.67106	2.10117
60	1.95681	1.88792	1.79745	1.66659	2.09339
80	1.94897	1.88073	1.79100	1.66103	2.08374
100	1.94430	1.87645	1.78715	1.65772	2.07798
120	1.94120	1.87360	1.78459	1.65551	2.07416
200	1.93501	1.86793	1.77950	1.65112	2.06654
∞	1.92581	1.85949	1.77191	1.64458	2.05520

TABLE A.2

One-sided Percentage Point (g) for Equal
Correlations $\left(\text{Case I with Correlation } \rho\right)$

$p = 7, \quad 1-\alpha = 0.90$

$\nu \downarrow$	ρ					
	0.0	0.1	0.2	0.3	0.4	0.5
2	4.50006	4.35890	4.21149	4.05613	3.89067	3.71208
3	3.46760	3.38422	3.29462	3.19764	3.09169	2.97449
4	3.06428	3.00278	2.93500	2.86000	2.77643	2.68233
5	2.85141	2.80131	2.74487	2.68128	2.60933	2.52719
6	2.72024	2.67712	2.62762	2.57101	2.50614	2.43130
7	2.63141	2.59300	2.54818	2.49627	2.43618	2.36625
8	2.56730	2.53228	2.49084	2.44231	2.38565	2.31925
9	2.51887	2.48640	2.44751	2.40154	2.34746	2.28371
10	2.48099	2.45052	2.41362	2.36965	2.31759	2.25592
11	2.45057	2.42169	2.38639	2.34402	2.29359	2.23357
12	2.42558	2.39802	2.36404	2.32299	2.27388	2.21523
13	2.40471	2.37825	2.34536	2.30541	2.25742	2.19990
14	2.38701	2.36147	2.32952	2.29050	2.24345	2.18690
15	2.37180	2.34707	2.31591	2.27769	2.23146	2.17573
16	2.35860	2.33456	2.30410	2.26658	2.22104	2.16604
17	2.34704	2.32360	2.29375	2.25684	2.21192	2.15754
18	2.33682	2.31392	2.28461	2.24824	2.20386	2.15003
19	2.32773	2.30530	2.27647	2.24058	2.19669	2.14336
20	2.31958	2.29758	2.26918	2.23372	2.19026	2.13737
21	2.31225	2.29063	2.26262	2.22754	2.18447	2.13198
22	2.30561	2.28434	2.25667	2.22195	2.17924	2.12711
23	2.29956	2.27861	2.25127	2.21686	2.17447	2.12267
24	2.29404	2.27338	2.24632	2.21221	2.17011	2.11861
25	2.28898	2.26858	2.24179	2.20795	2.16612	2.11489
26	2.28432	2.26416	2.23762	2.20402	2.16244	2.11147
27	2.28001	2.26008	2.23377	2.20040	2.15905	2.10831
28	2.27602	2.25630	2.23020	2.19704	2.15590	2.10538
29	2.27232	2.25279	2.22688	2.19392	2.15298	2.10266
30	2.26887	2.24952	2.22379	2.19102	2.15026	2.10012
35	2.25464	2.23603	2.21107	2.17904	2.13904	2.08968
40	2.24405	2.22599	2.20159	2.17013	2.13069	2.08190
45	2.23586	2.21823	2.19425	2.16323	2.12423	2.07588
50	2.22934	2.21204	2.18841	2.15774	2.11909	2.07109
60	2.21959	2.20280	2.17969	2.14954	2.11141	2.06394
80	2.20750	2.19133	2.16886	2.13935	2.10187	2.05506
100	2.20028	2.18449	2.16240	2.13327	2.09618	2.04976
120	2.19548	2.17994	2.15811	2.12924	2.09240	2.04624
200	2.18593	2.17088	2.14956	2.12120	2.08487	2.03923
∞	2.17171	2.15739	2.13683	2.10923	2.07366	2.02879

TABLE A.2

One-sided Percentage Point (g) for Equal
Correlations $\left(Case\ I\ with\ Correlation\ \rho\right)$

p = 7, 1-α = 0.90

ν ↓	ρ				
	0.6	0.7	0.8	0.9	$1/(1+\sqrt{7})$
2	3.51584	3.29443	3.03328	2.69491	4.09694
3	2.84254	2.68995	2.50523	2.25869	3.22336
4	2.57461	2.44802	2.29229	2.08084	2.88004
5	2.43201	2.31890	2.17822	1.98502	2.69837
6	2.34376	2.23885	2.10734	1.92527	2.58630
7	2.28385	2.18444	2.05908	1.88451	2.51035
8	2.24053	2.14508	2.02414	1.85493	2.45552
9	2.20778	2.11530	1.99767	1.83251	2.41408
10	2.18214	2.09198	1.97694	1.81492	2.38167
11	2.16153	2.07322	1.96026	1.80076	2.35564
12	2.14461	2.05782	1.94655	1.78912	2.33426
13	2.13046	2.04494	1.93508	1.77937	2.31640
14	2.11846	2.03401	1.92535	1.77110	2.30125
15	2.10816	2.02463	1.91699	1.76399	2.28824
16	2.09921	2.01648	1.90973	1.75781	2.27695
17	2.09137	2.00933	1.90336	1.75239	2.26705
18	2.08444	2.00302	1.89774	1.74760	2.25831
19	2.07827	1.99740	1.89273	1.74334	2.25053
20	2.07275	1.99237	1.88824	1.73951	2.24356
21	2.06777	1.98783	1.88420	1.73607	2.23728
22	2.06327	1.98373	1.88054	1.73295	2.23160
23	2.05917	1.97999	1.87721	1.73011	2.22643
24	2.05542	1.97658	1.87417	1.72752	2.22170
25	2.05199	1.97345	1.87137	1.72514	2.21737
26	2.04883	1.97057	1.86880	1.72295	2.21338
27	2.04591	1.96791	1.86643	1.72092	2.20970
28	2.04320	1.96544	1.86423	1.71905	2.20629
29	2.04069	1.96315	1.86218	1.71730	2.20312
30	2.03835	1.96101	1.86028	1.71568	2.20016
35	2.02870	1.95222	1.85243	1.70898	2.18799
40	2.02152	1.94567	1.84659	1.70400	2.17893
45	2.01596	1.94060	1.84206	1.70014	2.17192
50	2.01154	1.93656	1.83846	1.69706	2.16634
60	2.00493	1.93054	1.83308	1.69247	2.15801
80	1.99673	1.92305	1.82639	1.68676	2.14766
100	1.99183	1.91859	1.82241	1.68335	2.14148
120	1.98858	1.91562	1.81976	1.68108	2.13738
200	1.98210	1.90971	1.81448	1.67657	2.12921
∞	1.97246	1.90091	1.80662	1.66985	2.11704

TABLE A.2

One-sided Percentage Point (g) for Equal
Correlations ⎡*Case I with Correlation* ρ⎤

p = 8, 1-α = 0.90

ν ↓	0.0	0.1	0.2	0.3	0.4	0.5
				ρ		
2	4.68887	4.53468	4.37402	4.20512	4.02571	3.83260
3	3.59635	3.50479	3.40687	3.30133	3.18648	3.05989
4	3.16958	3.10198	3.02790	2.94633	2.85583	2.75432
5	2.94426	2.88923	2.82760	2.75852	2.68069	2.59219
6	2.80538	2.75808	2.70410	2.64266	2.57257	2.49202
7	2.71130	2.66923	2.62043	2.56416	2.49929	2.42409
8	2.64339	2.60509	2.56002	2.50748	2.44637	2.37502
9	2.59206	2.55662	2.51438	2.46465	2.40638	2.33792
10	2.55192	2.51871	2.47868	2.43115	2.37510	2.30890
11	2.51966	2.48825	2.44999	2.40424	2.34997	2.28559
12	2.49318	2.46324	2.42644	2.38214	2.32933	2.26644
13	2.47104	2.44233	2.40676	2.36368	2.31209	2.25044
14	2.45226	2.42460	2.39007	2.34802	2.29747	2.23687
15	2.43614	2.40937	2.37573	2.33457	2.28491	2.22522
16	2.42214	2.39615	2.36329	2.32290	2.27401	2.21510
17	2.40987	2.38456	2.35238	2.31267	2.26445	2.20624
18	2.39903	2.37433	2.34274	2.30363	2.25601	2.19840
19	2.38938	2.36522	2.33417	2.29558	2.24850	2.19143
20	2.38074	2.35705	2.32649	2.28838	2.24178	2.18519
21	2.37295	2.34970	2.31957	2.28189	2.23572	2.17957
22	2.36590	2.34305	2.31330	2.27602	2.23023	2.17448
23	2.35949	2.33699	2.30760	2.27067	2.22524	2.16985
24	2.35363	2.33145	2.30240	2.26579	2.22068	2.16562
25	2.34826	2.32638	2.29762	2.26131	2.21650	2.16174
26	2.34331	2.32170	2.29322	2.25719	2.21265	2.15816
27	2.33874	2.31739	2.28916	2.25338	2.20910	2.15486
28	2.33450	2.31339	2.28540	2.24985	2.20580	2.15181
29	2.33057	2.30967	2.28190	2.24657	2.20275	2.14897
30	2.32691	2.30621	2.27865	2.24352	2.19990	2.14633
35	2.31181	2.29195	2.26523	2.23094	2.18815	2.13543
40	2.30056	2.28133	2.25524	2.22157	2.17941	2.12731
45	2.29186	2.27311	2.24750	2.21433	2.17265	2.12104
50	2.28493	2.26656	2.24135	2.20856	2.16726	2.11604
60	2.27459	2.25679	2.23215	2.19994	2.15922	2.10858
80	2.26174	2.24465	2.22073	2.18924	2.14923	2.09931
100	2.25407	2.23740	2.21392	2.18285	2.14327	2.09378
120	2.24898	2.23259	2.20939	2.17861	2.13932	2.09011
200	2.23883	2.22300	2.20037	2.17017	2.13143	2.08280
∞	2.22372	2.20871	2.18694	2.15758	2.11970	2.07191

TABLE A.2

One-sided Percentage Point (g) for Equal
Correlations [*Case I with Correlation* ρ]

$p = 8, \quad 1-\alpha = 0.90$

ν ↓	ρ				
	0.6	0.7	0.8	0.9	$1/(1+\sqrt{8})$
2	3.62102	3.38304	3.10330	2.74234	4.27175
3	2.91786	2.75420	2.55678	2.29434	3.34328
4	2.63854	2.50296	2.33677	2.11194	2.97894
5	2.49002	2.36900	2.21900	2.01374	2.78628
6	2.39814	2.28599	2.14586	1.95254	2.66745
7	2.33578	2.22959	2.09609	1.91080	2.58693
8	2.29071	2.18880	2.06006	1.88052	2.52881
9	2.25663	2.15794	2.03277	1.85757	2.48489
10	2.22997	2.13378	2.01140	1.83957	2.45054
11	2.20853	2.11435	1.99421	1.82508	2.42294
12	2.19093	2.09840	1.98008	1.81316	2.40028
13	2.17622	2.08506	1.96826	1.80319	2.38134
14	2.16374	2.07374	1.95824	1.79473	2.36528
15	2.15303	2.06402	1.94962	1.78745	2.35149
16	2.14372	2.05558	1.94214	1.78113	2.33951
17	2.13557	2.04818	1.93558	1.77558	2.32902
18	2.12836	2.04165	1.92979	1.77068	2.31975
19	2.12195	2.03583	1.92463	1.76632	2.31150
20	2.11621	2.03062	1.92000	1.76241	2.30411
21	2.11104	2.02592	1.91584	1.75889	2.29746
22	2.10636	2.02167	1.91207	1.75570	2.29143
23	2.10210	2.01781	1.90864	1.75280	2.28595
24	2.09820	2.01427	1.90550	1.75014	2.28094
25	2.09463	2.01103	1.90263	1.74771	2.27634
26	2.09135	2.00805	1.89998	1.74547	2.27211
27	2.08831	2.00529	1.89753	1.74340	2.26821
28	2.08550	2.00274	1.89527	1.74148	2.26459
29	2.08289	2.00037	1.89316	1.73970	2.26123
30	2.08045	1.99816	1.89120	1.73804	2.25810
35	2.07043	1.98905	1.88312	1.73119	2.24519
40	2.06296	1.98227	1.87710	1.72609	2.23558
45	2.05719	1.97703	1.87244	1.72214	2.22815
50	2.05259	1.97285	1.86873	1.71899	2.22223
60	2.04572	1.96661	1.86319	1.71430	2.21339
80	2.03719	1.95887	1.85630	1.70846	2.20241
100	2.03210	1.95425	1.85220	1.70497	2.19586
120	2.02872	1.95118	1.84947	1.70266	2.19151
200	2.02199	1.94506	1.84403	1.69804	2.18284
∞	2.01197	1.93595	1.83594	1.69118	2.16993

TABLE A.2

One-sided Percentage Point (g) for Equal
Correlations $\left(Case\ I\ with\ Correlation\ \rho\right)$

p = 9, 1-α = 0.90

ν ↓	ρ					
	0.0	0.1	0.2	0.3	0.4	0.5
2	4.85401	4.68849	4.51622	4.33544	4.14376	3.93787
3	3.70933	3.61049	3.50517	3.39202	3.26928	3.13438
4	3.26204	3.18895	3.10923	3.02179	2.92513	2.81706
5	3.02579	2.96629	2.89999	2.82599	2.74293	2.64880
6	2.88011	2.82900	2.77098	2.70522	2.63048	2.54487
7	2.78138	2.73598	2.68357	2.62339	2.55427	2.47440
8	2.71009	2.66881	2.62046	2.56431	2.49925	2.42351
9	2.65620	2.61805	2.57277	2.51967	2.45767	2.38504
10	2.61404	2.57834	2.53547	2.48476	2.42514	2.35495
11	2.58015	2.54642	2.50549	2.45670	2.39901	2.33077
12	2.55232	2.52021	2.48088	2.43367	2.37756	2.31092
13	2.52906	2.49831	2.46031	2.41442	2.35963	2.29433
14	2.50933	2.47972	2.44286	2.39810	2.34443	2.28026
15	2.49237	2.46376	2.42788	2.38408	2.33137	2.26818
16	2.47766	2.44990	2.41487	2.37191	2.32004	2.25769
17	2.46475	2.43775	2.40347	2.36124	2.31011	2.24850
18	2.45336	2.42702	2.39339	2.35182	2.30133	2.24038
19	2.44321	2.41747	2.38443	2.34344	2.29352	2.23316
20	2.43412	2.40891	2.37640	2.33593	2.28653	2.22668
21	2.42593	2.40120	2.36916	2.32916	2.28023	2.22086
22	2.41852	2.39422	2.36261	2.32303	2.27453	2.21558
23	2.41177	2.38787	2.35665	2.31746	2.26934	2.21078
24	2.40561	2.38207	2.35121	2.31237	2.26460	2.20639
25	2.39996	2.37674	2.34621	2.30770	2.26025	2.20237
26	2.39475	2.37184	2.34161	2.30340	2.25625	2.19866
27	2.38994	2.36731	2.33737	2.29943	2.25255	2.19524
28	2.38549	2.36312	2.33343	2.29575	2.24913	2.19208
29	2.38135	2.35922	2.32978	2.29234	2.24595	2.18913
30	2.37750	2.35559	2.32637	2.28915	2.24299	2.18639
35	2.36161	2.34063	2.31234	2.27604	2.23078	2.17509
40	2.34977	2.32948	2.30189	2.26627	2.22169	2.16668
45	2.34061	2.32086	2.29380	2.25871	2.21466	2.16018
50	2.33332	2.31399	2.28736	2.25269	2.20905	2.15500
60	2.32243	2.30373	2.27774	2.24371	2.20069	2.14726
80	2.30890	2.29099	2.26579	2.23254	2.19031	2.13765
100	2.30082	2.28338	2.25866	2.22589	2.18412	2.13192
120	2.29546	2.27833	2.25393	2.22146	2.18000	2.12812
200	2.28478	2.26826	2.24449	2.21265	2.17181	2.12054
∞	2.26886	2.25326	2.23044	2.19953	2.15960	2.10925

TABLE A.2

One-sided Percentage Point (g) for Equal Correlations (*Case I with Correlation* ρ)

$p = 9, \quad 1-\alpha = 0.90$

$\nu \downarrow$	ρ				
	0.6	0.7	0.8	0.9	$1/(1+\sqrt{9})$
2	3.71279	3.46024	3.16419	2.78347	4.42703
3	2.98348	2.81008	2.60152	2.32519	3.44967
4	2.69418	2.55069	2.37532	2.13884	3.06655
5	2.54044	2.41248	2.25433	2.03857	2.86403
6	2.44539	2.32688	2.17923	1.97611	2.73913
7	2.38089	2.26874	2.12813	1.93351	2.65451
8	2.33428	2.22670	2.09115	1.90262	2.59342
9	2.29904	2.19489	2.06314	1.87920	2.54726
10	2.27147	2.16999	2.04121	1.86085	2.51115
11	2.24931	2.14998	2.02357	1.84607	2.48213
12	2.23111	2.13354	2.00908	1.83391	2.45831
13	2.21590	2.11980	1.99696	1.82375	2.43841
14	2.20301	2.10814	1.98667	1.81512	2.42152
15	2.19193	2.09813	1.97783	1.80770	2.40702
16	2.18231	2.08943	1.97016	1.80125	2.39443
17	2.17388	2.08181	1.96343	1.79560	2.38340
18	2.16644	2.07508	1.95749	1.79061	2.37365
19	2.15981	2.06909	1.95219	1.78616	2.36498
20	2.15388	2.06372	1.94745	1.78217	2.35721
21	2.14853	2.05889	1.94318	1.77858	2.35021
22	2.14369	2.05451	1.93932	1.77533	2.34387
23	2.13929	2.05053	1.93580	1.77237	2.33810
24	2.13527	2.04689	1.93258	1.76966	2.33284
25	2.13158	2.04355	1.92963	1.76718	2.32800
26	2.12818	2.04048	1.92692	1.76490	2.32355
27	2.12504	2.03764	1.92441	1.76279	2.31945
28	2.12214	2.03501	1.92209	1.76083	2.31564
29	2.11944	2.03257	1.91993	1.75901	2.31211
30	2.11692	2.03029	1.91792	1.75732	2.30881
35	2.10656	2.02092	1.90963	1.75034	2.29524
40	2.09885	2.01394	1.90345	1.74514	2.28513
45	2.09288	2.00854	1.89868	1.74112	2.27731
50	2.08813	2.00423	1.89487	1.73791	2.27108
60	2.08103	1.99781	1.88919	1.73312	2.26178
80	2.07222	1.98983	1.88213	1.72717	2.25023
100	2.06696	1.98508	1.87792	1.72362	2.24333
120	2.06347	1.98192	1.87513	1.72126	2.23875
200	2.05652	1.97562	1.86956	1.71656	2.22963
∞	2.04616	1.96624	1.86126	1.70956	2.21604

TABLE A.2

One-sided Percentage Point (g) for Equal Correlations $\begin{pmatrix} Case\ I\ with\ Correlation\ \rho \end{pmatrix}$

$$p = 10, \quad 1-\alpha = 0.90$$

			ρ			
$\nu \downarrow$	0.0	0.1	0.2	0.3	0.4	0.5
2	5.00042	4.82492	4.64240	4.45105	4.24845	4.03118
3	3.80982	3.70444	3.59247	3.47250	3.34268	3.20036
4	3.34437	3.26629	3.18146	3.08873	2.98654	2.87259
5	3.09839	3.03480	2.96427	2.88581	2.79804	2.69886
6	2.94664	2.89204	2.83034	2.76067	2.68174	2.59159
7	2.84376	2.79528	2.73958	2.67587	2.60292	2.51886
8	2.76943	2.72541	2.67406	2.61465	2.54602	2.46634
9	2.71324	2.67259	2.62454	2.56839	2.50301	2.42665
10	2.66926	2.63126	2.58580	2.53220	2.46938	2.39560
11	2.63390	2.59803	2.55466	2.50313	2.44236	2.37065
12	2.60486	2.57075	2.52910	2.47926	2.42017	2.35017
13	2.58058	2.54794	2.50773	2.45931	2.40163	2.33306
14	2.55998	2.52859	2.48961	2.44239	2.38591	2.31854
15	2.54228	2.51197	2.47404	2.42786	2.37241	2.30608
16	2.52692	2.49754	2.46052	2.41524	2.36069	2.29526
17	2.51344	2.48488	2.44867	2.40418	2.35042	2.28578
18	2.50154	2.47370	2.43821	2.39442	2.34134	2.27740
19	2.49094	2.46375	2.42889	2.38572	2.33327	2.26995
20	2.48145	2.45484	2.42055	2.37794	2.32604	2.26327
21	2.47290	2.44681	2.41303	2.37093	2.31952	2.25726
22	2.46515	2.43953	2.40622	2.36458	2.31362	2.25181
23	2.45810	2.43291	2.40003	2.35880	2.30826	2.24686
24	2.45166	2.42687	2.39437	2.35352	2.30336	2.24233
25	2.44575	2.42132	2.38918	2.34868	2.29886	2.23818
26	2.44031	2.41621	2.38440	2.34422	2.29472	2.23436
27	2.43529	2.41149	2.37998	2.34010	2.29090	2.23084
28	2.43063	2.40712	2.37589	2.33629	2.28736	2.22757
29	2.42631	2.40306	2.37209	2.33275	2.28407	2.22453
30	2.42228	2.39928	2.36855	2.32945	2.28100	2.22170
35	2.40567	2.38368	2.35397	2.31585	2.26838	2.21005
40	2.39330	2.37206	2.34310	2.30572	2.25897	2.20137
45	2.38373	2.36307	2.33469	2.29788	2.25170	2.19466
50	2.37610	2.35591	2.32799	2.29164	2.24591	2.18931
60	2.36471	2.34521	2.31799	2.28232	2.23726	2.18134
80	2.35056	2.33192	2.30557	2.27075	2.22652	2.17143
100	2.34212	2.32399	2.29815	2.26384	2.22011	2.16551
120	2.33651	2.31872	2.29323	2.25926	2.21586	2.16159
200	2.32533	2.30822	2.28342	2.25012	2.20738	2.15377
∞	2.30868	2.29257	2.26879	2.23651	2.19476	2.14212

TABLE A.2

One-sided Percentage Point (g) for Equal
Correlations   (*Case I with Correlation* ρ)

p = 10, 1-α = 0.90

ν ↓	\(\rho\) 0.6	0.7	0.8	0.9	1/(1+√10)
2	3.79407	3.52854	3.21798	2.81972	4.56656
3	3.04153	2.85944	2.64098	2.35235	3.54523
4	2.74335	2.59282	2.40930	2.16249	3.14516
5	2.58499	2.45084	2.28545	2.06040	2.93371
6	2.48710	2.36294	2.20860	1.99681	2.80331
7	2.42069	2.30324	2.15633	1.95346	2.71496
8	2.37271	2.26008	2.11850	1.92203	2.65117
9	2.33644	2.22744	2.08986	1.89821	2.60296
10	2.30806	2.20189	2.06743	1.87953	2.56525
11	2.28525	2.18135	2.04939	1.86450	2.53495
12	2.26653	2.16448	2.03457	1.85214	2.51007
13	2.25088	2.15038	2.02218	1.84180	2.48927
14	2.23761	2.13842	2.01167	1.83302	2.47164
15	2.22621	2.12815	2.00263	1.82547	2.45649
16	2.21631	2.11923	1.99479	1.81892	2.44333
17	2.20764	2.11141	1.98791	1.81317	2.43181
18	2.19998	2.10450	1.98183	1.80809	2.42162
19	2.19316	2.09836	1.97642	1.80357	2.41256
20	2.18706	2.09285	1.97158	1.79951	2.40444
21	2.18156	2.08789	1.96721	1.79586	2.39713
22	2.17658	2.08340	1.96326	1.79255	2.39050
23	2.17204	2.07931	1.95966	1.78954	2.38448
24	2.16791	2.07558	1.95638	1.78679	2.37897
25	2.16411	2.07216	1.95336	1.78427	2.37392
26	2.16061	2.06900	1.95059	1.78195	2.36927
27	2.15739	2.06609	1.94802	1.77980	2.36498
28	2.15440	2.06340	1.94565	1.77781	2.36100
29	2.15162	2.06089	1.94344	1.77596	2.35731
30	2.14903	2.05856	1.94139	1.77424	2.35387
35	2.13837	2.04894	1.93292	1.76715	2.33968
40	2.13044	2.04178	1.92661	1.76186	2.32911
45	2.12430	2.03624	1.92173	1.75777	2.32093
50	2.11941	2.03183	1.91784	1.75451	2.31442
60	2.11211	2.02524	1.91203	1.74964	2.30470
80	2.10304	2.01706	1.90482	1.74359	2.29262
100	2.09763	2.01218	1.90052	1.73998	2.28541
120	2.09404	2.00894	1.89766	1.73758	2.28062
200	2.08689	2.00248	1.89197	1.73280	2.27108
∞	2.07624	1.99287	1.88349	1.72569	2.25687

BECHHOFER and DUNNETT

TABLE A.2

*One-sided Percentage Point (g) for Equal
Correlations* (*Case I with Correlation* ρ)

p = 11, 1-α = 0.90

	ρ					
ν ↓	0.0	0.1	0.2	0.3	0.4	0.5
2	5.13168	4.94732	4.75563	4.55481	4.34240	4.11488
3	3.90018	3.78889	3.67090	3.54476	3.40854	3.25951
4	3.41848	3.33585	3.24636	3.14882	3.04160	2.92233
5	3.16376	3.09643	3.02201	2.93950	2.84745	2.74369
6	3.00654	2.94873	2.88364	2.81041	2.72767	2.63340
7	2.89991	2.84860	2.78988	2.72293	2.64650	2.55864
8	2.82284	2.77627	2.72217	2.65978	2.58790	2.50466
9	2.76456	2.72159	2.67099	2.61205	2.54361	2.46386
10	2.71893	2.67879	2.63095	2.57472	2.50898	2.43195
11	2.68224	2.64439	2.59877	2.54472	2.48115	2.40631
12	2.65210	2.61613	2.57234	2.52009	2.45830	2.38526
13	2.62689	2.59250	2.55025	2.49950	2.43921	2.36767
14	2.60550	2.57245	2.53151	2.48204	2.42302	2.35275
15	2.58712	2.55523	2.51541	2.46705	2.40911	2.33994
16	2.57116	2.54027	2.50143	2.45403	2.39704	2.32882
17	2.55717	2.52716	2.48918	2.44262	2.38646	2.31908
18	2.54480	2.51557	2.47836	2.43254	2.37712	2.31047
19	2.53379	2.50526	2.46872	2.42357	2.36880	2.30281
20	2.52393	2.49602	2.46009	2.41554	2.36135	2.29595
21	2.51504	2.48769	2.45232	2.40830	2.35464	2.28977
22	2.50699	2.48015	2.44528	2.40174	2.34857	2.28417
23	2.49966	2.47329	2.43887	2.39578	2.34304	2.27908
24	2.49297	2.46702	2.43302	2.39033	2.33799	2.27443
25	2.48683	2.46126	2.42764	2.38533	2.33336	2.27017
26	2.48117	2.45597	2.42270	2.38073	2.32910	2.26624
27	2.47595	2.45108	2.41813	2.37648	2.32516	2.26261
28	2.47111	2.44654	2.41390	2.37255	2.32152	2.25926
29	2.46662	2.44233	2.40997	2.36889	2.31813	2.25614
30	2.46243	2.43841	2.40631	2.36549	2.31497	2.25323
35	2.44516	2.42223	2.39122	2.35145	2.30197	2.24125
40	2.43230	2.41018	2.37997	2.34099	2.29228	2.23233
45	2.42234	2.40086	2.37128	2.33290	2.28479	2.22544
50	2.41441	2.39343	2.36434	2.32646	2.27882	2.21994
60	2.40256	2.38233	2.35399	2.31684	2.26992	2.21174
80	2.38784	2.36854	2.34113	2.30489	2.25885	2.20156
100	2.37905	2.36030	2.33346	2.29776	2.25225	2.19548
120	2.37322	2.35483	2.32836	2.29302	2.24787	2.19145
200	2.36159	2.34394	2.31820	2.28358	2.23914	2.18341
∞	2.34426	2.32769	2.30307	2.26953	2.22613	2.17144

TABLE A.2

One-sided Percentage Point (g) for Equal
Correlations (Case I with Correlation ρ)

p = 11, 1-α = 0.90

ν ↓	ρ				
	0.6	0.7	0.8	0.9	1/(1+√11)
2	3.86693	3.58971	3.26610	2.85208	4.69313
3	3.09351	2.90360	2.67624	2.37656	3.63192
4	2.78735	2.63048	2.43963	2.18357	3.21641
5	2.62483	2.48511	2.31321	2.07984	2.99682
6	2.52439	2.39514	2.23479	2.01525	2.86139
7	2.45627	2.33404	2.18147	1.97123	2.76962
8	2.40705	2.28988	2.14288	1.93931	2.70336
9	2.36985	2.25648	2.11367	1.91512	2.65327
10	2.34074	2.23035	2.09080	1.89616	2.61409
11	2.31736	2.20933	2.07240	1.88090	2.58261
12	2.29815	2.19208	2.05729	1.86835	2.55675
13	2.28211	2.17766	2.04465	1.85785	2.53514
14	2.26850	2.16542	2.03393	1.84894	2.51681
15	2.25681	2.15492	2.02472	1.84128	2.50106
16	2.24666	2.14579	2.01672	1.83463	2.48739
17	2.23777	2.13780	2.00971	1.82880	2.47541
18	2.22991	2.13073	2.00352	1.82364	2.46482
19	2.22292	2.12445	1.99800	1.81905	2.45540
20	2.21666	2.11882	1.99306	1.81494	2.44696
21	2.21102	2.11374	1.98861	1.81123	2.43935
22	2.20591	2.10915	1.98458	1.80788	2.43247
23	2.20127	2.10497	1.98092	1.80482	2.42620
24	2.19703	2.10116	1.97757	1.80203	2.42048
25	2.19313	2.09765	1.97449	1.79947	2.41523
26	2.18955	2.09443	1.97166	1.79711	2.41039
27	2.18624	2.09145	1.96905	1.79493	2.40593
28	2.18318	2.08870	1.96663	1.79292	2.40179
29	2.18033	2.08614	1.96438	1.79104	2.39795
30	2.17768	2.08375	1.96229	1.78929	2.39437
35	2.16675	2.07391	1.95365	1.78209	2.37961
40	2.15861	2.06659	1.94722	1.77673	2.36862
45	2.15231	2.06093	1.94225	1.77258	2.36011
50	2.14730	2.05642	1.93828	1.76927	2.35334
60	2.13982	2.04968	1.93237	1.76433	2.34322
80	2.13052	2.04132	1.92502	1.75819	2.33065
100	2.12498	2.03633	1.92063	1.75453	2.32315
120	2.12130	2.03301	1.91772	1.75209	2.31817
200	2.11396	2.02641	1.91192	1.74724	2.30824
∞	2.10304	2.01658	1.90328	1.74003	2.29345

TABLE A.2

One-sided Percentage Point (g) for Equal
Correlations (*Case I with Correlation* ρ)

p = 12, 1-α = 0.90

ν ↓	ρ					
	0.0	0.1	0.2	0.3	0.4	0.5
2	5.25047	5.05815	4.85819	4.64882	4.42751	4.19069
3	3.98219	3.86551	3.74203	3.61026	3.46821	3.31306
4	3.48580	3.39899	3.30523	3.20329	3.09148	2.96735
5	3.22318	3.15237	3.07438	2.98815	2.89218	2.78424
6	3.06099	3.00019	2.93199	2.85547	2.76924	2.67121
7	2.95094	2.89699	2.83548	2.76556	2.68593	2.59460
8	2.87137	2.82243	2.76578	2.70064	2.62578	2.53929
9	2.81118	2.76605	2.71309	2.65158	2.58033	2.49748
10	2.76404	2.72191	2.67186	2.61320	2.54478	2.46478
11	2.72613	2.68642	2.63872	2.58236	2.51622	2.43851
12	2.69497	2.65727	2.61151	2.55703	2.49277	2.41695
13	2.66892	2.63289	2.58875	2.53587	2.47317	2.39893
14	2.64680	2.61221	2.56945	2.51792	2.45655	2.38364
15	2.62780	2.59443	2.55287	2.50250	2.44228	2.37052
16	2.61129	2.57899	2.53847	2.48911	2.42989	2.35912
17	2.59682	2.56546	2.52585	2.47738	2.41903	2.34914
18	2.58403	2.55350	2.51470	2.46701	2.40944	2.34032
19	2.57264	2.54285	2.50477	2.45779	2.40090	2.33247
20	2.56243	2.53331	2.49588	2.44953	2.39326	2.32544
21	2.55324	2.52472	2.48787	2.44208	2.38637	2.31911
22	2.54491	2.51693	2.48061	2.43534	2.38013	2.31338
23	2.53733	2.50984	2.47401	2.42921	2.37446	2.30816
24	2.53040	2.50337	2.46798	2.42360	2.36928	2.30340
25	2.52404	2.49743	2.46244	2.41846	2.36452	2.29903
26	2.51819	2.49196	2.45735	2.41373	2.36015	2.29500
27	2.51279	2.48691	2.45264	2.40936	2.35611	2.29129
28	2.50778	2.48223	2.44828	2.40531	2.35236	2.28785
29	2.50312	2.47788	2.44423	2.40155	2.34888	2.28465
30	2.49879	2.47383	2.44046	2.39805	2.34564	2.28167
35	2.48091	2.45712	2.42490	2.38361	2.33229	2.26940
40	2.46759	2.44467	2.41331	2.37285	2.32235	2.26026
45	2.45728	2.43503	2.40434	2.36453	2.31466	2.25320
50	2.44906	2.42735	2.39720	2.35790	2.30853	2.24757
60	2.43679	2.41589	2.38653	2.34800	2.29939	2.23917
80	2.42154	2.40163	2.37327	2.33571	2.28803	2.22873
100	2.41244	2.39312	2.36535	2.32837	2.28125	2.22251
120	2.40639	2.38747	2.36009	2.32350	2.27675	2.21837
200	2.39434	2.37620	2.34962	2.31379	2.26778	2.21014
∞	2.37638	2.35941	2.33401	2.29932	2.25443	2.19788

TABLE A.2

One-sided Percentage Point (g) for Equal
Correlations ⟨*Case I with Correlation* ρ⟩

p = 12, 1-α = 0.90

ν ↓	ρ				
	0.6	0.7	0.8	0.9	1/(1+√12)
2	3.93289	3.64506	3.30959	2.88128	4.80886
3	3.14054	2.94351	2.70807	2.39839	3.71122
4	2.82714	2.66449	2.46699	2.20256	3.28155
5	2.66083	2.51605	2.33825	2.09734	3.05448
6	2.55808	2.42419	2.25841	2.03185	2.91442
7	2.48840	2.36183	2.20412	1.98721	2.81950
8	2.43806	2.31676	2.16485	1.95486	2.75095
9	2.40001	2.28268	2.13512	1.93034	2.69913
10	2.37024	2.25601	2.11184	1.91112	2.65859
11	2.34633	2.23457	2.09313	1.89565	2.62601
12	2.32669	2.21696	2.07775	1.88294	2.59925
13	2.31028	2.20225	2.06489	1.87230	2.57689
14	2.29636	2.18976	2.05398	1.86327	2.55791
15	2.28441	2.17904	2.04461	1.85551	2.54161
16	2.27403	2.16973	2.03647	1.84877	2.52746
17	2.26494	2.16158	2.02934	1.84286	2.51505
18	2.25691	2.15437	2.02304	1.83763	2.50409
19	2.24976	2.14796	2.01743	1.83298	2.49433
20	2.24336	2.14221	2.01240	1.82881	2.48559
21	2.23759	2.13704	2.00788	1.82506	2.47772
22	2.23237	2.13235	2.00378	1.82166	2.47059
23	2.22762	2.12809	2.00005	1.81856	2.46410
24	2.22328	2.12420	1.99664	1.81573	2.45817
25	2.21930	2.12063	1.99351	1.81314	2.45273
26	2.21564	2.11734	1.99064	1.81075	2.44773
27	2.21226	2.11430	1.98798	1.80854	2.44310
28	2.20912	2.11149	1.98552	1.80650	2.43882
29	2.20621	2.10888	1.98323	1.80460	2.43484
30	2.20350	2.10644	1.98110	1.80283	2.43113
35	2.19232	2.09641	1.97231	1.79553	2.41584
40	2.18400	2.08894	1.96577	1.79010	2.40445
45	2.17757	2.08316	1.96071	1.78589	2.39564
50	2.17244	2.07856	1.95668	1.78254	2.38862
60	2.16479	2.07169	1.95066	1.77754	2.37814
80	2.15529	2.06316	1.94319	1.77132	2.36512
100	2.14962	2.05807	1.93873	1.76761	2.35734
120	2.14585	2.05469	1.93577	1.76514	2.35218
200	2.13836	2.04795	1.92987	1.76022	2.34189
∞	2.12719	2.03793	1.92108	1.75292	2.32656

BECHHOFER and DUNNETT

TABLE A.2

One-sided Percentage Point (g) for Equal
Correlations $\left[Case\ I\ with\ Correlation\ \rho \right]$

$p = 13,\ 1-\alpha = 0.90$

ν ↓	0.0	0.1	0.2	0.3	0.4	0.5
2	5.35882	5.15930	4.95184	4.73467	4.50524	4.25991
3	4.05718	3.93557	3.80705	3.67011	3.52270	3.36194
4	3.54744	3.45676	3.35906	3.25306	3.13702	3.00842
5	3.27760	3.20357	3.12227	3.03260	2.93302	2.82122
6	3.11086	3.04728	2.97619	2.89663	2.80718	2.70569
7	2.99768	2.94127	2.87716	2.80448	2.72191	2.62739
8	2.91582	2.86466	2.80563	2.73795	2.66034	2.57085
9	2.85387	2.80671	2.75155	2.68766	2.61382	2.52812
10	2.80534	2.76134	2.70923	2.64832	2.57743	2.49471
11	2.76630	2.72486	2.67522	2.61671	2.54820	2.46786
12	2.73422	2.69488	2.64728	2.59075	2.52419	2.44581
13	2.70738	2.66982	2.62392	2.56905	2.50413	2.42740
14	2.68459	2.64854	2.60410	2.55064	2.48712	2.41177
15	2.66501	2.63026	2.58707	2.53484	2.47251	2.39836
16	2.64800	2.61438	2.57228	2.52111	2.45982	2.38672
17	2.63308	2.60045	2.55932	2.50908	2.44871	2.37651
18	2.61989	2.58815	2.54787	2.49845	2.43889	2.36750
19	2.60815	2.57719	2.53767	2.48899	2.43015	2.35948
20	2.59763	2.56738	2.52854	2.48052	2.42232	2.35229
21	2.58815	2.55853	2.52031	2.47289	2.41527	2.34582
22	2.57956	2.55052	2.51285	2.46597	2.40889	2.33996
23	2.57174	2.54323	2.50607	2.45968	2.40308	2.33463
24	2.56460	2.53656	2.49987	2.45394	2.39777	2.32976
25	2.55804	2.53045	2.49419	2.44867	2.39291	2.32530
26	2.55201	2.52482	2.48895	2.44381	2.38843	2.32118
27	2.54643	2.51962	2.48412	2.43933	2.38429	2.31739
28	2.54127	2.51480	2.47964	2.43518	2.38046	2.31387
29	2.53647	2.51032	2.47548	2.43132	2.37690	2.31060
30	2.53199	2.50615	2.47160	2.42773	2.37358	2.30756
35	2.51355	2.48895	2.45561	2.41292	2.35991	2.29502
40	2.49980	2.47613	2.44370	2.40188	2.34973	2.28568
45	2.48916	2.46621	2.43449	2.39335	2.34185	2.27846
50	2.48068	2.45830	2.42714	2.38655	2.33558	2.27270
60	2.46802	2.44649	2.41618	2.37640	2.32622	2.26412
80	2.45228	2.43180	2.40255	2.36378	2.31459	2.25345
100	2.44288	2.42304	2.39441	2.35625	2.30765	2.24709
120	2.43664	2.41721	2.38901	2.35125	2.30304	2.24286
200	2.42419	2.40560	2.37824	2.34129	2.29386	2.23445
∞	2.40565	2.38830	2.36218	2.32645	2.28018	2.22191

TABLE A.2

One-sided Percentage Point (g) for Equal
Correlations (*Case I with Correlation* ρ)

p = 13, 1-α = 0.90

ν ↓	ρ				
	0.6	0.7	0.8	0.9	1/(1+√13)
2	3.99311	3.69555	3.34923	2.90786	4.91539
3	3.18344	2.97989	2.73705	2.41824	3.78424
4	2.86341	2.69548	2.49190	2.21982	3.34153
5	2.69365	2.54422	2.36102	2.11325	3.10754
6	2.58878	2.45064	2.27988	2.04693	2.96319
7	2.51766	2.38713	2.22472	2.00173	2.86535
8	2.46630	2.34122	2.18482	1.96898	2.79468
9	2.42747	2.30651	2.15462	1.94416	2.74126
10	2.39710	2.27935	2.13097	1.92470	2.69945
11	2.37270	2.25752	2.11196	1.90905	2.66585
12	2.35266	2.23959	2.09634	1.89618	2.63825
13	2.33592	2.22461	2.08328	1.88541	2.61518
14	2.32172	2.21190	2.07220	1.87627	2.59560
15	2.30953	2.20098	2.06269	1.86842	2.57879
16	2.29894	2.19151	2.05442	1.86160	2.56418
17	2.28967	2.18320	2.04718	1.85561	2.55138
18	2.28147	2.17586	2.04078	1.85033	2.54007
19	2.27418	2.16933	2.03508	1.84562	2.53000
20	2.26765	2.16349	2.02998	1.84140	2.52098
21	2.26177	2.15822	2.02538	1.83760	2.51286
22	2.25644	2.15345	2.02122	1.83416	2.50550
23	2.25159	2.14911	2.01743	1.83103	2.49880
24	2.24717	2.14514	2.01397	1.82817	2.49268
25	2.24311	2.14151	2.01079	1.82554	2.48707
26	2.23937	2.13816	2.00787	1.82313	2.48190
27	2.23592	2.13507	2.00517	1.82089	2.47713
28	2.23272	2.13220	2.00267	1.81882	2.47270
29	2.22975	2.12954	2.00035	1.81690	2.46859
30	2.22699	2.12707	1.99818	1.81511	2.46477
35	2.21558	2.11685	1.98926	1.80773	2.44898
40	2.20710	2.10925	1.98262	1.80223	2.43723
45	2.20053	2.10337	1.97748	1.79797	2.42813
50	2.19530	2.09868	1.97339	1.79458	2.42088
60	2.18750	2.09169	1.96728	1.78952	2.41005
80	2.17780	2.08300	1.95969	1.78323	2.39660
100	2.17202	2.07782	1.95516	1.77947	2.38857
120	2.16818	2.07438	1.95215	1.77698	2.38324
200	2.16053	2.06753	1.94616	1.77199	2.37261
∞	2.14914	2.05732	1.93724	1.76461	2.35677

TABLE A.2

One-sided Percentage Point (g) for Equal
Correlations (*Case I with Correlation* ρ)

p = 14, 1-α = 0.90

ν ↓	ρ					
	0.0	0.1	0.2	0.3	0.4	0.5
2	5.45832	5.25224	5.03793	4.81360	4.57672	4.32356
3	4.12621	4.00005	3.86689	3.72518	3.57282	3.40687
4	3.60424	3.50997	3.40861	3.29885	3.17889	3.04617
5	3.32776	3.25073	3.16635	3.07349	2.97056	2.85521
6	3.15685	3.09066	3.01687	2.93449	2.84205	2.73736
7	3.04078	2.98206	2.91552	2.84028	2.75497	2.65749
8	2.95680	2.90356	2.84231	2.77225	2.69210	2.59983
9	2.89322	2.84416	2.78695	2.72083	2.64459	2.55625
10	2.84341	2.79765	2.74362	2.68061	2.60742	2.52217
11	2.80333	2.76025	2.70879	2.64828	2.57756	2.49478
12	2.77038	2.72951	2.68018	2.62173	2.55305	2.47230
13	2.74281	2.70380	2.65625	2.59954	2.53256	2.45352
14	2.71940	2.68198	2.63595	2.58071	2.51519	2.43759
15	2.69928	2.66323	2.61851	2.56455	2.50026	2.42391
16	2.68180	2.64693	2.60337	2.55051	2.48731	2.41203
17	2.66647	2.63265	2.59009	2.53820	2.47595	2.40162
18	2.65292	2.62002	2.57836	2.52733	2.46592	2.39243
19	2.64085	2.60878	2.56791	2.51765	2.45700	2.38425
20	2.63003	2.59871	2.55855	2.50898	2.44900	2.37692
21	2.62028	2.58963	2.55012	2.50117	2.44180	2.37032
22	2.61145	2.58141	2.54248	2.49410	2.43528	2.36434
23	2.60342	2.57392	2.53553	2.48767	2.42934	2.35891
24	2.59607	2.56708	2.52918	2.48179	2.42392	2.35394
25	2.58933	2.56081	2.52336	2.47640	2.41895	2.34939
26	2.58312	2.55503	2.51799	2.47143	2.41438	2.34519
27	2.57739	2.54969	2.51304	2.46685	2.41015	2.34132
28	2.57207	2.54474	2.50845	2.46260	2.40623	2.33773
29	2.56714	2.54015	2.50418	2.45865	2.40260	2.33440
30	2.56254	2.53586	2.50021	2.45497	2.39921	2.33130
35	2.54356	2.51820	2.48382	2.43982	2.38524	2.31850
40	2.52942	2.50504	2.47162	2.42853	2.37484	2.30898
45	2.51847	2.49485	2.46217	2.41980	2.36680	2.30161
50	2.50975	2.48673	2.45464	2.41284	2.36039	2.29575
60	2.49671	2.47459	2.44340	2.40245	2.35082	2.28699
80	2.48051	2.45951	2.42942	2.38954	2.33894	2.27611
100	2.47084	2.45051	2.42108	2.38183	2.33185	2.26962
120	2.46441	2.44452	2.41554	2.37672	2.32714	2.26531
200	2.45160	2.43259	2.40450	2.36652	2.31776	2.25672
∞	2.43250	2.41480	2.38803	2.35133	2.30379	2.24394

TABLE A.2

One-sided Percentage Point (g) for Equal
Correlations (*Case I with Correlation* ρ)

p = 14, 1-α = 0.90

ν ↓	ρ				
	0.6	0.7	0.8	0.9	$1/(1+\sqrt{14})$
2	4.04846	3.74195	3.38563	2.93223	5.01401
3	3.22286	3.01329	2.76363	2.43642	3.85190
4	2.89673	2.72391	2.51473	2.23562	3.39709
5	2.72377	2.57006	2.38190	2.12781	3.15667
6	2.61695	2.47490	2.29956	2.06073	3.00833
7	2.54452	2.41032	2.24360	2.01502	2.90777
8	2.49220	2.36365	2.20311	1.98190	2.83512
9	2.45267	2.32836	2.17248	1.95680	2.78019
10	2.42174	2.30074	2.14849	1.93713	2.73721
11	2.39689	2.27855	2.12921	1.92130	2.70265
12	2.37648	2.26033	2.11337	1.90829	2.67426
13	2.35943	2.24510	2.10012	1.89741	2.65053
14	2.34497	2.23218	2.08888	1.88817	2.63040
15	2.33256	2.22108	2.07923	1.88023	2.61310
16	2.32178	2.21145	2.07085	1.87333	2.59807
17	2.31233	2.20301	2.06350	1.86728	2.58490
18	2.30399	2.19555	2.05701	1.86194	2.57326
19	2.29656	2.18891	2.05123	1.85718	2.56290
20	2.28991	2.18297	2.04606	1.85292	2.55362
21	2.28392	2.17762	2.04140	1.84908	2.54526
22	2.27850	2.17277	2.03717	1.84560	2.53768
23	2.27356	2.16836	2.03333	1.84243	2.53079
24	2.26906	2.16433	2.02982	1.83954	2.52449
25	2.26492	2.16063	2.02660	1.83689	2.51872
26	2.26112	2.15723	2.02364	1.83444	2.51340
27	2.25760	2.15409	2.02090	1.83219	2.50848
28	2.25435	2.15118	2.01837	1.83010	2.50393
29	2.25132	2.14847	2.01601	1.82815	2.49970
30	2.24851	2.14595	2.01382	1.82634	2.49576
35	2.23690	2.13558	2.00478	1.81888	2.47951
40	2.22826	2.12785	1.99804	1.81332	2.46740
45	2.22157	2.12187	1.99283	1.80902	2.45804
50	2.21625	2.11711	1.98868	1.80559	2.45057
60	2.20830	2.11000	1.98248	1.80048	2.43942
80	2.19843	2.10117	1.97479	1.79412	2.42557
100	2.19254	2.09591	1.97020	1.79032	2.41730
120	2.18863	2.09241	1.96715	1.78780	2.41180
200	2.18084	2.08545	1.96107	1.78276	2.40085
∞	2.16925	2.07508	1.95202	1.77530	2.38453

BECHHOFER and DUNNETT

TABLE A.2

*One-sided Percentage Point (g) for Equal
Correlations* (*Case I with Correlation* ρ)

p = 15, 1-α = 0.90

ν ↓	0.0	0.1	0.2	0.3	0.4	0.5
				ρ		
2	5.55023	5.33815	5.11752	4.88661	4.64283	4.38243
3	4.19012	4.05975	3.92228	3.77614	3.61919	3.44843
4	3.65687	3.55926	3.45450	3.34123	3.21763	3.08107
5	3.37427	3.29443	3.20717	3.11133	3.00528	2.88661
6	3.19949	3.13086	3.05454	2.96952	2.87430	2.76663
7	3.08074	3.01985	2.95104	2.87340	2.78554	2.68531
8	2.99480	2.93959	2.87626	2.80399	2.72145	2.62660
9	2.92971	2.87885	2.81971	2.75152	2.67302	2.58223
10	2.87871	2.83129	2.77544	2.71047	2.63514	2.54753
11	2.83765	2.79302	2.73986	2.67747	2.60470	2.51965
12	2.80390	2.76158	2.71062	2.65038	2.57971	2.49676
13	2.77565	2.73527	2.68617	2.62773	2.55883	2.47764
14	2.75166	2.71294	2.66543	2.60851	2.54111	2.46142
15	2.73104	2.69375	2.64760	2.59201	2.52590	2.44749
16	2.71312	2.67707	2.63212	2.57768	2.51269	2.43540
17	2.69740	2.66245	2.61855	2.56512	2.50112	2.42480
18	2.68351	2.64953	2.60655	2.55402	2.49089	2.41544
19	2.67113	2.63802	2.59587	2.54414	2.48179	2.40711
20	2.66004	2.62770	2.58631	2.53529	2.47364	2.39965
21	2.65004	2.61841	2.57769	2.52732	2.46629	2.39293
22	2.64098	2.60999	2.56988	2.52010	2.45965	2.38685
23	2.63274	2.60232	2.56277	2.51353	2.45360	2.38131
24	2.62520	2.59532	2.55628	2.50752	2.44807	2.37625
25	2.61829	2.58889	2.55032	2.50202	2.44300	2.37162
26	2.61192	2.58297	2.54484	2.49695	2.43833	2.36735
27	2.60604	2.57751	2.53977	2.49227	2.43402	2.36340
28	2.60059	2.57244	2.53508	2.48793	2.43003	2.35975
29	2.59552	2.56773	2.53072	2.48390	2.42632	2.35636
30	2.59080	2.56335	2.52665	2.48014	2.42287	2.35320
35	2.57133	2.54525	2.50990	2.46467	2.40863	2.34017
40	2.55681	2.53177	2.49741	2.45314	2.39802	2.33048
45	2.54558	2.52133	2.48775	2.44422	2.38982	2.32298
50	2.53662	2.51301	2.48005	2.43711	2.38329	2.31700
60	2.52324	2.50057	2.46855	2.42650	2.37353	2.30808
80	2.50660	2.48512	2.45425	2.41332	2.36141	2.29701
100	2.49667	2.47589	2.44572	2.40545	2.35418	2.29040
120	2.49007	2.46975	2.44005	2.40022	2.34938	2.28601
200	2.47691	2.45752	2.42874	2.38981	2.33981	2.27727
∞	2.45729	2.43928	2.41190	2.37429	2.32556	2.26425

TABLE A.2

One-sided Percentage Point (g) for Equal
Correlations (*Case I with Correlation* ρ)

p = 15, 1−α = 0.90

ν ↓	ρ				
	0.6	0.7	0.8	0.9	1/(1+√15)
2	4.09965	3.78483	3.41925	2.95473	5.10576
3	3.25929	3.04415	2.78817	2.45319	3.91489
4	2.92751	2.75017	2.53579	2.25019	3.44882
5	2.75160	2.59391	2.40115	2.14122	3.20240
6	2.64297	2.49729	2.31770	2.07344	3.05033
7	2.56931	2.43171	2.26100	2.02726	2.94722
8	2.51612	2.38433	2.21998	1.99380	2.87272
9	2.47592	2.34851	2.18894	1.96845	2.81639
10	2.44447	2.32048	2.16464	1.94858	2.77229
11	2.41921	2.29795	2.14510	1.93259	2.73684
12	2.39846	2.27945	2.12905	1.91945	2.70771
13	2.38113	2.26399	2.11564	1.90845	2.68336
14	2.36643	2.25088	2.10425	1.89912	2.66270
15	2.35380	2.23961	2.09448	1.89110	2.64494
16	2.34284	2.22984	2.08599	1.88414	2.62952
17	2.33324	2.22127	2.07854	1.87803	2.61600
18	2.32476	2.21370	2.07197	1.87263	2.60406
19	2.31721	2.20696	2.06612	1.86783	2.59342
20	2.31045	2.20093	2.06088	1.86352	2.58389
21	2.30436	2.19550	2.05615	1.85964	2.57531
22	2.29884	2.19058	2.05188	1.85613	2.56753
23	2.29383	2.18610	2.04799	1.85293	2.56045
24	2.28925	2.18201	2.04443	1.85001	2.55399
25	2.28504	2.17826	2.04117	1.84733	2.54805
26	2.28117	2.17481	2.03817	1.84486	2.54259
27	2.27760	2.17162	2.03540	1.84258	2.53754
28	2.27429	2.16866	2.03283	1.84047	2.53287
29	2.27122	2.16592	2.03044	1.83851	2.52852
30	2.26835	2.16336	2.02822	1.83668	2.52448
35	2.25655	2.15283	2.01906	1.82915	2.50779
40	2.24777	2.14499	2.01224	1.82353	2.49536
45	2.24097	2.13892	2.00696	1.81919	2.48573
50	2.23556	2.13409	2.00276	1.81573	2.47806
60	2.22748	2.12688	1.99649	1.81056	2.46661
80	2.21744	2.11792	1.98869	1.80414	2.45237
100	2.21146	2.11257	1.98404	1.80031	2.44387
120	2.20748	2.10902	1.98095	1.79776	2.43823
200	2.19956	2.10196	1.97480	1.79267	2.42697
∞	2.18777	2.09143	1.96564	1.78514	2.41019

BECHHOFER and DUNNETT

TABLE A.2

*One-sided Percentage Point (g) for Equal
Correlations* (*Case I with Correlation* ρ)

p = 16, 1-α = 0.90

ν ↓	ρ					
	0.0	0.1	0.2	0.3	0.4	0.5
2	5.63558	5.41795	5.19149	4.95447	4.70429	4.43716
3	4.24957	4.11528	3.97381	3.82354	3.66231	3.48706
4	3.70589	3.60515	3.49720	3.38066	3.25365	3.11350
5	3.41761	3.33512	3.24516	3.14653	3.03756	2.91580
6	3.23923	3.16829	3.08960	3.00210	2.90427	2.79381
7	3.11799	3.05505	2.98409	2.90420	2.81395	2.71115
8	3.03021	2.97315	2.90786	2.83350	2.74873	2.65146
9	2.96372	2.91116	2.85019	2.78005	2.69945	2.60636
10	2.91160	2.86260	2.80505	2.73823	2.66089	2.57108
11	2.86963	2.82354	2.76875	2.70461	2.62992	2.54274
12	2.83512	2.79143	2.73893	2.67700	2.60448	2.51947
13	2.80624	2.76456	2.71400	2.65393	2.58322	2.50003
14	2.78171	2.74175	2.69283	2.63435	2.56519	2.48354
15	2.76062	2.72215	2.67465	2.61753	2.54971	2.46938
16	2.74229	2.70512	2.65885	2.60292	2.53626	2.45708
17	2.72621	2.69018	2.64501	2.59012	2.52448	2.44631
18	2.71199	2.67697	2.63277	2.57881	2.51407	2.43680
19	2.69932	2.66521	2.62187	2.56874	2.50481	2.42833
20	2.68797	2.65468	2.61211	2.55973	2.49651	2.42074
21	2.67774	2.64518	2.60331	2.55160	2.48904	2.41391
22	2.66847	2.63657	2.59535	2.54424	2.48227	2.40773
23	2.66003	2.62874	2.58809	2.53755	2.47611	2.40210
24	2.65232	2.62158	2.58147	2.53143	2.47049	2.39696
25	2.64524	2.61501	2.57539	2.52582	2.46533	2.39224
26	2.63872	2.60896	2.56979	2.52065	2.46058	2.38790
27	2.63270	2.60337	2.56462	2.51588	2.45619	2.38389
28	2.62712	2.59819	2.55983	2.51145	2.45212	2.38018
29	2.62193	2.59338	2.55537	2.50735	2.44835	2.37673
30	2.61709	2.58890	2.55123	2.50352	2.44483	2.37352
35	2.59715	2.57040	2.53412	2.48774	2.43034	2.36028
40	2.58229	2.55661	2.52138	2.47599	2.41954	2.35042
45	2.57078	2.54594	2.51151	2.46690	2.41119	2.34279
50	2.56160	2.53743	2.50365	2.45965	2.40453	2.33672
60	2.54789	2.52471	2.49190	2.44883	2.39460	2.32765
80	2.53085	2.50890	2.47731	2.43539	2.38226	2.31639
100	2.52067	2.49946	2.46859	2.42737	2.37490	2.30967
120	2.51390	2.49318	2.46280	2.42204	2.37001	2.30521
200	2.50042	2.48067	2.45126	2.41142	2.36027	2.29632
∞	2.48031	2.46201	2.43405	2.39559	2.34576	2.28308

TABLE A.2

One-sided Percentage Point (g) for Equal
Correlations (Case I with Correlation ρ)

p = 16, 1-α = 0.90

ν ↓	ρ				
	0.6	0.7	0.8	0.9	$1/(1+\sqrt{16})$
2	4.14723	3.82469	3.45048	2.97559	5.19149
3	3.29315	3.07280	2.81095	2.46874	3.97381
4	2.95610	2.77454	2.55534	2.26369	3.49720
5	2.77745	2.61605	2.41901	2.15366	3.24516
6	2.66713	2.51806	2.33453	2.08522	3.08960
7	2.59233	2.45157	2.27713	2.03860	2.98409
8	2.53832	2.40352	2.23561	2.00482	2.90786
9	2.49750	2.36720	2.20420	1.97923	2.85019
10	2.46557	2.33878	2.17960	1.95918	2.80505
11	2.43992	2.31594	2.15983	1.94304	2.76875
12	2.41886	2.29718	2.14359	1.92978	2.73893
13	2.40126	2.28151	2.13002	1.91868	2.71400
14	2.38633	2.26821	2.11850	1.90926	2.69283
15	2.37351	2.25680	2.10860	1.90117	2.67465
16	2.36239	2.24689	2.10001	1.89414	2.65885
17	2.35264	2.23820	2.09248	1.88798	2.64501
18	2.34402	2.23053	2.08583	1.88253	2.63277
19	2.33636	2.22370	2.07991	1.87768	2.62187
20	2.32950	2.21758	2.07460	1.87334	2.61211
21	2.32331	2.21207	2.06983	1.86942	2.60331
22	2.31771	2.20708	2.06550	1.86588	2.59535
23	2.31262	2.20255	2.06156	1.86265	2.58809
24	2.30797	2.19840	2.05797	1.85970	2.58147
25	2.30370	2.19460	2.05467	1.85700	2.57539
26	2.29977	2.19110	2.05163	1.85451	2.56979
27	2.29615	2.18787	2.04882	1.85221	2.56462
28	2.29279	2.18487	2.04623	1.85008	2.55983
29	2.28966	2.18209	2.04381	1.84810	2.55537
30	2.28676	2.17950	2.04156	1.84625	2.55123
35	2.27478	2.16882	2.03230	1.83865	2.53412
40	2.26586	2.16087	2.02540	1.83299	2.52138
45	2.25896	2.15472	2.02006	1.82860	2.51151
50	2.25346	2.14982	2.01580	1.82511	2.50365
60	2.24526	2.14251	2.00946	1.81990	2.49190
80	2.23507	2.13343	2.00157	1.81342	2.47731
100	2.22899	2.12802	1.99687	1.80955	2.46859
120	2.22496	2.12442	1.99374	1.80698	2.46280
200	2.21692	2.11725	1.98752	1.80184	2.45126
∞	2.20495	2.10658	1.97825	1.79425	2.43405

TABLE A.2

One-sided Percentage Point (g) for Equal Correlations (*Case I with Correlation* ρ)

$$p = 18, \ 1-\alpha = 0.90$$

	ρ					
$\nu \downarrow$	0.0	0.1	0.2	0.3	0.4	0.5
2	5.78974	5.56218	5.32526	5.07724	4.81552	4.53621
3	4.35724	4.21587	4.06713	3.90936	3.74036	3.55696
4	3.79477	3.68833	3.57457	3.45206	3.31886	3.17218
5	3.49624	3.40891	3.31401	3.21028	3.09599	2.96859
6	3.31136	3.23618	3.15313	3.06111	2.95851	2.84298
7	3.18561	3.11888	3.04399	2.95997	2.86535	2.75785
8	3.09451	3.03400	2.96510	2.88692	2.79808	2.69640
9	3.02545	2.96973	2.90541	2.83169	2.74723	2.64996
10	2.97130	2.91938	2.85868	2.78846	2.70746	2.61363
11	2.92768	2.87885	2.82110	2.75372	2.67550	2.58445
12	2.89180	2.84553	2.79021	2.72518	2.64926	2.56049
13	2.86175	2.81765	2.76438	2.70132	2.62732	2.54047
14	2.83623	2.79398	2.74246	2.68108	2.60872	2.52349
15	2.81427	2.77362	2.72362	2.66369	2.59274	2.50891
16	2.79519	2.75593	2.70725	2.64859	2.57887	2.49625
17	2.77845	2.74042	2.69290	2.63535	2.56671	2.48516
18	2.76364	2.72670	2.68022	2.62365	2.55597	2.47536
19	2.75045	2.71448	2.66892	2.61324	2.54641	2.46664
20	2.73862	2.70353	2.65880	2.60391	2.53784	2.45883
21	2.72796	2.69366	2.64969	2.59551	2.53013	2.45179
22	2.71830	2.68472	2.64142	2.58790	2.52314	2.44542
23	2.70950	2.67658	2.63391	2.58097	2.51679	2.43963
24	2.70146	2.66914	2.62703	2.57464	2.51098	2.43433
25	2.69408	2.66231	2.62073	2.56884	2.50565	2.42948
26	2.68728	2.65602	2.61493	2.56349	2.50075	2.42501
27	2.68100	2.65021	2.60956	2.55855	2.49622	2.42088
28	2.67518	2.64482	2.60459	2.55398	2.49203	2.41705
29	2.66977	2.63982	2.59998	2.54973	2.48813	2.41350
30	2.66473	2.63515	2.59567	2.54577	2.48450	2.41019
35	2.64393	2.61591	2.57793	2.52945	2.46953	2.39655
40	2.62841	2.60157	2.56471	2.51728	2.45839	2.38640
45	2.61640	2.59046	2.55447	2.50787	2.44977	2.37855
50	2.60682	2.58160	2.54631	2.50037	2.44290	2.37229
60	2.59251	2.56837	2.53413	2.48918	2.43264	2.36295
80	2.57470	2.55191	2.51897	2.47526	2.41990	2.35135
100	2.56407	2.54208	2.50993	2.46696	2.41230	2.34443
120	2.55700	2.53554	2.50391	2.46144	2.40725	2.33983
200	2.54291	2.52251	2.49193	2.45044	2.39719	2.33068
∞	2.52189	2.50306	2.47405	2.43405	2.38221	2.31705

TABLE A.2

One-sided Percentage Point (g) for Equal
Correlations (*Case I with Correlation* ρ)

p = 18, 1-α = 0.90

ν ↓	ρ				
	0.6	0.7	0.8	0.9	1/(1+√18)
2	4.23333	3.89677	3.50693	3.01327	5.34762
3	3.35438	3.12460	2.85208	2.49677	4.08125
4	3.00780	2.81857	2.59061	2.28802	3.58544
5	2.82415	2.65603	2.45122	2.17606	3.32314
6	2.71077	2.55556	2.36488	2.10644	3.16118
7	2.63391	2.48740	2.30622	2.05903	3.05129
8	2.57841	2.43815	2.26380	2.02468	2.97185
9	2.53647	2.40092	2.23170	1.99866	2.91175
10	2.50366	2.37179	2.20658	1.97826	2.86468
11	2.47731	2.34838	2.18638	1.96186	2.82683
12	2.45567	2.32917	2.16979	1.94837	2.79573
13	2.43759	2.31310	2.15593	1.93709	2.76971
14	2.42225	2.29948	2.14416	1.92752	2.74763
15	2.40908	2.28778	2.13406	1.91929	2.72865
16	2.39765	2.27763	2.12528	1.91215	2.71217
17	2.38763	2.26873	2.11759	1.90588	2.69771
18	2.37878	2.26086	2.11080	1.90035	2.68493
19	2.37091	2.25387	2.10475	1.89542	2.67356
20	2.36386	2.24760	2.09933	1.89100	2.66336
21	2.35751	2.24196	2.09446	1.88703	2.65417
22	2.35176	2.23685	2.09004	1.88342	2.64585
23	2.34652	2.23220	2.08602	1.88014	2.63828
24	2.34174	2.22795	2.08234	1.87715	2.63135
25	2.33736	2.22405	2.07898	1.87440	2.62500
26	2.33332	2.22047	2.07587	1.87187	2.61915
27	2.32960	2.21715	2.07301	1.86953	2.61375
28	2.32615	2.21409	2.07036	1.86737	2.60874
29	2.32294	2.21124	2.06789	1.86535	2.60409
30	2.31995	2.20858	2.06560	1.86348	2.59976
35	2.30764	2.19764	2.05613	1.85575	2.58188
40	2.29848	2.18950	2.04909	1.85000	2.56855
45	2.29139	2.18320	2.04364	1.84554	2.55824
50	2.28574	2.17818	2.03929	1.84200	2.55002
60	2.27732	2.17069	2.03281	1.83670	2.53773
80	2.26685	2.16139	2.02476	1.83011	2.52246
100	2.26061	2.15584	2.01996	1.82618	2.51335
120	2.25646	2.15215	2.01677	1.82357	2.50728
200	2.24820	2.14481	2.01041	1.81834	2.49520
∞	2.23591	2.13388	2.00095	1.81064	2.47719

TABLE A.2

One-sided Percentage Point (g) for Equal
Correlations $\left(\text{Case I with Correlation } \rho\right)$

p = 20, 1-α = 0.90

ν ↓	ρ					
	0.0	0.1	0.2	0.3	0.4	0.5
2	5.92588	5.68964	5.44355	5.18587	4.91396	4.62389
3	4.45261	4.30498	4.14980	3.98539	3.80948	3.61884
4	3.87362	3.76210	3.64317	3.51533	3.37660	3.22411
5	3.56605	3.47439	3.37506	3.26677	3.14772	3.01529
6	3.37544	3.29644	3.20947	3.11338	3.00653	2.88646
7	3.24569	3.17553	3.09710	3.00938	2.91084	2.79915
8	3.15164	3.08801	3.01585	2.93424	2.84174	2.73612
9	3.08031	3.02172	2.95437	2.87742	2.78951	2.68849
10	3.02435	2.96976	2.90622	2.83295	2.74865	2.65124
11	2.97926	2.92794	2.86749	2.79720	2.71582	2.62131
12	2.94214	2.89354	2.83565	2.76783	2.68886	2.59673
13	2.91106	2.86475	2.80903	2.74328	2.66632	2.57620
14	2.88465	2.84030	2.78643	2.72244	2.64720	2.55878
15	2.86192	2.81927	2.76700	2.70454	2.63078	2.54382
16	2.84217	2.80099	2.75012	2.68900	2.61653	2.53084
17	2.82483	2.78496	2.73532	2.67537	2.60403	2.51946
18	2.80949	2.77079	2.72224	2.66333	2.59299	2.50941
19	2.79582	2.75816	2.71059	2.65260	2.58317	2.50046
20	2.78357	2.74684	2.70015	2.64300	2.57437	2.49245
21	2.77252	2.73663	2.69074	2.63435	2.56644	2.48524
22	2.76251	2.72739	2.68222	2.62651	2.55926	2.47870
23	2.75339	2.71897	2.67446	2.61938	2.55273	2.47276
24	2.74505	2.71128	2.66737	2.61286	2.54676	2.46733
25	2.73740	2.70421	2.66087	2.60688	2.54129	2.46234
26	2.73035	2.69771	2.65488	2.60137	2.53625	2.45776
27	2.72384	2.69170	2.64934	2.59629	2.53159	2.45352
28	2.71780	2.68613	2.64421	2.59158	2.52728	2.44960
29	2.71219	2.68095	2.63945	2.58720	2.52327	2.44596
30	2.70696	2.67613	2.63501	2.58312	2.51954	2.44256
35	2.68538	2.65622	2.61669	2.56631	2.50416	2.42857
40	2.66928	2.64137	2.60304	2.55378	2.49270	2.41815
45	2.65681	2.62987	2.59247	2.54408	2.48383	2.41009
50	2.64687	2.62071	2.58405	2.53636	2.47677	2.40368
60	2.63201	2.60700	2.57146	2.52482	2.46623	2.39409
80	2.61352	2.58995	2.55581	2.51048	2.45313	2.38219
100	2.60247	2.57977	2.54646	2.50192	2.44531	2.37509
120	2.59512	2.57300	2.54025	2.49623	2.44012	2.37038
200	2.58048	2.55949	2.52786	2.48490	2.42978	2.36098
∞	2.55864	2.53934	2.50939	2.46801	2.41437	2.34699

TABLE A.2

One-sided Percentage Point (g) for Equal
Correlations ⎛*Case I with Correlation* ρ⎞

p = 20, 1-α = 0.90

ν ↓	ρ				
	0.6	0.7	0.8	0.9	$1/(1+\sqrt{20})$
2	4.30954	3.96056	3.55683	3.04652	5.48679
3	3.40855	3.17039	2.88841	2.52150	4.17719
4	3.05351	2.85748	2.62174	2.30947	3.66428
5	2.86544	2.69134	2.47964	2.19579	3.39280
6	2.74934	2.58867	2.39164	2.12513	3.22509
7	2.67065	2.51902	2.33186	2.07701	3.11126
8	2.61383	2.46870	2.28864	2.04216	3.02894
9	2.57088	2.43067	2.25594	2.01575	2.96664
10	2.53730	2.40091	2.23035	1.99507	2.91785
11	2.51031	2.37700	2.20978	1.97842	2.87859
12	2.48816	2.35737	2.19288	1.96474	2.84633
13	2.46965	2.34097	2.17876	1.95330	2.81934
14	2.45395	2.32705	2.16677	1.94359	2.79642
15	2.44047	2.31510	2.15648	1.93524	2.77672
16	2.42877	2.30473	2.14754	1.92800	2.75961
17	2.41851	2.29564	2.13971	1.92164	2.74460
18	2.40946	2.28761	2.13279	1.91603	2.73134
19	2.40139	2.28046	2.12663	1.91103	2.71952
20	2.39418	2.27406	2.12112	1.90655	2.70893
21	2.38767	2.26830	2.11615	1.90251	2.69939
22	2.38179	2.26308	2.11165	1.89886	2.69075
23	2.37643	2.25833	2.10755	1.89553	2.68288
24	2.37154	2.25399	2.10381	1.89249	2.67569
25	2.36705	2.25001	2.10038	1.88971	2.66909
26	2.36292	2.24635	2.09722	1.88714	2.66301
27	2.35910	2.24297	2.09431	1.88477	2.65740
28	2.35557	2.23983	2.09160	1.88257	2.65219
29	2.35228	2.23692	2.08909	1.88053	2.64736
30	2.34923	2.23421	2.08676	1.87863	2.64285
35	2.33663	2.22304	2.07712	1.87080	2.62427
40	2.32724	2.21472	2.06994	1.86496	2.61042
45	2.31999	2.20829	2.06439	1.86044	2.59969
50	2.31421	2.20316	2.05997	1.85685	2.59114
60	2.30558	2.19551	2.05337	1.85147	2.57837
80	2.29487	2.18601	2.04517	1.84480	2.56248
100	2.28847	2.18034	2.04028	1.84081	2.55299
120	2.28423	2.17658	2.03703	1.83816	2.54669
200	2.27578	2.16909	2.03056	1.83285	2.53411
∞	2.26319	2.15792	2.02093	1.82505	2.51536

TABLE A.3

One-sided Percentage Point (g) for Equal
Correlations (*Case I with Correlation* ρ)

p = 2, 1-α = 0.95

ν ↓	ρ					
	0.0	0.1	0.2	0.3	0.4	0.5
1	10.83390	10.60194	10.35669	10.09558	9.81504	9.50998
2	4.07505	4.03252	3.98517	3.93219	3.87249	3.80445
3	3.08960	3.06750	3.04201	3.01257	2.97838	2.93832
4	2.72153	2.70615	2.68794	2.66642	2.64090	2.61042
5	2.53197	2.51980	2.50511	2.48743	2.46615	2.44038
6	2.41699	2.40668	2.39403	2.37860	2.35981	2.33680
7	2.33998	2.33087	2.31955	2.30559	2.28843	2.26724
8	2.28487	2.27660	2.26620	2.25327	2.23725	2.21735
9	2.24349	2.23584	2.22613	2.21396	2.19879	2.17983
10	2.21130	2.20412	2.19494	2.18336	2.16884	2.15061
11	2.18555	2.17874	2.16998	2.15886	2.14486	2.12721
12	2.16448	2.15797	2.14955	2.13881	2.12522	2.10806
13	2.14693	2.14067	2.13252	2.12210	2.10886	2.09209
14	2.13208	2.12603	2.11812	2.10795	2.09501	2.07857
15	2.11936	2.11348	2.10577	2.09583	2.08314	2.06698
16	2.10833	2.10261	2.09507	2.08533	2.07285	2.05693
17	2.09869	2.09310	2.08571	2.07613	2.06385	2.04814
18	2.09018	2.08471	2.07746	2.06802	2.05590	2.04038
19	2.08262	2.07726	2.07012	2.06082	2.04884	2.03349
20	2.07586	2.07059	2.06355	2.05437	2.04253	2.02732
21	2.06978	2.06458	2.05764	2.04856	2.03684	2.02177
22	2.06427	2.05915	2.05230	2.04331	2.03170	2.01674
23	2.05927	2.05422	2.04744	2.03854	2.02702	2.01217
24	2.05470	2.04971	2.04300	2.03418	2.02275	2.00800
25	2.05052	2.04558	2.03894	2.03019	2.01884	2.00418
26	2.04667	2.04179	2.03520	2.02651	2.01524	2.00067
27	2.04312	2.03828	2.03175	2.02312	2.01191	1.99742
28	2.03983	2.03504	2.02855	2.01998	2.00884	1.99441
29	2.03677	2.03202	2.02558	2.01707	2.00598	1.99162
30	2.03393	2.02922	2.02282	2.01436	2.00332	1.98903
35	2.02223	2.01767	2.01145	2.00318	1.99237	1.97833
40	2.01353	2.00909	2.00300	1.99488	1.98423	1.97038
45	2.00681	2.00246	1.99647	1.98846	1.97794	1.96423
50	2.00147	1.99718	1.99127	1.98335	1.97294	1.95934
60	1.99350	1.98931	1.98352	1.97574	1.96547	1.95205
80	1.98362	1.97956	1.97392	1.96630	1.95622	1.94300
100	1.97773	1.97375	1.96819	1.96067	1.95070	1.93761
120	1.97383	1.96990	1.96439	1.95694	1.94704	1.93403
200	1.96606	1.96222	1.95684	1.94951	1.93976	1.92692
∞	1.95450	1.95082	1.94560	1.93846	1.92893	1.91633

TABLE A.3

One-sided Percentage Point (g) for Equal
Correlations $\bigl[$ *Case I with Correlation* ρ $\bigr]$

p = 2, 1-α = 0.95

ν ↓	ρ				
	0.6	0.7	0.8	0.9	$1/(1+\sqrt{2})$
1	9.17254	8.78954	8.33522	7.74315	9.77332
2	3.72560	3.63169	3.51447	3.35253	3.86337
3	2.89060	2.83226	2.75745	2.65105	2.97308
4	2.57347	2.52752	2.46761	2.38091	2.63690
5	2.40874	2.36893	2.31643	2.23959	2.46279
6	2.30830	2.27212	2.22403	2.15304	2.35682
7	2.24079	2.20701	2.16182	2.09471	2.28568
8	2.19235	2.16026	2.11712	2.05276	2.23468
9	2.15592	2.12508	2.08347	2.02116	2.19635
10	2.12754	2.09767	2.05724	1.99650	2.16649
11	2.10480	2.07570	2.03621	1.97674	2.14259
12	2.08618	2.05771	2.01899	1.96054	2.12302
13	2.07066	2.04271	2.00462	1.94702	2.10671
14	2.05751	2.03001	1.99245	1.93558	2.09291
15	2.04624	2.01911	1.98202	1.92575	2.08108
16	2.03648	2.00967	1.97297	1.91724	2.07082
17	2.02793	2.00140	1.96505	1.90978	2.06184
18	2.02038	1.99411	1.95806	1.90320	2.05393
19	2.01368	1.98762	1.95184	1.89734	2.04689
20	2.00768	1.98182	1.94628	1.89211	2.04059
21	2.00228	1.97659	1.94127	1.88739	2.03492
22	1.99739	1.97187	1.93674	1.88312	2.02979
23	1.99295	1.96757	1.93262	1.87924	2.02513
24	1.98889	1.96365	1.92886	1.87570	2.02088
25	1.98517	1.96005	1.92541	1.87245	2.01698
26	1.98175	1.95674	1.92223	1.86946	2.01339
27	1.97859	1.95368	1.91931	1.86670	2.01007
28	1.97567	1.95086	1.91659	1.86414	2.00701
29	1.97295	1.94823	1.91407	1.86176	2.00416
30	1.97042	1.94578	1.91173	1.85955	2.00151
35	1.96002	1.93571	1.90207	1.85045	1.99059
40	1.95228	1.92822	1.89488	1.84368	1.98248
45	1.94630	1.92243	1.88933	1.83844	1.97621
50	1.94154	1.91783	1.88491	1.83428	1.97122
60	1.93444	1.91096	1.87832	1.82806	1.96378
80	1.92563	1.90244	1.87014	1.82035	1.95455
100	1.92039	1.89736	1.86527	1.81575	1.94905
120	1.91691	1.89399	1.86204	1.81270	1.94540
200	1.90998	1.88728	1.85560	1.80662	1.93814
∞	1.89967	1.87729	1.84601	1.79758	1.92734

BECHHOFER and DUNNETT

TABLE A.3

One-sided Percentage Point (g) for Equal
Correlations (*Case I with Correlation* ρ)

p = 3, 1-α = 0.95

ν ↓	ρ					
	0.0	0.1	0.2	0.3	0.4	0.5
2	4.83395	4.75118	4.66141	4.56345	4.45562	4.33548
3	3.55084	3.50914	3.46213	3.40901	3.34861	3.27924
4	3.08042	3.05211	3.01925	2.98111	2.93673	2.88467
5	2.84045	2.81852	2.79245	2.76158	2.72501	2.68142
6	2.69575	2.67750	2.65537	2.62874	2.59674	2.55814
7	2.59922	2.58334	2.56377	2.53990	2.51090	2.47556
8	2.53032	2.51609	2.49831	2.47639	2.44949	2.41646
9	2.47872	2.46570	2.44924	2.42875	2.40341	2.37208
10	2.43864	2.42654	2.41110	2.39170	2.36757	2.33756
11	2.40662	2.39525	2.38060	2.36208	2.33890	2.30994
12	2.38045	2.36967	2.35567	2.33786	2.31545	2.28734
13	2.35867	2.34837	2.33491	2.31769	2.29592	2.26852
14	2.34026	2.33037	2.31736	2.30063	2.27940	2.25259
15	2.32450	2.31495	2.30232	2.28601	2.26525	2.23894
16	2.31085	2.30160	2.28930	2.27335	2.25298	2.22712
17	2.29891	2.28992	2.27791	2.26228	2.24226	2.21677
18	2.28839	2.27962	2.26786	2.25251	2.23279	2.20765
19	2.27904	2.27047	2.25894	2.24383	2.22439	2.19954
20	2.27068	2.26229	2.25096	2.23607	2.21687	2.19228
21	2.26316	2.25493	2.24378	2.22909	2.21010	2.18576
22	2.25637	2.24828	2.23728	2.22277	2.20398	2.17985
23	2.25019	2.24223	2.23138	2.21703	2.19842	2.17449
24	2.24455	2.23671	2.22599	2.21179	2.19334	2.16959
25	2.23938	2.23165	2.22106	2.20699	2.18869	2.16510
26	2.23463	2.22700	2.21652	2.20257	2.18441	2.16097
27	2.23025	2.22271	2.21233	2.19850	2.18046	2.15715
28	2.22619	2.21874	2.20845	2.19473	2.17680	2.15363
29	2.22242	2.21505	2.20485	2.19122	2.17340	2.15035
30	2.21892	2.21161	2.20150	2.18796	2.17024	2.14730
35	2.20449	2.19748	2.18770	2.17454	2.15723	2.13474
40	2.19378	2.18699	2.17746	2.16457	2.14757	2.12541
45	2.18551	2.17888	2.16954	2.15687	2.14010	2.11821
50	2.17893	2.17244	2.16325	2.15074	2.13416	2.11247
60	2.16912	2.16283	2.15386	2.14161	2.12530	2.10392
80	2.15697	2.15091	2.14223	2.13028	2.11432	2.09332
100	2.14973	2.14382	2.13530	2.12354	2.10778	2.08700
120	2.14493	2.13912	2.13070	2.11906	2.10344	2.08281
200	2.13538	2.12976	2.12156	2.11016	2.09481	2.07448
∞	2.12120	2.11585	2.10797	2.09693	2.08197	2.06208

TABLE A.3

One-sided Percentage Point (g) for Equal Correlations ⌈*Case I with Correlation* ρ⌉

p = 3, 1-α = 0.95

ν ↓	ρ				
	0.6	0.7	0.8	0.9	1/(1+√3)
2	4.19926	4.04061	3.84708	3.58653	4.49349
3	3.19827	3.10123	2.97936	2.80995	3.37004
4	2.82268	2.74699	2.65015	2.51292	2.95259
5	2.62878	2.56364	2.47924	2.35807	2.73815
6	2.51101	2.45210	2.37507	2.26345	2.60828
7	2.43205	2.37725	2.30508	2.19976	2.52139
8	2.37550	2.32360	2.25486	2.15401	2.45925
9	2.33302	2.28328	2.21710	2.11957	2.41262
10	2.29996	2.25189	2.18768	2.09273	2.37636
11	2.27351	2.22675	2.16412	2.07121	2.34736
12	2.25186	2.20618	2.14483	2.05359	2.32364
13	2.23381	2.18903	2.12874	2.03889	2.30389
14	2.21855	2.17452	2.11512	2.02644	2.28718
15	2.20547	2.16208	2.10345	2.01576	2.27286
16	2.19413	2.15130	2.09333	2.00650	2.26046
17	2.18421	2.14187	2.08447	1.99840	2.24961
18	2.17546	2.13354	2.07666	1.99125	2.24004
19	2.16768	2.12615	2.06971	1.98489	2.23154
20	2.16073	2.11953	2.06349	1.97920	2.22393
21	2.15446	2.11357	2.05790	1.97408	2.21709
22	2.14880	2.10818	2.05284	1.96944	2.21090
23	2.14365	2.10328	2.04824	1.96523	2.20528
24	2.13895	2.09881	2.04403	1.96138	2.20014
25	2.13465	2.09471	2.04018	1.95785	2.19544
26	2.13068	2.09094	2.03664	1.95461	2.19111
27	2.12703	2.08746	2.03337	1.95161	2.18712
28	2.12364	2.08424	2.03034	1.94884	2.18342
29	2.12050	2.08124	2.02753	1.94626	2.17999
30	2.11757	2.07846	2.02491	1.94386	2.17679
35	2.10552	2.06699	2.01413	1.93398	2.16364
40	2.09657	2.05846	2.00611	1.92663	2.15386
45	2.08965	2.05188	1.99992	1.92096	2.14631
50	2.08415	2.04663	1.99499	1.91644	2.14031
60	2.07594	2.03882	1.98764	1.90970	2.13135
80	2.06576	2.02912	1.97852	1.90133	2.12025
100	2.05970	2.02335	1.97309	1.89635	2.11364
120	2.05567	2.01952	1.96949	1.89304	2.10925
200	2.04767	2.01189	1.96231	1.88646	2.10052
∞	2.03577	2.00055	1.95164	1.87666	2.08755

TABLE A.3

One-sided Percentage Point (g) for Equal
Correlations $\left(Case\ I\ with\ Correlation\ \rho\right)$

p = 4, 1-α = 0.95

ν ↓	ρ					
	0.0	0.1	0.2	0.3	0.4	0.5
2	5.39748	5.27985	5.15439	5.01952	4.87310	4.71211
3	3.88834	3.82955	3.76439	3.69184	3.61050	3.51831
4	3.34010	3.30059	3.25542	3.20374	3.14439	3.07562
5	3.06173	3.03143	2.99589	2.95435	2.90572	2.84845
6	2.89435	2.86937	2.83943	2.80380	2.76147	2.71095
7	2.78291	2.76136	2.73505	2.70329	2.66508	2.61899
8	2.70348	2.68432	2.66056	2.63151	2.59621	2.55324
9	2.64405	2.62664	2.60476	2.57772	2.54456	2.50392
10	2.59793	2.58186	2.56142	2.53593	2.50442	2.46557
11	2.56111	2.54609	2.52679	2.50252	2.47234	2.43490
12	2.53104	2.51687	2.49849	2.47522	2.44610	2.40983
13	2.50602	2.49255	2.47494	2.45248	2.42425	2.38894
14	2.48488	2.47199	2.45503	2.43326	2.40578	2.37127
15	2.46679	2.45440	2.43798	2.41680	2.38996	2.35614
16	2.45112	2.43916	2.42321	2.40255	2.37625	2.34303
17	2.43743	2.42584	2.41030	2.39008	2.36426	2.33156
18	2.42536	2.41410	2.39891	2.37909	2.35369	2.32145
19	2.41464	2.40367	2.38880	2.36932	2.34430	2.31246
20	2.40506	2.39434	2.37976	2.36059	2.33590	2.30443
21	2.39644	2.38595	2.37163	2.35273	2.32834	2.29720
22	2.38865	2.37836	2.36427	2.34562	2.32151	2.29066
23	2.38157	2.37147	2.35758	2.33917	2.31530	2.28471
24	2.37511	2.36518	2.35148	2.33327	2.30963	2.27929
25	2.36919	2.35942	2.34589	2.32787	2.30443	2.27431
26	2.36375	2.35412	2.34075	2.32291	2.29965	2.26974
27	2.35873	2.34923	2.33601	2.31832	2.29525	2.26552
28	2.35408	2.34470	2.33162	2.31408	2.29116	2.26161
29	2.34977	2.34050	2.32754	2.31014	2.28737	2.25798
30	2.34576	2.33659	2.32375	2.30647	2.28385	2.25461
35	2.32924	2.32050	2.30814	2.29139	2.26933	2.24071
40	2.31698	2.30855	2.29654	2.28018	2.25854	2.23038
45	2.30751	2.29932	2.28758	2.27152	2.25021	2.22241
50	2.29999	2.29199	2.28046	2.26464	2.24358	2.21606
60	2.28877	2.28105	2.26984	2.25437	2.23371	2.20660
80	2.27487	2.26750	2.25669	2.24165	2.22146	2.19487
100	2.26660	2.25943	2.24885	2.23408	2.21417	2.18789
120	2.26111	2.25408	2.24365	2.22905	2.20933	2.18325
200	2.25020	2.24343	2.23332	2.21905	2.19970	2.17404
∞	2.23400	2.22762	2.21796	2.20420	2.18540	2.16033

TABLE A.3

One-sided Percentage Point (g) for Equal Correlations (*Case I with Correlation* ρ)

$$p = 4, \quad 1-\alpha = 0.95$$

$\nu \downarrow$	ρ				
	0.6	0.7	0.8	0.9	$1/(1+\sqrt{4})$
2	4.53190	4.32469	4.07529	3.74453	4.97212
3	3.41205	3.28628	3.13032	2.91653	3.66579
4	2.99474	2.89712	2.77374	2.60115	3.18488
5	2.78005	2.69635	2.58913	2.43708	2.93899
6	2.64992	2.57445	2.47681	2.33696	2.79050
7	2.56280	2.49274	2.40142	2.26963	2.69133
8	2.50047	2.43424	2.34737	2.22130	2.62050
9	2.45369	2.39030	2.30676	2.18493	2.56741
10	2.41730	2.35610	2.27513	2.15659	2.52615
11	2.38819	2.32874	2.24981	2.13389	2.49317
12	2.36438	2.30635	2.22908	2.11529	2.46622
13	2.34454	2.28770	2.21181	2.09979	2.44377
14	2.32776	2.27191	2.19719	2.08666	2.42479
15	2.31339	2.25839	2.18465	2.07540	2.40854
16	2.30093	2.24666	2.17379	2.06564	2.39446
17	2.29004	2.23641	2.16429	2.05710	2.38215
18	2.28043	2.22736	2.15590	2.04956	2.37130
19	2.27189	2.21932	2.14845	2.04286	2.36165
20	2.26425	2.21213	2.14178	2.03686	2.35302
21	2.25737	2.20566	2.13578	2.03146	2.34527
22	2.25116	2.19980	2.13035	2.02658	2.33825
23	2.24550	2.19448	2.12541	2.02214	2.33187
24	2.24035	2.18963	2.12091	2.01808	2.32605
25	2.23562	2.18517	2.11678	2.01437	2.32072
26	2.23127	2.18108	2.11298	2.01095	2.31581
27	2.22726	2.17730	2.10947	2.00779	2.31128
28	2.22354	2.17380	2.10622	2.00487	2.30709
29	2.22009	2.17055	2.10321	2.00215	2.30320
30	2.21688	2.16752	2.10040	1.99963	2.29958
35	2.20366	2.15507	2.08884	1.98922	2.28468
40	2.19384	2.14581	2.08025	1.98148	2.27361
45	2.18625	2.13866	2.07361	1.97550	2.26506
50	2.18022	2.13298	2.06833	1.97075	2.25825
60	2.17122	2.12449	2.06046	1.96365	2.24811
80	2.16006	2.11398	2.05069	1.95484	2.23555
100	2.15342	2.10771	2.04487	1.94960	2.22806
120	2.14900	2.10355	2.04100	1.94611	2.22310
200	2.14023	2.09528	2.03332	1.93918	2.21322
∞	2.12719	2.08298	2.02189	1.92888	2.19855

TABLE A.3

One-sided Percentage Point (g) for Equal
Correlations $\left[\text{Case I with Correlation } \rho\right]$

$p = 5, \ 1-\alpha = 0.95$

ν ↓	ρ					
	0.0	0.1	0.2	0.3	0.4	0.5
2	5.84237	5.69520	5.53984	5.37438	5.19633	5.00220
3	4.15394	4.08030	3.99964	3.91081	3.81219	3.70142
4	3.54353	3.49421	3.43848	3.37541	3.30369	3.22134
5	3.23435	3.19671	3.15304	3.10252	3.04395	2.97555
6	3.04871	3.01783	2.98120	2.93802	2.88718	2.82700
7	2.92523	2.89872	2.86667	2.82830	2.78252	2.72774
8	2.83728	2.81383	2.78499	2.75000	2.70779	2.65682
9	2.77151	2.75030	2.72383	2.69134	2.65178	2.60364
10	2.72049	2.70099	2.67634	2.64578	2.60827	2.56231
11	2.67978	2.66162	2.63842	2.60938	2.57349	2.52926
12	2.64653	2.62947	2.60743	2.57963	2.54506	2.50225
13	2.61888	2.60271	2.58164	2.55487	2.52140	2.47975
14	2.59552	2.58010	2.55984	2.53394	2.50139	2.46073
15	2.57553	2.56075	2.54118	2.51602	2.48425	2.44444
16	2.55822	2.54399	2.52502	2.50049	2.46941	2.43033
17	2.54310	2.52935	2.51089	2.48692	2.45643	2.41798
18	2.52977	2.51643	2.49843	2.47496	2.44499	2.40710
19	2.51794	2.50497	2.48737	2.46433	2.43482	2.39743
20	2.50736	2.49471	2.47747	2.45482	2.42573	2.38878
21	2.49784	2.48549	2.46858	2.44627	2.41755	2.38100
22	2.48924	2.47715	2.46053	2.43853	2.41015	2.37396
23	2.48142	2.46958	2.45322	2.43151	2.40343	2.36757
24	2.47429	2.46267	2.44654	2.42509	2.39729	2.36173
25	2.46776	2.45633	2.44043	2.41922	2.39167	2.35638
26	2.46176	2.45051	2.43481	2.41381	2.38650	2.35146
27	2.45622	2.44513	2.42962	2.40882	2.38172	2.34692
28	2.45109	2.44016	2.42482	2.40421	2.37731	2.34271
29	2.44633	2.43554	2.42036	2.39992	2.37321	2.33881
30	2.44190	2.43125	2.41621	2.39593	2.36939	2.33518
35	2.42368	2.41357	2.39914	2.37952	2.35368	2.32023
40	2.41015	2.40044	2.38646	2.36733	2.34201	2.30913
45	2.39971	2.39030	2.37667	2.35791	2.33300	2.30055
50	2.39141	2.38224	2.36888	2.35042	2.32583	2.29373
60	2.37904	2.37023	2.35727	2.33926	2.31515	2.28355
80	2.36372	2.35535	2.34289	2.32542	2.30190	2.27095
100	2.35460	2.34649	2.33432	2.31719	2.29402	2.26344
120	2.34855	2.34061	2.32864	2.31172	2.28878	2.25845
200	2.33653	2.32892	2.31734	2.30085	2.27838	2.24854
∞	2.31868	2.31157	2.30056	2.28470	2.26291	2.23382

TABLE A.3

One-sided Percentage Point (g) for Equal
Correlations (*Case I with Correlation* ρ)

p = 5, 1-α = 0.95

ν ↓	ρ				
	0.6	0.7	0.8	0.9	1/(1+√5)
2	4.78668	4.54092	4.24773	3.86277	5.35888
3	3.57487	3.42636	3.24381	2.99596	3.90234
4	3.12532	3.01040	2.86638	2.66675	3.36932
5	2.89457	2.79626	2.67135	2.49575	3.09759
6	2.75489	2.66640	2.55281	2.39149	2.93377
7	2.66147	2.57944	2.47332	2.32142	2.82450
8	2.59466	2.51721	2.41636	2.27114	2.74650
9	2.54455	2.47049	2.37358	2.23332	2.68808
10	2.50558	2.43415	2.34027	2.20386	2.64270
11	2.47442	2.40508	2.31361	2.18026	2.60644
12	2.44894	2.38129	2.29180	2.16094	2.57681
13	2.42771	2.36148	2.27362	2.14482	2.55215
14	2.40976	2.34472	2.25823	2.13118	2.53130
15	2.39438	2.33035	2.24505	2.11949	2.51344
16	2.38106	2.31791	2.23362	2.10935	2.49798
17	2.36941	2.30703	2.22362	2.10048	2.48446
18	2.35913	2.29742	2.21480	2.09265	2.47254
19	2.35000	2.28889	2.20696	2.08569	2.46195
20	2.34183	2.28126	2.19995	2.07946	2.45248
21	2.33448	2.27439	2.19364	2.07386	2.44396
22	2.32784	2.26818	2.18793	2.06879	2.43625
23	2.32180	2.26253	2.18274	2.06418	2.42925
24	2.31628	2.25738	2.17800	2.05997	2.42286
25	2.31123	2.25265	2.17366	2.05611	2.41701
26	2.30658	2.24831	2.16966	2.05256	2.41163
27	2.30229	2.24430	2.16598	2.04928	2.40666
28	2.29832	2.24059	2.16256	2.04625	2.40206
29	2.29464	2.23714	2.15939	2.04343	2.39779
30	2.29120	2.23393	2.15644	2.04080	2.39381
35	2.27708	2.22072	2.14430	2.03000	2.37746
40	2.26659	2.21091	2.13527	2.02197	2.36531
45	2.25848	2.20332	2.12829	2.01577	2.35594
50	2.25203	2.19729	2.12274	2.01083	2.34847
60	2.24242	2.18830	2.11446	2.00346	2.33735
80	2.23050	2.17715	2.10420	1.99433	2.32357
100	2.22340	2.17050	2.09809	1.98888	2.31536
120	2.21869	2.16610	2.09403	1.98527	2.30992
200	2.20932	2.15733	2.08596	1.97808	2.29909
∞	2.19540	2.14429	2.07395	1.96738	2.28299

TABLE A.3

One-sided Percentage Point (g) for Equal
Correlations   (*Case I with Correlation* ρ)

$p = 6, \ 1-\alpha = 0.95$

$\nu \downarrow$	ρ 0.0	0.1	0.2	0.3	0.4	0.5
2	6.20753	6.03533	5.85470	5.66348	5.45891	5.23712
3	4.37229	4.28566	4.19162	4.08887	3.97561	3.84926
4	3.71054	3.65253	3.58759	3.51473	3.43249	3.33873
5	3.37576	3.33158	3.28080	3.22254	3.15551	3.07777
6	3.17489	3.13875	3.09626	3.04657	2.98849	2.92020
7	3.04134	3.01042	2.97334	2.92927	2.87706	2.81498
8	2.94625	2.91898	2.88569	2.84558	2.79752	2.73983
9	2.87516	2.85057	2.82009	2.78291	2.73793	2.68350
10	2.82002	2.79749	2.76916	2.73425	2.69164	2.63973
11	2.77602	2.75511	2.72849	2.69537	2.65465	2.60475
12	2.74010	2.72049	2.69527	2.66361	2.62443	2.57615
13	2.71023	2.69170	2.66762	2.63717	2.59926	2.55234
14	2.68500	2.66737	2.64425	2.61482	2.57799	2.53221
15	2.66340	2.64654	2.62425	2.59568	2.55977	2.51496
16	2.64471	2.62851	2.60692	2.57911	2.54399	2.50003
17	2.62838	2.61275	2.59178	2.56463	2.53020	2.48697
18	2.61399	2.59886	2.57843	2.55185	2.51803	2.47545
19	2.60121	2.58652	2.56657	2.54050	2.50723	2.46522
20	2.58978	2.57549	2.55597	2.53036	2.49756	2.45607
21	2.57951	2.56557	2.54643	2.52123	2.48887	2.44784
22	2.57022	2.55660	2.53781	2.51298	2.48101	2.44039
23	2.56179	2.54845	2.52997	2.50548	2.47387	2.43363
24	2.55409	2.54101	2.52282	2.49863	2.46734	2.42745
25	2.54704	2.53419	2.51627	2.49236	2.46137	2.42179
26	2.54055	2.52793	2.51024	2.48659	2.45588	2.41659
27	2.53457	2.52215	2.50468	2.48127	2.45080	2.41179
28	2.52903	2.51680	2.49954	2.47634	2.44611	2.40734
29	2.52390	2.51183	2.49476	2.47177	2.44175	2.40322
30	2.51912	2.50721	2.49032	2.46751	2.43770	2.39937
35	2.49945	2.48819	2.47202	2.45000	2.42101	2.38356
40	2.48485	2.47407	2.45843	2.43699	2.40861	2.37182
45	2.47358	2.46317	2.44794	2.42694	2.39904	2.36275
50	2.46463	2.45450	2.43960	2.41895	2.39142	2.35553
60	2.45128	2.44158	2.42717	2.40704	2.38007	2.34478
80	2.43475	2.42558	2.41176	2.39228	2.36600	2.33145
100	2.42492	2.41605	2.40258	2.38349	2.35762	2.32351
120	2.41839	2.40973	2.39649	2.37765	2.35207	2.31824
200	2.40542	2.39716	2.38439	2.36606	2.34101	2.30776
∞	2.38617	2.37849	2.36640	2.34882	2.32458	2.29219

TABLE A.3

*One-sided Percentage Point (g) for Equal
Correlations* $\left(\text{Case I with Correlation } \rho\right)$

$$p = 6, \quad 1-\alpha = 0.95$$

$\nu \downarrow$	ρ				
	0.6	0.7	0.8	0.9	$1/(1+\sqrt{6})$
2	4.99229	4.71473	4.38565	3.95672	5.68336
3	3.70582	3.53855	3.33427	3.05889	4.09969
4	3.23010	3.10093	2.94007	2.71864	3.52249
5	2.98631	2.87598	2.73666	2.54210	3.22880
6	2.83888	2.73969	2.61314	2.43454	3.05195
7	2.74033	2.64848	2.53034	2.36228	2.93407
8	2.66989	2.58323	2.47105	2.31045	2.84997
9	2.61707	2.53427	2.42652	2.27148	2.78700
10	2.57601	2.49619	2.39186	2.24112	2.73810
11	2.54318	2.46574	2.36413	2.21681	2.69904
12	2.51633	2.44083	2.34144	2.19690	2.66713
13	2.49398	2.42007	2.32253	2.18031	2.64056
14	2.47508	2.40252	2.30653	2.16626	2.61811
15	2.45888	2.38748	2.29282	2.15422	2.59888
16	2.44485	2.37446	2.28094	2.14378	2.58223
17	2.43259	2.36306	2.27054	2.13464	2.56768
18	2.42177	2.35301	2.26137	2.12658	2.55484
19	2.41215	2.34408	2.25323	2.11942	2.54344
20	2.40356	2.33609	2.24593	2.11300	2.53325
21	2.39582	2.32890	2.23937	2.10723	2.52408
22	2.38883	2.32240	2.23344	2.10201	2.51579
23	2.38247	2.31649	2.22805	2.09727	2.50825
24	2.37666	2.31110	2.22312	2.09293	2.50138
25	2.37135	2.30615	2.21861	2.08896	2.49508
26	2.36645	2.30160	2.21446	2.08531	2.48928
27	2.36194	2.29741	2.21063	2.08193	2.48393
28	2.35776	2.29352	2.20708	2.07881	2.47899
29	2.35388	2.28992	2.20378	2.07591	2.47439
30	2.35027	2.28656	2.20072	2.07321	2.47011
35	2.33541	2.27274	2.18810	2.06209	2.45252
40	2.32437	2.26247	2.17871	2.05383	2.43945
45	2.31584	2.25454	2.17147	2.04744	2.42935
50	2.30905	2.24823	2.16570	2.04236	2.42133
60	2.29894	2.23882	2.15710	2.03478	2.40936
80	2.28640	2.22715	2.14644	2.02538	2.39453
100	2.27893	2.22021	2.14009	2.01978	2.38570
120	2.27398	2.21560	2.13587	2.01606	2.37984
200	2.26412	2.20643	2.12749	2.00866	2.37984...
∞	2.24948	2.19280	2.11502	1.99766	2.35088

BECHHOFER and DUNNETT

TABLE A.3

One-sided Percentage Point (g) for Equal
Correlations (*Case I with Correlation* ρ)

p = 7, 1-α = 0.95

ν ↓	ρ					
	0.0	0.1	0.2	0.3	0.4	0.5
2	6.51568	6.32211	6.11984	5.90658	5.67932	5.43392
3	4.55718	4.45912	4.35338	4.23853	4.11263	3.97289
4	3.85198	3.78620	3.71314	3.63170	3.54034	3.43675
5	3.49541	3.44534	3.38826	3.32321	3.24881	3.16302
6	3.28152	3.24064	3.19293	3.13752	3.07313	2.99785
7	3.13934	3.10443	3.06286	3.01378	2.95598	2.88761
8	3.03812	3.00740	2.97014	2.92553	2.87237	2.80890
9	2.96244	2.93481	2.90075	2.85945	2.80975	2.74991
10	2.90375	2.87848	2.84689	2.80815	2.76111	2.70408
11	2.85692	2.83351	2.80387	2.76716	2.72225	2.66746
12	2.81869	2.79679	2.76873	2.73368	2.69050	2.63753
13	2.78690	2.76624	2.73949	2.70581	2.66406	2.61260
14	2.76004	2.74043	2.71478	2.68226	2.64172	2.59153
15	2.73706	2.71833	2.69363	2.66209	2.62258	2.57349
16	2.71717	2.69920	2.67531	2.64462	2.60601	2.55786
17	2.69979	2.68248	2.65930	2.62936	2.59152	2.54419
18	2.68448	2.66775	2.64518	2.61589	2.57875	2.53214
19	2.67088	2.65466	2.63264	2.60393	2.56740	2.52144
20	2.65872	2.64295	2.62143	2.59324	2.55725	2.51186
21	2.64779	2.63243	2.61135	2.58362	2.54812	2.50325
22	2.63790	2.62291	2.60223	2.57493	2.53987	2.49546
23	2.62893	2.61427	2.59394	2.56702	2.53236	2.48838
24	2.62074	2.60638	2.58638	2.55981	2.52552	2.48192
25	2.61323	2.59915	2.57945	2.55320	2.51924	2.47600
26	2.60633	2.59250	2.57308	2.54712	2.51347	2.47056
27	2.59997	2.58637	2.56720	2.54151	2.50815	2.46553
28	2.59408	2.58069	2.56176	2.53632	2.50322	2.46088
29	2.58861	2.57542	2.55671	2.53150	2.49864	2.45657
30	2.58353	2.57052	2.55201	2.52701	2.49439	2.45255
35	2.56260	2.55035	2.53267	2.50856	2.47686	2.43601
40	2.54707	2.53537	2.51830	2.49485	2.46384	2.42372
45	2.53509	2.52380	2.50721	2.48426	2.45379	2.41423
50	2.52556	2.51461	2.49838	2.47584	2.44580	2.40669
60	2.51136	2.50090	2.48523	2.46329	2.43388	2.39544
80	2.49378	2.48392	2.46894	2.44773	2.41911	2.38149
100	2.48332	2.47382	2.45924	2.43847	2.41031	2.37319
120	2.47638	2.46711	2.45280	2.43233	2.40447	2.36768
200	2.46259	2.45378	2.44000	2.42010	2.39287	2.35672
∞	2.44211	2.43398	2.42098	2.40194	2.37562	2.34044

TABLE A.3

One-sided Percentage Point (g) for Equal
Correlations (*Case I with Correlation* ρ)

p = 7, 1-α = 0.95

ν ↓	ρ				
	0.6	0.7	0.8	0.9	$1/(1+\sqrt{7})$
2	5.16413	4.85958	4.50021	4.03437	5.96261
3	3.81503	3.63183	3.40921	3.11078	4.26902
4	3.31735	3.17608	3.00104	2.76138	3.65351
5	3.06261	2.94208	2.79063	2.58024	3.34076
6	2.90866	2.80041	2.66295	2.46995	3.15256
7	2.80581	2.70564	2.57740	2.39586	3.02718
8	2.73232	2.63787	2.51616	2.34276	2.93776
9	2.67722	2.58703	2.47017	2.30283	2.87082
10	2.63440	2.54750	2.43439	2.27173	2.81885
11	2.60016	2.51588	2.40576	2.24683	2.77733
12	2.57218	2.49002	2.38234	2.22644	2.74342
13	2.54887	2.46849	2.36283	2.20944	2.71519
14	2.52916	2.45027	2.34631	2.19506	2.69133
15	2.51228	2.43467	2.33217	2.18273	2.67091
16	2.49766	2.42115	2.31991	2.17204	2.65322
17	2.48488	2.40933	2.30918	2.16268	2.63776
18	2.47360	2.39890	2.29972	2.15443	2.62412
19	2.46358	2.38963	2.29131	2.14710	2.61201
20	2.45462	2.38135	2.28379	2.14053	2.60118
21	2.44656	2.37389	2.27702	2.13462	2.59144
22	2.43927	2.36714	2.27090	2.12928	2.58263
23	2.43265	2.36101	2.26534	2.12442	2.57463
24	2.42660	2.35542	2.26026	2.11998	2.56733
25	2.42106	2.35029	2.25560	2.11592	2.56063
26	2.41596	2.34557	2.25132	2.11218	2.55448
27	2.41126	2.34122	2.24737	2.10872	2.54880
28	2.40690	2.33719	2.24371	2.10553	2.54354
29	2.40286	2.33345	2.24031	2.10256	2.53866
30	2.39910	2.32997	2.23715	2.09980	2.53412
35	2.38361	2.31564	2.22413	2.08842	2.51543
40	2.37211	2.30499	2.21445	2.07996	2.50154
45	2.36322	2.29676	2.20698	2.07342	2.49083
50	2.35616	2.29022	2.20103	2.06822	2.48230
60	2.34562	2.28046	2.19217	2.06046	2.46959
80	2.33256	2.26837	2.18117	2.05084	2.45384
100	2.32478	2.26116	2.17462	2.04511	2.44446
120	2.31962	2.25638	2.17028	2.04131	2.43824
200	2.30936	2.24688	2.16163	2.03374	2.42587
∞	2.29410	2.23274	2.14878	2.02248	2.40748

TABLE A.3

One-sided Percentage Point (g) for Equal
Correlations ⟮*Case I with Correlation* ρ⟯

p = 8, 1-α = 0.95

ν ↓	ρ					
	0.0	0.1	0.2	0.3	0.4	0.5
2	6.78121	6.56919	6.34818	6.11576	5.86878	5.60286
3	4.71716	4.60895	4.49286	4.36735	4.23035	4.07891
4	3.97447	3.90169	3.82138	3.73234	3.63293	3.52072
5	3.59900	3.54360	3.48084	3.40975	3.32885	3.23599
6	3.37378	3.32857	3.27617	3.21566	3.14570	3.06428
7	3.22407	3.18552	3.13990	3.08635	3.02359	2.94971
8	3.11747	3.08360	3.04277	2.99413	2.93647	2.86791
9	3.03778	3.00736	2.97008	2.92510	2.87122	2.80663
10	2.97597	2.94820	2.91365	2.87150	2.82055	2.75903
11	2.92665	2.90097	2.86859	2.82868	2.78006	2.72099
12	2.88639	2.86240	2.83179	2.79371	2.74698	2.68990
13	2.85291	2.83031	2.80116	2.76459	2.71945	2.66401
14	2.82462	2.80320	2.77528	2.73999	2.69617	2.64213
15	2.80042	2.77999	2.75311	2.71892	2.67624	2.62339
16	2.77947	2.75989	2.73393	2.70068	2.65898	2.60716
17	2.76116	2.74233	2.71715	2.68473	2.64389	2.59297
18	2.74503	2.72685	2.70237	2.67067	2.63059	2.58046
19	2.73071	2.71310	2.68923	2.65818	2.61877	2.56935
20	2.71790	2.70081	2.67749	2.64701	2.60820	2.55940
21	2.70639	2.68975	2.66692	2.63696	2.59869	2.55046
22	2.69598	2.67976	2.65737	2.62787	2.59009	2.54237
23	2.68652	2.67067	2.64869	2.61962	2.58228	2.53503
24	2.67790	2.66239	2.64077	2.61208	2.57515	2.52832
25	2.66999	2.65479	2.63351	2.60518	2.56861	2.52217
26	2.66273	2.64781	2.62684	2.59883	2.56260	2.51652
27	2.65602	2.64137	2.62068	2.59297	2.55706	2.51130
28	2.64982	2.63540	2.61498	2.58755	2.55193	2.50647
29	2.64406	2.62987	2.60969	2.58251	2.54716	2.50199
30	2.63870	2.62472	2.60476	2.57783	2.54273	2.49782
35	2.61667	2.60353	2.58450	2.55855	2.52448	2.48065
40	2.60031	2.58779	2.56945	2.54423	2.51092	2.46789
45	2.58769	2.57564	2.55783	2.53317	2.50045	2.45805
50	2.57765	2.56598	2.54858	2.52437	2.49212	2.45021
60	2.56270	2.55158	2.53481	2.51126	2.47971	2.43853
80	2.54418	2.53374	2.51773	2.49501	2.46433	2.42406
100	2.53317	2.52312	2.50757	2.48534	2.45517	2.41544
120	2.52586	2.51608	2.50082	2.47892	2.44909	2.40972
200	2.51134	2.50207	2.48741	2.46615	2.43700	2.39834
∞	2.48978	2.48126	2.46749	2.44718	2.41904	2.38144

TABLE A.3

One-sided Percentage Point (g) for Equal
Correlations (*Case I with Correlation* ρ)

p = 8, 1-α = 0.95

ν ↓	ρ				
	0.6	0.7	0.8	0.9	1/(1+√8)
2	5.31141	4.98349	4.59794	4.10038	6.20746
3	3.90848	3.71147	3.47302	3.15481	4.41728
4	3.39193	3.24017	3.05288	2.79760	3.76801
5	3.12778	2.99840	2.83650	2.61255	3.43840
6	2.96823	2.85211	2.70525	2.49992	3.24017
7	2.86167	2.75429	2.61735	2.42431	3.10814
8	2.78555	2.68435	2.55444	2.37010	3.01400
9	2.72849	2.63190	2.50721	2.32935	2.94353
10	2.68415	2.59111	2.47047	2.29762	2.88882
11	2.64870	2.55850	2.44107	2.27221	2.84513
12	2.61973	2.53183	2.41702	2.25141	2.80943
13	2.59560	2.50962	2.39699	2.23408	2.77973
14	2.57520	2.49084	2.38004	2.21940	2.75462
15	2.55773	2.47475	2.36551	2.20683	2.73312
16	2.54260	2.46081	2.35293	2.19592	2.71451
17	2.52937	2.44862	2.34192	2.18639	2.69824
18	2.51770	2.43786	2.33221	2.17797	2.68389
19	2.50733	2.42831	2.32358	2.17049	2.67114
20	2.49806	2.41977	2.31586	2.16379	2.65975
21	2.48971	2.41208	2.30892	2.15777	2.64950
22	2.48217	2.40512	2.30264	2.15232	2.64023
23	2.47531	2.39881	2.29693	2.14736	2.63181
24	2.46906	2.39304	2.29171	2.14284	2.62412
25	2.46332	2.38775	2.28694	2.13869	2.61708
26	2.45805	2.38289	2.28254	2.13488	2.61060
27	2.45318	2.37840	2.27849	2.13136	2.60462
28	2.44868	2.37425	2.27473	2.12810	2.59909
29	2.44449	2.37039	2.27124	2.12507	2.59396
30	2.44060	2.36680	2.26800	2.12226	2.58918
35	2.42458	2.35203	2.25464	2.11065	2.56951
40	2.41268	2.34105	2.24471	2.10203	2.55490
45	2.40348	2.33257	2.23705	2.09537	2.54362
50	2.39617	2.32583	2.23094	2.09006	2.53465
60	2.38527	2.31577	2.22185	2.08216	2.52127
80	2.37176	2.30331	2.21057	2.07235	2.50470
100	2.36371	2.29589	2.20385	2.06651	2.49483
120	2.35837	2.29096	2.19939	2.06263	2.48828
200	2.34776	2.28116	2.19053	2.05491	2.47526
∞	2.33198	2.26660	2.17734	2.04344	2.45591

TABLE A.3

*One-sided Percentage   Point (g) for Equal
Correlations   (Case I vith Correlation ρ)*

p = 9, 1-α = 0.95

ν ↓	0.0	0.1	0.2	0.3	0.4	0.5
2	7.01380	6.78569	6.54824	6.29897	6.03461	5.75060
3	4.85789	4.74060	4.61527	4.48026	4.33339	4.17156
4	4.08235	4.00324	3.91638	3.82052	3.71393	3.59406
5	3.69025	3.62998	3.56209	3.48555	3.39884	3.29969
6	3.45503	3.40585	3.34918	3.28406	3.20911	3.12222
7	3.29864	3.25674	3.20744	3.14985	3.08265	3.00384
8	3.18729	3.15050	3.10641	3.05414	2.99243	2.91935
9	3.10402	3.07102	3.03080	2.98249	2.92487	2.85605
10	3.03944	3.00935	2.97211	2.92687	2.87240	2.80688
11	2.98790	2.96011	2.92524	2.88244	2.83049	2.76760
12	2.94582	2.91990	2.88696	2.84614	2.79624	2.73549
13	2.91082	2.88644	2.85510	2.81593	2.76774	2.70876
14	2.88126	2.85817	2.82817	2.79039	2.74364	2.68617
15	2.85596	2.83397	2.80512	2.76853	2.72301	2.66682
16	2.83407	2.81302	2.78516	2.74960	2.70514	2.65006
17	2.81493	2.79470	2.76771	2.73305	2.68952	2.63541
18	2.79807	2.77856	2.75233	2.71845	2.67575	2.62249
19	2.78310	2.76422	2.73866	2.70549	2.66351	2.61101
20	2.76971	2.75140	2.72644	2.69390	2.65257	2.60075
21	2.75767	2.73987	2.71545	2.68347	2.64273	2.59151
22	2.74679	2.72945	2.70552	2.67404	2.63383	2.58316
23	2.73691	2.71997	2.69649	2.66548	2.62574	2.57558
24	2.72789	2.71133	2.68825	2.65766	2.61836	2.56865
25	2.71963	2.70341	2.68070	2.65049	2.61159	2.56231
26	2.71203	2.69613	2.67375	2.64390	2.60537	2.55647
27	2.70503	2.68941	2.66734	2.63782	2.59963	2.55108
28	2.69854	2.68319	2.66141	2.63219	2.59432	2.54610
29	2.69252	2.67742	2.65591	2.62697	2.58939	2.54147
30	2.68692	2.67204	2.65078	2.62211	2.58479	2.53716
35	2.66388	2.64994	2.62970	2.60210	2.56590	2.51944
40	2.64678	2.63352	2.61404	2.58723	2.55187	2.50627
45	2.63359	2.62085	2.60194	2.57575	2.54103	2.49610
50	2.62309	2.61077	2.59233	2.56662	2.53241	2.48801
60	2.60746	2.59575	2.57799	2.55302	2.51956	2.47596
80	2.58811	2.57714	2.56023	2.53615	2.50364	2.46101
100	2.57659	2.56607	2.54965	2.52611	2.49415	2.45211
120	2.56895	2.55872	2.54263	2.51944	2.48786	2.44621
200	2.55377	2.54410	2.52867	2.50619	2.47535	2.43447
∞	2.53124	2.52240	2.50793	2.48650	2.45675	2.41702

TABLE A.3

One-sided Percentage Point (g) for Equal
Correlations $\left(Case\ I\ with\ Correlation\ \rho\right)$

$$p = 9, \quad 1-\alpha = 0.95$$

$\nu \downarrow$	ρ				
	0.6	0.7	0.8	0.9	$1/(1+\sqrt{9})$
2	5.44006	5.09156	4.68301	4.15766	6.42526
3	3.99003	3.78083	3.52848	3.19296	4.54909
4	3.45695	3.29593	3.09790	2.82896	3.86967
5	3.18455	3.04738	2.87629	2.64051	3.52498
6	3.02009	2.89704	2.74194	2.52585	3.31775
7	2.91028	2.79655	2.65199	2.44890	3.17975
8	2.83186	2.72472	2.58762	2.39373	3.08136
9	2.77308	2.67085	2.53930	2.35228	3.00772
10	2.72740	2.62897	2.50171	2.31999	2.95055
11	2.69090	2.59548	2.47165	2.29415	2.90490
12	2.66106	2.56810	2.44705	2.27299	2.86760
13	2.63621	2.54530	2.42657	2.25536	2.83655
14	2.61521	2.52602	2.40923	2.24044	2.81032
15	2.59722	2.50950	2.39438	2.22764	2.78786
16	2.58164	2.49520	2.38152	2.21656	2.76841
17	2.56801	2.48268	2.37026	2.20686	2.75140
18	2.55599	2.47165	2.36033	2.19830	2.73641
19	2.54532	2.46184	2.35151	2.19069	2.72309
20	2.53577	2.45307	2.34362	2.18388	2.71119
21	2.52718	2.44518	2.33652	2.17776	2.70048
22	2.51942	2.43805	2.33010	2.17222	2.69079
23	2.51236	2.43156	2.32426	2.16718	2.68199
24	2.50592	2.42564	2.31893	2.16258	2.67396
25	2.50001	2.42022	2.31405	2.15836	2.66660
26	2.49458	2.41523	2.30955	2.15448	2.65983
27	2.48957	2.41062	2.30541	2.15090	2.65358
28	2.48493	2.40636	2.30157	2.14759	2.64780
29	2.48063	2.40240	2.29801	2.14451	2.64244
30	2.47662	2.39872	2.29469	2.14165	2.63744
35	2.46013	2.38356	2.28104	2.12985	2.61689
40	2.44787	2.37230	2.27089	2.12108	2.60162
45	2.43841	2.36360	2.26305	2.11431	2.58983
50	2.43088	2.35668	2.25681	2.10892	2.58046
60	2.41966	2.34636	2.24752	2.10088	2.56648
80	2.40575	2.33357	2.23599	2.09091	2.54916
100	2.39747	2.32596	2.22912	2.08497	2.53884
120	2.39198	2.32090	2.22457	2.08103	2.53200
200	2.38105	2.31085	2.21550	2.07319	2.51839
∞	2.36480	2.29591	2.20203	2.06152	2.49817

BECHHOFER and DUNNETT

TABLE A.3

One-sided Percentage Point (g) for Equal
Correlations   (*Case I with Correlation* ρ)

p = 10, 1-α = 0.95

ν ↓	ρ					
	0.0	0.1	0.2	0.3	0.4	0.5
2	7.22028	6.97798	6.72594	6.46168	6.18184	5.88169
3	4.98332	4.85785	4.72419	4.58062	4.42489	4.25374
4	4.17865	4.09376	4.00095	3.89891	3.78584	3.65907
5	3.77174	3.70698	3.63440	3.55292	3.46094	3.35612
6	3.52757	3.47473	3.41414	3.34483	3.26536	3.17355
7	3.36520	3.32019	3.26751	3.20624	3.13501	3.05178
8	3.24957	3.21009	3.16299	3.10741	3.04204	2.96488
9	3.16309	3.12771	3.08477	3.03343	2.97241	2.89978
10	3.09601	3.06377	3.02405	2.97599	2.91835	2.84922
11	3.04247	3.01273	2.97556	2.93012	2.87516	2.80882
12	2.99876	2.97104	2.93595	2.89264	2.83987	2.77581
13	2.96240	2.93635	2.90299	2.86144	2.81049	2.74833
14	2.93168	2.90704	2.87513	2.83507	2.78566	2.72510
15	2.90539	2.88194	2.85127	2.81250	2.76440	2.70520
16	2.88264	2.86022	2.83062	2.79295	2.74599	2.68797
17	2.86275	2.84123	2.81256	2.77586	2.72990	2.67291
18	2.84523	2.82448	2.79665	2.76079	2.71570	2.65963
19	2.82967	2.80961	2.78250	2.74740	2.70309	2.64783
20	2.81576	2.79632	2.76986	2.73543	2.69182	2.63727
21	2.80325	2.78436	2.75849	2.72466	2.68168	2.62778
22	2.79194	2.77355	2.74820	2.71493	2.67251	2.61920
23	2.78166	2.76372	2.73886	2.70608	2.66417	2.61140
24	2.77229	2.75476	2.73033	2.69800	2.65657	2.60428
25	2.76370	2.74654	2.72251	2.69061	2.64960	2.59775
26	2.75581	2.73899	2.71533	2.68380	2.64319	2.59175
27	2.74852	2.73202	2.70869	2.67752	2.63727	2.58622
28	2.74178	2.72557	2.70256	2.67171	2.63180	2.58109
29	2.73553	2.71958	2.69686	2.66631	2.62671	2.57633
30	2.72970	2.71401	2.69155	2.66129	2.62198	2.57190
35	2.70576	2.69108	2.66973	2.64063	2.60252	2.55368
40	2.68798	2.67405	2.65352	2.62527	2.58805	2.54014
45	2.67427	2.66090	2.64100	2.61342	2.57689	2.52969
50	2.66336	2.65045	2.63104	2.60399	2.56800	2.52137
60	2.64711	2.63486	2.61620	2.58993	2.55476	2.50898
80	2.62699	2.61556	2.59781	2.57251	2.53835	2.49362
100	2.61502	2.60406	2.58686	2.56214	2.52858	2.48447
120	2.60708	2.59643	2.57959	2.55526	2.52210	2.47840
200	2.59130	2.58127	2.56514	2.54157	2.50920	2.46633
∞	2.56788	2.55875	2.54367	2.52123	2.49004	2.44839

TABLE A.3

One-sided Percentage Point (g) for Equal
Correlations (*Case I with Correlation* ρ)

p = 10, 1-α = 0.95

ν ↓	ρ				
	0.6	0.7	0.8	0.9	$1/(1+\sqrt{10})$
2	5.55411	5.18725	4.75822	4.20817	6.62120
3	4.06226	3.84219	3.57745	3.22657	4.66770
4	3.51451	3.34522	3.13761	2.85656	3.96107
5	3.23478	3.09064	2.91138	2.66511	3.60275
6	3.06595	2.93672	2.77428	2.54866	3.38736
7	2.95326	2.83385	2.68250	2.47051	3.24394
8	2.87278	2.76033	2.61684	2.41451	3.14170
9	2.81247	2.70521	2.56756	2.37243	3.06518
10	2.76561	2.66236	2.52923	2.33966	3.00577
11	2.72817	2.62810	2.49857	2.31343	2.95832
12	2.69756	2.60009	2.47349	2.29195	2.91956
13	2.67207	2.57676	2.45260	2.27406	2.88731
14	2.65052	2.55704	2.43493	2.25891	2.86004
15	2.63207	2.54014	2.41979	2.24593	2.83670
16	2.61609	2.52551	2.40667	2.23468	2.81649
17	2.60211	2.51271	2.39520	2.22484	2.79882
18	2.58979	2.50142	2.38508	2.21615	2.78324
19	2.57884	2.49139	2.37609	2.20843	2.76940
20	2.56905	2.48242	2.36804	2.20153	2.75702
21	2.56024	2.47435	2.36080	2.19531	2.74589
22	2.55227	2.46705	2.35426	2.18969	2.73582
23	2.54504	2.46042	2.34831	2.18458	2.72667
24	2.53843	2.45437	2.34287	2.17991	2.71833
25	2.53237	2.44882	2.33790	2.17564	2.71068
26	2.52681	2.44372	2.33332	2.17170	2.70364
27	2.52167	2.43901	2.32909	2.16807	2.69715
28	2.51691	2.43465	2.32518	2.16471	2.69114
29	2.51249	2.43060	2.32155	2.16158	2.68556
30	2.50838	2.42683	2.31816	2.15868	2.68037
35	2.49147	2.41133	2.30425	2.14671	2.65901
40	2.47890	2.39981	2.29391	2.13782	2.64314
45	2.46920	2.39092	2.28592	2.13094	2.63089
50	2.46148	2.38384	2.27956	2.12547	2.62114
60	2.44998	2.37329	2.27009	2.11732	2.60661
80	2.43571	2.36021	2.25834	2.10721	2.58861
100	2.42722	2.35242	2.25134	2.10118	2.57788
120	2.42159	2.34726	2.24670	2.09718	2.57077
200	2.41038	2.33698	2.23746	2.08923	2.55662
∞	2.39372	2.32170	2.22373	2.07740	2.53559

TABLE A.3

One-sided Percentage Point (g) for Equal
Correlations $\left(Case\ I\ with\ Correlation\ \rho\right)$

p = 11, 1-α = 0.95

ν ↓	ρ					
	0.0	0.1	0.2	0.3	0.4	0.5
2	7.40558	7.15064	6.88556	6.60784	6.31406	5.99937
3	5.09631	4.96341	4.82219	4.67086	4.50708	4.32750
4	4.26552	4.17532	4.07707	3.96941	3.85043	3.71740
5	3.84529	3.77638	3.69950	3.61348	3.51671	3.40674
6	3.59306	3.53681	3.47261	3.39946	3.31586	3.21956
7	3.42528	3.37738	3.32157	3.25691	3.18201	3.09474
8	3.30577	3.26377	3.21390	3.15527	3.08654	3.00567
9	3.21638	3.17876	3.13332	3.07918	3.01506	2.93896
10	3.14703	3.11278	3.07076	3.02011	2.95955	2.88714
11	3.09167	3.06010	3.02080	2.97292	2.91521	2.84574
12	3.04646	3.01706	2.97999	2.93437	2.87897	2.81191
13	3.00886	2.98125	2.94602	2.90229	2.84882	2.78375
14	2.97709	2.95099	2.91731	2.87517	2.82332	2.75994
15	2.94990	2.92509	2.89273	2.85195	2.80150	2.73956
16	2.92636	2.90266	2.87145	2.83184	2.78259	2.72190
17	2.90579	2.88305	2.85284	2.81426	2.76607	2.70647
18	2.88766	2.86576	2.83644	2.79876	2.75150	2.69286
19	2.87156	2.85041	2.82186	2.78499	2.73855	2.68076
20	2.85717	2.83668	2.80883	2.77267	2.72697	2.66995
21	2.84423	2.82433	2.79711	2.76160	2.71656	2.66022
22	2.83253	2.81317	2.78651	2.75158	2.70715	2.65143
23	2.82190	2.80302	2.77688	2.74248	2.69859	2.64344
24	2.81220	2.79376	2.76809	2.73417	2.69078	2.63614
25	2.80331	2.78528	2.76003	2.72656	2.68362	2.62946
26	2.79514	2.77748	2.75262	2.71956	2.67704	2.62331
27	2.78761	2.77028	2.74579	2.71310	2.67097	2.61763
28	2.78063	2.76362	2.73946	2.70712	2.66535	2.61238
29	2.77416	2.75743	2.73359	2.70157	2.66013	2.60751
30	2.76813	2.75168	2.72812	2.69640	2.65527	2.60297
35	2.74336	2.72799	2.70562	2.67514	2.63528	2.58430
40	2.72496	2.71040	2.68891	2.65935	2.62043	2.57043
45	2.71076	2.69682	2.67601	2.64715	2.60897	2.55972
50	2.69948	2.68602	2.66574	2.63745	2.59985	2.55120
60	2.68266	2.66992	2.65044	2.62299	2.58625	2.53850
80	2.66184	2.64998	2.63148	2.60506	2.56940	2.52275
100	2.64945	2.63810	2.62018	2.59439	2.55937	2.51338
120	2.64123	2.63022	2.61269	2.58731	2.55271	2.50716
200	2.62490	2.61456	2.59779	2.57322	2.53947	2.49479
∞	2.60066	2.59128	2.57564	2.55229	2.51979	2.47641

TABLE A.3

One-sided Percentage Point (g) for Equal
Correlations (*Case I with Correlation* ρ)

p = 11, 1-α = 0.95

ν ↓	ρ				
	0.6	0.7	0.8	0.9	1/(1+√11)
2	5.65643	5.27303	4.82555	4.25330	6.79913
3	4.12702	3.89713	3.62123	3.25655	4.77546
4	3.56609	3.38932	3.17309	2.88118	4.04408
5	3.27977	3.12933	2.94272	2.68703	3.67332
6	3.10702	2.97219	2.80315	2.56898	3.45048
7	2.99173	2.86719	2.70974	2.48978	3.30211
8	2.90941	2.79216	2.64292	2.43302	3.19634
9	2.84772	2.73591	2.59278	2.39037	3.11717
10	2.79980	2.69219	2.55378	2.35717	3.05571
11	2.76150	2.65723	2.52259	2.33059	3.00662
12	2.73019	2.62865	2.49707	2.30884	2.96652
13	2.70413	2.60486	2.47582	2.29071	2.93314
14	2.68210	2.58473	2.45785	2.27537	2.90493
15	2.66323	2.56750	2.44245	2.26222	2.88078
16	2.64688	2.55257	2.42910	2.25082	2.85987
17	2.63259	2.53951	2.41743	2.24085	2.84158
18	2.61999	2.52800	2.40714	2.23205	2.82546
19	2.60880	2.51777	2.39799	2.22423	2.81114
20	2.59879	2.50862	2.38981	2.21723	2.79833
21	2.58978	2.50039	2.38245	2.21094	2.78681
22	2.58163	2.49294	2.37579	2.20524	2.77640
23	2.57423	2.48618	2.36974	2.20007	2.76693
24	2.56748	2.48000	2.36421	2.19534	2.75829
25	2.56129	2.47434	2.35915	2.19101	2.75037
26	2.55559	2.46914	2.35450	2.18702	2.74309
27	2.55034	2.46433	2.35020	2.18334	2.73638
28	2.54548	2.45989	2.34622	2.17994	2.73016
29	2.54096	2.45576	2.34253	2.17678	2.72439
30	2.53676	2.45192	2.33909	2.17383	2.71901
35	2.51947	2.43611	2.32493	2.16171	2.69690
40	2.50662	2.42436	2.31442	2.15270	2.68048
45	2.49670	2.41529	2.30629	2.14574	2.66779
50	2.48881	2.40807	2.29983	2.14021	2.65771
60	2.47704	2.39731	2.29020	2.13195	2.64267
80	2.46246	2.38397	2.27825	2.12171	2.62403
100	2.45378	2.37603	2.27114	2.11560	2.61293
120	2.44802	2.37076	2.26642	2.11156	2.60556
200	2.43656	2.36028	2.25702	2.10350	2.59092
∞	2.41953	2.34470	2.24307	2.09152	2.56915

BECHHOFER and DUNNETT

TABLE A.3

One-sided Percentage Point (g) for Equal
Correlations (*Case I with Correlation* ρ)

p = 12, 1-α = 0.95

ν ↓	ρ					
	0.0	0.1	0.2	0.3	0.4	0.5
2	7.57341	7.30712	7.03026	6.74036	6.43393	6.10603
3	5.19901	5.05932	4.91118	4.75277	4.58164	4.39435
4	4.34460	4.24950	4.14624	4.03340	3.90901	3.77026
5	3.91227	3.83952	3.75865	3.66846	3.56728	3.45259
6	3.65270	3.59328	3.52574	3.44904	3.36164	3.26123
7	3.48000	3.42939	3.37067	3.30289	3.22461	3.13364
8	3.35695	3.31259	3.26013	3.19869	3.12687	3.04260
9	3.26490	3.22518	3.17740	3.12068	3.05369	2.97441
10	3.19346	3.15733	3.11317	3.06011	2.99687	2.92145
11	3.13643	3.10314	3.06187	3.01173	2.95148	2.87914
12	3.08986	3.05888	3.01995	2.97221	2.91439	2.84457
13	3.05112	3.02205	2.98507	2.93931	2.88351	2.81578
14	3.01838	2.99092	2.95559	2.91150	2.85742	2.79146
15	2.99036	2.96427	2.93034	2.88769	2.83507	2.77062
16	2.96610	2.94119	2.90848	2.86707	2.81572	2.75258
17	2.94490	2.92102	2.88937	2.84904	2.79880	2.73681
18	2.92622	2.90323	2.87252	2.83315	2.78388	2.72290
19	2.90962	2.88743	2.85755	2.81903	2.77063	2.71054
20	2.89479	2.87331	2.84416	2.80640	2.75878	2.69949
21	2.88145	2.86060	2.83212	2.79504	2.74812	2.68955
22	2.86938	2.84911	2.82123	2.78477	2.73848	2.68056
23	2.85843	2.83867	2.81134	2.77544	2.72972	2.67239
24	2.84843	2.82914	2.80231	2.76692	2.72173	2.66494
25	2.83927	2.82041	2.79403	2.75911	2.71440	2.65810
26	2.83085	2.81238	2.78642	2.75193	2.70766	2.65182
27	2.82308	2.80498	2.77940	2.74531	2.70145	2.64602
28	2.81588	2.79812	2.77290	2.73917	2.69569	2.64066
29	2.80921	2.79175	2.76686	2.73348	2.69035	2.63568
30	2.80300	2.78583	2.76125	2.72818	2.68538	2.63104
35	2.77745	2.76145	2.73813	2.70638	2.66491	2.61196
40	2.75848	2.74334	2.72096	2.69018	2.64971	2.59778
45	2.74384	2.72937	2.70770	2.67767	2.63797	2.58684
50	2.73221	2.71825	2.69715	2.66772	2.62863	2.57813
60	2.71487	2.70167	2.68143	2.65288	2.61471	2.56515
80	2.69340	2.68114	2.66194	2.63450	2.59746	2.54906
100	2.68062	2.66891	2.65033	2.62355	2.58719	2.53948
120	2.67215	2.66080	2.64263	2.61628	2.58037	2.53313
200	2.65531	2.64467	2.62732	2.60183	2.56681	2.52049
∞	2.63031	2.62070	2.60455	2.58035	2.54666	2.50170

TABLE A.3

One-sided Percentage Point (g) for Equal
Correlations (*Case I with Correlation* ρ)

p = 12, 1-α = 0.95

ν ↓	ρ				
	0.6	0.7	0.8	0.9	$1/(1+\sqrt{12})$
2	5.74912	5.35068	4.88643	4.29404	6.96196
3	4.18567	3.94683	3.66078	3.28360	4.87417
4	3.61278	3.42920	3.20513	2.90336	4.12009
5	3.32048	3.16430	2.97101	2.70679	3.73790
6	3.14417	3.00424	2.82920	2.58728	3.50821
7	3.02652	2.89731	2.73431	2.50712	3.35527
8	2.94252	2.82091	2.66644	2.44968	3.24625
9	2.87958	2.76363	2.61552	2.40653	3.16464
10	2.83069	2.71911	2.57591	2.37294	3.10128
11	2.79162	2.68353	2.54424	2.34604	3.05068
12	2.75968	2.65444	2.51833	2.32404	3.00933
13	2.73310	2.63021	2.49675	2.30569	2.97492
14	2.71062	2.60973	2.47850	2.29017	2.94584
15	2.69137	2.59218	2.46287	2.27687	2.92094
16	2.67470	2.57698	2.44932	2.26534	2.89937
17	2.66012	2.56369	2.43747	2.25525	2.88052
18	2.64727	2.55197	2.42702	2.24636	2.86390
19	2.63585	2.54156	2.41773	2.23845	2.84913
20	2.62564	2.53225	2.40943	2.23137	2.83592
21	2.61645	2.52387	2.40195	2.22500	2.82404
22	2.60814	2.51629	2.39519	2.21924	2.81330
23	2.60059	2.50941	2.38905	2.21400	2.80354
24	2.59370	2.50312	2.38344	0.00000	2.79463
25	2.58739	2.49736	2.37830	0.00000	2.78647
26	2.58158	2.49207	2.37358	2.20081	2.77896
27	2.57622	2.48718	2.36921	2.19709	2.77203
28	2.57126	2.48265	2.36517	2.19364	2.76562
29	2.56666	2.47845	2.36142	2.19045	2.75966
30	2.56237	2.47454	2.35793	2.18747	2.75412
35	2.54473	2.45845	2.34357	2.17521	2.73132
40	2.53163	2.44649	2.33289	2.16610	2.71438
45	2.52151	2.43726	2.32465	2.15906	2.70130
50	2.51346	2.42991	2.31809	2.15346	2.69089
60	2.50146	2.41896	2.30831	2.14511	2.67538
80	2.48659	2.40539	2.29618	2.13475	2.65615
100	2.47774	2.39731	2.28896	2.12858	2.64470
120	2.47186	2.39194	2.28417	2.12448	2.63710
200	2.46018	2.38128	2.27464	2.11634	2.62199
∞	2.44281	2.36542	2.26047	2.10422	2.59953

BECHHOFER and DUNNETT

TABLE A.3

One-sided Percentage Point (g) for Equal Correlations (Case I with Correlation ρ)

$p = 13, \ 1-\alpha = 0.95$

$\nu \downarrow$	ρ					
	0.0	0.1	0.2	0.3	0.4	0.5
2	7.72661	7.45004	7.16248	6.86146	6.54348	6.20349
3	5.29304	5.14713	4.99263	4.82769	4.64980	4.45543
4	4.41711	4.31748	4.20958	4.09195	3.96257	3.81854
5	3.97374	3.89740	3.81282	3.71877	3.61351	3.49446
6	3.70745	3.64505	3.57439	3.49439	3.40348	3.29927
7	3.53022	3.47707	3.41564	3.34495	3.26353	3.16914
8	3.40392	3.35733	3.30247	3.23839	3.16372	3.07630
9	3.30941	3.26772	3.21775	3.15862	3.08899	3.00676
10	3.23606	3.19814	3.15198	3.09669	3.03095	2.95276
11	3.17749	3.14258	3.09944	3.04721	2.98460	2.90961
12	3.12966	3.09719	3.05652	3.00679	2.94672	2.87435
13	3.08986	3.05941	3.02079	2.97314	2.91519	2.84500
14	3.05623	3.02748	2.99060	2.94470	2.88854	2.82020
15	3.02744	3.00014	2.96474	2.92035	2.86572	2.79895
16	3.00252	2.97647	2.94235	2.89926	2.84596	2.78055
17	2.98073	2.95577	2.92277	2.88082	2.82868	2.76447
18	2.96153	2.93752	2.90551	2.86456	2.81344	2.75028
19	2.94448	2.92131	2.89017	2.85012	2.79991	2.73768
20	2.92923	2.90682	2.87646	2.83720	2.78781	2.72642
21	2.91552	2.89378	2.86412	2.82559	2.77692	2.71628
22	2.90312	2.88199	2.85297	2.81508	2.76707	2.70711
23	2.89186	2.87128	2.84283	2.80553	2.75813	2.69878
24	2.88159	2.86150	2.83358	2.79682	2.74996	2.69118
25	2.87217	2.85254	2.82510	2.78883	2.74248	2.68422
26	2.86351	2.84430	2.81730	2.78149	2.73560	2.67781
27	2.85553	2.83670	2.81011	2.77471	2.72925	2.67190
28	2.84813	2.82966	2.80345	2.76844	2.72337	2.66643
29	2.84127	2.82313	2.79726	2.76261	2.71791	2.66135
30	2.83489	2.81705	2.79151	2.75719	2.71283	2.65662
35	2.80862	2.79203	2.76782	2.73488	2.69193	2.63716
40	2.78913	2.77345	2.75023	2.71831	2.67640	2.62270
45	2.77408	2.75910	2.73664	2.70551	2.66441	2.61154
50	2.76211	2.74768	2.72583	2.69533	2.65487	2.60266
60	2.74429	2.73067	2.70971	2.68015	2.64066	2.58943
80	2.72221	2.70959	2.68974	2.66134	2.62303	2.57302
100	2.70908	2.69703	2.67784	2.65014	2.61254	2.56326
120	2.70036	2.68870	2.66995	2.64270	2.60558	2.55677
200	2.68305	2.67214	2.65425	2.62792	2.59172	2.54388
∞	2.65735	2.64753	2.63092	2.60594	2.57114	2.52473

TABLE A.3

One-sided Percentage Point (g) for Equal
Correlations (*Case I with Correlation* ρ)

$p = 13, \ 1-\alpha = 0.95$

ν ↓	ρ				
	0.6	0.7	0.8	0.9	$1/(1+\sqrt{13})$
2	5.83379	5.42156	4.94195	4.33113	7.11195
3	4.23921	3.99217	3.69683	3.30820	4.96518
4	3.65539	3.46556	3.23432	2.92354	4.19017
5	3.35763	3.19618	2.99676	2.72475	3.79742
6	3.17806	3.03345	2.85291	2.60392	3.56139
7	3.05824	2.92475	2.75667	2.52289	3.40422
8	2.97271	2.84709	2.68784	2.46483	3.29218
9	2.90863	2.78888	2.63620	2.42121	3.20830
10	2.85885	2.74363	2.59604	2.38726	3.14318
11	2.81907	2.70747	2.56393	2.36008	3.09117
12	2.78656	2.67791	2.53766	2.33784	3.04867
13	2.75949	2.65329	2.51579	2.31931	3.01330
14	2.73661	2.63248	2.49728	2.30362	2.98340
15	2.71701	2.61465	2.48143	2.29018	2.95780
16	2.70004	2.59921	2.46770	2.27853	2.93563
17	2.68520	2.58570	2.45569	2.26834	2.91625
18	2.67212	2.57379	2.44509	2.25935	2.89916
19	2.66049	2.56321	2.43568	2.25135	2.88397
20	2.65010	2.55375	2.42726	2.24420	2.87040
21	2.64074	2.54524	2.41968	2.23777	2.85818
22	2.63229	2.53754	2.41283	2.23195	2.84714
23	2.62460	2.53054	2.40661	2.22666	2.83710
24	2.61759	2.52416	2.40092	2.22183	2.82794
25	2.61116	2.51831	2.39571	2.21740	2.81954
26	2.60525	2.51292	2.39092	2.21333	2.81182
27	2.59979	2.50796	2.38650	2.20957	2.80470
28	2.59474	2.50336	2.38240	2.20609	2.79810
29	2.59006	2.49909	2.37860	2.20286	2.79198
30	2.58569	2.49512	2.37506	2.19985	2.78628
35	2.56774	2.47877	2.36050	2.18746	2.76283
40	2.55440	2.46662	2.34968	2.17826	2.74541
45	2.54410	2.45724	2.34133	2.17115	2.73195
50	2.53591	2.44978	2.33468	2.16549	2.72125
60	2.52369	2.43865	2.32476	2.15705	2.70529
80	2.50856	2.42487	2.31248	2.14659	2.68552
100	2.49954	2.41666	2.30516	2.14035	2.67374
120	2.49356	2.41121	2.30030	2.13622	2.66592
200	2.48167	2.40037	2.29064	2.12799	2.65038
∞	2.46399	2.38427	2.27628	2.11575	2.62727

BECHHOFER and DUNNETT

TABLE A.3

*One-sided Percentage Point (g) for Equal
Correlations* (*Case I with Correlation* ρ)

p = 14, 1-α = 0.95

ν ↓	ρ					
	0.0	0.1	0.2	0.3	0.4	0.5
2	7.86740	7.58144	7.28408	6.97287	6.64427	6.29314
3	5.37971	5.22803	5.06766	4.89668	4.71254	4.51163
4	4.48402	4.38017	4.26796	4.14588	4.01187	3.86296
5	4.03049	3.95080	3.86276	3.76510	3.65605	3.53297
6	3.75801	3.69282	3.61924	3.53617	3.44198	3.33426
7	3.57661	3.52107	3.45709	3.38369	3.29935	3.20179
8	3.44730	3.39862	3.34148	3.27496	3.19762	3.10728
9	3.35052	3.30696	3.25494	3.19356	3.12145	3.03649
10	3.27540	3.23579	3.18774	3.13036	3.06231	2.98152
11	3.21540	3.17895	3.13406	3.07987	3.01506	2.93761
12	3.16640	3.13251	3.09020	3.03862	2.97645	2.90172
13	3.12561	3.09386	3.05370	3.00428	2.94432	2.87185
14	3.09115	3.06118	3.02284	2.97525	2.91715	2.84660
15	3.06165	3.03321	2.99642	2.95040	2.89389	2.82497
16	3.03611	3.00898	2.97353	2.92888	2.87375	2.80625
17	3.01378	2.98780	2.95352	2.91006	2.85614	2.78987
18	2.99410	2.96912	2.93588	2.89346	2.84061	2.77544
19	2.97662	2.95253	2.92021	2.87872	2.82681	2.76261
20	2.96099	2.93769	2.90619	2.86554	2.81448	2.75114
21	2.94693	2.92435	2.89358	2.85368	2.80338	2.74082
22	2.93423	2.91228	2.88218	2.84295	2.79334	2.73149
23	2.92268	2.90131	2.87181	2.83321	2.78422	2.72302
24	2.91214	2.89130	2.86236	2.82431	2.77590	2.71528
25	2.90249	2.88213	2.85369	2.81616	2.76827	2.70819
26	2.89361	2.87369	2.84572	2.80866	2.76126	2.70167
27	2.88542	2.86591	2.83836	2.80174	2.75479	2.69565
28	2.87785	2.85871	2.83155	2.79534	2.74879	2.69008
29	2.87081	2.85202	2.82523	2.78939	2.74323	2.68491
30	2.86426	2.84579	2.81935	2.78386	2.73805	2.68010
35	2.83733	2.82017	2.79513	2.76109	2.71675	2.66029
40	2.81734	2.80114	2.77714	2.74417	2.70092	2.64558
45	2.80190	2.78645	2.76325	2.73110	2.68870	2.63422
50	2.78963	2.77476	2.75219	2.72071	2.67897	2.62518
60	2.77135	2.75734	2.73572	2.70521	2.66448	2.61171
80	2.74871	2.73574	2.71529	2.68600	2.64651	2.59501
100	2.73524	2.72289	2.70312	2.67456	2.63581	2.58507
120	2.72631	2.71435	2.69505	2.66697	2.62871	2.57847
200	2.70855	2.69739	2.67899	2.65187	2.61459	2.56535
∞	2.68219	2.67218	2.65513	2.62943	2.59361	2.54585

TABLE A.3

One-sided Percentage Point (g) for Equal
Correlations (Case I with Correlation ρ)

p = 14, 1-α = 0.95

ν ↓	ρ				
	0.6	0.7	0.8	0.9	1/(1+√14)
2	5.91166	5.48672	4.99295	4.36516	7.25090
3	4.28845	4.03382	3.72991	3.33076	5.04958
4	3.69456	3.49896	3.26109	2.94203	4.25517
5	3.39176	3.22545	3.02038	2.74120	3.85261
6	3.20919	3.06025	2.87465	2.61916	3.61067
7	3.08738	2.94993	2.77717	2.53732	3.44957
8	3.00044	2.87111	2.70746	2.47869	3.33471
9	2.93530	2.81204	2.65516	2.43465	3.24872
10	2.88470	2.76613	2.61449	2.40037	3.18196
11	2.84427	2.72943	2.58197	2.37293	3.12863
12	2.81122	2.69943	2.55538	2.35048	3.08505
13	2.78372	2.67446	2.53323	2.33176	3.04878
14	2.76046	2.65334	2.51449	2.31593	3.01812
15	2.74054	2.63525	2.49844	2.30236	2.99187
16	2.72330	2.61958	2.48454	2.29060	2.96913
17	2.70821	2.60588	2.47238	2.28031	2.94925
18	2.69492	2.59380	2.46165	2.27123	2.93172
19	2.68310	2.58306	2.45212	2.26317	2.91615
20	2.67254	2.57347	2.44359	2.25595	2.90222
21	2.66303	2.56483	2.43592	2.24945	2.88969
22	2.65444	2.55702	2.42899	2.24358	2.87836
23	2.64663	2.54992	2.42268	2.23824	2.86807
24	2.63950	2.54344	2.41693	2.23336	2.85867
25	2.63297	2.53751	2.41165	2.22889	2.85006
26	2.62696	2.53205	2.40680	2.22478	2.84214
27	2.62141	2.52701	2.40233	2.22099	2.83483
28	2.61628	2.52234	2.39818	2.21747	2.82807
29	2.61152	2.51801	2.39433	2.21421	2.82179
30	2.60709	2.51398	2.39075	2.21118	2.81594
35	2.58884	2.49739	2.37601	2.19868	2.79188
40	2.57528	2.48507	2.36506	2.18938	2.77400
45	2.56481	2.47556	2.35660	2.18221	2.76020
50	2.55649	2.46798	2.34987	2.17650	2.74922
60	2.54408	2.45670	2.33983	2.16798	2.73285
80	2.52869	2.44271	2.32739	2.15742	2.71255
100	2.51953	2.43438	2.31999	2.15113	2.70046
120	2.51345	2.42886	2.31507	2.14695	2.69244
200	2.50137	2.41786	2.30529	2.13865	2.67649
∞	2.48340	2.40153	2.29076	2.12629	2.65277

TABLE A.3

One-sided Percentage Point (g) for Equal
Correlations   [*Case I with Correlation* ρ]

p = 15,   1-α = 0.95

ν ↓	ρ					
	0.0	0.1	0.2	0.3	0.4	0.5
2	7.99752	7.70296	7.39658	7.07597	6.73754	6.37611
3	5.46002	5.30300	5.13717	4.96058	4.77064	4.56364
4	4.54611	4.43831	4.32207	4.19585	4.05752	3.90406
5	4.08319	4.00035	3.90906	3.80804	3.69545	3.56861
6	3.80497	3.73715	3.66083	3.57487	3.47763	3.36662
7	3.61970	3.56189	3.49552	3.41957	3.33250	3.23198
8	3.48760	3.43693	3.37765	3.30883	3.22899	3.13592
9	3.38871	3.34337	3.28941	3.22592	3.15149	3.06399
10	3.31193	3.27072	3.22089	3.16153	3.09131	3.00812
11	3.25060	3.21269	3.16615	3.11011	3.04324	2.96349
12	3.20050	3.16527	3.12142	3.06808	3.00396	2.92702
13	3.15881	3.12580	3.08419	3.03310	2.97126	2.89666
14	3.12357	3.09244	3.05271	3.00353	2.94362	2.87099
15	3.09340	3.06386	3.02576	2.97821	2.91995	2.84902
16	3.06728	3.03912	3.00242	2.95628	2.89945	2.82999
17	3.04444	3.01749	2.98201	2.93711	2.88153	2.81335
18	3.02431	2.99841	2.96401	2.92020	2.86573	2.79868
19	3.00643	2.98146	2.94802	2.90518	2.85169	2.78564
20	2.99044	2.96630	2.93372	2.89175	2.83914	2.77399
21	2.97606	2.95267	2.92086	2.87967	2.82784	2.76350
22	2.96306	2.94034	2.90922	2.86874	2.81763	2.75402
23	2.95125	2.92914	2.89865	2.85881	2.80835	2.74540
24	2.94047	2.91891	2.88900	2.84974	2.79988	2.73754
25	2.93060	2.90954	2.88016	2.84144	2.79212	2.73033
26	2.92151	2.90092	2.87202	2.83380	2.78498	2.72371
27	2.91314	2.89297	2.86452	2.82675	2.77839	2.71759
28	2.90538	2.88561	2.85757	2.82022	2.77230	2.71193
29	2.89818	2.87877	2.85112	2.81417	2.76664	2.70667
30	2.89148	2.87241	2.84511	2.80853	2.76137	2.70178
35	2.86393	2.84623	2.82040	2.78532	2.73968	2.68165
40	2.84346	2.82678	2.80204	2.76808	2.72357	2.66670
45	2.82767	2.81177	2.78786	2.75476	2.71113	2.65515
50	2.81511	2.79982	2.77658	2.74417	2.70124	2.64597
60	2.79640	2.78201	2.75977	2.72837	2.68649	2.63228
80	2.77323	2.75994	2.73892	2.70880	2.66820	2.61531
100	2.75944	2.74680	2.72650	2.69714	2.65731	2.60520
120	2.75030	2.73808	2.71826	2.68940	2.65008	2.59850
200	2.73212	2.72073	2.70187	2.67401	2.63571	2.58516
∞	2.70515	2.69496	2.67750	2.65113	2.61435	2.56534

TABLE A.3

*One-sided Percentage Point (g) for Equal
Correlations* (*Case I with Correlation* ρ)

$$p = 15, \quad 1-\alpha = 0.95$$

ν ↓	ρ				
	0.6	0.7	0.8	0.9	$1/(1+\sqrt{15})$
2	5.98370	5.54698	5.04009	4.39657	7.38025
3	4.33399	4.07233	3.76047	3.35156	5.12825
4	3.73078	3.52981	3.28581	2.95908	4.31576
5	3.42332	3.25248	3.04218	2.75636	3.90405
6	3.23796	3.08501	2.89471	2.63320	3.65660
7	3.11432	2.97318	2.79608	2.55062	3.49181
8	3.02606	2.89329	2.72555	2.49147	3.37431
9	2.95995	2.83342	2.67265	2.44703	3.28634
10	2.90859	2.78689	2.63151	2.41245	3.21804
11	2.86755	2.74970	2.59861	2.38477	3.16347
12	2.83401	2.71930	2.57171	2.36212	3.11888
13	2.80609	2.69399	2.54931	2.34324	3.08177
14	2.78249	2.67259	2.53036	2.32727	3.05039
15	2.76227	2.65426	2.51412	2.31358	3.02352
16	2.74477	2.63838	2.50006	2.30172	3.00025
17	2.72946	2.62450	2.48776	2.29134	2.97991
18	2.71597	2.61226	2.47691	2.28218	2.96197
19	2.70398	2.60138	2.46727	2.27404	2.94603
20	2.69325	2.59165	2.45865	2.26676	2.93177
21	2.68361	2.58290	2.45090	2.26021	2.91895
22	2.67489	2.57499	2.44388	2.25429	2.90735
23	2.66696	2.56779	2.43750	2.24890	2.89681
24	2.65972	2.56123	2.43168	2.24398	2.88719
25	2.65310	2.55521	2.42635	2.23947	2.87838
26	2.64700	2.54968	2.42144	2.23533	2.87027
27	2.64137	2.54458	2.41692	2.23150	2.86279
28	2.63616	2.53985	2.41273	2.22796	2.85586
29	2.63133	2.53546	2.40883	2.22467	2.84943
30	2.62683	2.53138	2.40521	2.22161	2.84345
35	2.60831	2.51457	2.39030	2.20900	2.81881
40	2.59455	2.50209	2.37923	2.19963	2.80051
45	2.58393	2.49244	2.37067	2.19239	2.78638
50	2.57548	2.48477	2.36387	2.18663	2.77513
60	2.56289	2.47334	2.35372	2.17804	2.75837
80	2.54727	2.45917	2.34114	2.16739	2.73758
100	2.53798	2.45073	2.33365	2.16104	2.72520
120	2.53181	2.44513	2.32868	2.15683	2.71699
200	2.51954	2.43399	2.31879	2.14846	2.70065
∞	2.50131	2.41744	2.30409	2.13600	2.67636

BECHHOFER and DUNNETT

One-sided Percentage Point (g) for Equal
Correlations (*Case I with Correlation* ρ)

p = 16, 1-α = 0.95

ν ↓	0.0	0.1	0.2	0.3	0.4	0.5
				ρ		
2	8.11841	7.81589	7.50118	7.17185	6.82430	6.45327
3	5.53480	5.37281	5.20189	5.02007	4.82470	4.61202
4	4.60400	4.49250	4.37249	4.24238	4.10001	3.94229
5	4.13235	4.04654	3.95220	3.84802	3.73211	3.60174
6	3.84880	3.77849	3.69958	3.61091	3.51080	3.39672
7	3.65991	3.59996	3.53133	3.45298	3.36335	3.26005
8	3.52521	3.47265	3.41135	3.34036	3.25818	3.16256
9	3.42434	3.37732	3.32152	3.25604	3.17944	3.08954
10	3.34602	3.30329	3.25176	3.19056	3.11829	3.03284
11	3.28345	3.24414	3.19603	3.13825	3.06945	2.98754
12	3.23232	3.19581	3.15049	3.09551	3.02953	2.95052
13	3.18977	3.15558	3.11258	3.05993	2.99631	2.91971
14	3.15381	3.12156	3.08053	3.02985	2.96823	2.89367
15	3.12301	3.09243	3.05308	3.00409	2.94418	2.87136
16	3.09634	3.06721	3.02931	2.98178	2.92335	2.85205
17	3.07303	3.04515	3.00853	2.96228	2.90514	2.83516
18	3.05248	3.02569	2.99019	2.94508	2.88908	2.82027
19	3.03422	3.00841	2.97391	2.92980	2.87482	2.80704
20	3.01790	2.99296	2.95934	2.91613	2.86206	2.79521
21	3.00322	2.97905	2.94624	2.90384	2.85058	2.78456
22	2.98994	2.96648	2.93439	2.89272	2.84021	2.77494
23	2.97788	2.95505	2.92362	2.88262	2.83078	2.76620
24	2.96687	2.94462	2.91379	2.87340	2.82217	2.75821
25	2.95678	2.93506	2.90478	2.86495	2.81428	2.75090
26	2.94751	2.92627	2.89650	2.85717	2.80703	2.74417
27	2.93895	2.91816	2.88885	2.85000	2.80033	2.73797
28	2.93103	2.91065	2.88178	2.84336	2.79414	2.73222
29	2.92368	2.90368	2.87520	2.83720	2.78838	2.72689
30	2.91684	2.89719	2.86909	2.83146	2.78303	2.72192
35	2.88869	2.87049	2.84391	2.80785	2.76099	2.70149
40	2.86779	2.85065	2.82521	2.79030	2.74462	2.68631
45	2.85165	2.83533	2.81076	2.77675	2.73198	2.67459
50	2.83883	2.82314	2.79927	2.76597	2.72192	2.66527
60	2.81971	2.80497	2.78213	2.74990	2.70693	2.65137
80	2.79604	2.78245	2.76088	2.72998	2.68834	2.63415
100	2.78195	2.76903	2.74823	2.71811	2.67728	2.62389
120	2.77261	2.76013	2.73983	2.71023	2.66993	2.61708
200	2.75404	2.74243	2.72313	2.69457	2.65532	2.60355
∞	2.72648	2.71613	2.69829	2.67128	2.63361	2.58343

TABLE A.3

One-sided Percentage Point (g) for Equal
Correlations   (Case I with Correlation ρ)

p = 16, 1-α = 0.95

ν ↓	ρ				
	0.6	0.7	0.8	0.9	1/(1+√16)
2	6.05069	5.60300	5.08388	4.42573	7.50118
3	4.37633	4.10811	3.78884	3.37086	5.20189
4	3.76445	3.55848	3.30875	2.97488	4.37249
5	3.45265	3.27759	3.06241	2.77042	3.95220
6	3.26470	3.10800	2.91332	2.64622	3.69958
7	3.13934	2.99476	2.81362	2.56295	3.53133
8	3.04986	2.91388	2.74234	2.50330	3.41135
9	2.98284	2.85327	2.68887	2.45851	3.32152
10	2.93077	2.80616	2.64729	2.42364	3.25176
11	2.88917	2.76851	2.61405	2.39573	3.19603
12	2.85517	2.73774	2.58686	2.37290	3.15049
13	2.82687	2.71212	2.56422	2.35387	3.11258
14	2.80294	2.69045	2.54507	2.33777	3.08053
15	2.78245	2.67190	2.52866	2.32397	3.05308
16	2.76471	2.65583	2.51445	2.31201	3.02931
17	2.74919	2.64177	2.50202	2.30155	3.00853
18	2.73551	2.62938	2.49106	2.29232	2.99019
19	2.72335	2.61837	2.48132	2.28412	2.97391
20	2.71249	2.60852	2.47261	2.27678	2.95934
21	2.70271	2.59967	2.46477	2.27018	2.94624
22	2.69387	2.59165	2.45768	2.26421	2.93439
23	2.68583	2.58437	2.45124	2.25878	2.92362
24	2.67850	2.57773	2.44536	2.25382	2.91379
25	2.67178	2.57164	2.43997	2.24928	2.90478
26	2.66560	2.56604	2.43501	2.24510	2.89650
27	2.65989	2.56087	2.43044	2.24124	2.88885
28	2.65462	2.55609	2.42620	2.23767	2.88178
29	2.64972	2.55165	2.42227	2.23435	2.87520
30	2.64515	2.54751	2.41861	2.23127	2.86909
35	2.62638	2.53050	2.40355	2.21856	2.84391
40	2.61244	2.51787	2.39236	2.20911	2.82521
45	2.60167	2.50811	2.38371	2.20182	2.81076
50	2.59310	2.50034	2.37684	2.19601	2.79927
60	2.58034	2.48877	2.36659	2.18736	2.78213
80	2.56451	2.47443	2.35388	2.17662	2.76088
100	2.55509	2.46588	2.34631	2.17023	2.74823
120	2.54883	2.46022	2.34129	2.16598	2.73983
200	2.53640	2.44894	2.33130	2.15755	2.72313
∞	2.51792	2.43219	2.31645	2.14499	2.69829

TABLE A.3

One-sided Percentage Point (g) for Equal
Correlations (*Case I with Correlation* ρ)

$$p = 18, \quad 1-\alpha = 0.95$$

ν ↓	ρ 0.0	0.1	0.2	0.3	0.4	0.5
2	8.33691	8.02015	7.69045	7.34540	6.98137	6.59299
3	5.67039	5.49938	5.31922	5.12789	4.92265	4.69964
4	4.70914	4.59087	4.46395	4.32675	4.17701	4.01152
5	4.22172	4.13044	4.03050	3.92052	3.79855	3.66175
6	3.92850	3.85359	3.76991	3.67627	3.57090	3.45120
7	3.73307	3.66914	3.59633	3.51356	3.41923	3.31086
8	3.59362	3.53755	3.47252	3.39753	3.31105	3.21075
9	3.48916	3.43900	3.37980	3.31064	3.23006	3.13579
10	3.40802	3.36245	3.30779	3.24317	3.16716	3.07757
11	3.34318	3.30128	3.25026	3.18926	3.11691	3.03105
12	3.29019	3.25128	3.20323	3.14521	3.07584	2.99305
13	3.24607	3.20965	3.16409	3.10853	3.04166	2.96141
14	3.20877	3.17445	3.13099	3.07753	3.01277	2.93466
15	3.17683	3.14430	3.10264	3.05098	2.98803	2.91176
16	3.14917	3.11819	3.07808	3.02798	2.96660	2.89193
17	3.12498	3.09535	3.05661	3.00787	2.94786	2.87459
18	3.10365	3.07521	3.03767	2.99014	2.93134	2.85930
19	3.08471	3.05732	3.02084	2.97438	2.91666	2.84571
20	3.06776	3.04132	3.00580	2.96029	2.90353	2.83357
21	3.05252	3.02692	2.99226	2.94762	2.89173	2.82264
22	3.03874	3.01389	2.98001	2.93615	2.88105	2.81276
23	3.02622	3.00206	2.96888	2.92573	2.87134	2.80378
24	3.01479	2.99126	2.95872	2.91622	2.86248	2.79558
25	3.00432	2.98136	2.94941	2.90751	2.85437	2.78807
26	2.99469	2.97225	2.94085	2.89949	2.84690	2.78116
27	2.98580	2.96385	2.93295	2.89210	2.84001	2.77479
28	2.97758	2.95607	2.92563	2.88525	2.83364	2.76889
29	2.96994	2.94885	2.91884	2.87889	2.82772	2.76341
30	2.96284	2.94212	2.91252	2.87297	2.82221	2.75831
35	2.93360	2.91445	2.88650	2.84862	2.79953	2.73733
40	2.91189	2.89389	2.86716	2.83052	2.78268	2.72175
45	2.89513	2.87801	2.85222	2.81654	2.76966	2.70971
50	2.88181	2.86538	2.84033	2.80542	2.75931	2.70014
60	2.86195	2.84654	2.82261	2.78884	2.74388	2.68587
80	2.83735	2.82319	2.80064	2.76828	2.72475	2.66818
100	2.82271	2.80929	2.78755	2.75604	2.71336	2.65764
120	2.81301	2.80006	2.77886	2.74791	2.70580	2.65065
200	2.79371	2.78170	2.76158	2.73175	2.69076	2.63675
∞	2.76507	2.75442	2.73588	2.70771	2.66840	2.61609

TABLE A.3

One-sided Percentage Point (g) for Equal
Correlations   (Case I with Correlation ρ)

p = 18, 1-α = 0.95

ν ↓	ρ				
	0.6	0.7	0.8	0.9	1/(1+√18)
2	6.17198	5.70437	5.16307	4.47838	7.72157
3	4.45298	4.17283	3.84012	3.40568	5.33633
4	3.82538	3.61031	3.35019	3.00339	4.47611
5	3.50571	3.32298	3.09893	2.79577	4.04015
6	3.31307	3.14954	2.94692	2.66968	3.77806
7	3.18459	3.03376	2.84528	2.58517	3.60347
8	3.09290	2.95108	2.77262	2.52464	3.47894
9	3.02422	2.88911	2.71812	2.47918	3.38568
10	2.97087	2.84096	2.67576	2.44380	3.31326
11	2.92824	2.80248	2.64188	2.41549	3.25538
12	2.89341	2.77102	2.61418	2.39232	3.20808
13	2.86441	2.74484	2.59111	2.37302	3.16871
14	2.83990	2.72269	2.57159	2.35668	3.13541
15	2.81890	2.70373	2.55488	2.34269	3.10690
16	2.80072	2.68731	2.54040	2.33056	3.08220
17	2.78483	2.67295	2.52774	2.31995	3.06060
18	2.77081	2.66028	2.51657	2.31058	3.04155
19	2.75836	2.64903	2.50665	2.30227	3.02462
20	2.74722	2.63897	2.49777	2.29482	3.00949
21	2.73720	2.62992	2.48979	2.28813	2.99587
22	2.72815	2.62173	2.48257	2.28207	2.98355
23	2.71991	2.61429	2.47601	2.27656	2.97236
24	2.71240	2.60750	2.47001	2.27153	2.96214
25	2.70552	2.60128	2.46452	2.26692	2.95277
26	2.69918	2.59556	2.45947	2.26269	2.94416
27	2.69334	2.59027	2.45481	2.25877	2.93621
28	2.68793	2.58539	2.45050	2.25515	2.92885
29	2.68291	2.58085	2.44650	2.25179	2.92202
30	2.67824	2.57662	2.44277	2.24866	2.91566
35	2.65901	2.55924	2.42743	2.23577	2.88948
40	2.64472	2.54633	2.41602	2.22619	2.87003
45	2.63369	2.53636	2.40722	2.21880	2.85500
50	2.62491	2.52842	2.40022	2.21291	2.84305
60	2.61183	2.51660	2.38977	2.20413	2.82522
80	2.59562	2.50194	2.37683	2.19324	2.80312
100	2.58597	2.49321	2.36912	2.18676	2.78995
120	2.57956	2.48742	2.36400	2.18246	2.78121
200	2.56682	2.47590	2.35383	2.17390	2.76383
∞	2.54789	2.45879	2.33871	2.16117	2.73798

BECHHOFER and DUNNETT

TABLE A.3

*One-sided Percentage Point (g) for Equal
Correlations* $\left(Case\ I\ with\ Correlation\ \rho\right)$

p = 20, 1-α = 0.95

ν ↓	ρ					
	0.0	0.1	0.2	0.3	0.4	0.5
2	8.53001	8.20079	7.85795	7.49907	7.12049	6.71676
3	5.79066	5.61166	5.42330	5.22350	5.00949	4.77728
4	4.80258	4.67826	4.54517	4.40161	4.24528	4.07287
5	4.30124	4.20503	4.10006	3.98487	3.85746	3.71491
6	3.99946	3.92038	3.83240	3.73427	3.62419	3.49945
7	3.79822	3.73067	3.65407	3.56733	3.46877	3.35586
8	3.65456	3.59529	3.52686	3.44826	3.35792	3.25343
9	3.54691	3.49387	3.43158	3.35909	3.27491	3.17672
10	3.46325	3.41507	3.35756	3.28984	3.21045	3.11715
11	3.39638	3.35208	3.29842	3.23451	3.15895	3.06956
12	3.34172	3.30060	3.25007	3.18928	3.11687	3.03067
13	3.29620	3.25772	3.20982	3.15164	3.08184	2.99830
14	3.25770	3.22146	3.17578	3.11981	3.05222	2.97094
15	3.22473	3.19040	3.14662	3.09254	3.02686	2.94750
16	3.19618	3.16350	3.12137	3.06893	3.00489	2.92721
17	3.17120	3.13997	3.09928	3.04829	2.98569	2.90946
18	3.14918	3.11921	3.07980	3.03008	2.96875	2.89382
19	3.12961	3.10077	3.06249	3.01390	2.95371	2.87992
20	3.11211	3.08427	3.04701	2.99943	2.94025	2.86749
21	3.09637	3.06943	3.03308	2.98641	2.92814	2.85631
22	3.08214	3.05601	3.02048	2.97464	2.91720	2.84619
23	3.06920	3.04381	3.00903	2.96394	2.90725	2.83700
24	3.05739	3.03267	2.99858	2.95417	2.89817	2.82862
25	3.04657	3.02246	2.98900	2.94522	2.88984	2.82093
26	3.03662	3.01307	2.98019	2.93698	2.88219	2.81386
27	3.02744	3.00441	2.97205	2.92939	2.87513	2.80734
28	3.01894	2.99638	2.96452	2.92235	2.86859	2.80130
29	3.01105	2.98894	2.95753	2.91582	2.86252	2.79570
30	3.00370	2.98200	2.95103	2.90974	2.85687	2.79048
35	2.97349	2.95346	2.92424	2.88472	2.83362	2.76901
40	2.95104	2.93225	2.90433	2.86613	2.81634	2.75306
45	2.93372	2.91587	2.88895	2.85177	2.80300	2.74074
50	2.91994	2.90283	2.87672	2.84034	2.79238	2.73094
60	2.89941	2.88340	2.85847	2.82331	2.77656	2.71634
80	2.87397	2.85930	2.83584	2.80218	2.75694	2.69823
100	2.85883	2.84495	2.82236	2.78960	2.74525	2.68745
120	2.84879	2.83542	2.81341	2.78125	2.73750	2.68030
200	2.82884	2.81647	2.79561	2.76463	2.72207	2.66607
∞	2.79921	2.78831	2.76914	2.73993	2.69915	2.64492

TABLE A.3

One-sided Percentage Point (g) for Equal
Correlations (*Case I with Correlation* ρ)

p = 20, 1-α = 0.95

ν ↓	ρ				
	0.6	0.7	0.8	0.9	1/(1+√20)
2	6.27940	5.79412	5.23314	4.52490	7.91817
3	4.52085	4.23010	3.88543	3.43641	5.45655
4	3.87933	3.65615	3.38679	3.02853	4.56883
5	3.55268	3.36310	3.13118	2.81811	4.11886
6	3.35587	3.18626	2.97657	2.69036	3.84827
7	3.22462	3.06822	2.87322	2.60474	3.66798
8	3.13096	2.98393	2.79934	2.54343	3.53936
9	3.06081	2.92077	2.74393	2.49739	3.44302
10	3.00632	2.87169	2.70086	2.46156	3.36818
11	2.96278	2.83247	2.66643	2.43289	3.30838
12	2.92721	2.80041	2.63827	2.40943	3.25949
13	2.89759	2.77372	2.61482	2.38988	3.21879
14	2.87255	2.75116	2.59498	2.37334	3.18437
15	2.85111	2.73183	2.57800	2.35917	3.15489
16	2.83255	2.71509	2.56328	2.34689	3.12935
17	2.81631	2.70046	2.55041	2.33614	3.10701
18	2.80200	2.68755	2.53906	2.32666	3.08731
19	2.78928	2.67609	2.52898	2.31824	3.06981
20	2.77791	2.66583	2.51996	2.31070	3.05415
21	2.76768	2.65661	2.51184	2.30392	3.04006
22	2.75842	2.64827	2.50450	2.29779	3.02732
23	2.75002	2.64068	2.49783	2.29221	3.01574
24	2.74234	2.63377	2.49174	2.28712	3.00517
25	2.73531	2.62743	2.48616	2.28246	2.99548
26	2.72885	2.62159	2.48103	2.27817	2.98657
27	2.72288	2.61621	2.47630	2.27421	2.97834
28	2.71735	2.61123	2.47191	2.27054	2.97073
29	2.71223	2.60661	2.46784	2.26714	2.96366
30	2.70745	2.60230	2.46405	2.26397	2.95708
35	2.68781	2.58459	2.44846	2.25092	2.92999
40	2.67322	2.57143	2.43688	2.24123	2.90985
45	2.66196	2.56127	2.42793	2.23374	2.89430
50	2.65299	2.55319	2.42081	2.22778	2.88192
60	2.63963	2.54114	2.41020	2.21889	2.86347
80	2.62308	2.52620	2.39705	2.20787	2.84058
100	2.61322	2.51731	2.38922	2.20131	2.82695
120	2.60667	2.51141	2.38402	2.19695	2.81790
200	2.59366	2.49967	2.37368	2.18829	2.79990
∞	2.57433	2.48223	2.35831	2.17540	2.77312

TABLE A.4

One-sided Percentage Point (g) for Equal
Correlations (*Case I with Correlation* ρ)

$$p = 2, \ 1-\alpha = 0.99$$

$\nu \downarrow$	ρ 0.0	0.1	0.2	0.3	0.4	0.5
1	54.33212	53.17691	51.95551	50.65508	49.25793	47.73863
2	9.45145	9.36194	9.26198	9.14984	9.02311	8.87828
3	5.70979	5.67730	5.63946	5.59530	5.54352	5.48221
4	4.54411	4.52605	4.50426	4.47801	4.44631	4.40775
5	3.99737	3.98522	3.97012	3.95145	3.92835	3.89963
6	3.68431	3.67523	3.66366	3.64901	3.63054	3.60716
7	3.48271	3.47546	3.46601	3.45382	3.43820	3.41813
8	3.34245	3.33639	3.32835	3.31781	3.30409	3.28625
9	3.23940	3.23418	3.22713	3.21775	3.20540	3.18916
10	3.16058	3.15597	3.14964	3.14113	3.12980	3.11476
11	3.09837	3.09422	3.08846	3.08061	3.07007	3.05596
12	3.04805	3.04426	3.03894	3.03162	3.02171	3.00835
13	3.00651	3.00302	2.99805	2.99116	2.98176	2.96901
14	2.97165	2.96840	2.96373	2.95719	2.94821	2.93596
15	2.94199	2.93893	2.93450	2.92827	2.91964	2.90782
16	2.91644	2.91355	2.90933	2.90335	2.89502	2.88357
17	2.89421	2.89146	2.88742	2.88165	2.87359	2.86245
18	2.87469	2.87206	2.86818	2.86260	2.85477	2.84390
19	2.85741	2.85489	2.85114	2.84573	2.83810	2.82747
20	2.84201	2.83959	2.83596	2.83069	2.82324	2.81283
21	2.82821	2.82587	2.82234	2.81720	2.80991	2.79969
22	2.81575	2.81349	2.81006	2.80504	2.79789	2.78784
23	2.80447	2.80227	2.79892	2.79401	2.78698	2.77709
24	2.79419	2.79205	2.78878	2.78396	2.77706	2.76730
25	2.78479	2.78271	2.77951	2.77478	2.76797	2.75835
26	2.77617	2.77413	2.77099	2.76634	2.75964	2.75013
27	2.76822	2.76623	2.76315	2.75857	2.75196	2.74256
28	2.76088	2.75893	2.75590	2.75139	2.74486	2.73556
29	2.75407	2.75216	2.74919	2.74473	2.73828	2.72907
30	2.74775	2.74587	2.74294	2.73855	2.73216	2.72304
35	2.72181	2.72008	2.71733	2.71317	2.70707	2.69829
40	2.70264	2.70101	2.69840	2.69441	2.68851	2.67999
45	2.68790	2.68635	2.68383	2.67997	2.67423	2.66591
50	2.67621	2.67471	2.67228	2.66852	2.66291	2.65474
60	2.65884	2.65743	2.65511	2.65150	2.64607	2.63813
80	2.63741	2.63611	2.63393	2.63050	2.62530	2.61763
100	2.62470	2.62346	2.62137	2.61804	2.61297	2.60547
120	2.61629	2.61509	2.61305	2.60979	2.60481	2.59742
200	2.59961	2.59849	2.59655	2.59343	2.58863	2.58145
∞	2.57496	2.57394	2.57216	2.56923	2.56468	2.55781

TABLE A.4

One-sided Percentage Point (g) for Equal
Correlations (*Case I with Correlation* ρ)

p = 2, 1-α = 0.99

ν ↓	ρ				
	0.6	0.7	0.8	0.9	1/(1+√2)
1	46.05811	44.15063	41.88802	38.93932	49.05016
2	8.70987	8.50867	8.25657	7.90674	9.00373
3	5.40846	5.31731	5.19908	5.02869	5.53544
4	4.36014	4.29981	4.21958	4.10089	4.44129
5	3.86344	3.81671	3.75340	3.65792	3.92464
6	3.57723	3.53797	3.48402	3.40144	3.62755
7	3.39208	3.35750	3.30943	3.23500	3.43565
8	3.26284	3.23145	3.18739	3.11855	3.30183
9	3.16765	3.13856	3.09741	3.03263	3.20335
10	3.09468	3.06733	3.02838	2.96668	3.12791
11	3.03700	3.01100	2.97378	2.91448	3.06831
12	2.99027	2.96536	2.92952	2.87216	3.02004
13	2.95166	2.92764	2.89293	2.83717	2.98017
14	2.91922	2.89595	2.86219	2.80775	2.94669
15	2.89159	2.86895	2.83599	2.78267	2.91818
16	2.86778	2.84567	2.81340	2.76105	2.89361
17	2.84704	2.82540	2.79372	2.74221	2.87222
18	2.82882	2.80759	2.77643	2.72566	2.85343
19	2.81269	2.79181	2.76112	2.71099	2.83679
20	2.79830	2.77775	2.74746	2.69791	2.82196
21	2.78540	2.76513	2.73521	2.68618	2.80866
22	2.77375	2.75374	2.72415	2.67558	2.79666
23	2.76319	2.74341	2.71412	2.66598	2.78577
24	2.75358	2.73401	2.70499	2.65722	2.77586
25	2.74478	2.72540	2.69663	2.64922	2.76680
26	2.73670	2.71750	2.68896	2.64186	2.75848
27	2.72926	2.71022	2.68188	2.63509	2.75081
28	2.72238	2.70349	2.67535	2.62882	2.74373
29	2.71601	2.69726	2.66929	2.62301	2.73716
30	2.71008	2.69146	2.66365	2.61761	2.73105
35	2.68576	2.66766	2.64054	2.59545	2.70600
40	2.66777	2.65006	2.62343	2.57905	2.68748
45	2.65392	2.63651	2.61026	2.56643	2.67323
50	2.64294	2.62576	2.59982	2.55641	2.66192
60	2.62661	2.60978	2.58428	2.54151	2.64512
80	2.60646	2.59005	2.56510	2.52311	2.62438
100	2.59450	2.57834	2.55372	2.51219	2.61207
120	2.58658	2.57059	2.54618	2.50495	2.60393
200	2.57087	2.55521	2.53122	2.49060	2.58777
∞	2.54762	2.53244	2.50908	2.46935	2.56386

TABLE A.4

One-sided Percentage Point (g) for Equal
Correlations $\left(Case\ I\ with\ Correlation\ \rho\right)$

p = 3, 1-α = 0.99

ν ↓	ρ					
	0.0	0.1	0.2	0.3	0.4	0.5
2	11.11255	10.93543	10.74305	10.53276	10.30089	10.04211
3	6.46163	6.39832	6.32643	6.24457	6.15082	6.04237
4	5.04428	5.00959	4.96867	4.92048	4.86356	4.79581
5	4.38708	4.36408	4.33605	4.30208	4.26091	4.21076
6	4.01350	3.99654	3.97528	3.94887	3.91618	3.87558
7	3.77412	3.76075	3.74357	3.72178	3.69429	3.65960
8	3.60820	3.59717	3.58269	3.56396	3.53997	3.50927
9	3.48666	3.47726	3.46466	3.44812	3.42662	3.39879
10	3.39390	3.38568	3.37449	3.35956	3.33993	3.31424
11	3.32084	3.31352	3.30338	3.28970	3.27151	3.24750
12	3.26183	3.25521	3.24591	3.23321	3.21617	3.19349
13	3.21319	3.20712	3.19851	3.18661	3.17050	3.14890
14	3.17242	3.16681	3.15874	3.14751	3.13217	3.11148
15	3.13775	3.13252	3.12492	3.11424	3.09955	3.07962
16	3.10792	3.10301	3.09580	3.08559	3.07146	3.05218
17	3.08199	3.07735	3.07047	3.06066	3.04701	3.02830
18	3.05923	3.05482	3.04824	3.03878	3.02555	3.00732
19	3.03910	3.03490	3.02857	3.01942	3.00655	2.98876
20	3.02118	3.01715	3.01104	3.00216	2.98962	2.97221
21	3.00511	3.00124	2.99533	2.98669	2.97444	2.95737
22	2.99062	2.98689	2.98116	2.97274	2.96075	2.94399
23	2.97750	2.97389	2.96832	2.96010	2.94834	2.93185
24	2.96556	2.96206	2.95663	2.94858	2.93703	2.92080
25	2.95464	2.95124	2.94595	2.93806	2.92670	2.91070
26	2.94462	2.94132	2.93614	2.92839	2.91722	2.90142
27	2.93539	2.93218	2.92710	2.91949	2.90848	2.89288
28	2.92687	2.92373	2.91876	2.91127	2.90040	2.88498
29	2.91898	2.91591	2.91102	2.90365	2.89292	2.87766
30	2.91164	2.90863	2.90383	2.89657	2.88597	2.87086
35	2.88157	2.87883	2.87437	2.86753	2.85745	2.84296
40	2.85937	2.85681	2.85260	2.84607	2.83637	2.82234
45	2.84230	2.83988	2.83586	2.82957	2.82015	2.80647
50	2.82878	2.82647	2.82259	2.81648	2.80729	2.79389
60	2.80869	2.80654	2.80288	2.79704	2.78819	2.77519
80	2.78394	2.78198	2.77857	2.77306	2.76462	2.75212
100	2.76927	2.76742	2.76416	2.75884	2.75064	2.73844
120	2.75957	2.75778	2.75462	2.74943	2.74139	2.72938
200	2.74034	2.73869	2.73572	2.73078	2.72305	2.71142
∞	2.71194	2.71048	2.70778	2.70321	2.69593	2.68485

TABLE A.4

One-sided Percentage Point (g) for Equal
Correlations (*Case I with Correlation* ρ)

p = 3, 1-α = 0.99

ν ↓	ρ				
	0.6	0.7	0.8	0.9	1/(1+√3)
2	9.74816	9.40513	8.98579	8.41973	10.38238
3	5.91482	5.76076	5.56563	5.29174	6.18416
4	4.71397	4.61254	4.48076	4.29068	4.88399
5	4.14888	4.07064	3.96699	3.81448	4.27580
6	3.82460	3.75911	3.67101	3.53938	3.92808
7	3.61542	3.55790	3.47958	3.36113	3.70435
8	3.46970	3.41761	3.34597	3.23655	3.54879
9	3.36253	3.31437	3.24757	3.14470	3.43456
10	3.28048	3.23529	3.17214	3.07425	3.34720
11	3.21568	3.17280	3.11252	3.01852	3.27827
12	3.16323	3.12221	3.06422	2.97336	3.22252
13	3.11992	3.08041	3.02432	2.93602	3.17651
14	3.08356	3.04531	2.99079	2.90464	3.13791
15	3.05259	3.01542	2.96223	2.87791	3.10506
16	3.02592	2.98967	2.93762	2.85486	3.07677
17	3.00270	2.96724	2.91619	2.83478	3.05215
18	2.98231	2.94755	2.89736	2.81714	3.03053
19	2.96425	2.93011	2.88069	2.80151	3.01141
20	2.94816	2.91456	2.86582	2.78758	2.99436
21	2.93373	2.90061	2.85248	2.77508	2.97907
22	2.92070	2.88803	2.84045	2.76380	2.96529
23	2.90890	2.87662	2.82954	2.75356	2.95279
24	2.89815	2.86623	2.81960	2.74424	2.94142
25	2.88832	2.85673	2.81051	2.73572	2.93101
26	2.87929	2.84801	2.80216	2.72789	2.92146
27	2.87098	2.83997	2.79447	2.72068	2.91267
28	2.86329	2.83254	2.78736	2.71401	2.90454
29	2.85617	2.82565	2.78077	2.70782	2.89701
30	2.84955	2.81925	2.77465	2.70208	2.89001
35	2.82239	2.79299	2.74951	2.67849	2.86130
40	2.80232	2.77358	2.73092	2.66104	2.84008
45	2.78687	2.75863	2.71662	2.64761	2.82376
50	2.77461	2.74678	2.70527	2.63695	2.81082
60	2.75640	2.72916	2.68839	2.62111	2.79160
80	2.73393	2.70742	2.66757	2.60154	2.76788
100	2.72060	2.69452	2.65521	2.58993	2.75382
120	2.71177	2.68598	2.64702	2.58224	2.74451
200	2.69427	2.66903	2.63079	2.56698	2.72605
∞	2.66839	2.64397	2.60677	2.54441	2.69877

TABLE A.4

One-sided Percentage Point (g) for Equal
Correlations $\left[Case\ I\ with\ Correlation\ \rho \right]$

$p = 4,\ 1-\alpha = 0.99$

$\nu \downarrow$	0.0	0.1	0.2	0.3	0.4	0.5
				ρ		
2	12.35454	12.10072	11.82969	11.53801	11.22101	10.87205
3	7.01867	6.92779	6.82642	6.71291	6.58490	6.43896
4	5.41182	5.36214	5.30457	5.23786	5.16029	5.06930
5	4.67137	4.63860	4.59931	4.55243	4.49646	4.42926
6	4.25214	4.22812	4.19846	4.16215	4.11782	4.06354
7	3.98426	3.96545	3.94160	3.91174	3.87458	3.82831
8	3.79898	3.78356	3.76355	3.73799	3.70565	3.66479
9	3.66348	3.65042	3.63310	3.61060	3.58169	3.54472
10	3.56021	3.54886	3.53353	3.51329	3.48696	3.45291
11	3.47895	3.46891	3.45509	3.43660	3.41226	3.38047
12	3.41339	3.40435	3.39173	3.37462	3.35187	3.32189
13	3.35939	3.35116	3.33950	3.32351	3.30205	3.27355
14	3.31416	3.30659	3.29572	3.28065	3.26026	3.23298
15	3.27573	3.26870	3.25849	3.24420	3.22470	3.19846
16	3.24268	3.23610	3.22645	3.21282	3.19409	3.16874
17	3.21396	3.20777	3.19859	3.18553	3.16746	3.14287
18	3.18876	3.18291	3.17414	3.16157	3.14408	3.12016
19	3.16649	3.16093	3.15252	3.14038	3.12339	3.10006
20	3.14666	3.14135	3.13326	3.12150	3.10496	3.08215
21	3.12889	3.12380	3.11599	3.10457	3.08843	3.06609
22	3.11288	3.10799	3.10043	3.08931	3.07353	3.05160
23	3.09838	3.09366	3.08633	3.07548	3.06002	3.03848
24	3.08518	3.08062	3.07349	3.06289	3.04773	3.02652
25	3.07312	3.06870	3.06175	3.05138	3.03648	3.01559
26	3.06205	3.05777	3.05099	3.04082	3.02616	3.00556
27	3.05187	3.04770	3.04107	3.03109	3.01666	2.99632
28	3.04246	3.03840	3.03191	3.02210	3.00788	2.98778
29	3.03374	3.02979	3.02342	3.01377	2.99974	2.97986
30	3.02564	3.02178	3.01554	3.00603	2.99217	2.97251
35	2.99248	2.98898	2.98321	2.97431	2.96117	2.94234
40	2.96800	2.96476	2.95934	2.95087	2.93825	2.92005
45	2.94919	2.94615	2.94099	2.93285	2.92063	2.90290
50	2.93429	2.93140	2.92645	2.91856	2.90666	2.88931
60	2.91218	2.90951	2.90485	2.89735	2.88591	2.86911
80	2.88495	2.88253	2.87823	2.87118	2.86031	2.84419
100	2.86881	2.86654	2.86245	2.85567	2.84513	2.82941
120	2.85814	2.85597	2.85201	2.84541	2.83509	2.81963
200	2.83701	2.83502	2.83133	2.82507	2.81518	2.80024
∞	2.80582	2.80409	2.80077	2.79501	2.78575	2.77156

TABLE A.4

*One-sided Percentage Point (g) for Equal
Correlations* (*Case I with Correlation* ρ)

$$p = 4, \quad 1-\alpha = 0.99$$

$\nu \downarrow$	ρ				
	0.6	0.7	0.8	0.9	$1/(1+\sqrt{4})$
2	10.48098	10.03073	9.48805	8.76712	11.43543
3	6.26974	6.06820	5.81661	5.46903	6.67199
4	4.96095	4.82855	4.65899	4.41827	5.21332
5	4.34751	4.24557	4.11243	3.91955	4.53487
6	3.99632	3.91113	3.79813	3.63184	4.14834
7	3.77015	3.69544	3.59510	3.44559	3.90024
8	3.61278	3.54521	3.45352	3.31550	3.72803
9	3.49714	3.43474	3.34931	3.21966	3.60174
10	3.40867	3.35017	3.26948	3.14615	3.50526
11	3.33884	3.28338	3.20640	3.08803	3.42921
12	3.28234	3.22932	3.15532	3.04094	3.36773
13	3.23570	3.18468	3.11313	3.00202	3.31703
14	3.19655	3.14721	3.07769	2.96932	3.27451
15	3.16323	3.11530	3.04751	2.94146	3.23834
16	3.13453	3.08781	3.02151	2.91744	3.20721
17	3.10955	3.06388	2.99886	2.89653	3.18012
18	3.08762	3.04286	2.97897	2.87815	3.15635
19	3.06821	3.02426	2.96136	2.86188	3.13532
20	3.05090	3.00768	2.94566	2.84737	3.11658
21	3.03539	2.99280	2.93158	2.83435	3.09978
22	3.02139	2.97939	2.91887	2.82260	3.08463
23	3.00870	2.96722	2.90735	2.81195	3.07090
24	2.99715	2.95614	2.89686	2.80224	3.05841
25	2.98659	2.94601	2.88726	2.79336	3.04698
26	2.97689	2.93671	2.87845	2.78521	3.03650
27	2.96795	2.92815	2.87034	2.77770	3.02684
28	2.95970	2.92023	2.86283	2.77076	3.01791
29	2.95205	2.91289	2.85588	2.76433	3.00965
30	2.94493	2.90607	2.84941	2.75834	3.00196
35	2.91577	2.87808	2.82289	2.73379	2.97046
40	2.89421	2.85740	2.80328	2.71564	2.94719
45	2.87762	2.84148	2.78819	2.70166	2.92929
50	2.86447	2.82885	2.77621	2.69057	2.91510
60	2.84492	2.81009	2.75842	2.67408	2.89403
80	2.82081	2.78693	2.73645	2.65373	2.86805
100	2.80650	2.77320	2.72342	2.64165	2.85264
120	2.79703	2.76411	2.71479	2.63365	2.84244
200	2.77826	2.74607	2.69768	2.61778	2.82224
∞	2.75050	2.71940	2.67236	2.59430	2.79237

BECHHOFER and DUNNETT

TABLE A.4

One-sided Percentage Point (g) for Equal
Correlations $\left(\textit{Case I with Correlation } \rho \right)$

$p = 5, \quad 1-\alpha = 0.99$

ν ↓	0.0	0.1	0.2	0.3	0.4	0.5
			ρ			
2	13.33896	13.01972	12.68244	12.32295	11.93579	11.51332
3	7.46045	7.34522	7.21836	7.07795	6.92131	6.74451
4	5.70264	5.63952	5.56735	5.48476	5.38979	5.27958
5	4.89562	4.85402	4.80480	4.74678	4.67832	4.59698
6	4.43977	4.40937	4.37228	4.32740	4.27324	4.20761
7	4.14899	4.12526	4.09551	4.05867	4.01333	3.95745
8	3.94813	3.92876	3.90386	3.87239	3.83298	3.78369
9	3.80139	3.78504	3.76356	3.73591	3.70074	3.65619
10	3.68965	3.67550	3.65654	3.63172	3.59973	3.55874
11	3.60179	3.58931	3.57227	3.54964	3.52011	3.48189
12	3.53094	3.51976	3.50423	3.48333	3.45576	3.41975
13	3.47262	3.46247	3.44817	3.42867	3.40270	3.36849
14	3.42380	3.41449	3.40118	3.38284	3.35819	3.32548
15	3.38233	3.37371	3.36124	3.34388	3.32034	3.28890
16	3.34668	3.33865	3.32688	3.31034	3.28775	3.25739
17	3.31570	3.30817	3.29700	3.28117	3.25940	3.22998
18	3.28854	3.28144	3.27079	3.25558	3.23452	3.20592
19	3.26454	3.25780	3.24762	3.23294	3.21251	3.18463
20	3.24318	3.23676	3.22697	3.21278	3.19290	3.16566
21	3.22404	3.21790	3.20847	3.19470	3.17532	3.14865
22	3.20679	3.20091	3.19179	3.17840	3.15947	3.13331
23	3.19117	3.18552	3.17669	3.16364	3.14510	3.11941
24	3.17697	3.17151	3.16293	3.15020	3.13202	3.10675
25	3.16398	3.15871	3.15037	3.13791	3.12007	3.09517
26	3.15208	3.14697	3.13883	3.12663	3.10909	3.08455
27	3.14112	3.13616	3.12822	3.11625	3.09899	3.07477
28	3.13099	3.12617	3.11841	3.10666	3.08965	3.06573
29	3.12162	3.11692	3.10932	3.09777	3.08100	3.05735
30	3.11291	3.10833	3.10087	3.08951	3.07295	3.04956
35	3.07724	3.07312	3.06627	3.05565	3.03999	3.01763
40	3.05093	3.04714	3.04073	3.03065	3.01564	2.99404
45	3.03073	3.02718	3.02109	3.01143	2.99691	2.97590
50	3.01472	3.01136	3.00553	2.99619	2.98206	2.96151
60	2.99098	2.98789	2.98243	2.97357	2.96001	2.94014
80	2.96174	2.95897	2.95396	2.94567	2.93282	2.91378
100	2.94443	2.94185	2.93710	2.92914	2.91670	2.89815
120	2.93299	2.93052	2.92594	2.91820	2.90603	2.88780
200	2.91032	2.90808	2.90383	2.89652	2.88489	2.86729
∞	2.87689	2.87497	2.87118	2.86449	2.85364	2.83698

TABLE A.4

One-sided Percentage Point (g) for Equal
Correlations (*Case I with Correlation* ρ)

p = 5, 1-α = 0.99

ν ↓	ρ				
	0.6	0.7	0.8	0.9	1/(1+√5)
2	11.04392	10.50819	9.86845	9.02756	12.28927
3	6.54151	6.30204	6.00603	5.60154	7.06454
4	5.14968	4.99252	4.79326	4.51347	5.47674
5	4.49904	4.37815	4.22185	3.99786	4.74107
6	4.12716	4.02622	3.89367	3.70070	4.32293
7	3.88791	3.79947	3.68184	3.50846	4.05496
8	3.72157	3.64164	3.53423	3.37424	3.86919
9	3.59941	3.52564	3.42563	3.27540	3.73307
10	3.50599	3.43688	3.34246	3.19960	3.62915
11	3.43228	3.36680	3.27676	3.13970	3.54728
12	3.37266	3.31009	3.22357	3.09116	3.48114
13	3.32345	3.26328	3.17964	3.05106	3.42662
14	3.28216	3.22398	3.14276	3.01736	3.38090
15	3.24702	3.19053	3.11135	2.98866	3.34203
16	3.21676	3.16172	3.08428	2.96392	3.30857
17	3.19043	3.13664	3.06073	2.94238	3.27948
18	3.16731	3.11461	3.04003	2.92345	3.25394
19	3.14684	3.09512	3.02171	2.90669	3.23136
20	3.12861	3.07774	3.00538	2.89174	3.21124
21	3.11225	3.06216	2.99073	2.87833	3.19320
22	3.09750	3.04810	2.97751	2.86624	3.17694
23	3.08414	3.03536	2.96553	2.85527	3.16221
24	3.07196	3.02376	2.95462	2.84528	3.14880
25	3.06083	3.01314	2.94464	2.83614	3.13654
26	3.05062	3.00340	2.93548	2.82774	3.12529
27	3.04120	2.99443	2.92703	2.82001	3.11493
28	3.03251	2.98614	2.91923	2.81286	3.10536
29	3.02445	2.97845	2.91200	2.80624	3.09649
30	3.01695	2.97130	2.90528	2.80008	3.08825
35	2.98624	2.94200	2.87771	2.77481	3.05447
40	2.96353	2.92034	2.85732	2.75612	3.02952
45	2.94606	2.90368	2.84163	2.74173	3.01034
50	2.93221	2.89046	2.82918	2.73032	2.99513
60	2.91164	2.87082	2.81069	2.71335	2.97256
80	2.88625	2.84659	2.78786	2.69241	2.94472
100	2.87119	2.83221	2.77432	2.67998	2.92822
120	2.86123	2.82269	2.76535	2.67175	2.91731
200	2.84147	2.80382	2.74757	2.65542	2.89567
∞	2.81226	2.77592	2.72127	2.63126	2.86371

BECHHOFER and DUNNETT

TABLE A.4

One-sided Percentage Point (g) for Equal
Correlations (Case I with Correlation ρ)

p = 6, 1-α = 0.99

ν ↓	ρ					
	0.0	0.1	0.2	0.3	0.4	0.5
2	14.14916	13.77424	13.38074	12.96393	12.51775	12.03373
3	7.82561	7.68888	7.53978	7.37618	7.19508	6.99220
4	5.94309	5.86789	5.78281	5.68636	5.57643	5.44988
5	5.08082	5.03120	4.97313	4.90534	4.82607	4.73268
6	4.59448	4.55824	4.51450	4.46208	4.39939	4.32407
7	4.28458	4.25634	4.22129	4.17831	4.12586	4.06176
8	4.07069	4.04769	4.01840	3.98172	3.93617	3.87967
9	3.91453	3.89517	3.86994	3.83775	3.79715	3.74611
10	3.79568	3.77898	3.75675	3.72789	3.69099	3.64407
11	3.70228	3.68759	3.66765	3.64137	3.60734	3.56361
12	3.62700	3.61386	3.59573	3.57149	3.53975	3.49858
13	3.56505	3.55316	3.53648	3.51390	3.48401	3.44493
14	3.51320	3.50232	3.48684	3.46562	3.43728	3.39994
15	3.46918	3.45914	3.44465	3.42458	3.39754	3.36166
16	3.43134	3.42200	3.40835	3.38926	3.36333	3.32871
17	3.39848	3.38973	3.37680	3.35855	3.33357	3.30004
18	3.36967	3.36144	3.34913	3.33160	3.30746	3.27487
19	3.34421	3.33642	3.32466	3.30777	3.28436	3.25261
20	3.32155	3.31415	3.30286	3.28654	3.26378	3.23277
21	3.30126	3.29420	3.28333	3.26751	3.24533	3.21498
22	3.28298	3.27622	3.26573	3.25036	3.22870	3.19894
23	3.26643	3.25994	3.24979	3.23482	3.21363	3.18441
24	3.25137	3.24512	3.23527	3.22067	3.19991	3.17117
25	3.23761	3.23158	3.22201	3.20774	3.18736	3.15907
26	3.22500	3.21916	3.20984	3.19587	3.17585	3.14797
27	3.21338	3.20773	3.19864	3.18495	3.16525	3.13774
28	3.20266	3.19717	3.18829	3.17485	3.15546	3.12829
29	3.19273	3.18739	3.17870	3.16550	3.14638	3.11953
30	3.18350	3.17830	3.16979	3.15681	3.13794	3.11139
35	3.14574	3.14109	3.13330	3.12120	3.10337	3.07802
40	3.11789	3.11363	3.10635	3.09489	3.07783	3.05337
45	3.09651	3.09253	3.08565	3.07468	3.05820	3.03441
50	3.07958	3.07582	3.06924	3.05865	3.04263	3.01937
60	3.05446	3.05103	3.04489	3.03486	3.01951	2.99704
80	3.02355	3.02049	3.01488	3.00553	2.99100	2.96950
100	3.00525	3.00241	2.99710	2.98815	2.97410	2.95317
120	2.99315	2.99045	2.98535	2.97665	2.96292	2.94236
200	2.96921	2.96677	2.96205	2.95386	2.94076	2.92094
∞	2.93390	2.93183	2.92766	2.92020	2.90801	2.88927

TABLE A.4

One-sided Percentage Point (g) for Equal
Correlations (*Case I with Correlation* ρ)

p = 6, 1-α = 0.99

ν ↓	ρ				
	0.6	0.7	0.8	0.9	1/(1+√6)
2	11.49911	10.89266	10.17319	9.23476	13.00725
3	6.76089	6.48995	6.15745	5.70674	7.39344
4	5.30185	5.12410	4.90045	4.58896	5.69668
5	4.62111	4.48445	4.30911	4.05992	4.91268
6	4.23247	4.11842	3.96981	3.75524	4.46781
7	3.98263	3.88275	3.75094	3.55822	4.18305
8	3.80902	3.71880	3.59848	3.42072	3.98580
9	3.68157	3.59835	3.48636	3.31950	3.84135
10	3.58414	3.50620	3.40052	3.24186	3.73114
11	3.50728	3.43347	3.33272	3.18055	3.64435
12	3.44513	3.37463	3.27784	3.13086	3.57425
13	3.39384	3.32606	3.23252	3.08981	3.51648
14	3.35081	3.28530	3.19448	3.05533	3.46806
15	3.31420	3.25060	3.16208	3.02596	3.42689
16	3.28267	3.22072	3.13417	3.00064	3.39147
17	3.25523	3.19471	3.10987	2.97860	3.36067
18	3.23115	3.17187	3.08854	2.95923	3.33365
19	3.20983	3.15166	3.06965	2.94208	3.30974
20	3.19084	3.13364	3.05281	2.92679	3.28845
21	3.17380	3.11748	3.03770	2.91308	3.26937
22	3.15844	3.10291	3.02408	2.90070	3.25217
23	3.14452	3.08970	3.01173	2.88948	3.23658
24	3.13184	3.07767	3.00048	2.87926	3.22240
25	3.12025	3.06667	2.99019	2.86991	3.20943
26	3.10961	3.05657	2.98075	2.86133	3.19753
27	3.09982	3.04727	2.97204	2.85342	3.18658
28	3.09076	3.03868	2.96400	2.84611	3.17646
29	3.08237	3.03071	2.95655	2.83933	3.16708
30	3.07457	3.02330	2.94962	2.83303	3.15836
35	3.04259	2.99294	2.92120	2.80719	3.12265
40	3.01895	2.97049	2.90019	2.78807	3.09628
45	3.00077	2.95322	2.88403	2.77336	3.07601
50	2.98635	2.93953	2.87120	2.76169	3.05994
60	2.96494	2.91918	2.85215	2.74434	3.03609
80	2.93852	2.89407	2.82863	2.72292	3.00668
100	2.92285	2.87918	2.81468	2.71022	2.98926
120	2.91248	2.86932	2.80544	2.70180	2.97773
200	2.89192	2.84977	2.78712	2.68511	2.95488
∞	2.86152	2.82087	2.76003	2.66042	2.92114

BECHHOFER and DUNNETT

TABLE A.4

One-sided Percentage Point (g) for Equal
Correlations   (Case I with Correlation ρ)

$p = 7$, $1-\alpha = 0.99$

$\nu \downarrow$			ρ			
	0.0	0.1	0.2	0.3	0.4	0.5
2	14.83417	14.41156	13.96980	13.50380	13.00702	12.47034
3	8.13608	7.98027	7.81160	7.62770	7.42536	7.19994
4	6.14786	6.06172	5.96511	5.85642	5.73338	5.59264
5	5.23851	5.18157	5.11552	5.03905	4.95027	4.84638
6	4.72610	4.68450	4.63473	4.57559	4.50538	4.42160
7	4.39981	4.36742	4.32756	4.27907	4.22035	4.14907
8	4.17473	4.14838	4.11509	4.07374	4.02277	3.95997
9	4.01048	3.98833	3.95969	3.92342	3.87801	3.82131
10	3.88551	3.86643	3.84123	3.80875	3.76750	3.71540
11	3.78734	3.77058	3.74801	3.71846	3.68044	3.63190
12	3.70822	3.69328	3.67278	3.64554	3.61010	3.56442
13	3.64314	3.62965	3.61081	3.58546	3.55211	3.50877
14	3.58868	3.57636	3.55889	3.53510	3.50349	3.46210
15	3.54246	3.53110	3.51478	3.49228	3.46215	3.42239
16	3.50273	3.49219	3.47683	3.45545	3.42657	3.38822
17	3.46823	3.45838	3.44385	3.42342	3.39562	3.35849
18	3.43800	3.42874	3.41492	3.39532	3.36846	3.33239
19	3.41128	3.40253	3.38934	3.37047	3.34444	3.30930
20	3.38751	3.37921	3.36656	3.34834	3.32304	3.28873
21	3.36622	3.35831	3.34615	3.32850	3.30385	3.27029
22	3.34704	3.33948	3.32776	3.31061	3.28656	3.25366
23	3.32968	3.32243	3.31110	3.29441	3.27089	3.23859
24	3.31389	3.30692	3.29593	3.27966	3.25662	3.22487
25	3.29946	3.29274	3.28207	3.26619	3.24358	3.21232
26	3.28623	3.27974	3.26936	3.25382	3.23161	3.20081
27	3.27406	3.26777	3.25766	3.24243	3.22059	3.19021
28	3.26281	3.25672	3.24684	3.23191	3.21041	3.18041
29	3.25240	3.24648	3.23683	3.22216	3.20097	3.17133
30	3.24273	3.23697	3.22752	3.21310	3.19220	3.16289
35	3.20315	3.19802	3.18940	3.17598	3.15626	3.12831
40	3.17397	3.16929	3.16125	3.14857	3.12971	3.10275
45	3.15157	3.14722	3.13963	3.12751	3.10930	3.08310
50	3.13384	3.12974	3.12250	3.11081	3.09312	3.06752
60	3.10754	3.10381	3.09707	3.08602	3.06910	3.04438
80	3.07517	3.07187	3.06574	3.05546	3.03947	3.01584
100	3.05602	3.05296	3.04718	3.03735	3.02191	2.99892
120	3.04336	3.04046	3.03491	3.02537	3.01029	2.98772
200	3.01831	3.01571	3.01059	3.00164	2.98726	2.96552
∞	2.98139	2.97920	2.97470	2.96658	2.95323	2.93271

TABLE A.4

One-sided Percentage Point (g) for Equal
Correlations (*Case I with Correlation* ρ)

$$p = 7, \quad 1-\alpha = 0.99$$

ν ↓	0.6	0.7	0.8	0.9	1/(1+√7)
			ρ		
2	11.88008	11.21351	10.42661	9.40619	13.62625
3	6.94434	6.64655	6.28315	5.79364	7.67659
4	5.42901	5.23368	4.98935	4.65127	5.88563
5	4.72306	4.57290	4.38144	4.11110	5.05980
6	4.32037	4.19510	4.03288	3.80019	4.59178
7	4.06165	3.95199	3.80816	3.59923	4.29244
8	3.88194	3.78292	3.65167	3.45901	4.08522
9	3.75006	3.65875	3.53662	3.35583	3.93355
10	3.64926	3.56377	3.44855	3.27666	3.81787
11	3.56976	3.48882	3.37901	3.21420	3.72680
12	3.50548	3.42820	3.32272	3.16354	3.65326
13	3.45245	3.37816	3.27625	3.12172	3.59267
14	3.40795	3.33617	3.23723	3.08658	3.54189
15	3.37010	3.30043	3.20402	3.05665	3.49873
16	3.33750	3.26965	3.17540	3.03086	3.46160
17	3.30915	3.24287	3.15049	3.00840	3.42931
18	3.28425	3.21935	3.12862	2.98868	3.40099
19	3.26222	3.19853	3.10925	2.97121	3.37594
20	3.24259	3.17998	3.09199	2.95563	3.35363
21	3.22498	3.16334	3.07651	2.94166	3.33364
22	3.20911	3.14834	3.06255	2.92905	3.31562
23	3.19472	3.13474	3.04989	2.91762	3.29929
24	3.18162	3.12235	3.03836	2.90721	3.28443
25	3.16964	3.11103	3.02781	2.89769	3.27085
26	3.15865	3.10063	3.01813	2.88895	3.25838
27	3.14852	3.09106	3.00922	2.88089	3.24691
28	3.13917	3.08221	3.00098	2.87345	3.23631
29	3.13049	3.07401	2.99334	2.86655	3.22649
30	3.12244	3.06639	2.98624	2.86013	3.21736
35	3.08939	3.03513	2.95712	2.83381	3.17997
40	3.06497	3.01202	2.93559	2.81435	3.15236
45	3.04619	2.99425	2.91902	2.79937	3.13114
50	3.03130	2.98015	2.90588	2.78748	3.11432
60	3.00918	2.95921	2.88636	2.76982	3.08935
80	2.98188	2.93337	2.86226	2.74801	3.05858
100	2.96570	2.91805	2.84797	2.73507	3.04035
120	2.95499	2.90790	2.83851	2.72650	3.02829
200	2.93376	2.88779	2.81975	2.70951	3.00439
∞	2.90236	2.85805	2.79199	2.68437	2.96910

TABLE A.4

One-sided Percentage Point (g) for Equal Correlations (Case I with Correlation ρ)

p = 8, 1-α = 0.99

ν ↓	ρ					
	0.0	0.1	0.2	0.3	0.4	0.5
2	15.42531	14.96145	14.47777	13.96896	13.42810	12.84559
3	8.40557	8.23273	8.04665	7.84478	7.62371	7.37851
4	6.32597	6.22990	6.12291	6.00327	5.86859	5.71534
5	5.37575	5.31207	5.23878	5.15450	5.05724	4.94407
6	4.84062	4.79406	4.73879	4.67358	4.59664	4.50536
7	4.50001	4.46376	4.41949	4.36602	4.30168	4.22403
8	4.26513	4.23565	4.19871	4.15312	4.09728	4.02888
9	4.09377	4.06903	4.03726	3.99729	3.94756	3.88583
10	3.96344	3.94215	3.91421	3.87844	3.83329	3.77657
11	3.86106	3.84240	3.81740	3.78487	3.74327	3.69046
12	3.77859	3.76196	3.73928	3.70932	3.67055	3.62087
13	3.71075	3.69576	3.67494	3.64707	3.61061	3.56348
14	3.65399	3.64032	3.62104	3.59489	3.56036	3.51536
15	3.60582	3.59325	3.57524	3.55055	3.51763	3.47443
16	3.56443	3.55278	3.53585	3.51239	3.48085	3.43919
17	3.52850	3.51762	3.50162	3.47922	3.44887	3.40854
18	3.49700	3.48679	3.47159	3.45011	3.42081	3.38164
19	3.46918	3.45954	3.44504	3.42437	3.39599	3.35784
20	3.44442	3.43529	3.42141	3.40145	3.37387	3.33664
21	3.42225	3.41357	3.40022	3.38091	3.35405	3.31763
22	3.40228	3.39399	3.38114	3.36239	3.33618	3.30049
23	3.38421	3.37627	3.36385	3.34561	3.31999	3.28496
24	3.36777	3.36014	3.34811	3.33034	3.30525	3.27081
25	3.35275	3.34541	3.33373	3.31638	3.29177	3.25789
26	3.33898	3.33189	3.32054	3.30357	3.27941	3.24602
27	3.32631	3.31945	3.30840	3.29178	3.26802	3.23510
28	3.31461	3.30797	3.29718	3.28089	3.25750	3.22500
29	3.30377	3.29732	3.28679	3.27079	3.24775	3.21564
30	3.29371	3.28744	3.27713	3.26141	3.23869	3.20695
35	3.25253	3.24697	3.23758	3.22299	3.20157	3.17130
40	3.22217	3.21711	3.20839	3.19461	3.17415	3.14497
45	3.19888	3.19419	3.18596	3.17280	3.15306	3.12472
50	3.18043	3.17603	3.16820	3.15552	3.13635	3.10867
60	3.15309	3.14910	3.14183	3.12986	3.11154	3.08483
80	3.11945	3.11593	3.10934	3.09823	3.08090	3.05542
100	3.09954	3.09630	3.09009	3.07948	3.06280	3.03799
120	3.08639	3.08332	3.07737	3.06709	3.05080	3.02645
200	3.06036	3.05762	3.05216	3.04252	3.02702	3.00358
∞	3.02201	3.01972	3.01495	3.00624	2.99188	2.96978

TABLE A.4

One-sided Percentage Point (g) for Equal
Correlations (*Case I with Correlation* ρ)

p = 8, 1-α = 0.99

ν ↓	ρ				
	0.6	0.7	0.8	0.9	1/(1+√8)
2	12.20696	11.48824	10.64300	9.55200	14.16973
3	7.10166	6.78051	6.39036	5.86745	7.92516
4	5.53802	5.32735	5.06511	4.70415	6.05132
5	4.81041	4.64849	4.44305	4.15452	5.18862
6	4.39565	4.26060	4.08659	3.83832	4.70016
7	4.12930	4.01110	3.85686	3.63400	4.38795
8	3.94435	3.83765	3.69693	3.49147	4.17192
9	3.80866	3.71029	3.57938	3.38662	4.01386
10	3.70497	3.61289	3.48940	3.30615	3.89334
11	3.62320	3.53604	3.41837	3.24270	3.79847
12	3.55709	3.47388	3.36088	3.19123	3.72189
13	3.50255	3.42259	3.31342	3.14874	3.65880
14	3.45680	3.37954	3.27358	3.11305	3.60594
15	3.41788	3.34291	3.23966	3.08265	3.56101
16	3.38437	3.31136	3.21044	3.05645	3.52236
17	3.35521	3.28391	3.18501	3.03365	3.48876
18	3.32962	3.25981	3.16268	3.01361	3.45928
19	3.30697	3.23848	3.14291	2.99587	3.43322
20	3.28679	3.21947	3.12529	2.98005	3.41000
21	3.26869	3.20242	3.10948	2.96586	3.38920
22	3.25238	3.18704	3.09523	2.95306	3.37045
23	3.23759	3.17311	3.08230	2.94145	3.35347
24	3.22412	3.16042	3.07053	2.93088	3.33801
25	3.21181	3.14881	3.05977	2.92121	3.32388
26	3.20051	3.13816	3.04989	2.91233	3.31092
27	3.19010	3.12835	3.04079	2.90415	3.29898
28	3.18049	3.11928	3.03238	2.89659	3.28796
29	3.17158	3.11088	3.02458	2.88959	3.27774
30	3.16329	3.10307	3.01733	2.88307	3.26825
35	3.12933	3.07105	2.98761	2.85635	3.22936
40	3.10424	3.04738	2.96564	2.83658	3.20065
45	3.08494	3.02917	2.94873	2.82137	3.17859
50	3.06964	3.01473	2.93532	2.80931	3.16111
60	3.04690	2.99328	2.91540	2.79137	3.13516
80	3.01886	2.96681	2.89081	2.76924	3.10317
100	3.00223	2.95111	2.87623	2.75610	3.08422
120	2.99123	2.94072	2.86657	2.74740	3.07169
200	2.96941	2.92012	2.84743	2.73016	3.04686
∞	2.93716	2.88967	2.81911	2.70464	3.01019

TABLE A.4

One-sided Percentage Point (g) for Equal
Correlations   (Case I with Correlation ρ)

p = 9, 1-α = 0.99

ν ↓	ρ					
	0.0	0.1	0.2	0.3	0.4	0.5
2	15.94375	15.44382	14.92330	14.37676	13.79702	13.17404
3	8.64326	8.45510	8.25339	8.03545	7.79766	7.53485
4	6.48342	6.37828	6.26185	6.13234	5.98721	5.82276
5	5.49716	5.42727	5.34736	5.25599	5.15108	5.02958
6	4.94194	4.89076	4.83045	4.75970	4.67668	4.57866
7	4.58863	4.54877	4.50045	4.44243	4.37299	4.28961
8	4.34504	4.31264	4.27231	4.22285	4.16260	4.08916
9	4.16737	4.14019	4.10552	4.06217	4.00851	3.94224
10	4.03225	4.00889	3.97842	3.93963	3.89093	3.83006
11	3.92613	3.90567	3.87842	3.84316	3.79831	3.74164
12	3.84065	3.82245	3.79774	3.76528	3.72350	3.67020
13	3.77035	3.75395	3.73129	3.70111	3.66183	3.61129
14	3.71154	3.69661	3.67564	3.64734	3.61014	3.56189
15	3.66164	3.64792	3.62835	3.60163	3.56619	3.51988
16	3.61876	3.60606	3.58768	3.56231	3.52836	3.48371
17	3.58153	3.56969	3.55233	3.52812	3.49547	3.45226
18	3.54891	3.53781	3.52133	3.49813	3.46661	3.42465
19	3.52010	3.50964	3.49392	3.47161	3.44108	3.40022
20	3.49446	3.48456	3.46952	3.44798	3.41833	3.37846
21	3.47150	3.46209	3.44765	3.42681	3.39795	3.35895
22	3.45083	3.44185	3.42795	3.40773	3.37957	3.34137
23	3.43211	3.42353	3.41010	3.39044	3.36292	3.32543
24	3.41509	3.40685	3.39386	3.37471	3.34776	3.31092
25	3.39955	3.39162	3.37902	3.36033	3.33390	3.29765
26	3.38529	3.37765	3.36540	3.34713	3.32119	3.28547
27	3.37218	3.36479	3.35287	3.33498	3.30948	3.27426
28	3.36007	3.35291	3.34129	3.32376	3.29866	3.26390
29	3.34885	3.34191	3.33056	3.31336	3.28864	3.25430
30	3.33844	3.33169	3.32060	3.30369	3.27932	3.24538
35	3.29582	3.28986	3.27978	3.26410	3.24115	3.20881
40	3.26441	3.25901	3.24965	3.23487	3.21295	3.18179
45	3.24031	3.23531	3.22651	3.21240	3.19127	3.16101
50	3.22124	3.21655	3.20817	3.19459	3.17409	3.14454
60	3.19296	3.18872	3.18096	3.16816	3.14858	3.12008
80	3.15817	3.15446	3.14744	3.13558	3.11712	3.08991
100	3.13759	3.13418	3.12759	3.11628	3.09847	3.07203
120	3.12400	3.12077	3.11446	3.10351	3.08614	3.06019
200	3.09710	3.09423	3.08845	3.07820	3.06169	3.03673
∞	3.05747	3.05509	3.05007	3.04084	3.02556	3.00206

TABLE A.4

One-sided Percentage Point (g) for Equal
Correlations (*Case I with Correlation* ρ)

p = 9, 1-α = 0.99

ν ↓	ρ				
	0.6	0.7	0.8	0.9	1/(1+√9)
2	12.49274	11.72804	10.83150	9.67862	14.65368
3	7.23917	6.89736	6.48365	5.93147	8.14665
4	5.63327	5.40903	5.13101	4.74999	6.19887
5	4.88671	4.71437	4.49661	4.19215	5.30322
6	4.46140	4.31767	4.13326	3.87135	4.79648
7	4.18837	4.06259	3.89917	3.66411	4.47274
8	3.99883	3.88531	3.73625	3.51958	4.24881
9	3.85980	3.75516	3.61651	3.41328	4.08502
10	3.75357	3.65564	3.52488	3.33167	3.96015
11	3.66981	3.57713	3.45254	3.26738	3.86188
12	3.60210	3.51364	3.39401	3.21520	3.78257
13	3.54624	3.46124	3.34568	3.17213	3.71723
14	3.49939	3.41727	3.30512	3.13595	3.66249
15	3.45953	3.37986	3.27059	3.10515	3.61597
16	3.42521	3.34764	3.24085	3.07860	3.57596
17	3.39536	3.31961	3.21496	3.05548	3.54117
18	3.36915	3.29499	3.19223	3.03518	3.51066
19	3.34596	3.27321	3.17210	3.01720	3.48368
20	3.32530	3.25380	3.15417	3.00117	3.45966
21	3.30678	3.23639	3.13808	2.98679	3.43813
22	3.29007	3.22069	3.12357	2.97382	3.41872
23	3.27493	3.20646	3.11042	2.96206	3.40114
24	3.26115	3.19350	3.09844	2.95135	3.38515
25	3.24854	3.18166	3.08749	2.94155	3.37053
26	3.23697	3.17078	3.07743	2.93256	3.35711
27	3.22632	3.16076	3.06817	2.92427	3.34476
28	3.21648	3.15151	3.05961	2.91661	3.33336
29	3.20735	3.14293	3.05168	2.90951	3.32279
30	3.19887	3.13495	3.04430	2.90291	3.31297
35	3.16411	3.10226	3.01405	2.87584	3.27274
40	3.13842	3.07809	2.99169	2.85581	3.24304
45	3.11867	3.05951	2.97449	2.84040	3.22022
50	3.10300	3.04477	2.96085	2.82818	3.20214
60	3.07974	3.02287	2.94058	2.81001	3.17530
80	3.05104	2.99585	2.91556	2.78759	3.14222
100	3.03402	2.97983	2.90072	2.77428	3.12263
120	3.02276	2.96923	2.89090	2.76547	3.10967
200	3.00043	2.94820	2.87142	2.74800	3.08399
∞	2.96742	2.91711	2.84261	2.72215	3.04609

TABLE A.4

One-sided Percentage Point (g) for Equal
Correlations (*Case I with Correlation* ρ)

p = 10, 1-α = 0.99

ν ↓	ρ					
	0.0	0.1	0.2	0.3	0.4	0.5
2	16.40440	15.87262	15.31940	14.73926	14.12482	13.46570
3	8.85557	8.65354	8.43771	8.20524	7.95238	7.67374
4	6.62439	6.51091	6.38587	6.24736	6.09275	5.91820
5	5.60596	5.53031	5.44431	5.34645	5.23458	5.10554
6	5.03275	4.97728	4.91229	4.83647	4.74790	4.64377
7	4.66805	4.62482	4.57273	4.51052	4.43643	4.34785
8	4.41664	4.38150	4.33802	4.28498	4.22069	4.14268
9	4.23328	4.20381	4.16644	4.11996	4.06272	3.99233
10	4.09385	4.06853	4.03570	3.99412	3.94217	3.87753
11	3.98436	3.96220	3.93285	3.89507	3.84723	3.78706
12	3.89616	3.87647	3.84987	3.81510	3.77055	3.71397
13	3.82364	3.80591	3.78154	3.74922	3.70735	3.65370
14	3.76298	3.74685	3.72430	3.69401	3.65437	3.60317
15	3.71150	3.69670	3.67567	3.64709	3.60932	3.56019
16	3.66728	3.65359	3.63385	3.60672	3.57056	3.52319
17	3.62889	3.61614	3.59750	3.57162	3.53685	3.49102
18	3.59525	3.58330	3.56563	3.54083	3.50727	3.46278
19	3.56553	3.55429	3.53745	3.51360	3.48111	3.43780
20	3.53910	3.52846	3.51236	3.48936	3.45781	3.41554
21	3.51543	3.50533	3.48987	3.46762	3.43692	3.39558
22	3.49411	3.48449	3.46961	3.44804	3.41808	3.37759
23	3.47482	3.46562	3.45126	3.43029	3.40102	3.36129
24	3.45727	3.44846	3.43457	3.41414	3.38549	3.34645
25	3.44125	3.43277	3.41931	3.39938	3.37129	3.33288
26	3.42655	3.41839	3.40531	3.38583	3.35826	3.32043
27	3.41303	3.40515	3.39242	3.37336	3.34627	3.30896
28	3.40055	3.39292	3.38052	3.36184	3.33518	3.29837
29	3.38899	3.38160	3.36949	3.35117	3.32491	3.28855
30	3.37826	3.37108	3.35925	3.34125	3.31537	3.27943
35	3.33434	3.32801	3.31728	3.30061	3.27625	3.24203
40	3.30198	3.29626	3.28632	3.27061	3.24736	3.21439
45	3.27716	3.27187	3.26253	3.24755	3.22515	3.19315
50	3.25750	3.25256	3.24368	3.22928	3.20755	3.17631
60	3.22838	3.22392	3.21571	3.20215	3.18141	3.15129
80	3.19255	3.18866	3.18126	3.16871	3.14918	3.12044
100	3.17136	3.16780	3.16085	3.14890	3.13008	3.10215
120	3.15737	3.15400	3.14736	3.13580	3.11744	3.09005
200	3.12967	3.12670	3.12064	3.10983	3.09239	3.06606
∞	3.08889	3.08644	3.08120	3.07149	3.05538	3.03061

TABLE A.4

One-sided Percentage Point (g) for Equal
Correlations (Case I with Correlation ρ)

p = 10, 1-α = 0.99

ν ↓	ρ				
	0.6	0.7	0.8	0.9	$1/(1+\sqrt{10})$
2	12.74627	11.94053	10.99824	9.79034	15.08947
3	7.36115	7.00085	6.56611	5.98792	8.34632
4	5.71775	5.48134	5.18922	4.79039	6.33186
5	4.95438	4.77267	4.54391	4.22529	5.40645
6	4.51969	4.36816	4.17447	3.90043	4.88316
7	4.24073	4.10814	3.93652	3.69062	4.54899
8	4.04711	3.92746	3.77094	3.54432	4.31790
9	3.90511	3.79484	3.64927	3.43674	4.14891
10	3.79662	3.69344	3.55617	3.35413	4.02010
11	3.71109	3.61346	3.48269	3.28909	3.91874
12	3.64196	3.54877	3.42323	3.23628	3.83694
13	3.58493	3.49540	3.37414	3.19271	3.76956
14	3.53710	3.45062	3.33294	3.15610	3.71312
15	3.49641	3.41251	3.29787	3.12494	3.66515
16	3.46137	3.37970	3.26766	3.09808	3.62390
17	3.43090	3.35115	3.24137	3.07469	3.58804
18	3.40415	3.32608	3.21828	3.05415	3.55659
19	3.38048	3.30389	3.19785	3.03597	3.52878
20	3.35939	3.28412	3.17963	3.01975	3.50401
21	3.34048	3.26640	3.16329	3.00521	3.48182
22	3.32342	3.25041	3.14856	2.99208	3.46182
23	3.30797	3.23592	3.13520	2.98019	3.44370
24	3.29390	3.22272	3.12304	2.96935	3.42722
25	3.28104	3.21066	3.11192	2.95944	3.41215
26	3.26923	3.19959	3.10171	2.95034	3.39833
27	3.25836	3.18939	3.09230	2.94196	3.38560
28	3.24831	3.17996	3.08361	2.93421	3.37384
29	3.23900	3.17123	3.07555	2.92703	3.36295
30	3.23034	3.16311	3.06807	2.92036	3.35283
35	3.19487	3.12982	3.03736	2.89297	3.31138
40	3.16865	3.10521	3.01465	2.87272	3.28078
45	3.14849	3.08629	2.99719	2.85713	3.25727
50	3.13250	3.07128	2.98334	2.84477	3.23864
60	3.10876	3.04899	2.96275	2.82640	3.21099
80	3.07947	3.02148	2.93736	2.80372	3.17693
100	3.06211	3.00517	2.92229	2.79026	3.15675
120	3.05061	2.99438	2.91232	2.78135	3.14340
200	3.02783	2.97297	2.89255	2.76369	3.11696
∞	2.99415	2.94132	2.86330	2.73755	3.07793

TABLE A.4

One-sided Percentage Point (g) for Equal Correlations (*Case I with Correlation* ρ)

$p = 11,\ 1-\alpha = 0.99$

$\nu \downarrow$	ρ					
	0.0	0.1	0.2	0.3	0.4	0.5
2	16.81815	16.25797	15.67544	15.06509	14.41940	13.72769
3	9.04718	8.83252	8.60382	8.35814	8.09158	7.79857
4	6.75191	6.63074	6.49776	6.35101	6.18774	6.00399
5	5.70448	5.62347	5.53182	5.42799	5.30975	5.17382
6	5.11502	5.05552	4.98619	4.90566	4.81200	4.70228
7	4.74000	4.69359	4.63798	4.57190	4.49352	4.40018
8	4.48149	4.44376	4.39733	4.34098	4.27297	4.19076
9	4.29296	4.26132	4.22142	4.17204	4.11148	4.03731
10	4.14960	4.12244	4.08739	4.04322	3.98827	3.92016
11	4.03704	4.01327	3.98196	3.94183	3.89123	3.82785
12	3.94637	3.92527	3.89689	3.85998	3.81287	3.75327
13	3.87182	3.85284	3.82685	3.79254	3.74828	3.69178
14	3.80947	3.79221	3.76818	3.73603	3.69413	3.64022
15	3.75656	3.74073	3.71833	3.68801	3.64810	3.59637
16	3.71111	3.69648	3.67546	3.64669	3.60849	3.55863
17	3.67165	3.65804	3.63821	3.61077	3.57404	3.52580
18	3.63708	3.62434	3.60554	3.57926	3.54382	3.49699
19	3.60654	3.59456	3.57666	3.55139	3.51709	3.47151
20	3.57938	3.56806	3.55094	3.52658	3.49327	3.44880
21	3.55506	3.54432	3.52790	3.50434	3.47193	3.42845
22	3.53316	3.52293	3.50713	3.48429	3.45268	3.41010
23	3.51333	3.50356	3.48833	3.46613	3.43525	3.39347
24	3.49531	3.48595	3.47121	3.44960	3.41938	3.37833
25	3.47884	3.46985	3.45558	3.43450	3.40487	3.36449
26	3.46375	3.45509	3.44123	3.42064	3.39156	3.35179
27	3.44986	3.44150	3.42802	3.40787	3.37930	3.34009
28	3.43704	3.42896	3.41582	3.39608	3.36798	3.32928
29	3.42516	3.41734	3.40452	3.38516	3.35748	3.31927
30	3.41414	3.40654	3.39402	3.37501	3.34773	3.30996
35	3.36903	3.36236	3.35102	3.33343	3.30777	3.27181
40	3.33580	3.32977	3.31929	3.30272	3.27825	3.24363
45	3.31031	3.30476	3.29491	3.27913	3.25556	3.22196
50	3.29013	3.28494	3.27560	3.26043	3.23757	3.20478
60	3.26022	3.25556	3.24694	3.23267	3.21087	3.17926
80	3.22345	3.21940	3.21164	3.19846	3.17794	3.14780
100	3.20170	3.19800	3.19073	3.17818	3.15842	3.12914
120	3.18734	3.18385	3.17691	3.16478	3.14551	3.11680
200	3.15893	3.15585	3.14953	3.13821	3.11992	3.09233
∞	3.11708	3.11457	3.10912	3.09897	3.08211	3.05618

TABLE A.4

One-sided Percentage Point (g) for Equal
Correlations (Case I with Correlation ρ)

p = 11, 1-α = 0.99

ν ↓	ρ				
	0.6	0.7	0.8	0.9	1/(1+√11)
2	12.97386	12.13110	11.14758	9.89019	15.48550
3	7.47065	7.09363	6.63991	6.03832	8.52803
4	5.79359	5.54614	5.24130	4.82644	6.45290
5	5.01510	4.82491	4.58621	4.25486	5.50036
6	4.57199	4.41339	4.21131	3.92638	4.96198
7	4.28770	4.14894	3.96991	3.71427	4.61827
8	4.09042	3.96521	3.80195	3.56639	4.38064
9	3.94575	3.83036	3.67855	3.45767	4.20688
10	3.83523	3.72728	3.58414	3.37415	4.07446
11	3.74811	3.64597	3.50962	3.30845	3.97027
12	3.67769	3.58022	3.44933	3.25508	3.88620
13	3.61961	3.52597	3.39956	3.21105	3.81695
14	3.57089	3.48046	3.35779	3.17407	3.75895
15	3.52945	3.44173	3.32223	3.14258	3.70966
16	3.49378	3.40838	3.29161	3.11544	3.66726
17	3.46274	3.37936	3.26495	3.09181	3.63042
18	3.43550	3.35389	3.24155	3.07106	3.59810
19	3.41140	3.33134	3.22083	3.05269	3.56953
20	3.38992	3.31125	3.20237	3.03631	3.54408
21	3.37067	3.29324	3.18581	3.02161	3.52129
22	3.35330	3.27699	3.17087	3.00836	3.50074
23	3.33757	3.26227	3.15734	2.99634	3.48213
24	3.32324	3.24886	3.14501	2.98539	3.46519
25	3.31014	3.23660	3.13373	2.97538	3.44971
26	3.29812	3.22535	3.12338	2.96619	3.43551
27	3.28705	3.21499	3.11385	2.95773	3.42244
28	3.27682	3.20541	3.10504	2.94990	3.41037
29	3.26734	3.19653	3.09687	2.94264	3.39918
30	3.25853	3.18828	3.08928	2.93590	3.38878
35	3.22241	3.15446	3.05816	2.90824	3.34621
40	3.19571	3.12946	3.03515	2.88778	3.31478
45	3.17519	3.11023	3.01745	2.87204	3.29064
50	3.15891	3.09498	3.00341	2.85955	3.27151
60	3.13474	3.07233	2.98255	2.84100	3.24312
80	3.10492	3.04439	2.95681	2.81809	3.20814
100	3.08724	3.02782	2.94154	2.80450	3.18742
120	3.07554	3.01685	2.93144	2.79550	3.17372
200	3.05235	2.99510	2.91140	2.77766	3.14657
∞	3.01806	2.96296	2.88177	2.75126	3.10651

TABLE A.4

One-sided Percentage Point (g) for Equal
Correlations ⟮*Case I with Correlation* ρ⟯

p = 12, 1-α = 0.99

ν ↓	ρ					
	0.0	0.1	0.2	0.3	0.4	0.5
2	17.19315	16.60741	15.99841	15.36069	14.68662	13.96528
3	9.22160	8.99537	8.75487	8.49709	8.21798	7.91182
4	6.86824	6.73994	6.59963	6.44527	6.27404	6.08183
5	5.79445	5.70843	5.61154	5.50217	5.37804	5.23578
6	5.19018	5.12689	5.05350	4.96862	4.87024	4.75538
7	4.80574	4.75634	4.69744	4.62774	4.54540	4.44767
8	4.54074	4.50056	4.45137	4.39192	4.32046	4.23438
9	4.34748	4.31378	4.27151	4.21941	4.15578	4.07812
10	4.20052	4.17160	4.13447	4.08788	4.03014	3.95883
11	4.08514	4.05984	4.02667	3.98434	3.93120	3.86484
12	3.99220	3.96975	3.93970	3.90077	3.85129	3.78891
13	3.91579	3.89561	3.86809	3.83192	3.78543	3.72631
14	3.85188	3.83354	3.80811	3.77423	3.73023	3.67381
15	3.79765	3.78085	3.75715	3.72520	3.68330	3.62917
16	3.75107	3.73555	3.71333	3.68301	3.64291	3.59075
17	3.71063	3.69620	3.67524	3.64634	3.60780	3.55733
18	3.67520	3.66171	3.64185	3.61417	3.57698	3.52801
19	3.64391	3.63123	3.61232	3.58573	3.54973	3.50206
20	3.61607	3.60410	3.58603	3.56039	3.52545	3.47895
21	3.59115	3.57980	3.56247	3.53768	3.50369	3.45823
22	3.56871	3.55791	3.54125	3.51722	3.48407	3.43955
23	3.54840	3.53809	3.52202	3.49868	3.46630	3.42262
24	3.52993	3.52005	3.50453	3.48181	3.45012	3.40721
25	3.51306	3.50358	3.48854	3.46638	3.43533	3.39312
26	3.49760	3.48847	3.47388	3.45223	3.42176	3.38019
27	3.48337	3.47457	3.46038	3.43921	3.40926	3.36828
28	3.47024	3.46173	3.44791	3.42717	3.39772	3.35728
29	3.45807	3.44984	3.43636	3.41602	3.38702	3.34709
30	3.44678	3.43879	3.42562	3.40566	3.37708	3.33761
35	3.40057	3.39357	3.38167	3.36321	3.33634	3.29878
40	3.36654	3.36023	3.34923	3.33186	3.30625	3.27010
45	3.34043	3.33463	3.32431	3.30778	3.28312	3.24804
50	3.31977	3.31436	3.30457	3.28869	3.26478	3.23055
60	3.28915	3.28429	3.27528	3.26035	3.23756	3.20458
80	3.25149	3.24729	3.23920	3.22542	3.20399	3.17256
100	3.22923	3.22540	3.21783	3.20473	3.18410	3.15357
120	3.21453	3.21092	3.20370	3.19104	3.17093	3.14101
200	3.18545	3.18228	3.17572	3.16392	3.14485	3.11610
∞	3.14263	3.14006	3.13443	3.12387	3.10631	3.07930

TABLE A.4

One-sided Percentage Point (g) for Equal
Correlations (*Case I with Correlation* ρ)

p = 12, 1−α = 0.99

ν ↓	ρ 0.6	0.7	0.8	0.9	1/(1+√12)
2	13.18015	12.30371	11.28269	9.98036	15.84817
3	7.56990	7.17764	6.70665	6.08381	8.69469
4	5.86232	5.60481	5.28838	4.85897	6.56394
5	5.07014	4.87220	4.62444	4.28153	5.58649
6	4.61939	4.45432	4.24460	3.94977	5.03424
7	4.33025	4.18585	4.00007	3.73559	4.68175
8	4.12965	3.99936	3.82996	3.58628	4.43809
9	3.98256	3.86250	3.70499	3.47652	4.25995
10	3.87020	3.75789	3.60939	3.39220	4.12420
11	3.78163	3.67538	3.53394	3.32590	4.01740
12	3.71005	3.60866	3.47290	3.27202	3.93122
13	3.65101	3.55362	3.42251	3.22758	3.86025
14	3.60149	3.50743	3.38022	3.19025	3.80080
15	3.55937	3.46814	3.34422	3.15847	3.75029
16	3.52311	3.43431	3.31322	3.13108	3.70685
17	3.49157	3.40487	3.28624	3.10724	3.66909
18	3.46388	3.37902	3.26255	3.08629	3.63597
19	3.43938	3.35615	3.24158	3.06775	3.60669
20	3.41756	3.33577	3.22289	3.05122	3.58062
21	3.39799	3.31750	3.20613	3.03639	3.55726
22	3.38034	3.30102	3.19101	3.02302	3.53621
23	3.36435	3.28608	3.17730	3.01089	3.51714
24	3.34979	3.27248	3.16483	2.99984	3.49979
25	3.33648	3.26004	3.15341	2.98974	3.48393
26	3.32426	3.24863	3.14294	2.98047	3.46939
27	3.31301	3.23812	3.13329	2.97192	3.45599
28	3.30261	3.22840	3.12437	2.96403	3.44362
29	3.29298	3.21940	3.11611	2.95670	3.43216
30	3.28402	3.21103	3.10843	2.94990	3.42151
35	3.24732	3.17672	3.07692	2.92199	3.37790
40	3.22019	3.15136	3.05363	2.90135	3.34571
45	3.19934	3.13186	3.03572	2.88546	3.32098
50	3.18280	3.11639	3.02151	2.87286	3.30138
60	3.15823	3.09342	3.00040	2.85414	3.27231
80	3.12793	3.06508	2.97435	2.83103	3.23648
100	3.10997	3.04827	2.95890	2.81732	3.21526
120	3.09808	3.03715	2.94868	2.80824	3.20123
200	3.07451	3.01509	2.92840	2.79023	3.17344
∞	3.03968	2.98248	2.89841	2.76360	3.13242

BECHHOFER and DUNNETT

TABLE A.4

One-sided Percentage Point (g) for Equal
Correlations [*Case I with Correlation* ρ]

p = 13, 1-α = 0.99

$\nu \downarrow$	ρ					
	0.0	0.1	0.2	0.3	0.4	0.5
2	17.53566	16.92674	16.29366	15.63097	14.93094	14.18246
3	9.38154	9.14464	8.89328	8.62434	8.33367	8.01541
4	6.97513	6.84019	6.69307	6.53166	6.35306	6.15305
5	5.87721	5.78649	5.68470	5.57018	5.44058	5.29246
6	5.25935	5.19249	5.11530	5.02635	4.92358	4.80395
7	4.86625	4.81401	4.75201	4.67894	4.59290	4.49110
8	4.59528	4.55276	4.50097	4.43863	4.36395	4.27428
9	4.39765	4.36199	4.31748	4.26284	4.19634	4.11544
10	4.24738	4.21678	4.17767	4.12881	4.06847	3.99419
11	4.12939	4.10263	4.06770	4.02332	3.96778	3.89866
12	4.03435	4.01061	3.97899	3.93816	3.88646	3.82149
13	3.95622	3.93489	3.90593	3.86801	3.81944	3.75787
14	3.89087	3.87150	3.84474	3.80923	3.76327	3.70452
15	3.83541	3.81768	3.79276	3.75927	3.71551	3.65915
16	3.78779	3.77141	3.74805	3.71629	3.67441	3.62011
17	3.74644	3.73123	3.70920	3.67893	3.63868	3.58615
18	3.71022	3.69600	3.67513	3.64615	3.60732	3.55634
19	3.67822	3.66487	3.64501	3.61717	3.57959	3.52998
20	3.64977	3.63716	3.61819	3.59135	3.55489	3.50649
21	3.62429	3.61235	3.59416	3.56822	3.53274	3.48544
22	3.60135	3.58999	3.57251	3.54737	3.51278	3.46645
23	3.58058	3.56975	3.55290	3.52848	3.49470	3.44925
24	3.56170	3.55133	3.53506	3.51129	3.47823	3.43359
25	3.54446	3.53451	3.51875	3.49558	3.46319	3.41928
26	3.52865	3.51908	3.50379	3.48116	3.44938	3.40614
27	3.51411	3.50488	3.49002	3.46788	3.43666	3.39404
28	3.50068	3.49177	3.47730	3.45562	3.42491	3.38286
29	3.48825	3.47963	3.46551	3.44426	3.41403	3.37250
30	3.47671	3.46835	3.45457	3.43370	3.40391	3.36287
35	3.42948	3.42217	3.40973	3.39045	3.36246	3.32342
40	3.39470	3.38812	3.37664	3.35852	3.33184	3.29427
45	3.36802	3.36198	3.35123	3.33398	3.30830	3.27185
50	3.34691	3.34128	3.33109	3.31453	3.28965	3.25408
60	3.31562	3.31059	3.30121	3.28566	3.26195	3.22770
80	3.27716	3.27281	3.26441	3.25008	3.22779	3.19516
100	3.25442	3.25046	3.24261	3.22900	3.20755	3.17586
120	3.23940	3.23569	3.22820	3.21505	3.19416	3.16310
200	3.20970	3.20645	3.19966	3.18742	3.16761	3.13780
∞	3.16598	3.16335	3.15755	3.14662	3.12840	3.10040

TABLE A.4

One-sided Percentage Point (g) for Equal
Correlations (*Case I with Correlation* ρ)

p = 13, 1-α = 0.99

ν ↓	ρ				
	0.6	0.7	0.8	0.9	$1/(1+\sqrt{13})$
2	13.36865	12.46133	11.40596	10.06249	16.18243
3	7.66061	7.25433	6.76750	6.12521	8.84856
4	5.92514	5.65836	5.33129	4.88857	6.66650
5	5.12043	4.91535	4.65928	4.30580	5.66604
6	4.66269	4.49168	4.27493	3.97106	5.10095
7	4.36913	4.21953	4.02754	3.75498	4.74033
8	4.16549	4.03051	3.85548	3.60438	4.49108
9	4.01618	3.89181	3.72908	3.49367	4.30887
10	3.90214	3.78580	3.63239	3.40861	4.17004
11	3.81225	3.70219	3.55609	3.34176	4.06081
12	3.73960	3.63459	3.49436	3.28742	3.97269
13	3.67968	3.57882	3.44340	3.24261	3.90011
14	3.62943	3.53203	3.40064	3.20496	3.83932
15	3.58669	3.49222	3.36425	3.17292	3.78767
16	3.54989	3.45794	3.33290	3.14530	3.74325
17	3.51788	3.42812	3.30562	3.12126	3.70465
18	3.48978	3.40194	3.28166	3.10014	3.67079
19	3.46493	3.37877	3.26046	3.08144	3.64085
20	3.44278	3.35812	3.24157	3.06478	3.61420
21	3.42292	3.33961	3.22462	3.04983	3.59032
22	3.40502	3.32291	3.20934	3.03634	3.56879
23	3.38879	3.30778	3.19548	3.02411	3.54930
24	3.37402	3.29400	3.18287	3.01297	3.53156
25	3.36051	3.28141	3.17133	3.00279	3.51535
26	3.34811	3.26984	3.16074	2.99344	3.50048
27	3.33670	3.25919	3.15098	2.98482	3.48679
28	3.32615	3.24935	3.14197	2.97686	3.47414
29	3.31637	3.24023	3.13361	2.96948	3.46243
30	3.30729	3.23176	3.12585	2.96262	3.45154
35	3.27004	3.19700	3.09400	2.93448	3.40696
40	3.24252	3.17132	3.07046	2.91367	3.37406
45	3.22136	3.15156	3.05235	2.89766	3.34878
50	3.20458	3.13590	3.03799	2.88496	3.32876
60	3.17966	3.11263	3.01665	2.86608	3.29904
80	3.14892	3.08392	2.99032	2.84278	3.26243
100	3.13069	3.06690	2.97470	2.82896	3.24075
120	3.11863	3.05563	2.96436	2.81981	3.22641
200	3.09472	3.03329	2.94386	2.80166	3.19801
∞	3.05938	3.00027	2.91355	2.77481	3.15610

TABLE A.4

One-sided Percentage Point (g) for Equal
Correlations (*Case I with Correlation* ρ)

p = 14, 1-α = 0.99

ν ↓	ρ 0.0	0.1	0.2	0.3	0.4	0.5
2	17.85057	17.22050	16.56535	15.87973	15.15582	14.38235
3	9.52911	9.28235	9.02091	8.74163	8.44026	8.11079
4	7.07395	6.93280	6.77933	6.61136	6.42590	6.21864
5	5.95381	5.85866	5.75227	5.63293	5.49824	5.34467
6	5.32340	5.25317	5.17239	5.07962	4.97275	4.84868
7	4.92229	4.86736	4.80244	4.72619	4.63670	4.53109
8	4.64578	4.60105	4.54680	4.48173	4.40404	4.31102
9	4.44411	4.40659	4.35995	4.30291	4.23373	4.14980
10	4.29076	4.25856	4.21759	4.16658	4.10380	4.02675
11	4.17035	4.14220	4.10560	4.05927	4.00149	3.92980
12	4.07337	4.04840	4.01526	3.97266	3.91887	3.85148
13	3.99363	3.97120	3.94087	3.90130	3.85078	3.78692
14	3.92694	3.90659	3.87857	3.84151	3.79371	3.73278
15	3.87035	3.85172	3.82563	3.79069	3.74519	3.68674
16	3.82174	3.80456	3.78011	3.74698	3.70343	3.64712
17	3.77955	3.76359	3.74055	3.70898	3.66713	3.61266
18	3.74259	3.72768	3.70586	3.67564	3.63527	3.58242
19	3.70994	3.69595	3.67518	3.64616	3.60710	3.55567
20	3.68090	3.66770	3.64788	3.61990	3.58200	3.53184
21	3.65491	3.64240	3.62341	3.59637	3.55950	3.51047
22	3.63150	3.61961	3.60136	3.57516	3.53922	3.49121
23	3.61031	3.59898	3.58139	3.55595	3.52085	3.47375
24	3.59105	3.58021	3.56322	3.53846	3.50412	3.45786
25	3.57346	3.56306	3.54661	3.52248	3.48884	3.44334
26	3.55733	3.54734	3.53138	3.50782	3.47481	3.43001
27	3.54249	3.53286	3.51736	3.49431	3.46189	3.41773
28	3.52879	3.51950	3.50440	3.48184	3.44995	3.40639
29	3.51611	3.50712	3.49240	3.47029	3.43889	3.39588
30	3.50433	3.49562	3.48126	3.45955	3.42862	3.38611
35	3.45616	3.44855	3.43560	3.41556	3.38650	3.34607
40	3.42068	3.41385	3.40191	3.38308	3.35540	3.31649
45	3.39347	3.38721	3.37603	3.35811	3.33149	3.29375
50	3.37193	3.36611	3.35553	3.33833	3.31253	3.27572
60	3.34003	3.33483	3.32510	3.30897	3.28439	3.24895
80	3.30081	3.29633	3.28763	3.27278	3.24969	3.21593
100	3.27762	3.27355	3.26544	3.25133	3.22912	3.19636
120	3.26231	3.25849	3.25076	3.23715	3.21552	3.18341
200	3.23203	3.22870	3.22170	3.20904	3.18855	3.15773
∞	3.18747	3.18479	3.17882	3.16754	3.14871	3.11979

TABLE A.4

One-sided Percentage Point (g) for Equal
Correlations   (Case I with Correlation ρ)

p = 14, 1−α = 0.99

ν ↓	ρ				
	0.6	0.7	0.8	0.9	1/(1+√14)
2	13.54209	12.60628	11.51922	10.13786	16.49223
3	7.74408	7.32485	6.82339	6.16319	8.99141
4	5.98294	5.70759	5.37070	4.91570	6.76178
5	5.16671	4.95501	4.69127	4.32804	5.73993
6	4.70253	4.52600	4.30278	3.99056	5.16290
7	4.40490	4.25047	4.05276	3.77274	4.79471
8	4.19845	4.05913	3.87889	3.62094	4.54026
9	4.04710	3.91874	3.75118	3.50938	4.35426
10	3.93151	3.81144	3.65349	3.42364	4.21254
11	3.84040	3.72683	3.57640	3.35628	4.10105
12	3.76677	3.65841	3.51404	3.30153	4.01111
13	3.70604	3.60197	3.46257	3.25637	3.93704
14	3.65511	3.55462	3.41938	3.21844	3.87499
15	3.61180	3.51434	3.38261	3.18615	3.82228
16	3.57451	3.47965	3.35095	3.15832	3.77695
17	3.54207	3.44947	3.32339	3.13410	3.73755
18	3.51360	3.42298	3.29920	3.11282	3.70300
19	3.48841	3.39953	3.27778	3.09398	3.67245
20	3.46596	3.37864	3.25870	3.07719	3.64525
21	3.44584	3.35991	3.24158	3.06213	3.62088
22	3.42769	3.34301	3.22614	3.04854	3.59892
23	3.41125	3.32770	3.21215	3.03622	3.57903
24	3.39628	3.31376	3.19941	3.02500	3.56093
25	3.38259	3.30102	3.18775	3.01474	3.54439
26	3.37003	3.28932	3.17706	3.00532	3.52921
27	3.35846	3.27854	3.16721	2.99664	3.51524
28	3.34778	3.26859	3.15810	2.98862	3.50234
29	3.33787	3.25936	3.14966	2.98118	3.49038
30	3.32866	3.25078	3.14182	2.97427	3.47928
35	3.29092	3.21562	3.10966	2.94592	3.43379
40	3.26304	3.18963	3.08588	2.92496	3.40023
45	3.24159	3.16965	3.06759	2.90883	3.37444
50	3.22459	3.15380	3.05309	2.89603	3.35401
60	3.19934	3.13025	3.03154	2.87702	3.32369
80	3.16819	3.10121	3.00495	2.85355	3.28634
100	3.14972	3.08399	2.98918	2.83962	3.26423
120	3.13750	3.07259	2.97874	2.83040	3.24960
200	3.11327	3.04999	2.95804	2.81212	3.22063
∞	3.07746	3.01658	2.92743	2.78508	3.17789

TABLE A.4

One-sided Percentage Point (g) for Equal
Correlations $\begin{pmatrix} Case\ I\ with\ Correlation\ \rho \end{pmatrix}$

$p = 15, \ 1-\alpha = 0.99$

$\nu \downarrow$	ρ 0.0	0.1	0.2	0.3	0.4	0.5
2	18.14175	17.49225	16.81680	16.11002	15.36401	14.56738
3	9.66601	9.41008	9.13928	8.85037	8.53904	8.19914
4	7.16579	7.01883	6.85940	6.68529	6.49342	6.27940
5	6.02506	5.92575	5.81503	5.69117	5.55170	5.39303
6	5.38302	5.30958	5.22543	5.12906	5.01835	4.89012
7	4.97447	4.91698	4.84929	4.77005	4.67731	4.56815
8	4.69281	4.64597	4.58937	4.52174	4.44121	4.34505
9	4.48737	4.44807	4.39940	4.34011	4.26839	4.18163
10	4.33115	4.29741	4.25466	4.20163	4.13656	4.05690
11	4.20848	4.17900	4.14081	4.09264	4.03275	3.95864
12	4.10968	4.08353	4.04896	4.00467	3.94892	3.87926
13	4.02844	4.00496	3.97333	3.93219	3.87983	3.81382
14	3.96050	3.93920	3.90998	3.87146	3.82192	3.75895
15	3.90285	3.88336	3.85616	3.81985	3.77269	3.71229
16	3.85333	3.83536	3.80987	3.77544	3.73033	3.67213
17	3.81035	3.79366	3.76965	3.73685	3.69349	3.63721
18	3.77269	3.75711	3.73438	3.70298	3.66117	3.60656
19	3.73943	3.72481	3.70319	3.67304	3.63258	3.57945
20	3.70985	3.69607	3.67542	3.64637	3.60712	3.55529
21	3.68336	3.67032	3.65055	3.62247	3.58429	3.53364
22	3.65952	3.64713	3.62813	3.60093	3.56372	3.51412
23	3.63794	3.62613	3.60782	3.58142	3.54507	3.49643
24	3.61832	3.60702	3.58935	3.56366	3.52811	3.48033
25	3.60040	3.58957	3.57246	3.54742	3.51259	3.46561
26	3.58397	3.57357	3.55698	3.53253	3.49836	3.45210
27	3.56885	3.55883	3.54272	3.51882	3.48525	3.43965
28	3.55490	3.54523	3.52955	3.50615	3.47314	3.42816
29	3.54198	3.53263	3.51735	3.49441	3.46192	3.41751
30	3.52999	3.52093	3.50601	3.48350	3.45150	3.40761
35	3.48092	3.47303	3.45959	3.43882	3.40877	3.36703
40	3.44479	3.43771	3.42534	3.40583	3.37721	3.33706
45	3.41708	3.41060	3.39902	3.38048	3.35295	3.31401
50	3.39515	3.38913	3.37818	3.36038	3.33372	3.29574
60	3.36266	3.35730	3.34724	3.33056	3.30516	3.26861
80	3.32273	3.31812	3.30914	3.29380	3.26996	3.23515
100	3.29912	3.29494	3.28658	3.27201	3.24909	3.21531
120	3.28354	3.27963	3.27166	3.25761	3.23529	3.20218
200	3.25271	3.24931	3.24211	3.22906	3.20793	3.17617
∞	3.20736	3.20464	3.19852	3.18691	3.16750	3.13772

TABLE A.4

One-sided Percentage Point (g) for Equal
Correlations (*Case I with Correlation* ρ)

p = 15, 1−α = 0.99

ν ↓	ρ				
	0.6	0.7	0.8	0.9	$1/(1+\sqrt{15})$
2	13.70260	12.74038	11.62393	10.20745	16.78079
3	7.82135	7.39008	6.87504	6.19824	9.12469
4	6.03644	5.75312	5.40711	4.94074	6.85071
5	5.20954	4.99169	4.72082	4.34856	5.80891
6	4.73941	4.55774	4.32849	4.00855	5.22072
7	4.43800	4.27908	4.07605	3.78913	4.84546
8	4.22896	4.08559	3.90051	3.63622	4.58613
9	4.07572	3.94363	3.77159	3.52386	4.39659
10	3.95868	3.83514	3.67298	3.43750	4.25217
11	3.86644	3.74959	3.59516	3.36968	4.13856
12	3.79190	3.68042	3.53222	3.31454	4.04691
13	3.73043	3.62336	3.48027	3.26905	3.97143
14	3.67887	3.57550	3.43667	3.23086	3.90821
15	3.63502	3.53477	3.39956	3.19835	3.85450
16	3.59727	3.49970	3.36761	3.17032	3.80831
17	3.56444	3.46920	3.33980	3.14593	3.76817
18	3.53562	3.44241	3.31538	3.12451	3.73297
19	3.51012	3.41871	3.29377	3.10554	3.70184
20	3.48740	3.39760	3.27450	3.08863	3.67413
21	3.46703	3.37866	3.25723	3.07346	3.64930
22	3.44866	3.36158	3.24165	3.05978	3.62693
23	3.43202	3.34611	3.22753	3.04738	3.60666
24	3.41687	3.33202	3.21467	3.03608	3.58822
25	3.40301	3.31913	3.20291	3.02575	3.57137
26	3.39030	3.30731	3.19212	3.01626	3.55591
27	3.37859	3.29642	3.18218	3.00753	3.54168
28	3.36777	3.28635	3.17299	2.99945	3.52853
29	3.35774	3.27702	3.16447	2.99196	3.51636
30	3.34842	3.26835	3.15656	2.98500	3.50504
35	3.31022	3.23281	3.12410	2.95646	3.45871
40	3.28200	3.20655	3.10011	2.93536	3.42451
45	3.26030	3.18635	3.08166	2.91912	3.39824
50	3.24309	3.17033	3.06702	2.90623	3.37743
60	3.21753	3.14653	3.04527	2.88709	3.34655
80	3.18600	3.11718	3.01844	2.86346	3.30852
100	3.16731	3.09977	3.00253	2.84944	3.28599
120	3.15494	3.08825	2.99200	2.84016	3.27110
200	3.13042	3.06541	2.97111	2.82176	3.24160
∞	3.09417	3.03164	2.94023	2.79453	3.19807

TABLE A.4

*One-sided Percentage Point (g) for Equal
Correlations* (*Case I with Correlation ρ*)

p = 16, 1-α = 0.99

ν ↓	ρ					
	0.0	0.1	0.2	0.3	0.4	0.5
2	18.41238	17.74493	17.05067	16.32423	15.55772	14.73954
3	9.79361	9.52912	9.24957	8.95167	8.63102	8.28137
4	7.25154	7.09911	6.93409	6.75420	6.55633	6.33597
5	6.09165	5.98839	5.87359	5.74547	5.60152	5.43807
6	5.43876	5.36229	5.27493	5.17518	5.06085	4.92871
7	5.02327	4.96334	4.89302	4.81095	4.71515	4.60265
8	4.73680	4.68794	4.62912	4.55905	4.47585	4.37673
9	4.52784	4.48683	4.43624	4.37479	4.30069	4.21126
10	4.36893	4.33372	4.28928	4.23432	4.16708	4.08497
11	4.24414	4.21337	4.17368	4.12376	4.06187	3.98548
12	4.14363	4.11635	4.08041	4.03451	3.97691	3.90511
13	4.06099	4.03650	4.00361	3.96099	3.90689	3.83885
14	3.99187	3.96966	3.93929	3.89938	3.84820	3.78330
15	3.93322	3.91290	3.88464	3.84703	3.79831	3.73607
16	3.88285	3.86412	3.83764	3.80198	3.75537	3.69541
17	3.83912	3.82174	3.79680	3.76283	3.71804	3.66005
18	3.80081	3.78459	3.76098	3.72847	3.68529	3.62902
19	3.76698	3.75177	3.72932	3.69810	3.65632	3.60157
20	3.73688	3.72255	3.70112	3.67105	3.63051	3.57712
21	3.70994	3.69638	3.67586	3.64680	3.60738	3.55520
22	3.68569	3.67281	3.65310	3.62495	3.58653	3.53544
23	3.66373	3.65146	3.63248	3.60515	3.56763	3.51753
24	3.64377	3.63204	3.61372	3.58713	3.55043	3.50122
25	3.62554	3.61431	3.59657	3.57067	3.53471	3.48632
26	3.60883	3.59804	3.58084	3.55556	3.52029	3.47264
27	3.59346	3.58307	3.56636	3.54164	3.50701	3.46005
28	3.57927	3.56924	3.55299	3.52879	3.49473	3.44841
29	3.56613	3.55643	3.54060	3.51688	3.48336	3.43763
30	3.55392	3.54454	3.52909	3.50582	3.47279	3.42760
35	3.50402	3.49586	3.48195	3.46049	3.42949	3.38653
40	3.46727	3.45996	3.44717	3.42702	3.39751	3.35618
45	3.43909	3.43241	3.42045	3.40130	3.37292	3.33285
50	3.41679	3.41059	3.39928	3.38092	3.35343	3.31435
60	3.38375	3.37824	3.36787	3.35066	3.32449	3.28689
80	3.34315	3.33842	3.32918	3.31336	3.28881	3.25301
100	3.31915	3.31487	3.30627	3.29127	3.26767	3.23293
120	3.30330	3.29930	3.29112	3.27665	3.25367	3.21964
200	3.27197	3.26850	3.26111	3.24769	3.22595	3.19330
∞	3.22587	3.22311	3.21685	3.20492	3.18498	3.15438

TABLE A.4

One-sided Percentage Point (g) for Equal
Correlations (*Case I with Correlation* ρ)

p = 16, 1-α = 0.99

ν ↓	ρ				
	0.6	0.7	0.8	0.9	1/(1+√16)
2	13.85192	12.86508	11.72125	10.27205	17.05067
3	7.89324	7.45072	6.92303	6.23076	9.24957
4	6.08622	5.79545	5.44092	4.96397	6.93409
5	5.24939	5.02579	4.74826	4.36759	5.87359
6	4.77372	4.58724	4.35237	4.02523	5.27493
7	4.46880	4.30567	4.09767	3.80432	4.89302
8	4.25733	4.11018	3.92058	3.65039	4.62912
9	4.10233	3.96676	3.79053	3.53719	4.43624
10	3.98396	3.85717	3.69106	3.45035	4.28928
11	3.89067	3.77074	3.61257	3.38209	4.17368
12	3.81527	3.70088	3.54908	3.32660	4.08041
13	3.75310	3.64324	3.49669	3.28080	4.00361
14	3.70096	3.59489	3.45272	3.24238	3.93929
15	3.65662	3.55375	3.41529	3.20965	3.88464
16	3.61844	3.51833	3.38306	3.18145	3.83764
17	3.58524	3.48752	3.35502	3.15690	3.79680
18	3.55609	3.46047	3.33039	3.13534	3.76098
19	3.53030	3.43653	3.30860	3.11625	3.72932
20	3.50733	3.41520	3.28917	3.09923	3.70112
21	3.48673	3.39608	3.27175	3.08397	3.67586
22	3.46816	3.37883	3.25604	3.07020	3.65310
23	3.45133	3.36320	3.24180	3.05772	3.63248
24	3.43600	3.34897	3.22883	3.04635	3.61372
25	3.42199	3.33596	3.21697	3.03595	3.59657
26	3.40914	3.32401	3.20609	3.02641	3.58084
27	3.39729	3.31301	3.19606	3.01762	3.56636
28	3.38635	3.30285	3.18680	3.00949	3.55299
29	3.37621	3.29343	3.17821	3.00196	3.54060
30	3.36679	3.28467	3.17023	2.99495	3.52909
35	3.32816	3.24878	3.13750	2.96623	3.48195
40	3.29962	3.22225	3.11331	2.94499	3.44717
45	3.27767	3.20185	3.09470	2.92865	3.42045
50	3.26027	3.18567	3.07994	2.91569	3.39928
60	3.23443	3.16164	3.05802	2.89642	3.36787
80	3.20255	3.13200	3.03096	2.87265	3.32918
100	3.18365	3.11442	3.01492	2.85854	3.30627
120	3.17114	3.10278	3.00430	2.84920	3.29112
200	3.14634	3.07972	2.98324	2.83069	3.26111
∞	3.10969	3.04562	2.95210	2.80329	3.21685

TABLE A.4

One-sided Percentage Point (g) for Equal
Correlations $\left(\text{Case } I \text{ with Correlation } \rho\right)$

$p = 18, \ 1-\alpha = 0.99$

$\nu \downarrow$	0.0	0.1	0.2	0.3	0.4	0.5
2	18.90178	18.20213	17.47406	16.71220	15.90859	15.05138
3	10.02526	9.74522	9.44974	9.13547	8.79786	8.43045
4	7.40757	7.24509	7.06981	6.87936	6.67049	6.43855
5	6.21299	6.10243	5.98009	5.84413	5.69195	5.51974
6	5.54042	5.45829	5.36499	5.25898	5.13799	4.99869
7	5.11230	5.04780	4.97260	4.88530	4.78385	4.66521
8	4.81706	4.76441	4.70145	4.62687	4.53873	4.43419
9	4.60167	4.55745	4.50326	4.43784	4.35933	4.26499
10	4.43785	4.39987	4.35226	4.29373	4.22248	4.13586
11	4.30920	4.27600	4.23347	4.18030	4.11472	4.03414
12	4.20556	4.17613	4.13763	4.08875	4.02771	3.95197
13	4.12035	4.09394	4.05871	4.01332	3.95600	3.88423
14	4.04907	4.02513	3.99261	3.95012	3.89589	3.82744
15	3.98859	3.96671	3.93645	3.89640	3.84479	3.77915
16	3.93664	3.91648	3.88815	3.85019	3.80082	3.73759
17	3.89155	3.87286	3.84617	3.81001	3.76259	3.70144
18	3.85205	3.83461	3.80936	3.77477	3.72904	3.66972
19	3.81716	3.80082	3.77681	3.74360	3.69937	3.64166
20	3.78612	3.77074	3.74784	3.71585	3.67294	3.61667
21	3.75834	3.74379	3.72187	3.69097	3.64925	3.59426
22	3.73333	3.71952	3.69847	3.66855	3.62789	3.57405
23	3.71069	3.69755	3.67728	3.64824	3.60854	3.55575
24	3.69011	3.67756	3.65800	3.62975	3.59093	3.53908
25	3.67131	3.65929	3.64038	3.61286	3.57483	3.52384
26	3.65408	3.64254	3.62421	3.59735	3.56005	3.50986
27	3.63823	3.62713	3.60933	3.58308	3.54645	3.49699
28	3.62360	3.61289	3.59558	3.56989	3.53388	3.48509
29	3.61005	3.59971	3.58285	3.55768	3.52223	3.47407
30	3.59747	3.58746	3.57102	3.54632	3.51141	3.46382
35	3.54602	3.53734	3.52257	3.49981	3.46706	3.42183
40	3.50814	3.50039	3.48681	3.46547	3.43430	3.39081
45	3.47909	3.47202	3.45934	3.43908	3.40912	3.36696
50	3.45610	3.44956	3.43758	3.41816	3.38916	3.34805
60	3.42206	3.41625	3.40530	3.38711	3.35952	3.31997
80	3.38022	3.37527	3.36552	3.34884	3.32297	3.28534
100	3.35549	3.35102	3.34197	3.32616	3.30131	3.26481
120	3.33917	3.33500	3.32640	3.31116	3.28698	3.25123
200	3.30690	3.30330	3.29556	3.28144	3.25858	3.22430
∞	3.25942	3.25658	3.25006	3.23756	3.21661	3.18451

TABLE A.4

One-sided Percentage Point (g) for Equal
Correlations (*Case I with Correlation* ρ)

p = 18, 1-α = 0.99

ν ↓	ρ				
	0.6	0.7	0.8	0.9	1/(1+√18)
2	14.12235	13.09082	11.89729	10.38876	17.54279
3	8.02347	7.56051	7.00980	6.28947	9.47782
4	6.17642	5.87206	5.50205	5.00589	7.08663
5	5.32159	5.08749	4.79785	4.40192	5.99196
6	4.83586	4.64062	4.39552	4.05532	5.37414
7	4.52457	4.35378	4.13674	3.83173	4.98005
8	4.30872	4.15466	3.95684	3.67595	4.70775
9	4.15053	4.00860	3.82474	3.56149	4.50873
10	4.02973	3.89699	3.72372	3.47353	4.35710
11	3.93452	3.80899	3.64401	3.40446	4.23783
12	3.85759	3.73786	3.57954	3.34836	4.14161
13	3.79415	3.67917	3.52634	3.30199	4.06238
14	3.74096	3.62995	3.48169	3.26314	3.99602
15	3.69571	3.58807	3.44370	3.23003	3.93964
16	3.65676	3.55201	3.41097	3.20150	3.89115
17	3.62288	3.52064	3.38250	3.17667	3.84902
18	3.59314	3.49310	3.35750	3.15486	3.81206
19	3.56683	3.46874	3.33537	3.13556	3.77940
20	3.54339	3.44703	3.31565	3.11835	3.75031
21	3.52238	3.42756	3.29796	3.10291	3.72425
22	3.50343	3.41000	3.28201	3.08898	3.70077
23	3.48626	3.39409	3.26756	3.07636	3.67950
24	3.47063	3.37960	3.25439	3.06486	3.66015
25	3.45633	3.36636	3.24236	3.05435	3.64246
26	3.44322	3.35420	3.23131	3.04469	3.62624
27	3.43114	3.34301	3.22113	3.03580	3.61130
28	3.41997	3.33266	3.21173	3.02758	3.59751
29	3.40963	3.32307	3.20301	3.01996	3.58473
30	3.40002	3.31416	3.19491	3.01288	3.57286
35	3.36061	3.27762	3.16168	2.98383	3.52424
40	3.33149	3.25062	3.13713	2.96236	3.48836
45	3.30910	3.22986	3.11824	2.94583	3.46081
50	3.29135	3.21339	3.10326	2.93272	3.43897
60	3.26499	3.18894	3.08100	2.91324	3.40658
80	3.23247	3.15877	3.05355	2.88920	3.36669
100	3.21318	3.14088	3.03726	2.87494	3.34306
120	3.20042	3.12904	3.02648	2.86550	3.32744
200	3.17513	3.10556	3.00511	2.84677	3.29651
∞	3.13774	3.07086	2.97351	2.81907	3.25088

TABLE A.4

One-sided Percentage Point (g) for Equal
Correlations   (*Case I with Correlation* ρ)

$p = 20$, $1-\alpha = 0.99$

ν ↓	ρ					
	0.0	0.1	0.2	0.3	0.4	0.5
2	19.33453	18.60670	17.84895	17.05589	16.21951	15.32775
3	10.23102	9.93717	9.62753	9.29866	8.94592	8.56269
4	7.54656	7.37505	7.19055	6.99061	6.77188	6.52958
5	6.32127	6.20407	6.07492	5.93189	5.77230	5.59222
6	5.63122	5.54391	5.44523	5.33354	5.20654	5.06080
7	5.19187	5.12317	5.04352	4.95146	4.84491	4.72073
8	4.88882	4.83267	4.76592	4.68722	4.59461	4.48517
9	4.66769	4.62048	4.56300	4.49394	4.41143	4.31266
10	4.49947	4.45892	4.40839	4.34659	4.27170	4.18101
11	4.36735	4.33190	4.28676	4.23061	4.16168	4.07731
12	4.26092	4.22949	4.18862	4.13700	4.07284	3.99354
13	4.17339	4.14520	4.10780	4.05987	3.99962	3.92449
14	4.10018	4.07463	4.04011	3.99524	3.93825	3.86660
15	4.03806	4.01470	3.98259	3.94031	3.88608	3.81737
16	3.98469	3.96319	3.93313	3.89306	3.84118	3.77500
17	3.93837	3.91844	3.89014	3.85198	3.80214	3.73815
18	3.89779	3.87921	3.85244	3.81594	3.76789	3.70581
19	3.86195	3.84454	3.81910	3.78407	3.73760	3.67721
20	3.83007	3.81369	3.78942	3.75568	3.71061	3.65173
21	3.80153	3.78605	3.76283	3.73024	3.68642	3.62888
22	3.77583	3.76115	3.73887	3.70731	3.66461	3.60829
23	3.75257	3.73861	3.71716	3.68654	3.64486	3.58963
24	3.73143	3.71810	3.69741	3.66764	3.62687	3.57264
25	3.71212	3.69937	3.67936	3.65036	3.61043	3.55711
26	3.69442	3.68219	3.66280	3.63450	3.59535	3.54285
27	3.67814	3.66637	3.64756	3.61990	3.58146	3.52973
28	3.66311	3.65177	3.63348	3.60642	3.56862	3.51760
29	3.64919	3.63825	3.62043	3.59392	3.55673	3.50636
30	3.63627	3.62568	3.60831	3.58231	3.54568	3.49592
35	3.58342	3.57426	3.55868	3.53474	3.50039	3.45311
40	3.54451	3.53635	3.52205	3.49961	3.46694	3.42149
45	3.51468	3.50725	3.49392	3.47262	3.44122	3.39718
50	3.49108	3.48421	3.47162	3.45122	3.42084	3.37790
60	3.45612	3.45005	3.43855	3.41947	3.39057	3.34928
80	3.41316	3.40801	3.39780	3.38032	3.35325	3.31398
100	3.38778	3.38314	3.37368	3.35713	3.33113	3.29305
120	3.37102	3.36670	3.35772	3.34178	3.31650	3.27920
200	3.33790	3.33418	3.32613	3.31138	3.28749	3.25175
∞	3.28918	3.28628	3.27952	3.26649	3.24464	3.21119

TABLE A.4

One-sided Percentage Point (g) for Equal
Correlations (*Case I with Correlation* ρ)

p = 20, 1-α = 0.99

ν ↓	ρ				
	0.6	0.7	0.8	0.9	$1/(1+\sqrt{20})$
2	14.36198	13.29078	12.05308	10.49188	17.98206
3	8.13890	7.65774	7.08655	6.34131	9.68222
4	6.25637	5.93990	5.55611	5.04288	7.22341
5	5.38559	5.14213	4.84169	4.43222	6.09814
6	4.89095	4.68788	4.43366	4.08187	5.46313
7	4.57401	4.39636	4.17126	3.85590	5.05810
8	4.35427	4.19402	3.98888	3.69849	4.77824
9	4.19324	4.04562	3.85497	3.58284	4.57370
10	4.07028	3.93223	3.75257	3.49398	4.41787
11	3.97338	3.84283	3.67179	3.42418	4.29529
12	3.89508	3.77057	3.60644	3.36754	4.19640
13	3.83052	3.71096	3.55253	3.32066	4.11497
14	3.77638	3.66095	3.50728	3.28144	4.04676
15	3.73033	3.61842	3.46877	3.24799	3.98882
16	3.69069	3.58179	3.43561	3.21917	3.93899
17	3.65621	3.54993	3.40676	3.19410	3.89568
18	3.62595	3.52196	3.38142	3.17207	3.85771
19	3.59918	3.49721	3.35900	3.15257	3.82413
20	3.57532	3.47516	3.33902	3.13519	3.79424
21	3.55394	3.45539	3.32110	3.11960	3.76746
22	3.53466	3.43756	3.30494	3.10553	3.74333
23	3.51718	3.42140	3.29029	3.09278	3.72147
24	3.50127	3.40669	3.27696	3.08117	3.70158
25	3.48673	3.39324	3.26476	3.07055	3.68340
26	3.47338	3.38089	3.25357	3.06080	3.66673
27	3.46109	3.36952	3.24325	3.05182	3.65138
28	3.44973	3.35901	3.23372	3.04352	3.63721
29	3.43921	3.34927	3.22489	3.03583	3.62407
30	3.42942	3.34022	3.21668	3.02867	3.61187
35	3.38932	3.30312	3.18302	2.99934	3.56191
40	3.35970	3.27570	3.15815	2.97765	3.52505
45	3.33691	3.25461	3.13901	2.96096	3.49673
50	3.31885	3.23789	3.12383	2.94773	3.47430
60	3.29202	3.21305	3.10129	2.92806	3.44101
80	3.25893	3.18242	3.07347	2.90378	3.40002
100	3.23931	3.16425	3.05697	2.88938	3.37575
120	3.22633	3.15222	3.04605	2.87984	3.35970
200	3.20059	3.12839	3.02440	2.86094	3.32793
∞	3.16255	3.09315	2.99239	2.83297	3.28106

TABLES B.1 to B.4

These tables give the two-sided upper equicoordinate $100(1-\alpha)$ percentage point $h = h(p,v,\rho;\alpha)$ of a central p-variate Student t-distribution based on v d.f. with a correlation matrix $\underline{P} = \{\rho_{ij}\}$ where $\rho_{ij} = \begin{cases} 1 & \text{if } i = j \\ \rho & \text{if } i \neq j \end{cases}$

$(1 \leq i,j \leq p)$. That is, it is the solution of the equation

$$P\{ \max_{1\leq i\leq p} |T_i| \leq h\} = \int_{-h}^{h} \cdots \int_{-h}^{h} f_v(t_1,\ldots,t_p;\underline{P})dt_1\ldots dt_p$$

$$= \int_0^{\infty} \left[\int_{-\infty}^{+\infty} \left\{ \Phi\left[\frac{x\sqrt{\rho} + hy}{\sqrt{1-\rho}} \right] - \Phi\left[\frac{x\sqrt{\rho} - hy}{\sqrt{1-\rho}} \right] \right\}^p d\Phi(x) \right] q_v(y)dy = 1-\alpha$$

where $f_v(t_1,\ldots,t_p;\underline{P})$ is the p-variate Student t density function, $\Phi(\bullet)$ is the cdf of a standard normal r.v., and $q_v(\bullet)$ is the density function of $\chi_v/\sqrt{v}$, i.e..

$$q_v(y) = \frac{2}{\Gamma(v/2)} (v/2)^{v/2} y^{v-1} \exp(-vy^2/2).$$

The constant h is tabulated to 5 decimal places for

$$p = 2(1)16(2)20$$

$$\rho = 0.0(0.1)0.9, \ 1/(1+\sqrt{p})$$

$$1-\alpha = 0.80 \text{ (Table B.1)}, \ 0.90 \text{ (Table B.2)},$$
$$0.95 \text{ (Table B.3)}, \ 0.99 \text{ (Table B.4)}.$$

For $p = 2$: $v = 1(1)30(5)50,60(20)120,200,\infty$

$p > 2$: $v = 2(1)30(5)50,60(20)120,200,\infty$.

TABLE B.1

Two-sided Percentage Point (h) for Equal
Correlations [*Case I with Correlation* ρ]

p = 2, 1-α = 0.80

ν ↓	ρ					
	0.0	0.1	0.2	0.3	0.4	0.5
1	4.40814	4.40248	4.38526	4.35580	4.31280	4.25410
2	2.54931	2.54635	2.53737	2.52202	2.49965	2.46918
3	2.16661	2.16421	2.15693	2.14449	2.12638	2.10176
4	2.00568	2.00352	1.99695	1.98574	1.96945	1.94731
5	1.91752	1.91548	1.90931	1.89879	1.88349	1.86274
6	1.86197	1.86001	1.85410	1.84400	1.82934	1.80946
7	1.82378	1.82189	1.81614	1.80634	1.79212	1.77286
8	1.79593	1.79408	1.78846	1.77888	1.76499	1.74618
9	1.77473	1.77291	1.76738	1.75797	1.74433	1.72586
10	1.75804	1.75625	1.75080	1.74152	1.72808	1.70989
11	1.74457	1.74280	1.73742	1.72825	1.71496	1.69699
12	1.73347	1.73172	1.72638	1.71730	1.70415	1.68637
13	1.72417	1.72242	1.71714	1.70813	1.69509	1.67746
14	1.71625	1.71452	1.70927	1.70033	1.68738	1.66989
15	1.70944	1.70772	1.70250	1.69361	1.68075	1.66338
16	1.70352	1.70180	1.69661	1.68777	1.67498	1.65771
17	1.69831	1.69661	1.69144	1.68265	1.66992	1.65273
18	1.69371	1.69201	1.68687	1.67811	1.66544	1.64833
19	1.68961	1.68792	1.68279	1.67407	1.66145	1.64441
20	1.68593	1.68425	1.67914	1.67045	1.65787	1.64090
21	1.68262	1.68094	1.67584	1.66718	1.65464	1.63773
22	1.67961	1.67793	1.67286	1.66421	1.65172	1.63486
23	1.67687	1.67520	1.67014	1.66152	1.64905	1.63224
24	1.67437	1.67270	1.66765	1.65905	1.64662	1.62985
25	1.67207	1.67041	1.66537	1.65679	1.64438	1.62766
26	1.66996	1.66830	1.66326	1.65470	1.64232	1.62563
27	1.66800	1.66634	1.66132	1.65278	1.64042	1.62377
28	1.66619	1.66453	1.65952	1.65099	1.63866	1.62204
29	1.66450	1.66285	1.65784	1.64933	1.63702	1.62043
30	1.66293	1.66128	1.65628	1.64778	1.63549	1.61893
35	1.65646	1.65482	1.64985	1.64140	1.62919	1.61274
40	1.65162	1.64999	1.64504	1.63664	1.62449	1.60813
45	1.64788	1.64625	1.64133	1.63295	1.62085	1.60456
50	1.64489	1.64327	1.63836	1.63001	1.61795	1.60171
60	1.64043	1.63882	1.63393	1.62562	1.61361	1.59745
80	1.63488	1.63328	1.62841	1.62015	1.60822	1.59216
100	1.63157	1.62997	1.62512	1.61688	1.60499	1.58900
120	1.62937	1.62777	1.62293	1.61471	1.60285	1.58690
200	1.62497	1.62338	1.61857	1.61039	1.59858	1.58271
∞	1.61842	1.61684	1.61205	1.60393	1.59221	1.57646

TABLE B.1

Two-sided Percentage Point (h) for Equal
Correlations $\left(\textit{Case I with Correlation } \rho\right)$

p = 2, 1-α = 0.80

ν ↓	ρ				
	0.6	0.7	0.8	0.9	1/(1+√2)
1	4.17610	4.07243	3.93038	3.71664	4.30548
2	2.42883	2.37546	2.30285	2.19494	2.49584
3	2.06923	2.02632	1.96819	1.88228	2.12331
4	1.91811	1.87968	1.82775	1.75123	1.96668
5	1.83539	1.79946	1.75099	1.67968	1.88089
6	1.78330	1.74898	1.70273	1.63472	1.82685
7	1.74753	1.71432	1.66962	1.60390	1.78971
8	1.72145	1.68907	1.64551	1.58146	1.76263
9	1.70161	1.66987	1.62718	1.56440	1.74201
10	1.68601	1.65477	1.61277	1.55100	1.72580
11	1.67342	1.64259	1.60116	1.54020	1.71271
12	1.66305	1.63256	1.59159	1.53130	1.70192
13	1.65436	1.62416	1.58358	1.52385	1.69288
14	1.64697	1.61701	1.57677	1.51752	1.68519
15	1.64061	1.61087	1.57092	1.51207	1.67857
16	1.63508	1.60553	1.56583	1.50734	1.67281
17	1.63023	1.60084	1.56136	1.50318	1.66776
18	1.62594	1.59669	1.55741	1.49951	1.66329
19	1.62211	1.59300	1.55390	1.49624	1.65931
20	1.61869	1.58969	1.55074	1.49330	1.65574
21	1.61559	1.58670	1.54790	1.49066	1.65252
22	1.61279	1.58400	1.54532	1.48826	1.64960
23	1.61024	1.58154	1.54298	1.48608	1.64694
24	1.60791	1.57929	1.54084	1.48409	1.64451
25	1.60577	1.57722	1.53887	1.48226	1.64228
26	1.60380	1.57532	1.53706	1.48057	1.64023
27	1.60198	1.57356	1.53539	1.47902	1.63833
28	1.60029	1.57193	1.53384	1.47757	1.63657
29	1.59873	1.57042	1.53240	1.47623	1.63494
30	1.59727	1.56901	1.53105	1.47498	1.63341
35	1.59124	1.56319	1.52552	1.46983	1.62713
40	1.58674	1.55886	1.52139	1.46599	1.62244
45	1.58326	1.55550	1.51819	1.46302	1.61880
50	1.58049	1.55282	1.51565	1.46064	1.61591
60	1.57634	1.54883	1.51184	1.45710	1.61158
80	1.57119	1.54386	1.50711	1.45270	1.60620
100	1.56811	1.54089	1.50429	1.45007	1.60299
120	1.56606	1.53892	1.50241	1.44832	1.60085
200	1.56199	1.53499	1.49868	1.44484	1.59659
∞	1.55591	1.52913	1.49310	1.43965	1.59024

BECHHOFER and DUNNETT

TABLE B.1

Two-sided Percentage Point (h) for Equal
Correlations (*Case I with Correlation* ρ)

p = 3, 1−α = 0.80

ν ↓	ρ					
	0.0	0.1	0.2	0.3	0.4	0.5
2	2.94190	2.93614	2.91928	2.89145	2.85210	2.79989
3	2.47575	2.47116	2.45773	2.43551	2.40407	2.36235
4	2.27908	2.27501	2.26305	2.24326	2.21523	2.17803
5	2.17104	2.16725	2.15612	2.13769	2.11157	2.07691
6	2.10281	2.09920	2.08861	2.07104	2.04616	2.01313
7	2.05582	2.05234	2.04212	2.02517	2.00114	1.96926
8	2.02149	2.01811	2.00817	1.99167	1.96828	1.93725
9	1.99532	1.99202	1.98228	1.96613	1.94324	1.91286
10	1.97471	1.97146	1.96190	1.94602	1.92352	1.89367
11	1.95805	1.95485	1.94543	1.92978	1.90760	1.87817
12	1.94431	1.94115	1.93184	1.91639	1.89447	1.86540
13	1.93278	1.92966	1.92045	1.90515	1.88346	1.85470
14	1.92297	1.91988	1.91075	1.89559	1.87410	1.84559
15	1.91452	1.91146	1.90240	1.88736	1.86604	1.83775
16	1.90717	1.90413	1.89513	1.88020	1.85902	1.83093
17	1.90071	1.89769	1.88875	1.87391	1.85286	1.82495
18	1.89500	1.89199	1.88310	1.86834	1.84741	1.81965
19	1.88990	1.88691	1.87807	1.86338	1.84255	1.81493
20	1.88533	1.88235	1.87355	1.85893	1.83820	1.81070
21	1.88121	1.87824	1.86948	1.85492	1.83427	1.80688
22	1.87747	1.87452	1.86579	1.85128	1.83071	1.80342
23	1.87407	1.87112	1.86242	1.84796	1.82746	1.80027
24	1.87095	1.86802	1.85935	1.84493	1.82450	1.79739
25	1.86810	1.86517	1.85652	1.84215	1.82177	1.79475
26	1.86546	1.86254	1.85392	1.83959	1.81927	1.79232
27	1.86303	1.86012	1.85152	1.83722	1.81695	1.79007
28	1.86077	1.85787	1.84929	1.83502	1.81480	1.78798
29	1.85867	1.85577	1.84721	1.83298	1.81280	1.78604
30	1.85672	1.85383	1.84528	1.83108	1.81094	1.78424
35	1.84864	1.84578	1.83731	1.82323	1.80326	1.77678
40	1.84262	1.83977	1.83136	1.81737	1.79753	1.77123
45	1.83795	1.83512	1.82675	1.81283	1.79309	1.76692
50	1.83422	1.83141	1.82307	1.80921	1.78955	1.76349
60	1.82866	1.82585	1.81757	1.80379	1.78426	1.75836
80	1.82172	1.81895	1.81073	1.79706	1.77768	1.75198
100	1.81758	1.81482	1.80664	1.79304	1.77375	1.74817
120	1.81483	1.81207	1.80392	1.79036	1.77113	1.74564
200	1.80933	1.80660	1.79850	1.78502	1.76592	1.74059
∞	1.80113	1.79842	1.79040	1.77706	1.75814	1.73306

TABLE B.1

*Two-sided Percentage Point (h) for Equal
Correlations* $\left(Case\ I\ with\ Correlation\ \rho\right)$

$$p = 3, \quad 1-\alpha = 0.80$$

$\nu \downarrow$	0.6	0.7	0.8	0.9	$1/(1+\sqrt{3})$
			ρ		
2	2.73235	2.64492	2.52834	2.35848	2.86682
3	2.30840	2.23863	2.14579	2.01102	2.41584
4	2.12994	2.06779	1.98521	1.86553	2.22572
5	2.03211	1.97426	1.89744	1.78616	2.12135
6	1.97045	1.91536	1.84225	1.73634	2.05547
7	1.92807	1.87492	1.80440	1.70220	2.01013
8	1.89716	1.84545	1.77684	1.67737	1.97703
9	1.87363	1.82303	1.75590	1.65851	1.95181
10	1.85512	1.80541	1.73945	1.64369	1.93195
11	1.84018	1.79119	1.72618	1.63175	1.91590
12	1.82788	1.77949	1.71527	1.62193	1.90268
13	1.81756	1.76968	1.70613	1.61370	1.89158
14	1.80879	1.76134	1.69836	1.60671	1.88214
15	1.80124	1.75417	1.69168	1.60070	1.87402
16	1.79468	1.74794	1.68587	1.59548	1.86695
17	1.78892	1.74247	1.68078	1.59089	1.86074
18	1.78383	1.73763	1.67628	1.58684	1.85525
19	1.77929	1.73333	1.67227	1.58323	1.85035
20	1.77522	1.72947	1.66868	1.58000	1.84596
21	1.77155	1.72598	1.66544	1.57708	1.84200
22	1.76822	1.72283	1.66250	1.57444	1.83841
23	1.76519	1.71996	1.65983	1.57204	1.83514
24	1.76243	1.71734	1.65739	1.56984	1.83215
25	1.75989	1.71493	1.65515	1.56782	1.82940
26	1.75755	1.71271	1.65309	1.56597	1.82687
27	1.75539	1.71066	1.65118	1.56425	1.82454
28	1.75338	1.70876	1.64942	1.56266	1.82237
29	1.75152	1.70700	1.64778	1.56118	1.82036
30	1.74979	1.70535	1.64625	1.55981	1.81848
35	1.74263	1.69857	1.63994	1.55413	1.81073
40	1.73729	1.69352	1.63525	1.54990	1.80495
45	1.73316	1.68961	1.63161	1.54662	1.80048
50	1.72987	1.68649	1.62871	1.54401	1.79691
60	1.72495	1.68184	1.62438	1.54012	1.79157
80	1.71883	1.67605	1.61901	1.53527	1.78493
100	1.71518	1.67259	1.61580	1.53238	1.78097
120	1.71276	1.67030	1.61366	1.53045	1.77833
200	1.70792	1.66573	1.60941	1.52662	1.77307
∞	1.70071	1.65891	1.60308	1.52091	1.76522

BECHHOFER and DUNNETT

TABLE B.1

Two-sided Percentage Point (h) for Equal
Correlations $\left(Case\ I\ with\ Correlation\ \rho\right)$

$p = 4,\ 1-\alpha = 0.80$

ν ↓	ρ					
	0.0	0.1	0.2	0.3	0.4	0.5
2	3.21733	3.20907	3.18554	3.14762	3.09505	3.02646
3	2.69246	2.68594	2.66733	2.63725	2.59547	2.54089
4	2.47029	2.46454	2.44807	2.42141	2.38433	2.33584
5	2.34789	2.34258	2.32733	2.30261	2.26818	2.22313
6	2.27041	2.26538	2.25093	2.22746	2.19475	2.15193
7	2.21695	2.21212	2.19823	2.17564	2.14414	2.10290
8	2.17783	2.17316	2.15968	2.13775	2.10716	2.06709
9	2.14797	2.14341	2.13026	2.10884	2.07895	2.03978
10	2.12442	2.11995	2.10706	2.08605	2.05672	2.01828
11	2.10536	2.10098	2.08830	2.06763	2.03876	2.00092
12	2.08964	2.08531	2.07281	2.05242	2.02394	1.98659
13	2.07643	2.07216	2.05981	2.03966	2.01150	1.97458
14	2.06519	2.06096	2.04875	2.02880	2.00092	1.96436
15	2.05550	2.05131	2.03921	2.01944	1.99181	1.95556
16	2.04706	2.04291	2.03091	2.01130	1.98387	1.94791
17	2.03964	2.03553	2.02361	2.00414	1.97691	1.94119
18	2.03307	2.02899	2.01715	1.99780	1.97074	1.93524
19	2.02722	2.02316	2.01139	1.99216	1.96524	1.92994
20	2.02196	2.01793	2.00622	1.98709	1.96031	1.92518
21	2.01722	2.01321	2.00156	1.98252	1.95586	1.92090
22	2.01292	2.00893	1.99733	1.97837	1.95183	1.91701
23	2.00900	2.00503	1.99348	1.97459	1.94816	1.91347
24	2.00542	2.00146	1.98996	1.97114	1.94480	1.91024
25	2.00213	1.99818	1.98672	1.96797	1.94172	1.90726
26	1.99909	1.99516	1.98374	1.96505	1.93887	1.90453
27	1.99629	1.99237	1.98098	1.96234	1.93625	1.90200
28	1.99369	1.98978	1.97843	1.95984	1.93381	1.89966
29	1.99127	1.98737	1.97605	1.95751	1.93155	1.89748
30	1.98902	1.98513	1.97384	1.95534	1.92944	1.89545
35	1.97971	1.97586	1.96469	1.94638	1.92073	1.88707
40	1.97275	1.96894	1.95786	1.93969	1.91424	1.88082
45	1.96736	1.96358	1.95256	1.93451	1.90920	1.87597
50	1.96306	1.95930	1.94834	1.93037	1.90519	1.87211
60	1.95663	1.95289	1.94202	1.92419	1.89918	1.86635
80	1.94861	1.94492	1.93415	1.91649	1.89172	1.85917
100	1.94382	1.94015	1.92945	1.91189	1.88725	1.85488
120	1.94063	1.93698	1.92632	1.90883	1.88428	1.85204
200	1.93427	1.93065	1.92008	1.90272	1.87837	1.84636
∞	1.92477	1.92120	1.91076	1.89361	1.86953	1.83789

TABLE B.1

Two-sided Percentage Point (h) for Equal
Correlations $\left(Case\ I\ with\ Correlation\ \rho\right)$

$p = 4, \ 1-\alpha = 0.80$

$\nu \downarrow$	ρ				
	0.6	0.7	0.8	0.9	$1/(1+\sqrt{4})$
2	2.93896	2.82711	2.67968	2.46728	3.13177
3	2.47123	2.38221	2.26502	2.09667	2.62466
4	2.27394	2.19484	2.09078	1.94146	2.41024
5	2.16561	2.09212	1.99547	1.85680	2.29224
6	2.09725	2.02739	1.93553	1.80367	2.21761
7	2.05022	1.98291	1.89440	1.76728	2.16616
8	2.01590	1.95049	1.86447	1.74082	2.12854
9	1.98975	1.92582	1.84171	1.72072	2.09984
10	1.96918	1.90642	1.82384	1.70495	2.07723
11	1.95257	1.89077	1.80943	1.69223	2.05894
12	1.93888	1.87788	1.79757	1.68177	2.04385
13	1.92740	1.86708	1.78764	1.67302	2.03119
14	1.91764	1.85791	1.77921	1.66558	2.02041
15	1.90924	1.85001	1.77196	1.65918	2.01113
16	1.90194	1.84315	1.76565	1.65362	2.00305
17	1.89553	1.83713	1.76013	1.64875	1.99595
18	1.88985	1.83180	1.75524	1.64444	1.98966
19	1.88480	1.82706	1.75089	1.64060	1.98406
20	1.88027	1.82281	1.74699	1.63716	1.97904
21	1.87619	1.81897	1.74347	1.63406	1.97450
22	1.87248	1.81550	1.74029	1.63125	1.97039
23	1.86911	1.81234	1.73739	1.62869	1.96664
24	1.86603	1.80945	1.73474	1.62636	1.96322
25	1.86320	1.80680	1.73231	1.62421	1.96007
26	1.86060	1.80436	1.73007	1.62224	1.95718
27	1.85819	1.80211	1.72800	1.62041	1.95450
28	1.85596	1.80002	1.72609	1.61872	1.95201
29	1.85389	1.79808	1.72431	1.61715	1.94971
30	1.85196	1.79627	1.72265	1.61569	1.94755
35	1.84399	1.78881	1.71581	1.60966	1.93867
40	1.83805	1.78324	1.71072	1.60516	1.93204
45	1.83345	1.77894	1.70678	1.60168	1.92690
50	1.82978	1.77551	1.70364	1.59891	1.92280
60	1.82430	1.77039	1.69894	1.59477	1.91667
80	1.81749	1.76402	1.69311	1.58962	1.90904
100	1.81343	1.76022	1.68963	1.58654	1.90448
120	1.81073	1.75770	1.68732	1.58450	1.90145
200	1.80534	1.75267	1.68272	1.58043	1.89540
∞	1.79732	1.74518	1.67585	1.57437	1.88637

BECHHOFER and DUNNETT

TABLE B.1

Two-sided Percentage Point (h) for Equal
Correlations $\left(\text{Case } I \text{ with Correlation } \rho\right)$

p = 5, 1-α = 0.80

ν ↓	ρ					
	0.0	0.1	0.2	0.3	0.4	0.5
2	3.42772	3.41722	3.38799	3.34172	3.27848	3.19689
3	2.85823	2.84998	2.82694	2.79035	2.74019	2.67539
4	2.61650	2.60926	2.58895	2.55660	2.51218	2.45472
5	2.48302	2.47635	2.45760	2.42767	2.38652	2.33322
6	2.39834	2.39206	2.37432	2.34597	2.30694	2.25636
7	2.33982	2.33380	2.31679	2.28956	2.25204	2.20338
8	2.29694	2.29112	2.27466	2.24826	2.21186	2.16464
9	2.26415	2.25850	2.24246	2.21672	2.18119	2.13508
10	2.23827	2.23274	2.21705	2.19183	2.15701	2.11180
11	2.21732	2.21189	2.19648	2.17170	2.13745	2.09297
12	2.20000	2.19467	2.17949	2.15507	2.12131	2.07745
13	2.18546	2.18020	2.16522	2.14111	2.10776	2.06442
14	2.17306	2.16786	2.15306	2.12921	2.09622	2.05334
15	2.16237	2.15723	2.14258	2.11896	2.08628	2.04379
16	2.15306	2.14797	2.13345	2.11004	2.07763	2.03548
17	2.14487	2.13982	2.12543	2.10219	2.07003	2.02818
18	2.13762	2.13261	2.11832	2.09525	2.06330	2.02172
19	2.13114	2.12618	2.11198	2.08905	2.05729	2.01597
20	2.12533	2.12040	2.10629	2.08349	2.05191	2.01080
21	2.12009	2.11518	2.10115	2.07848	2.04705	2.00615
22	2.11533	2.11045	2.09649	2.07393	2.04265	2.00193
23	2.11100	2.10614	2.09225	2.06978	2.03863	1.99808
24	2.10703	2.10220	2.08836	2.06599	2.03496	1.99457
25	2.10339	2.09858	2.08479	2.06251	2.03159	1.99134
26	2.10003	2.09524	2.08151	2.05930	2.02849	1.98836
27	2.09692	2.09215	2.07847	2.05633	2.02562	1.98562
28	2.09404	2.08928	2.07565	2.05358	2.02296	1.98307
29	2.09136	2.08662	2.07303	2.05102	2.02048	1.98070
30	2.08886	2.08414	2.07058	2.04863	2.01818	1.97849
35	2.07854	2.07388	2.06048	2.03878	2.00865	1.96939
40	2.07083	2.06621	2.05294	2.03143	2.00155	1.96259
45	2.06484	2.06026	2.04709	2.02573	1.99604	1.95733
50	2.06007	2.05551	2.04242	2.02118	1.99165	1.95313
60	2.05292	2.04841	2.03544	2.01437	1.98508	1.94686
80	2.04401	2.03956	2.02673	2.00589	1.97690	1.93905
100	2.03868	2.03426	2.02153	2.00083	1.97201	1.93439
120	2.03513	2.03074	2.01807	1.99745	1.96876	1.93129
200	2.02805	2.02370	2.01115	1.99073	1.96228	1.92512
∞	2.01746	2.01319	2.00083	1.98068	1.95261	1.91590

TABLE B.1

Two-sided Percentage Point (h) for Equal
Correlations $\left(Case\ I\ with\ Correlation\ \rho \right)$

$p = 5, \quad 1-\alpha = 0.80$

$\nu \downarrow$	ρ 0.6	0.7	0.8	0.9	$1/(1+\sqrt{5})$
2	3.09382	2.96315	2.79224	2.54781	3.33673
3	2.59345	2.48957	2.35380	2.16011	2.78639
4	2.38202	2.28982	2.16937	1.99768	2.55310
5	2.26576	2.18019	2.06841	1.90906	2.42443
6	2.19230	2.11104	2.00487	1.85345	2.34290
7	2.14172	2.06349	1.96127	1.81537	2.28661
8	2.10478	2.02882	1.92953	1.78768	2.24540
9	2.07663	2.00243	1.90540	1.76666	2.21392
10	2.05446	1.98167	1.88644	1.75015	2.18909
11	2.03656	1.96492	1.87117	1.73686	2.16900
12	2.02180	1.95112	1.85859	1.72592	2.15241
13	2.00943	1.93957	1.84806	1.71676	2.13848
14	1.99891	1.92974	1.83912	1.70899	2.12662
15	1.98985	1.92129	1.83143	1.70230	2.11639
16	1.98197	1.91394	1.82475	1.69649	2.10749
17	1.97505	1.90750	1.81889	1.69139	2.09967
18	1.96893	1.90179	1.81371	1.68689	2.09274
19	1.96348	1.89672	1.80909	1.68288	2.08656
20	1.95859	1.89216	1.80496	1.67928	2.08101
21	1.95418	1.88806	1.80123	1.67604	2.07601
22	1.95019	1.88435	1.79786	1.67311	2.07147
23	1.94655	1.88096	1.79479	1.67044	2.06734
24	1.94323	1.87787	1.79198	1.66800	2.06355
25	1.94017	1.87503	1.78940	1.66576	2.06008
26	1.93736	1.87242	1.78703	1.66369	2.05688
27	1.93477	1.87000	1.78484	1.66179	2.05392
28	1.93236	1.86777	1.78281	1.66002	2.05117
29	1.93012	1.86569	1.78092	1.65838	2.04862
30	1.92804	1.86375	1.77917	1.65685	2.04624
35	1.91944	1.85577	1.77192	1.65055	2.03642
40	1.91302	1.84981	1.76653	1.64586	2.02908
45	1.90806	1.84521	1.76235	1.64222	2.02340
50	1.90410	1.84154	1.75902	1.63932	2.01886
60	1.89819	1.83606	1.75405	1.63500	2.01207
80	1.89084	1.82924	1.74788	1.62962	2.00362
100	1.88645	1.82518	1.74420	1.62641	1.99857
120	1.88353	1.82248	1.74175	1.62428	1.99520
200	1.87772	1.81710	1.73687	1.62003	1.98850
∞	1.86906	1.80908	1.72960	1.61370	1.97848

BECHHOFER and DUNNETT

TABLE B.1

Two-sided Percentage Point (h) for Equal
Correlations ⎡*Case I with Correlation* ρ⎤

p = 6, 1−α = 0.80

	ρ					
ν ↓	0.0	0.1	0.2	0.3	0.4	0.5
2	3.59697	3.58446	3.55029	3.49695	3.42481	3.33253
3	2.99182	2.98204	2.95514	2.91300	2.85586	2.78262
4	2.73439	2.72582	2.70215	2.66496	2.61441	2.54954
5	2.59195	2.58407	2.56227	2.52791	2.48113	2.42102
6	2.50144	2.49403	2.47344	2.44094	2.39662	2.33962
7	2.43879	2.43171	2.41199	2.38081	2.33824	2.28345
8	2.39282	2.38599	2.36693	2.33674	2.29549	2.24235
9	2.35764	2.35101	2.33246	2.30305	2.26282	2.21098
10	2.32984	2.32337	2.30524	2.27645	2.23705	2.18624
11	2.30732	2.30097	2.28319	2.25492	2.21619	2.16624
12	2.28869	2.28245	2.26496	2.23712	2.19897	2.14973
13	2.27303	2.26689	2.24964	2.22217	2.18451	2.13588
14	2.25967	2.25361	2.23658	2.20943	2.17219	2.12409
15	2.24815	2.24216	2.22532	2.19845	2.16157	2.11393
16	2.23811	2.23218	2.21550	2.18888	2.15232	2.10508
17	2.22927	2.22341	2.20687	2.18047	2.14420	2.09732
18	2.22144	2.21563	2.19923	2.17302	2.13700	2.09044
19	2.21446	2.20869	2.19240	2.16637	2.13059	2.08431
20	2.20818	2.20246	2.18628	2.16040	2.12483	2.07881
21	2.20252	2.19683	2.18075	2.15502	2.11963	2.07385
22	2.19737	2.19172	2.17573	2.15013	2.11492	2.06936
23	2.19269	2.18707	2.17116	2.14568	2.11063	2.06526
24	2.18840	2.18280	2.16697	2.14161	2.10670	2.06151
25	2.18446	2.17889	2.16313	2.13787	2.10309	2.05808
26	2.18082	2.17528	2.15958	2.13442	2.09977	2.05491
27	2.17746	2.17194	2.15630	2.13123	2.09670	2.05198
28	2.17434	2.16885	2.15326	2.12827	2.09385	2.04926
29	2.17144	2.16597	2.15044	2.12552	2.09120	2.04674
30	2.16874	2.16328	2.14780	2.12296	2.08873	2.04439
35	2.15756	2.15218	2.13691	2.11237	2.07853	2.03468
40	2.14920	2.14388	2.12876	2.10446	2.07092	2.02744
45	2.14271	2.13744	2.12245	2.09832	2.06502	2.02183
50	2.13753	2.13230	2.11740	2.09342	2.06031	2.01735
60	2.12977	2.12460	2.10985	2.08610	2.05327	2.01066
80	2.12010	2.11500	2.10045	2.07697	2.04451	2.00234
100	2.11431	2.10926	2.09482	2.07151	2.03927	1.99737
120	2.11045	2.10543	2.09107	2.06788	2.03578	1.99407
200	2.10275	2.09779	2.08359	2.06063	2.02883	1.98748
∞	2.09123	2.08637	2.07240	2.04980	2.01845	1.97765

TABLE B.1

Two-sided Percentage Point (h) for Equal
Correlations (*Case I with Correlation* ρ)

$p = 6,\ 1-\alpha = 0.80$

$\nu \downarrow$	0.6	0.7	0.8	0.9	$1/(1+\sqrt{6})$
2	3.21677	3.07091	2.88117	2.61125	3.50319
3	2.69065	2.57473	2.42404	2.21013	2.91794
4	2.46800	2.36519	2.23155	2.04200	2.66932
5	2.34542	2.25005	2.12610	1.95025	2.53194
6	2.26789	2.17738	2.05970	1.89267	2.44475
7	2.21446	2.12738	2.01412	1.85323	2.38447
8	2.17541	2.09090	1.98093	1.82457	2.34029
9	2.14564	2.06312	1.95570	1.80280	2.30651
10	2.12219	2.04126	1.93588	1.78572	2.27984
11	2.10324	2.02363	1.91990	1.77196	2.25824
12	2.08762	2.00910	1.90675	1.76064	2.24040
13	2.07452	1.99693	1.89574	1.75116	2.22540
14	2.06337	1.98658	1.88639	1.74311	2.21263
15	2.05378	1.97768	1.87835	1.73620	2.20161
16	2.04543	1.96994	1.87136	1.73019	2.19202
17	2.03810	1.96315	1.86523	1.72491	2.18358
18	2.03162	1.95714	1.85981	1.72025	2.17610
19	2.02584	1.95179	1.85499	1.71610	2.16944
20	2.02066	1.94700	1.85066	1.71239	2.16345
21	2.01599	1.94267	1.84677	1.70903	2.15805
22	2.01175	1.93876	1.84324	1.70600	2.15315
23	2.00790	1.93519	1.84003	1.70324	2.14868
24	2.00437	1.93194	1.83709	1.70071	2.14460
25	2.00114	1.92895	1.83440	1.69839	2.14084
26	1.99816	1.92619	1.83192	1.69626	2.13739
27	1.99541	1.92365	1.82963	1.69429	2.13419
28	1.99286	1.92129	1.82751	1.69247	2.13122
29	1.99048	1.91910	1.82554	1.69077	2.12846
30	1.98827	1.91706	1.82370	1.68919	2.12589
35	1.97916	1.90865	1.81613	1.68267	2.11526
40	1.97236	1.90238	1.81049	1.67782	2.10733
45	1.96710	1.89753	1.80612	1.67406	2.10117
50	1.96290	1.89366	1.80265	1.67106	2.09625
60	1.95663	1.88789	1.79745	1.66659	2.08890
80	1.94884	1.88072	1.79100	1.66103	2.07974
100	1.94419	1.87644	1.78715	1.65772	2.07426
120	1.94110	1.87359	1.78459	1.65551	2.07062
200	1.93494	1.86793	1.77950	1.65112	2.06334
∞	1.92576	1.85949	1.77191	1.64458	2.05247

BECHHOFER and DUNNETT

TABLE B.1

Two-sided Percentage Point (h) for Equal
Correlations (*Case I with Correlation* ρ)

p = 7, 1-α = 0.80

ν ↓	ρ					
	0.0	0.1	0.2	0.3	0.4	0.5
2	3.73799	3.72368	3.68515	3.62570	3.54597	3.44465
3	3.10333	3.09215	3.06186	3.01492	2.95178	2.87140
4	2.83286	2.82307	2.79645	2.75505	2.69925	2.62808
5	2.68295	2.67397	2.64947	2.61126	2.55965	2.49375
6	2.58755	2.57912	2.55601	2.51990	2.47104	2.40859
7	2.52144	2.51339	2.49128	2.45667	2.40977	2.34977
8	2.47287	2.46512	2.44377	2.41028	2.36486	2.30670
9	2.43567	2.42815	2.40740	2.37479	2.33052	2.27380
10	2.40625	2.39891	2.37864	2.34675	2.30341	2.24785
11	2.38238	2.37521	2.35534	2.32404	2.28147	2.22686
12	2.36264	2.35560	2.33606	2.30526	2.26334	2.20953
13	2.34603	2.33910	2.31985	2.28947	2.24810	2.19498
14	2.33186	2.32502	2.30603	2.27602	2.23512	2.18259
15	2.31962	2.31288	2.29410	2.26441	2.22393	2.17192
16	2.30895	2.30229	2.28370	2.25430	2.21418	2.16263
17	2.29957	2.29297	2.27455	2.24540	2.20562	2.15446
18	2.29124	2.28471	2.26645	2.23752	2.19803	2.14723
19	2.28381	2.27733	2.25921	2.23049	2.19126	2.14079
20	2.27714	2.27071	2.25271	2.22418	2.18518	2.13501
21	2.27111	2.26472	2.24685	2.21848	2.17970	2.12979
22	2.26564	2.25929	2.24152	2.21331	2.17473	2.12506
23	2.26065	2.25434	2.23667	2.20859	2.17020	2.12075
24	2.25608	2.24981	2.23222	2.20428	2.16605	2.11681
25	2.25188	2.24564	2.22814	2.20032	2.16224	2.11320
26	2.24801	2.24180	2.22438	2.19666	2.15874	2.10986
27	2.24443	2.23825	2.22089	2.19329	2.15549	2.10678
28	2.24111	2.23495	2.21766	2.19015	2.15248	2.10393
29	2.23801	2.23189	2.21466	2.18724	2.14969	2.10127
30	2.23513	2.22903	2.21186	2.18452	2.14708	2.09880
35	2.22321	2.21720	2.20028	2.17330	2.13631	2.08858
40	2.21429	2.20835	2.19162	2.16491	2.12826	2.08096
45	2.20736	2.20149	2.18490	2.15840	2.12203	2.07505
50	2.20182	2.19600	2.17953	2.15321	2.11705	2.07034
60	2.19354	2.18778	2.17149	2.14543	2.10961	2.06330
80	2.18319	2.17753	2.16147	2.13574	2.10034	2.05454
100	2.17700	2.17139	2.15548	2.12995	2.09480	2.04930
120	2.17287	2.16730	2.15148	2.12609	2.09111	2.04582
200	2.16463	2.15914	2.14351	2.11839	2.08376	2.03889
∞	2.15229	2.14691	2.13158	2.10688	2.07278	2.02854

TABLE B.1

Two-sided Percentage Point (h) for Equal
Correlations $\left(Case\ I\ with\ Correlation\ \rho\right)$

$p = 7,\ 1-\alpha = 0.80$

$\nu \downarrow$	ρ				
	0.6	0.7	0.8	0.9	$1/(1+\sqrt{7})$
2	3.31824	3.15969	2.95431	2.66333	3.64291
3	2.77099	2.64499	2.48187	2.25123	3.02852
4	2.53909	2.42739	2.28276	2.07841	2.76706
5	2.41129	2.30771	2.17360	1.98407	2.62235
6	2.33038	2.23212	2.10484	1.92486	2.53039
7	2.27458	2.18008	2.05763	1.88431	2.46673
8	2.23378	2.14209	2.02324	1.85483	2.42002
9	2.20266	2.11316	1.99708	1.83245	2.38428
10	2.17813	2.09039	1.97654	1.81489	2.35603
11	2.15831	2.07202	1.95998	1.80074	2.33315
12	2.14196	2.05688	1.94635	1.78911	2.31423
13	2.12825	2.04419	1.93493	1.77937	2.29832
14	2.11658	2.03340	1.92524	1.77109	2.28476
15	2.10654	2.02412	1.91691	1.76398	2.27306
16	2.09780	2.01605	1.90966	1.75781	2.26287
17	2.09012	2.00897	1.90331	1.75239	2.25390
18	2.08333	2.00271	1.89769	1.74760	2.24596
19	2.07728	1.99713	1.89269	1.74333	2.23887
20	2.07185	1.99213	1.88821	1.73951	2.23250
21	2.06696	1.98763	1.88418	1.73607	2.22675
22	2.06253	1.98355	1.88052	1.73295	2.22154
23	2.05849	1.97983	1.87719	1.73011	2.21678
24	2.05479	1.97643	1.87415	1.72752	2.21243
25	2.05141	1.97332	1.87136	1.72514	2.20844
26	2.04828	1.97045	1.86879	1.72295	2.20475
27	2.04540	1.96780	1.86642	1.72092	2.20135
28	2.04272	1.96534	1.86422	1.71905	2.19819
29	2.04024	1.96306	1.86218	1.71730	2.19525
30	2.03792	1.96093	1.86027	1.71568	2.19251
35	2.02837	1.95216	1.85243	1.70898	2.18118
40	2.02125	1.94562	1.84658	1.70400	2.17271
45	2.01573	1.94056	1.84206	1.70014	2.16614
50	2.01134	1.93653	1.83846	1.69706	2.16090
60	2.00477	1.93052	1.83308	1.69247	2.15305
80	1.99660	1.92304	1.82639	1.68676	2.14327
100	1.99173	1.91858	1.82241	1.68335	2.13742
120	1.98849	1.91561	1.81976	1.68108	2.13352
200	1.98203	1.90971	1.81448	1.67657	2.12574
∞	1.97242	1.90091	1.80662	1.66985	2.11412

TABLE B.1

Two-sided Percentage Point (h) for Equal
Correlations   (*Case I with Correlation* ρ)

p = 8, 1-α = 0.80

ν ↓	ρ					
	0.0	0.1	0.2	0.3	0.4	0.5
2	3.85852	3.84257	3.80016	3.73534	3.64900	3.53988
3	3.19879	3.18633	3.15300	3.10183	3.03348	2.94690
4	2.91720	2.90631	2.87704	2.83194	2.77154	2.69492
5	2.76092	2.75094	2.72402	2.68241	2.62658	2.55566
6	2.66134	2.65197	2.62660	2.58730	2.53448	2.46729
7	2.59225	2.58332	2.55907	2.52142	2.47074	2.40621
8	2.54145	2.53286	2.50946	2.47305	2.42398	2.36146
9	2.50251	2.49418	2.47144	2.43602	2.38821	2.32726
10	2.47168	2.46357	2.44137	2.40674	2.35996	2.30027
11	2.44666	2.43873	2.41698	2.38301	2.33707	2.27843
12	2.42595	2.41817	2.39680	2.36338	2.31816	2.26039
13	2.40851	2.40086	2.37982	2.34687	2.30226	2.24524
14	2.39363	2.38609	2.36533	2.33280	2.28871	2.23234
15	2.38077	2.37334	2.35283	2.32065	2.27703	2.22122
16	2.36956	2.36221	2.34192	2.31006	2.26685	2.21154
17	2.35969	2.35242	2.33233	2.30075	2.25790	2.20303
18	2.35093	2.34374	2.32382	2.29250	2.24997	2.19550
19	2.34312	2.33598	2.31623	2.28513	2.24289	2.18878
20	2.33609	2.32902	2.30940	2.27852	2.23654	2.18276
21	2.32974	2.32272	2.30324	2.27254	2.23081	2.17732
22	2.32398	2.31701	2.29765	2.26712	2.22561	2.17239
23	2.31873	2.31180	2.29255	2.26218	2.22088	2.16790
24	2.31391	2.30703	2.28788	2.25766	2.21654	2.16379
25	2.30949	2.30264	2.28359	2.25351	2.21256	2.16001
26	2.30541	2.29860	2.27963	2.24968	2.20889	2.15654
27	2.30163	2.29485	2.27597	2.24613	2.20549	2.15333
28	2.29813	2.29138	2.27258	2.24285	2.20235	2.15035
29	2.29487	2.28815	2.26942	2.23979	2.19942	2.14758
30	2.29183	2.28514	2.26647	2.23694	2.19669	2.14500
35	2.27925	2.27267	2.25429	2.22516	2.18542	2.13434
40	2.26983	2.26334	2.24517	2.21636	2.17700	2.12639
45	2.26251	2.25610	2.23810	2.20952	2.17047	2.12022
50	2.25667	2.25030	2.23245	2.20407	2.16526	2.11531
60	2.24791	2.24163	2.22398	2.19590	2.15747	2.10796
80	2.23697	2.23080	2.21342	2.18572	2.14776	2.09882
100	2.23041	2.22431	2.20710	2.17963	2.14196	2.09335
120	2.22605	2.21999	2.20289	2.17557	2.13809	2.08972
200	2.21732	2.21136	2.19448	2.16748	2.13039	2.08248
∞	2.20424	2.19842	2.18189	2.15537	2.11889	2.07168

TABLE B.1

Two-sided Percentage Point (h) for Equal
Correlations $\left(\text{Case } I \text{ with Correlation } \rho\right)$

p = 8, 1-α = 0.80

ν ↓	ρ				
	0.6	0.7	0.8	0.9	1/(1+√8)
2	3.40431	3.23491	3.01620	2.70731	3.76304
3	2.83922	2.70460	2.53086	2.28598	3.12372
4	2.59951	2.48017	2.32616	2.10920	2.85124
5	2.46727	2.35665	2.21386	2.01267	2.70023
6	2.38350	2.27857	2.14309	1.95207	2.60415
7	2.32568	2.22480	2.09448	1.91058	2.53757
8	2.28338	2.18553	2.05906	1.88041	2.48867
9	2.25110	2.15561	2.03212	1.85751	2.45122
10	2.22565	2.13206	2.01096	1.83953	2.42161
11	2.20508	2.11305	1.99390	1.82506	2.39760
12	2.18811	2.09738	1.97986	1.81315	2.37774
13	2.17387	2.08425	1.96810	1.80318	2.36103
14	2.16176	2.07309	1.95812	1.79472	2.34678
15	2.15132	2.06348	1.94953	1.78745	2.33448
16	2.14224	2.05513	1.94207	1.78113	2.32376
17	2.13427	2.04781	1.93552	1.77558	2.31433
18	2.12722	2.04132	1.92974	1.77068	2.30597
19	2.12093	2.03555	1.92459	1.76632	2.29851
20	2.11529	2.03038	1.91997	1.76241	2.29181
21	2.11021	2.02571	1.91582	1.75889	2.28575
22	2.10560	2.02149	1.91205	1.75570	2.28026
23	2.10140	2.01764	1.90862	1.75280	2.27525
24	2.09756	2.01412	1.90549	1.75014	2.27067
25	2.09404	2.01090	1.90261	1.74771	2.26646
26	2.09080	2.00793	1.89997	1.74547	2.26257
27	2.08780	2.00518	1.89752	1.74340	2.25898
28	2.08502	2.00264	1.89526	1.74148	2.25565
29	2.08244	2.00028	1.89315	1.73970	2.25255
30	2.08003	1.99807	1.89119	1.73804	2.24966
35	2.07010	1.98899	1.88311	1.73119	2.23771
40	2.06270	1.98223	1.87709	1.72609	2.22878
45	2.05696	1.97699	1.87244	1.72214	2.22184
50	2.05239	1.97282	1.86873	1.71899	2.21630
60	2.04556	1.96659	1.86319	1.71430	2.20801
80	2.03708	1.95885	1.85630	1.70846	2.19767
100	2.03201	1.95424	1.85220	1.70497	2.19148
120	2.02864	1.95117	1.84947	1.70266	2.18736
200	2.02193	1.94506	1.84403	1.69804	2.17913
∞	2.01193	1.93595	1.83594	1.69118	2.16683

TABLE B.1

Two-sided Percentage Point (h) for Equal
Correlations (Case I with Correlation ρ)

p = 9, 1-α = 0.80

ν ↓	ρ					
	0.0	0.1	0.2	0.3	0.4	0.5
2	3.96353	3.94607	3.90017	3.83057	3.73841	3.62242
3	3.28208	3.26844	3.23237	3.17743	3.10445	3.01243
4	2.99084	2.97893	2.94726	2.89885	2.83439	2.75296
5	2.82901	2.81811	2.78900	2.74436	2.68479	2.60943
6	2.72579	2.71557	2.68815	2.64600	2.58965	2.51828
7	2.65411	2.64437	2.61817	2.57780	2.52376	2.45524
8	2.60135	2.59199	2.56672	2.52770	2.47540	2.40902
9	2.56088	2.55180	2.52727	2.48932	2.43838	2.37368
10	2.52882	2.51998	2.49605	2.45896	2.40912	2.34579
11	2.50278	2.49415	2.47071	2.43434	2.38542	2.32320
12	2.48121	2.47275	2.44973	2.41396	2.36581	2.30454
13	2.46305	2.45474	2.43207	2.39682	2.34933	2.28887
14	2.44754	2.43935	2.41700	2.38220	2.33529	2.27552
15	2.43414	2.42606	2.40399	2.36958	2.32317	2.26401
16	2.42244	2.41447	2.39264	2.35858	2.31261	2.25399
17	2.41214	2.40426	2.38265	2.34890	2.30332	2.24518
18	2.40300	2.39520	2.37379	2.34032	2.29509	2.23738
19	2.39484	2.38711	2.36587	2.33266	2.28775	2.23042
20	2.38750	2.37984	2.35876	2.32578	2.28116	2.22418
21	2.38087	2.37327	2.35234	2.31956	2.27520	2.21855
22	2.37485	2.36730	2.34651	2.31392	2.26981	2.21344
23	2.36936	2.36186	2.34119	2.30878	2.26489	2.20879
24	2.36433	2.35688	2.33632	2.30407	2.26038	2.20453
25	2.35970	2.35230	2.33185	2.29975	2.25625	2.20062
26	2.35544	2.34807	2.32772	2.29576	2.25243	2.19702
27	2.35149	2.34416	2.32390	2.29207	2.24891	2.19369
28	2.34782	2.34053	2.32036	2.28865	2.24564	2.19060
29	2.34441	2.33715	2.31706	2.28547	2.24260	2.18773
30	2.34123	2.33400	2.31399	2.28250	2.23976	2.18506
35	2.32806	2.32097	2.30127	2.27023	2.22805	2.17401
40	2.31820	2.31121	2.29175	2.26105	2.21930	2.16577
45	2.31053	2.30362	2.28436	2.25393	2.21251	2.15938
50	2.30441	2.29756	2.27845	2.24824	2.20709	2.15428
60	2.29522	2.28847	2.26960	2.23972	2.19899	2.14666
80	2.28375	2.27712	2.25856	2.22910	2.18889	2.13718
100	2.27687	2.27032	2.25194	2.22275	2.18285	2.13152
120	2.27228	2.26579	2.24754	2.21852	2.17883	2.12775
200	2.26312	2.25673	2.23874	2.21007	2.17082	2.12025
∞	2.24937	2.24315	2.22555	2.19743	2.15885	2.10905

TABLE B.1

Two-sided Percentage Point (h) for Equal
Correlations $\left[Case\ I\ with\ Correlation\ \rho \right]$

p = 9, 1-α = 0.80

ν ↓	ρ				
	0.6	0.7	0.8	0.9	$1/(1+\sqrt{9})$
2	3.47885	3.30000	3.06971	2.74530	3.86820
3	2.89838	2.75622	2.57325	2.31601	3.20716
4	2.65192	2.52591	2.36371	2.13580	2.92507
5	2.51584	2.39906	2.24870	2.03738	2.76854
6	2.42958	2.31883	2.17619	1.97559	2.66885
7	2.37001	2.26354	2.12636	1.93326	2.59970
8	2.32641	2.22316	2.09005	1.90250	2.54887
9	2.29312	2.19238	2.06244	1.87914	2.50992
10	2.26687	2.16815	2.04074	1.86081	2.47909
11	2.24564	2.14858	2.02324	1.84604	2.45409
12	2.22813	2.13246	2.00884	1.83390	2.43339
13	2.21343	2.11894	1.99678	1.82374	2.41597
14	2.20092	2.10745	1.98654	1.81511	2.40111
15	2.19015	2.09756	1.97773	1.80769	2.38828
16	2.18077	2.08896	1.97008	1.80125	2.37709
17	2.17254	2.08142	1.96337	1.79560	2.36725
18	2.16525	2.07475	1.95744	1.79060	2.35852
19	2.15876	2.06880	1.95216	1.78616	2.35072
20	2.15293	2.06347	1.94742	1.78217	2.34372
21	2.14768	2.05867	1.94316	1.77858	2.33739
22	2.14292	2.05432	1.93929	1.77533	2.33165
23	2.13858	2.05036	1.93578	1.77237	2.32641
24	2.13462	2.04674	1.93257	1.76966	2.32162
25	2.13098	2.04341	1.92962	1.76718	2.31722
26	2.12762	2.04035	1.92691	1.76490	2.31315
27	2.12453	2.03753	1.92440	1.76279	2.30940
28	2.12165	2.03491	1.92208	1.76083	2.30591
29	2.11899	2.03248	1.91992	1.75901	2.30267
30	2.11650	2.03021	1.91791	1.75732	2.29964
35	2.10624	2.02086	1.90962	1.75034	2.28713
40	2.09859	2.01389	1.90345	1.74514	2.27777
45	2.09266	2.00850	1.89868	1.74112	2.27051
50	2.08794	2.00421	1.89487	1.73791	2.26470
60	2.08088	1.99779	1.88919	1.73312	2.25601
80	2.07211	1.98982	1.88213	1.72717	2.24517
100	2.06687	1.98507	1.87792	1.72362	2.23867
120	2.06339	1.98191	1.87513	1.72126	2.23435
200	2.05646	1.97562	1.86956	1.71656	2.22571
∞	2.04613	1.96624	1.86126	1.70956	2.21279

TABLE B.1

Two-sided Percentage Point (h) for Equal
Correlations   [*Case I with Correlation* ρ]

$$p = 10, \ 1-\alpha = 0.80$$

ν ↓	ρ 0.0	0.1	0.2	0.3	0.4	0.5
2	4.05641	4.03755	3.98848	3.91459	3.81721	3.69513
3	3.35583	3.34111	3.30254	3.24420	3.16709	3.07021
4	3.05610	3.04325	3.00939	2.95799	2.88988	2.80416
5	2.88937	2.87762	2.84651	2.79912	2.73619	2.65688
6	2.78293	2.77191	2.74262	2.69789	2.63838	2.56328
7	2.70895	2.69846	2.67048	2.62766	2.57059	2.49850
8	2.65446	2.64438	2.61741	2.57603	2.52081	2.45099
9	2.61263	2.60286	2.57669	2.53645	2.48269	2.41465
10	2.57947	2.56997	2.54444	2.50513	2.45254	2.38595
11	2.55253	2.54325	2.51826	2.47971	2.42811	2.36270
12	2.53020	2.52111	2.49657	2.45868	2.40790	2.34349
13	2.51139	2.50246	2.47831	2.44098	2.39090	2.32736
14	2.49531	2.48653	2.46272	2.42587	2.37641	2.31361
15	2.48142	2.47276	2.44925	2.41283	2.36390	2.30175
16	2.46929	2.46074	2.43750	2.40146	2.35300	2.29142
17	2.45861	2.45016	2.42716	2.39145	2.34342	2.28235
18	2.44913	2.44076	2.41798	2.38257	2.33492	2.27431
19	2.44066	2.43237	2.40978	2.37465	2.32734	2.26714
20	2.43304	2.42483	2.40242	2.36753	2.32053	2.26071
21	2.42615	2.41801	2.39576	2.36110	2.31438	2.25490
22	2.41990	2.41182	2.38972	2.35526	2.30880	2.24963
23	2.41419	2.40617	2.38420	2.34994	2.30372	2.24484
24	2.40897	2.40100	2.37915	2.34507	2.29907	2.24045
25	2.40416	2.39624	2.37451	2.34059	2.29479	2.23642
26	2.39973	2.39185	2.37023	2.33646	2.29085	2.23270
27	2.39562	2.38779	2.36627	2.33264	2.28721	2.22927
28	2.39181	2.38402	2.36260	2.32910	2.28383	2.22609
29	2.38826	2.38051	2.35918	2.32580	2.28069	2.22313
30	2.38495	2.37723	2.35598	2.32273	2.27776	2.22037
35	2.37126	2.36368	2.34278	2.31001	2.26565	2.20897
40	2.36099	2.35353	2.33290	2.30050	2.25660	2.20047
45	2.35301	2.34564	2.32522	2.29312	2.24958	2.19388
50	2.34662	2.33933	2.31908	2.28722	2.24397	2.18862
60	2.33705	2.32987	2.30989	2.27839	2.23559	2.18076
80	2.32509	2.31805	2.29841	2.26738	2.22514	2.17098
100	2.31791	2.31096	2.29153	2.26078	2.21890	2.16513
120	2.31313	2.30624	2.28695	2.25639	2.21474	2.16124
200	2.30356	2.29679	2.27779	2.24762	2.20644	2.15350
∞	2.28921	2.28263	2.26407	2.23451	2.19405	2.14194

TABLE B.1

Two-sided Percentage Point (h) for Equal Correlations (*Case I with Correlation* ρ)

p = 10, 1-α = 0.80

ν ↓	ρ				
	0.6	0.7	0.8	0.9	1/(1+√10)
2	3.54446	3.35724	3.11672	2.77866	3.96158
3	2.95050	2.80166	2.61053	2.34239	3.28133
4	2.69811	2.56619	2.39675	2.15918	2.99072
5	2.55866	2.43641	2.27935	2.05909	2.82931
6	2.47021	2.35428	2.20530	1.99624	2.72640
7	2.40909	2.29767	2.15441	1.95319	2.65496
8	2.36434	2.25630	2.11731	1.92189	2.60242
9	2.33015	2.22476	2.08909	1.89813	2.56212
10	2.30319	2.19992	2.06692	1.87949	2.53021
11	2.28139	2.17987	2.04904	1.86447	2.50432
12	2.26339	2.16334	2.03432	1.85212	2.48287
13	2.24829	2.14948	2.02200	1.84179	2.46482
14	2.23543	2.13770	2.01153	1.83301	2.44941
15	2.22435	2.12756	2.00253	1.82547	2.43610
16	2.21471	2.11874	1.99471	1.81891	2.42448
17	2.20625	2.11100	1.98785	1.81317	2.41426
18	2.19875	2.10415	1.98178	1.80809	2.40520
19	2.19207	2.09806	1.97638	1.80357	2.39710
20	2.18608	2.09259	1.97155	1.79951	2.38983
21	2.18068	2.08767	1.96719	1.79586	2.38325
22	2.17578	2.08320	1.96324	1.79255	2.37729
23	2.17132	2.07914	1.95964	1.78954	2.37185
24	2.16725	2.07543	1.95636	1.78679	2.36686
25	2.16350	2.07202	1.95335	1.78427	2.36228
26	2.16005	2.06888	1.95058	1.78195	2.35806
27	2.15687	2.06598	1.94801	1.77980	2.35415
28	2.15391	2.06329	1.94564	1.77781	2.35052
29	2.15117	2.06080	1.94343	1.77596	2.34715
30	2.14861	2.05847	1.94138	1.77424	2.34400
35	2.13805	2.04888	1.93291	1.76715	2.33098
40	2.13018	2.04174	1.92661	1.76186	2.32124
45	2.12408	2.03621	1.92173	1.75777	2.31367
50	2.11922	2.03180	1.91784	1.75451	2.30762
60	2.11196	2.02522	1.91203	1.74964	2.29856
80	2.10294	2.01705	1.90482	1.74359	2.28726
100	2.09755	2.01217	1.90052	1.73998	2.28049
120	2.09397	2.00893	1.89766	1.73758	2.27597
200	2.08684	2.00248	1.89197	1.73280	2.26696
∞	2.07621	1.99287	1.88349	1.72569	2.25347

TABLE B.1

Two-sided Percentage Point (h) for Equal
Correlations $\left(\text{Case I with Correlation } \rho\right)$

p = 11, 1–α = 0.80

ν ↓	ρ					
	0.0	0.1	0.2	0.3	0.4	0.5
2	4.13956	4.11941	4.06743	3.98964	3.88756	3.75998
3	3.42193	3.40620	3.36534	3.30391	3.22306	3.12180
4	3.11462	3.10089	3.06503	3.01090	2.93949	2.84990
5	2.94352	2.93097	2.89802	2.84813	2.78216	2.69927
6	2.83420	2.82244	2.79143	2.74434	2.68196	2.60349
7	2.75816	2.74696	2.71736	2.67229	2.61248	2.53717
8	2.70212	2.69136	2.66282	2.61929	2.56143	2.48850
9	2.65906	2.64865	2.62096	2.57864	2.52232	2.45126
10	2.62492	2.61480	2.58780	2.54646	2.49138	2.42183
11	2.59716	2.58728	2.56086	2.52033	2.46629	2.39800
12	2.57415	2.56447	2.53854	2.49870	2.44553	2.37830
13	2.55475	2.54525	2.51973	2.48050	2.42807	2.36174
14	2.53817	2.52882	2.50367	2.46495	2.41318	2.34763
15	2.52383	2.51462	2.48980	2.45153	2.40032	2.33546
16	2.51131	2.50222	2.47769	2.43982	2.38912	2.32486
17	2.50028	2.49130	2.46702	2.42952	2.37926	2.31555
18	2.49049	2.48160	2.45756	2.42038	2.37052	2.30729
19	2.48174	2.47294	2.44910	2.41221	2.36272	2.29993
20	2.47387	2.46515	2.44150	2.40488	2.35572	2.29332
21	2.46675	2.45811	2.43464	2.39826	2.34940	2.28736
22	2.46029	2.45171	2.42840	2.39224	2.34366	2.28195
23	2.45439	2.44587	2.42271	2.38676	2.33843	2.27703
24	2.44898	2.44053	2.41750	2.38173	2.33364	2.27252
25	2.44401	2.43561	2.41271	2.37712	2.32924	2.26838
26	2.43942	2.43107	2.40829	2.37286	2.32519	2.26456
27	2.43517	2.42687	2.40420	2.36892	2.32144	2.26103
28	2.43123	2.42297	2.40040	2.36527	2.31796	2.25776
29	2.42756	2.41934	2.39687	2.36187	2.31472	2.25472
30	2.42414	2.41595	2.39357	2.35870	2.31171	2.25189
35	2.40995	2.40193	2.37993	2.34558	2.29924	2.24018
40	2.39932	2.39142	2.36972	2.33577	2.28992	2.23144
45	2.39104	2.38325	2.36178	2.32815	2.28269	2.22466
50	2.38443	2.37672	2.35543	2.32207	2.27691	2.21926
60	2.37450	2.36692	2.34592	2.31295	2.26828	2.21118
80	2.36209	2.35467	2.33404	2.30158	2.25751	2.20113
100	2.35464	2.34732	2.32692	2.29476	2.25108	2.19512
120	2.34968	2.34242	2.32218	2.29023	2.24679	2.19112
200	2.33974	2.33262	2.31270	2.28117	2.23824	2.18316
∞	2.32482	2.31791	2.29848	2.26762	2.22547	2.17127

TABLE B.1

Two-sided Percentage Point (h) for Equal
Correlations   (Case I with Correlation ρ)

p = 11, 1-α = 0.80

ν ↓	ρ				
	0.6	0.7	0.8	0.9	1/(1+√11)
2	3.60295	3.40824	3.15859	2.80834	4.04546
3	2.99701	2.84219	2.64375	2.36588	3.34801
4	2.73935	2.60212	2.42620	2.18000	3.04978
5	2.59690	2.46973	2.30668	2.07843	2.88398
6	2.50648	2.38592	2.23126	2.01463	2.77819
7	2.44399	2.32812	2.17941	1.97093	2.70469
8	2.39820	2.28586	2.14161	1.93916	2.65060
9	2.36322	2.25364	2.11285	1.91504	2.60909
10	2.33563	2.22827	2.09025	1.89611	2.57620
11	2.31330	2.20777	2.07202	1.88087	2.54950
12	2.29487	2.19088	2.05702	1.86834	2.52737
13	2.27940	2.17671	2.04446	1.85784	2.50874
14	2.26623	2.16467	2.03379	1.84894	2.49283
15	2.25488	2.15430	2.02461	1.84128	2.47909
16	2.24500	2.14528	2.01664	1.83463	2.46709
17	2.23633	2.13737	2.00965	1.82880	2.45653
18	2.22865	2.13037	2.00346	1.82364	2.44716
19	2.22181	2.12414	1.99796	1.81905	2.43879
20	2.21567	2.11855	1.99303	1.81494	2.43127
21	2.21013	2.11351	1.98858	1.81123	2.42447
22	2.20511	2.10895	1.98456	1.80788	2.41830
23	2.20054	2.10480	1.98090	1.80482	2.41267
24	2.19636	2.10100	1.97755	1.80203	2.40751
25	2.19252	2.09751	1.97448	1.79947	2.40277
26	2.18898	2.09430	1.97165	1.79711	2.39840
27	2.18572	2.09134	1.96904	1.79493	2.39435
28	2.18269	2.08859	1.96662	1.79292	2.39060
29	2.17987	2.08604	1.96437	1.79104	2.38710
30	2.17725	2.08366	1.96228	1.78929	2.38384
35	2.16643	2.07386	1.95365	1.78209	2.37036
40	2.15836	2.06655	1.94722	1.77673	2.36026
45	2.15211	2.06090	1.94224	1.77258	2.35241
50	2.14712	2.05639	1.93828	1.76927	2.34614
60	2.13968	2.04966	1.93237	1.76433	2.33674
80	2.13043	2.04130	1.92502	1.75819	2.32501
100	2.12490	2.03632	1.92063	1.75453	2.31798
120	2.12123	2.03300	1.91772	1.75209	2.31330
200	2.11392	2.02641	1.91192	1.74724	2.30394
∞	2.10301	2.01658	1.90328	1.74003	2.28992

TABLE B.1

Two-sided Percentage Point (h) for Equal
Correlations $\left(\text{Case I with Correlation } \rho\right)$

$p = 12, \ 1-\alpha = 0.80$

$\nu \downarrow$	ρ					
	0.0	0.1	0.2	0.3	0.4	0.5
2	4.21475	4.19339	4.13873	4.05737	3.95101	3.81845
3	3.48176	3.46508	3.42211	3.35785	3.27358	3.16835
4	3.16761	3.15306	3.11534	3.05873	2.98429	2.89118
5	2.99258	2.97927	2.94462	2.89244	2.82369	2.73755
6	2.88066	2.86819	2.83559	2.78635	2.72134	2.63980
7	2.80275	2.79089	2.75977	2.71265	2.65034	2.57208
8	2.74530	2.73392	2.70393	2.65841	2.59814	2.52237
9	2.70114	2.69013	2.66104	2.61679	2.55813	2.48431
10	2.66610	2.65540	2.62704	2.58383	2.52647	2.45423
11	2.63761	2.62716	2.59941	2.55707	2.50079	2.42986
12	2.61397	2.60374	2.57651	2.53490	2.47953	2.40972
13	2.59404	2.58400	2.55722	2.51623	2.46165	2.39278
14	2.57700	2.56712	2.54073	2.50029	2.44639	2.37835
15	2.56226	2.55253	2.52648	2.48652	2.43323	2.36590
16	2.54938	2.53978	2.51404	2.47451	2.42174	2.35505
17	2.53803	2.52855	2.50309	2.46393	2.41164	2.34551
18	2.52796	2.51858	2.49336	2.45455	2.40268	2.33706
19	2.51895	2.50967	2.48467	2.44617	2.39469	2.32953
20	2.51085	2.50165	2.47686	2.43864	2.38751	2.32276
21	2.50352	2.49440	2.46980	2.43184	2.38102	2.31666
22	2.49686	2.48782	2.46339	2.42566	2.37514	2.31112
23	2.49078	2.48181	2.45754	2.42003	2.36977	2.30607
24	2.48521	2.47630	2.45218	2.41487	2.36486	2.30145
25	2.48009	2.47124	2.44725	2.41013	2.36035	2.29721
26	2.47536	2.46656	2.44270	2.40576	2.35619	2.29331
27	2.47098	2.46224	2.43849	2.40171	2.35234	2.28969
28	2.46692	2.45822	2.43459	2.39795	2.34877	2.28634
29	2.46313	2.45448	2.43095	2.39446	2.34545	2.28323
30	2.45960	2.45099	2.42756	2.39120	2.34236	2.28032
35	2.44497	2.43654	2.41352	2.37772	2.32956	2.26833
40	2.43400	2.42570	2.40300	2.36763	2.31999	2.25938
45	2.42546	2.41727	2.39482	2.35980	2.31257	2.25243
50	2.41863	2.41053	2.38828	2.35353	2.30664	2.24690
60	2.40837	2.40041	2.37848	2.34416	2.29777	2.23862
80	2.39555	2.38777	2.36624	2.33245	2.28672	2.22831
100	2.38785	2.38017	2.35890	2.32544	2.28011	2.22216
120	2.38271	2.37511	2.35400	2.32078	2.27571	2.21806
200	2.37244	2.36498	2.34422	2.31145	2.26693	2.20990
∞	2.35700	2.34978	2.32955	2.29749	2.25381	2.19772

TABLE B.1

Two-sided Percentage Point (h) for Equal Correlations (Case I with Correlation ρ)

$$p = 12, \quad 1-\alpha = 0.80$$

ν ↓	ρ 0.6	0.7	0.8	0.9	$1/(1+\sqrt{12})$
2	3.65565	3.45417	3.19628	2.83503	4.12151
3	3.03895	2.87871	2.67368	2.38702	3.40853
4	2.77655	2.63451	2.45274	2.19874	3.10340
5	2.63139	2.49978	2.33130	2.09583	2.93362
6	2.53922	2.41444	2.25465	2.03118	2.82522
7	2.47548	2.35557	2.20193	1.98689	2.74986
8	2.42876	2.31252	2.16350	1.95470	2.69436
9	2.39306	2.27968	2.13425	1.93025	2.65174
10	2.36489	2.25382	2.11126	1.91107	2.61797
11	2.34209	2.23293	2.09273	1.89563	2.59053
12	2.32327	2.21570	2.07747	1.88292	2.56778
13	2.30747	2.20126	2.06469	1.87229	2.54862
14	2.29401	2.18897	2.05383	1.86326	2.53226
15	2.28242	2.17840	2.04450	1.85550	2.51811
16	2.27232	2.16921	2.03639	1.84876	2.50577
17	2.26346	2.16114	2.02927	1.84285	2.49489
18	2.25561	2.15400	2.02299	1.83763	2.48524
19	2.24862	2.14764	2.01739	1.83298	2.47662
20	2.24234	2.14194	2.01237	1.82881	2.46887
21	2.23668	2.13681	2.00785	1.82506	2.46187
22	2.23155	2.13215	2.00375	1.82166	2.45551
23	2.22688	2.12791	2.00003	1.81856	2.44970
24	2.22260	2.12404	1.99662	1.81573	2.44438
25	2.21868	2.12048	1.99350	1.81314	2.43950
26	2.21507	2.11721	1.99062	1.81075	2.43499
27	2.21173	2.11419	1.98797	1.80854	2.43081
28	2.20863	2.11138	1.98551	1.80650	2.42694
29	2.20575	2.10878	1.98322	1.80460	2.42334
30	2.20307	2.10636	1.98109	1.80283	2.41997
35	2.19201	2.09635	1.97231	1.79553	2.40606
40	2.18375	2.08890	1.96577	1.79010	2.39563
45	2.17736	2.08313	1.96071	1.78589	2.38753
50	2.17227	2.07854	1.95668	1.78254	2.38105
60	2.16466	2.07167	1.95066	1.77754	2.37134
80	2.15519	2.06315	1.94319	1.77132	2.35922
100	2.14955	2.05806	1.93873	1.76761	2.35195
120	2.14579	2.05468	1.93577	1.76514	2.34710
200	2.13831	2.04795	1.92987	1.76022	2.33742
∞	2.12717	2.03793	1.92108	1.75292	2.32291

TABLE B.1

Two-sided Percentage Point (h) for Equal
Correlations   (*Case I with Correlation* ρ)

p = 13, 1-α = 0.80

ν ↓	ρ					
	0.0	0.1	0.2	0.3	0.4	0.5
2	4.28330	4.26081	4.20367	4.11903	4.00873	3.87162
3	3.53636	3.51879	3.47386	3.40698	3.31958	3.21071
4	3.21600	3.20067	3.16123	3.10231	3.02511	2.92877
5	3.03738	3.02337	2.98714	2.93283	2.86152	2.77240
6	2.92309	2.90997	2.87588	2.82464	2.75723	2.67287
7	2.84349	2.83101	2.79848	2.74945	2.68483	2.60388
8	2.78476	2.77278	2.74143	2.69408	2.63159	2.55321
9	2.73959	2.72801	2.69760	2.65158	2.59076	2.51441
10	2.70373	2.69248	2.66285	2.61791	2.55845	2.48374
11	2.67456	2.66358	2.63459	2.59056	2.53223	2.45888
12	2.65035	2.63960	2.61117	2.56790	2.51052	2.43833
13	2.62993	2.61938	2.59142	2.54881	2.49225	2.42105
14	2.61247	2.60210	2.57455	2.53251	2.47666	2.40631
15	2.59736	2.58714	2.55996	2.51842	2.46320	2.39361
16	2.58415	2.57408	2.54722	2.50613	2.45147	2.38253
17	2.57252	2.56256	2.53599	2.49531	2.44114	2.37279
18	2.56218	2.55234	2.52603	2.48571	2.43198	2.36417
19	2.55293	2.54320	2.51713	2.47713	2.42381	2.35647
20	2.54462	2.53497	2.50912	2.46942	2.41646	2.34956
21	2.53709	2.52754	2.50188	2.46246	2.40983	2.34333
22	2.53026	2.52078	2.49531	2.45613	2.40381	2.33767
23	2.52402	2.51461	2.48931	2.45036	2.39832	2.33251
24	2.51830	2.50896	2.48381	2.44508	2.39330	2.32780
25	2.51303	2.50376	2.47876	2.44022	2.38868	2.32346
26	2.50817	2.49896	2.47409	2.43574	2.38443	2.31947
27	2.50368	2.49452	2.46977	2.43159	2.38049	2.31578
28	2.49950	2.49039	2.46577	2.42774	2.37684	2.31236
29	2.49561	2.48655	2.46204	2.42416	2.37344	2.30917
30	2.49198	2.48297	2.45856	2.42082	2.37027	2.30621
35	2.47694	2.46811	2.44414	2.40701	2.35717	2.29395
40	2.46565	2.45697	2.43334	2.39666	2.34738	2.28480
45	2.45686	2.44831	2.42495	2.38863	2.33978	2.27770
50	2.44983	2.44137	2.41823	2.38220	2.33371	2.27204
60	2.43927	2.43096	2.40816	2.37258	2.32463	2.26358
80	2.42607	2.41795	2.39558	2.36058	2.31331	2.25305
100	2.41813	2.41013	2.38803	2.35338	2.30654	2.24675
120	2.41284	2.40492	2.38300	2.34859	2.30203	2.24256
200	2.40224	2.39448	2.37293	2.33902	2.29303	2.23422
∞	2.38631	2.37881	2.35784	2.32469	2.27959	2.22177

TABLE B.1

Two-sided Percentage Point (h) for Equal
Correlations (*Case I with Correlation* ρ)

p = 13, 1-α = 0.80

ν ↓	ρ				
	0.6	0.7	0.8	0.9	1/(1+√13)
2	3.70355	3.49590	3.23050	2.85927	4.19102
3	3.07709	2.91192	2.70087	2.40622	3.46389
4	2.81040	2.66397	2.47685	2.21576	3.15246
5	2.66279	2.52711	2.35368	2.11163	2.97906
6	2.56901	2.44038	2.27590	2.04621	2.86827
7	2.50414	2.38054	2.22241	2.00139	2.79120
8	2.45658	2.33677	2.18339	1.96881	2.73441
9	2.42022	2.30337	2.15370	1.94407	2.69079
10	2.39152	2.27706	2.13036	1.92465	2.65619
11	2.36829	2.25580	2.11154	1.90902	2.62808
12	2.34911	2.23828	2.09604	1.89616	2.60477
13	2.33301	2.22358	2.08307	1.88540	2.58512
14	2.31929	2.21108	2.07204	1.87627	2.56834
15	2.30747	2.20032	2.06257	1.86841	2.55382
16	2.29718	2.19096	2.05433	1.86159	2.54115
17	2.28814	2.18275	2.04711	1.85561	2.52999
18	2.28014	2.17548	2.04072	1.85033	2.52008
19	2.27301	2.16901	2.03503	1.84562	2.51123
20	2.26661	2.16321	2.02994	1.84140	2.50327
21	2.26084	2.15798	2.02535	1.83760	2.49607
22	2.25560	2.15324	2.02119	1.83416	2.48954
23	2.25084	2.14892	2.01741	1.83103	2.48357
24	2.24648	2.14498	2.01395	1.82817	2.47811
25	2.24248	2.14136	2.01078	1.82554	2.47308
26	2.23879	2.13803	2.00786	1.82313	2.46845
27	2.23538	2.13495	2.00516	1.82089	2.46416
28	2.23223	2.13210	2.00266	1.81882	2.46017
29	2.22929	2.12945	2.00034	1.81690	2.45647
30	2.22656	2.12698	1.99818	1.81511	2.45301
35	2.21527	2.11680	1.98926	1.80773	2.43869
40	2.20685	2.10921	1.98262	1.80223	2.42796
45	2.20033	2.10334	1.97748	1.79797	2.41962
50	2.19513	2.09866	1.97339	1.79458	2.41295
60	2.18737	2.09167	1.96728	1.78952	2.40295
80	2.17771	2.08300	1.95969	1.78323	2.39045
100	2.17195	2.07782	1.95516	1.77947	2.38296
120	2.16812	2.07438	1.95215	1.77698	2.37797
200	2.16049	2.06753	1.94616	1.77199	2.36798
∞	2.14912	2.05732	1.93724	1.76461	2.35301

TABLE B.1

Two-sided Percentage Point (h) for Equal
Correlations $\left(Case\ I\ with\ Correlation\ \rho\right)$

$$p = 14,\ 1-\alpha = 0.80$$

				ρ		
ν ↓	0.0	0.1	0.2	0.3	0.4	0.5
2	4.34624	4.32269	4.26324	4.17555	4.06164	3.92032
3	3.58653	3.56813	3.52137	3.45207	3.36177	3.24954
4	3.26048	3.24443	3.20338	3.14232	3.06255	2.96324
5	3.07858	3.06391	3.02620	2.96992	2.89624	2.80437
6	2.96212	2.94838	2.91291	2.85981	2.79016	2.70320
7	2.88096	2.86790	2.83405	2.78325	2.71649	2.63304
8	2.82106	2.80852	2.77591	2.72685	2.66229	2.58151
9	2.77496	2.76284	2.73121	2.68354	2.62072	2.54203
10	2.73835	2.72658	2.69576	2.64921	2.58780	2.51081
11	2.70856	2.69707	2.66693	2.62132	2.56108	2.48551
12	2.68383	2.67258	2.64302	2.59821	2.53896	2.46458
13	2.66295	2.65192	2.62285	2.57873	2.52034	2.44697
14	2.64510	2.63426	2.60562	2.56210	2.50444	2.43197
15	2.62965	2.61897	2.59072	2.54772	2.49072	2.41902
16	2.61614	2.60561	2.57770	2.53517	2.47875	2.40774
17	2.60423	2.59384	2.56623	2.52412	2.46821	2.39782
18	2.59365	2.58338	2.55605	2.51432	2.45887	2.38902
19	2.58419	2.57402	2.54694	2.50556	2.45053	2.38118
20	2.57568	2.56561	2.53876	2.49768	2.44303	2.37414
21	2.56797	2.55800	2.53136	2.49057	2.43627	2.36778
22	2.56097	2.55108	2.52463	2.48410	2.43012	2.36201
23	2.55458	2.54477	2.51850	2.47821	2.42452	2.35676
24	2.54872	2.53898	2.51287	2.47281	2.41939	2.35195
25	2.54333	2.53365	2.50770	2.46784	2.41468	2.34753
26	2.53835	2.52874	2.50293	2.46326	2.41033	2.34346
27	2.53374	2.52419	2.49851	2.45902	2.40631	2.33970
28	2.52946	2.51996	2.49441	2.45509	2.40258	2.33621
29	2.52547	2.51602	2.49059	2.45143	2.39911	2.33296
30	2.52175	2.51235	2.48703	2.44801	2.39588	2.32994
35	2.50632	2.49713	2.47227	2.43388	2.38250	2.31744
40	2.49474	2.48571	2.46121	2.42330	2.37250	2.30810
45	2.48572	2.47682	2.45261	2.41508	2.36474	2.30087
50	2.47850	2.46970	2.44573	2.40851	2.35854	2.29509
60	2.46767	2.45903	2.43541	2.39867	2.34926	2.28646
80	2.45410	2.44566	2.42251	2.38638	2.33769	2.27571
100	2.44594	2.43764	2.41477	2.37901	2.33077	2.26929
120	2.44050	2.43228	2.40961	2.37411	2.32616	2.26502
200	2.42960	2.42156	2.39928	2.36431	2.31696	2.25651
∞	2.41322	2.40545	2.38380	2.34963	2.30322	2.24380

TABLE B.1

Two-sided Percentage Point (h) for Equal Correlations $\left(Case\ I\ with\ Correlation\ \rho\right)$

$p = 14,\ 1-\alpha = 0.80$

$\nu\ \downarrow$	0.6	0.7	0.8	0.9	$1/(1+\sqrt{14})$
			ρ		
2	3.74742	3.53410	3.26182	2.88144	4.25499
3	3.11205	2.94233	2.72576	2.42377	3.51486
4	2.84142	2.69096	2.49894	2.23134	3.19765
5	2.69157	2.55215	2.37417	2.12610	3.02093
6	2.59633	2.46416	2.29537	2.05997	2.90794
7	2.53041	2.40342	2.24116	2.01466	2.82930
8	2.48208	2.35898	2.20161	1.98172	2.77132
9	2.44512	2.32507	2.17151	1.95670	2.72676
10	2.41594	2.29836	2.14785	1.93708	2.69142
11	2.39232	2.27676	2.12877	1.92127	2.66268
12	2.37281	2.25896	2.11305	1.90827	2.63884
13	2.35642	2.24403	2.09990	1.89740	2.61875
14	2.34247	2.23133	2.08872	1.88816	2.60157
15	2.33044	2.22040	2.07911	1.88022	2.58672
16	2.31997	2.21089	2.07075	1.87333	2.57375
17	2.31077	2.20254	2.06343	1.86728	2.56232
18	2.30263	2.19516	2.05695	1.86194	2.55217
19	2.29537	2.18858	2.05119	1.85718	2.54310
20	2.28885	2.18269	2.04602	1.85292	2.53495
21	2.28298	2.17737	2.04137	1.84908	2.52758
22	2.27765	2.17255	2.03715	1.84560	2.52088
23	2.27280	2.16817	2.03331	1.84243	2.51476
24	2.26836	2.16416	2.02981	1.83954	2.50916
25	2.26429	2.16049	2.02659	1.83689	2.50401
26	2.26053	2.15710	2.02363	1.83444	2.49926
27	2.25706	2.15397	2.02089	1.83219	2.49485
28	2.25385	2.15107	2.01836	1.83010	2.49077
29	2.25086	2.14838	2.01601	1.82815	2.48697
30	2.24808	2.14587	2.01381	1.82634	2.48342
35	2.23658	2.13552	2.00477	1.81888	2.46872
40	2.22801	2.12781	1.99804	1.81332	2.45771
45	2.22137	2.12184	1.99283	1.80902	2.44915
50	2.21608	2.11709	1.98868	1.80559	2.44230
60	2.20817	2.10999	1.98248	1.80048	2.43202
80	2.19834	2.10117	1.97479	1.79412	2.41918
100	2.19247	2.09590	1.97020	1.79032	2.41148
120	2.18857	2.09241	1.96715	1.78780	2.40634
200	2.18080	2.08544	1.96107	1.78276	2.39607
∞	2.16922	2.07507	1.95202	1.77530	2.38067

TABLE B.1

Two-sided Percentage Point (h) for Equal
Correlations (*Case I with Correlation* ρ)

p = 15, 1-α = 0.80

ν ↓	ρ					
	0.0	0.1	0.2	0.3	0.4	0.5
2	4.40439	4.37983	4.31822	4.22771	4.11042	3.96523
3	3.63291	3.61372	3.56524	3.49369	3.40071	3.28536
4	3.30162	3.28488	3.24233	3.17927	3.09712	2.99504
5	3.11669	3.10139	3.06230	3.00419	2.92830	2.83387
6	2.99823	2.98390	2.94713	2.89230	2.82057	2.73119
7	2.91564	2.90202	2.86693	2.81447	2.74573	2.65997
8	2.85465	2.84158	2.80778	2.75713	2.69065	2.60763
9	2.80769	2.79506	2.76229	2.71307	2.64838	2.56752
10	2.77039	2.75812	2.72619	2.67814	2.61491	2.53580
11	2.74002	2.72806	2.69683	2.64975	2.58773	2.51008
12	2.71480	2.70309	2.67246	2.62622	2.56522	2.48880
13	2.69351	2.68203	2.65192	2.60639	2.54627	2.47091
14	2.67530	2.66401	2.63435	2.58944	2.53010	2.45565
15	2.65953	2.64842	2.61916	2.57480	2.51613	2.44248
16	2.64574	2.63479	2.60588	2.56201	2.50394	2.43100
17	2.63358	2.62277	2.59419	2.55075	2.49321	2.42091
18	2.62278	2.61209	2.58380	2.54075	2.48370	2.41197
19	2.61311	2.60254	2.57451	2.53182	2.47520	2.40398
20	2.60442	2.59395	2.56616	2.52379	2.46757	2.39682
21	2.59655	2.58617	2.55860	2.51654	2.46067	2.39035
22	2.58939	2.57911	2.55174	2.50994	2.45442	2.38448
23	2.58286	2.57266	2.54547	2.50393	2.44871	2.37913
24	2.57686	2.56674	2.53973	2.49842	2.44348	2.37424
25	2.57135	2.56130	2.53445	2.49336	2.43868	2.36974
26	2.56626	2.55628	2.52958	2.48869	2.43425	2.36560
27	2.56155	2.55162	2.52507	2.48436	2.43015	2.36177
28	2.55717	2.54730	2.52088	2.48035	2.42635	2.35822
29	2.55309	2.54328	2.51698	2.47661	2.42282	2.35491
30	2.54928	2.53952	2.51334	2.47313	2.41952	2.35183
35	2.53350	2.52396	2.49826	2.45871	2.40588	2.33911
40	2.52164	2.51227	2.48696	2.44791	2.39569	2.32960
45	2.51241	2.50318	2.47817	2.43952	2.38777	2.32223
50	2.50502	2.49589	2.47113	2.43281	2.38145	2.31636
60	2.49391	2.48496	2.46058	2.42275	2.37198	2.30756
80	2.48001	2.47127	2.44738	2.41020	2.36018	2.29662
100	2.47164	2.46305	2.43946	2.40267	2.35312	2.29008
120	2.46606	2.45756	2.43418	2.39766	2.34842	2.28573
200	2.45488	2.44658	2.42362	2.38765	2.33904	2.27706
∞	2.43806	2.43006	2.40776	2.37264	2.32502	2.26412

TABLE B.1

Two-sided Percentage Point (h) for Equal
Correlations (*Case I with Correlation* ρ)

p = 15, 1-α = 0.80

ν ↓	ρ				
	0.6	0.7	0.8	0.9	$1/(1+\sqrt{15})$
2	3.78784	3.56930	3.29067	2.90185	4.31418
3	3.14428	2.97037	2.74871	2.43995	3.56206
4	2.87004	2.71585	2.51930	2.24569	3.23952
5	2.71812	2.57524	2.39306	2.13942	3.05972
6	2.62153	2.48609	2.31332	2.07264	2.94470
7	2.55466	2.42452	2.25844	2.02688	2.86461
8	2.50561	2.37947	2.21841	1.99361	2.80553
9	2.46809	2.34509	2.18793	1.96834	2.76011
10	2.43847	2.31799	2.16397	1.94852	2.72406
11	2.41448	2.29609	2.14464	1.93256	2.69475
12	2.39467	2.27803	2.12873	1.91943	2.67042
13	2.37802	2.26288	2.11540	1.90844	2.64991
14	2.36385	2.25000	2.10408	1.89911	2.63237
15	2.35163	2.23891	2.09435	1.89110	2.61720
16	2.34099	2.22926	2.08589	1.88413	2.60395
17	2.33164	2.22079	2.07847	1.87803	2.59227
18	2.32337	2.21330	2.07191	1.87263	2.58190
19	2.31599	2.20662	2.06607	1.86782	2.57263
20	2.30937	2.20064	2.06084	1.86352	2.56430
21	2.30340	2.19525	2.05612	1.85964	2.55676
22	2.29798	2.19036	2.05185	1.85613	2.54990
23	2.29305	2.18591	2.04797	1.85293	2.54365
24	2.28854	2.18184	2.04441	1.85001	2.53792
25	2.28440	2.17811	2.04116	1.84733	2.53265
26	2.28058	2.17467	2.03816	1.84486	2.52779
27	2.27706	2.17150	2.03539	1.84258	2.52328
28	2.27379	2.16856	2.03282	1.84047	2.51910
29	2.27075	2.16582	2.03044	1.83851	2.51521
30	2.26792	2.16328	2.02822	1.83668	2.51158
35	2.25624	2.15277	2.01906	1.82915	2.49653
40	2.24752	2.14495	2.01224	1.82353	2.48525
45	2.24077	2.13889	2.00696	1.81919	2.47648
50	2.23539	2.13406	2.00276	1.81573	2.46946
60	2.22735	2.12686	1.99649	1.81056	2.45893
80	2.21736	2.11791	1.98869	1.80414	2.44576
100	2.21139	2.11257	1.98404	1.80031	2.43786
120	2.20743	2.10902	1.98095	1.79776	2.43259
200	2.19953	2.10195	1.97480	1.79267	2.42205
∞	2.18776	2.09143	1.96564	1.78514	2.40624

TABLE B.1

Two-sided Percentage Point (h) for Equal
Correlations (*Case I with Correlation* ρ)

$p = 16$, $1-\alpha = 0.80$

$\nu \downarrow$	0.0	0.1	0.2	0.3	0.4	0.5
			ρ			
2	4.45839	4.43288	4.36924	4.27608	4.15566	4.00685
3	3.67601	3.65608	3.60599	3.53232	3.43683	3.31859
4	3.33987	3.32248	3.27850	3.21358	3.12920	3.02455
5	3.15213	3.13624	3.09584	3.03600	2.95806	2.86125
6	3.03182	3.01693	2.97894	2.92248	2.84881	2.75718
7	2.94790	2.93375	2.89749	2.84348	2.77288	2.68496
8	2.88589	2.87232	2.83740	2.78525	2.71698	2.63187
9	2.83814	2.82503	2.79117	2.74050	2.67407	2.59118
10	2.80020	2.78746	2.75447	2.70501	2.64008	2.55899
11	2.76930	2.75687	2.72461	2.67616	2.61248	2.53289
12	2.74362	2.73147	2.69983	2.65224	2.58961	2.51129
13	2.72194	2.71003	2.67893	2.63207	2.57036	2.49312
14	2.70339	2.69168	2.66106	2.61484	2.55392	2.47762
15	2.68733	2.67580	2.64559	2.59994	2.53972	2.46425
16	2.67328	2.66191	2.63208	2.58694	2.52733	2.45259
17	2.66089	2.64967	2.62017	2.57548	2.51642	2.44234
18	2.64987	2.63879	2.60959	2.56531	2.50675	2.43325
19	2.64002	2.62906	2.60013	2.55622	2.49811	2.42515
20	2.63115	2.62030	2.59162	2.54804	2.49035	2.41786
21	2.62312	2.61237	2.58392	2.54066	2.48333	2.41129
22	2.61582	2.60517	2.57693	2.53395	2.47697	2.40533
23	2.60915	2.59859	2.57055	2.52782	2.47116	2.39989
24	2.60304	2.59255	2.56469	2.52222	2.46584	2.39492
25	2.59742	2.58700	2.55931	2.51706	2.46096	2.39035
26	2.59222	2.58188	2.55434	2.51230	2.45645	2.38614
27	2.58741	2.57713	2.54975	2.50790	2.45228	2.38224
28	2.58294	2.57272	2.54547	2.50381	2.44842	2.37863
29	2.57877	2.56862	2.54150	2.50000	2.44482	2.37527
30	2.57489	2.56479	2.53779	2.49645	2.44146	2.37214
35	2.55877	2.54890	2.52241	2.48176	2.42759	2.35921
40	2.54665	2.53696	2.51088	2.47075	2.41720	2.34955
45	2.53722	2.52767	2.50191	2.46220	2.40915	2.34206
50	2.52966	2.52023	2.49473	2.45536	2.40271	2.33608
60	2.51831	2.50906	2.48395	2.44511	2.39307	2.32714
80	2.50408	2.49507	2.47048	2.43231	2.38106	2.31601
100	2.49553	2.48666	2.46239	2.42464	2.37387	2.30936
120	2.48981	2.48105	2.45700	2.41952	2.36908	2.30494
200	2.47837	2.46981	2.44621	2.40931	2.35953	2.29612
∞	2.46114	2.45290	2.43000	2.39400	2.34524	2.28296

TABLE B.1

Two-sided Percentage Point (h) for Equal
Correlations $\Big(Case\ I\ with\ Correlation\ \rho\Big)$

$$p = 16, \quad 1-\alpha = 0.80$$

$\nu \downarrow$	ρ				
	0.6	0.7	0.8	0.9	$1/(1+\sqrt{16})$
2	3.82531	3.60191	3.31739	2.92075	4.36924
3	3.17417	2.99635	2.76996	2.45494	3.60599
4	2.89658	2.73892	2.53816	2.25898	3.27850
5	2.74275	2.59666	2.41057	2.15177	3.09584
6	2.64491	2.50642	2.32995	2.08438	2.97894
7	2.57715	2.44409	2.27446	2.03820	2.89749
8	2.52744	2.39847	2.23397	2.00462	2.83740
9	2.48941	2.36365	2.20314	1.97912	2.79117
10	2.45937	2.33620	2.17890	1.95912	2.75447
11	2.43504	2.31402	2.15935	1.94301	2.72461
12	2.41494	2.29572	2.14325	1.92976	2.69983
13	2.39806	2.28037	2.12977	1.91867	2.67893
14	2.38368	2.26731	2.11832	1.90926	2.66106
15	2.37128	2.25607	2.10847	1.90116	2.64559
16	2.36049	2.24629	2.09991	1.89414	2.63208
17	2.35100	2.23771	2.09240	1.88797	2.62017
18	2.34260	2.23011	2.08577	1.88253	2.60959
19	2.33511	2.22335	2.07986	1.87768	2.60013
20	2.32839	2.21729	2.07456	1.87334	2.59162
21	2.32233	2.21182	2.06979	1.86942	2.58392
22	2.31684	2.20686	2.06547	1.86588	2.57693
23	2.31183	2.20235	2.06154	1.86265	2.57055
24	2.30725	2.19823	2.05795	1.85970	2.56469
25	2.30305	2.19445	2.05465	1.85700	2.55931
26	2.29918	2.19096	2.05162	1.85451	2.55434
27	2.29560	2.18775	2.04881	1.85221	2.54975
28	2.29228	2.18476	2.04622	1.85008	2.54547
29	2.28920	2.18199	2.04380	1.84810	2.54150
30	2.28632	2.17941	2.04156	1.84625	2.53779
35	2.27446	2.16876	2.03229	1.83865	2.52241
40	2.26561	2.16083	2.02539	1.83299	2.51088
45	2.25876	2.15469	2.02006	1.82860	2.50191
50	2.25330	2.14980	2.01580	1.82511	2.49473
60	2.24514	2.14250	2.00945	1.81990	2.48395
80	2.23499	2.13342	2.00157	1.81342	2.47048
100	2.22893	2.12801	1.99687	1.80955	2.46239
120	2.22490	2.12441	1.99374	1.80698	2.45700
200	2.21688	2.11725	1.98752	1.80184	2.44621
∞	2.20493	2.10658	1.97825	1.79425	2.43000

TABLE B.1

Two-sided Percentage Point (h) for Equal
Correlations (*Case I with Correlation* ρ)

p = 18, 1-α = 0.80

ν ↓	ρ					
	0.0	0.1	0.2	0.3	0.4	0.5
2	4.55597	4.52870	4.46134	4.36336	4.23725	4.08188
3	3.75396	3.73264	3.67959	3.60208	3.50202	3.37852
4	3.40908	3.39048	3.34389	3.27555	3.18713	3.07781
5	3.21629	3.19929	3.15649	3.09350	3.01182	2.91068
6	3.09263	3.07671	3.03645	2.97703	2.89981	2.80408
7	3.00631	2.99118	2.95277	2.89592	2.82192	2.73008
8	2.94248	2.92797	2.89098	2.83609	2.76455	2.67565
9	2.89330	2.87928	2.84342	2.79009	2.72048	2.63391
10	2.85419	2.84057	2.80564	2.75359	2.68556	2.60087
11	2.82231	2.80904	2.77489	2.72391	2.65719	2.57407
12	2.79582	2.78284	2.74935	2.69928	2.63368	2.55189
13	2.77344	2.76071	2.72780	2.67851	2.61387	2.53322
14	2.75428	2.74177	2.70937	2.66076	2.59696	2.51730
15	2.73767	2.72537	2.69341	2.64541	2.58234	2.50356
16	2.72315	2.71103	2.67946	2.63200	2.56958	2.49157
17	2.71034	2.69837	2.66717	2.62018	2.55835	2.48103
18	2.69894	2.68712	2.65624	2.60969	2.54839	2.47169
19	2.68874	2.67705	2.64647	2.60032	2.53949	2.46335
20	2.67956	2.66799	2.63768	2.59188	2.53149	2.45586
21	2.67125	2.65979	2.62972	2.58426	2.52426	2.44909
22	2.66368	2.65233	2.62249	2.57733	2.51770	2.44296
23	2.65678	2.64552	2.61589	2.57101	2.51172	2.43737
24	2.65044	2.63927	2.60984	2.56522	2.50624	2.43225
25	2.64461	2.63352	2.60427	2.55989	2.50120	2.42754
26	2.63922	2.62822	2.59913	2.55498	2.49655	2.42321
27	2.63423	2.62330	2.59438	2.55043	2.49225	2.41920
28	2.62960	2.61873	2.58996	2.54620	2.48826	2.41548
29	2.62528	2.61447	2.58584	2.54227	2.48455	2.41203
30	2.62124	2.61050	2.58200	2.53860	2.48109	2.40880
35	2.60450	2.59402	2.56608	2.52341	2.46677	2.39548
40	2.59192	2.58163	2.55413	2.51203	2.45606	2.38553
45	2.58211	2.57199	2.54483	2.50318	2.44774	2.37782
50	2.57425	2.56426	2.53738	2.49611	2.44109	2.37166
60	2.56244	2.55264	2.52621	2.48550	2.43114	2.36245
80	2.54763	2.53809	2.51223	2.47224	2.41873	2.35099
100	2.53871	2.52934	2.50383	2.46430	2.41131	2.34414
120	2.53276	2.52350	2.49823	2.45900	2.40636	2.33958
200	2.52082	2.51179	2.48701	2.44841	2.39649	2.33049
∞	2.50283	2.49417	2.47017	2.43255	2.38173	2.31693

TABLE B.1

Two-sided Percentage Point (h) for Equal
Correlations $\left(\text{Case } I \text{ with Correlation } \rho\right)$

$p = 18, \ 1-\alpha = 0.80$

$\nu \downarrow$	0.6	0.7	0.8	0.9	$1/(1+\sqrt{18})$
			ρ		
2	3.89282	3.66065	3.36550	2.95477	4.46894
3	3.22806	3.04320	2.80827	2.48194	3.68559
4	2.94446	2.78053	2.57217	2.28293	3.34917
5	2.78719	2.63527	2.44214	2.17400	3.16134
6	2.68710	2.54309	2.35994	2.10552	3.04103
7	2.61775	2.47939	2.30333	2.05858	2.95714
8	2.56684	2.43274	2.26202	2.02445	2.89519
9	2.52787	2.39712	2.23056	1.99853	2.84750
10	2.49708	2.36904	2.20582	1.97820	2.80963
11	2.47215	2.34634	2.18587	1.96182	2.77879
12	2.45154	2.32761	2.16943	1.94835	2.75318
13	2.43422	2.31189	2.15567	1.93708	2.73157
14	2.41946	2.29852	2.14397	1.92751	2.71308
15	2.40674	2.28701	2.13391	1.91929	2.69707
16	2.39566	2.27700	2.12517	1.91215	2.68308
17	2.38593	2.26821	2.11751	1.90588	2.67075
18	2.37731	2.26043	2.11073	1.90035	2.65979
19	2.36962	2.25350	2.10470	1.89542	2.64998
20	2.36272	2.24729	2.09929	1.89100	2.64116
21	2.35650	2.24169	2.09442	1.88703	2.63318
22	2.35085	2.23662	2.09001	1.88342	2.62593
23	2.34571	2.23200	2.08599	1.88014	2.61930
24	2.34101	2.22777	2.08233	1.87715	2.61323
25	2.33669	2.22390	2.07896	1.87440	2.60764
26	2.33272	2.22033	2.07586	1.87187	2.60249
27	2.32904	2.21703	2.07300	1.86953	2.59771
28	2.32563	2.21398	2.07035	1.86737	2.59327
29	2.32246	2.21114	2.06788	1.86535	2.58914
30	2.31951	2.20849	2.06559	1.86348	2.58529
35	2.30733	2.19758	2.05613	1.85575	2.56930
40	2.29824	2.18946	2.04908	1.85000	2.55731
45	2.29120	2.18317	2.04363	1.84554	2.54797
50	2.28559	2.17816	2.03929	1.84200	2.54050
60	2.27720	2.17067	2.03281	1.83670	2.52928
80	2.26677	2.16138	2.02476	1.83011	2.51523
100	2.26055	2.15583	2.01996	1.82618	2.50680
120	2.25641	2.15215	2.01677	1.82357	2.50117
200	2.24817	2.14481	2.01041	1.81834	2.48991
∞	2.23589	2.13388	2.00095	1.81064	2.47298

TABLE B.1

Two-sided Percentage Point (h) for Equal Correlations (Case I with Correlation ρ)

p = 20, 1-α = 0.80

ν ↓	ρ 0.0	0.1	0.2	0.3	0.4	0.5
2	4.64221	4.61334	4.54263	4.44035	4.30917	4.14801
3	3.82292	3.80034	3.74463	3.66367	3.55955	3.43139
4	3.47035	3.45063	3.40169	3.33031	3.23827	3.12480
5	3.27311	3.25509	3.21012	3.14431	3.05929	2.95430
6	3.14650	3.12963	3.08733	3.02524	2.94487	2.84550
7	3.05806	3.04203	3.00168	2.94228	2.86526	2.76992
8	2.99263	2.97725	2.93839	2.88105	2.80658	2.71430
9	2.94218	2.92732	2.88965	2.83395	2.76150	2.67164
10	2.90204	2.88761	2.85092	2.79655	2.72575	2.63786
11	2.86931	2.85525	2.81938	2.76613	2.69670	2.61044
12	2.84209	2.82834	2.79318	2.74088	2.67262	2.58774
13	2.81908	2.80561	2.77106	2.71958	2.65232	2.56864
14	2.79938	2.78615	2.75213	2.70137	2.63499	2.55234
15	2.78230	2.76928	2.73574	2.68561	2.62000	2.53826
16	2.76735	2.75453	2.72140	2.67185	2.60692	2.52599
17	2.75416	2.74151	2.70876	2.65972	2.59540	2.51519
18	2.74243	2.72993	2.69753	2.64894	2.58518	2.50562
19	2.73192	2.71957	2.68748	2.63931	2.57605	2.49707
20	2.72246	2.71024	2.67844	2.63064	2.56784	2.48940
21	2.71389	2.70179	2.67025	2.62281	2.56043	2.48247
22	2.70610	2.69411	2.66281	2.61569	2.55369	2.47618
23	2.69898	2.68709	2.65602	2.60919	2.54754	2.47044
24	2.69244	2.68065	2.64979	2.60324	2.54192	2.46520
25	2.68643	2.67472	2.64406	2.59776	2.53674	2.46038
26	2.68087	2.66925	2.63876	2.59270	2.53197	2.45593
27	2.67572	2.66418	2.63386	2.58802	2.52756	2.45182
28	2.67093	2.65946	2.62931	2.58368	2.52346	2.44801
29	2.66647	2.65507	2.62507	2.57964	2.51965	2.44446
30	2.66231	2.65097	2.62111	2.57586	2.51609	2.44116
35	2.64502	2.63396	2.60470	2.56023	2.50138	2.42750
40	2.63201	2.62117	2.59238	2.54851	2.49037	2.41729
45	2.62187	2.61120	2.58278	2.53940	2.48182	2.40938
50	2.61374	2.60322	2.57510	2.53211	2.47499	2.40306
60	2.60151	2.59121	2.56357	2.52118	2.46475	2.39361
80	2.58616	2.57615	2.54913	2.50752	2.45199	2.38185
100	2.57692	2.56709	2.54045	2.49933	2.44435	2.37481
120	2.57075	2.56104	2.53466	2.49386	2.43926	2.37014
200	2.55836	2.54891	2.52307	2.48295	2.42911	2.36081
∞	2.53969	2.53064	2.50565	2.46658	2.41392	2.34689

TABLE B.1

Two-sided Percentage Point (h) for Equal
Correlations (*Case I with Correlation* ρ)

p = 20, 1-α = 0.80

ν ↓	ρ				
	0.6	0.7	0.8	0.9	1/(1+√20)
2	3.95229	3.71237	3.40785	2.98470	4.55724
3	3.27557	3.08448	2.84200	2.50570	3.75616
4	2.98669	2.81721	2.60213	2.30401	3.41184
5	2.82640	2.66933	2.46995	2.19358	3.21946
6	2.72432	2.57543	2.38636	2.12414	3.09613
7	2.65357	2.51052	2.32878	2.07653	3.01008
8	2.60160	2.46297	2.28674	2.04191	2.94649
9	2.56181	2.42665	2.25473	2.01562	2.89751
10	2.53037	2.39800	2.22954	1.99499	2.85859
11	2.50489	2.37484	2.20923	1.97838	2.82688
12	2.48382	2.35572	2.19249	1.96472	2.80054
13	2.46612	2.33969	2.17848	1.95328	2.77829
14	2.45104	2.32604	2.16657	1.94358	2.75925
15	2.43803	2.31429	2.15633	1.93524	2.74277
16	2.42670	2.30407	2.14743	1.92799	2.72835
17	2.41674	2.29510	2.13962	1.92164	2.71564
18	2.40792	2.28716	2.13272	1.91602	2.70434
19	2.40006	2.28008	2.12658	1.91103	2.69423
20	2.39300	2.27374	2.12107	1.90655	2.68512
21	2.38663	2.26802	2.11611	1.90251	2.67689
22	2.38086	2.26284	2.11162	1.89886	2.66940
23	2.37560	2.25812	2.10753	1.89553	2.66256
24	2.37079	2.25381	2.10379	1.89249	2.65629
25	2.36637	2.24985	2.10037	1.88971	2.65052
26	2.36230	2.24621	2.09721	1.88714	2.64519
27	2.35854	2.24284	2.09429	1.88477	2.64025
28	2.35505	2.23972	2.09159	1.88257	2.63567
29	2.35181	2.23682	2.08909	1.88053	2.63140
30	2.34878	2.23412	2.08675	1.87863	2.62741
35	2.33631	2.22298	2.07712	1.87080	2.61088
40	2.32701	2.21468	2.06994	1.86496	2.59847
45	2.31980	2.20826	2.06439	1.86044	2.58880
50	2.31405	2.20314	2.05997	1.85685	2.58106
60	2.30547	2.19550	2.05337	1.85147	2.56943
80	2.29479	2.18600	2.04517	1.84480	2.55487
100	2.28842	2.18034	2.04028	1.84081	2.54612
120	2.28418	2.17658	2.03703	1.83816	2.54028
200	2.27575	2.16908	2.03056	1.83285	2.52859
∞	2.26317	2.15792	2.02093	1.82505	2.51100

TABLE B.2

Two-sided Percentage Point (h) for Equal
Correlations (*Case I with Correlation* ρ)

$p = 2, \ 1-\alpha = 0.90$

$\nu \downarrow$	0.0	0.1	0.2	0.3	0.4	0.5
			ρ			
1	8.95678	8.94547	8.91114	8.85239	8.76666	8.64966
2	3.83103	3.82696	3.81460	3.79347	3.76269	3.72078
3	2.98942	2.98656	2.97788	2.96305	2.94146	2.91211
4	2.66238	2.66001	2.65280	2.64049	2.62257	2.59822
5	2.49050	2.48839	2.48197	2.47102	2.45508	2.43343
6	2.38495	2.38300	2.37709	2.36698	2.35229	2.33232
7	2.31367	2.31183	2.30626	2.29675	2.28291	2.26410
8	2.26234	2.26059	2.25527	2.24619	2.23297	2.21501
9	2.22364	2.22195	2.21682	2.20807	2.19533	2.17802
10	2.19342	2.19178	2.18681	2.17831	2.16596	2.14915
11	2.16918	2.16758	2.16273	2.15445	2.14240	2.12601
12	2.14930	2.14774	2.14299	2.13488	2.12308	2.10703
13	2.13271	2.13117	2.12651	2.11855	2.10696	2.09120
14	2.11865	2.11714	2.11255	2.10471	2.09331	2.07780
15	2.10659	2.10509	2.10057	2.09284	2.08160	2.06629
16	2.09612	2.09465	2.09018	2.08254	2.07144	2.05632
17	2.08696	2.08550	2.08108	2.07353	2.06254	2.04758
18	2.07886	2.07742	2.07304	2.06557	2.05469	2.03988
19	2.07167	2.07024	2.06590	2.05849	2.04771	2.03302
20	2.06522	2.06381	2.05950	2.05215	2.04146	2.02689
21	2.05942	2.05802	2.05374	2.04645	2.03583	2.02136
22	2.05417	2.05277	2.04853	2.04128	2.03074	2.01636
23	2.04940	2.04801	2.04379	2.03659	2.02610	2.01182
24	2.04504	2.04366	2.03946	2.03230	2.02188	2.00767
25	2.04104	2.03966	2.03549	2.02837	2.01800	2.00387
26	2.03736	2.03599	2.03184	2.02475	2.01443	2.00036
27	2.03396	2.03260	2.02847	2.02141	2.01114	1.99713
28	2.03082	2.02946	2.02535	2.01832	2.00809	1.99414
29	2.02789	2.02654	2.02245	2.01545	2.00525	1.99136
30	2.02517	2.02383	2.01974	2.01277	2.00262	1.98877
35	2.01396	2.01264	2.00862	2.00175	1.99175	1.97811
40	2.00562	2.00431	2.00034	1.99356	1.98367	1.97019
45	1.99918	1.99788	1.99394	1.98722	1.97743	1.96406
50	1.99404	1.99276	1.98885	1.98218	1.97246	1.95919
60	1.98639	1.98511	1.98125	1.97465	1.96504	1.95191
80	1.97688	1.97563	1.97182	1.96532	1.95584	1.94289
100	1.97122	1.96997	1.96620	1.95975	1.95035	1.93751
120	1.96745	1.96622	1.96247	1.95606	1.94671	1.93394
200	1.95997	1.95874	1.95504	1.94870	1.93947	1.92684
∞	1.94882	1.94762	1.94398	1.93776	1.92869	1.91627

TABLE B.2

Two-sided Percentage Point (h) for Equal
Correlations $\left(\text{Case I with Correlation } \rho\right)$

$$p = 2, \quad 1-\alpha = 0.90$$

$\nu \downarrow$	ρ				
	0.6	0.7	0.8	0.9	$1/(1+\sqrt{2})$
1	8.49422	8.28773	8.00498	7.57998	8.75206
2	3.66532	3.59200	3.49235	3.34437	3.75745
3	2.87331	2.82211	2.75266	2.64974	2.93780
4	2.56605	2.52361	2.46604	2.38059	2.61953
5	2.40481	2.36706	2.31579	2.23949	2.45238
6	2.30593	2.27110	2.22372	2.15301	2.34980
7	2.23924	2.20639	2.16165	2.09469	2.28056
8	2.19126	2.15986	2.11703	2.05275	2.23073
9	2.15511	2.12481	2.08342	2.02115	2.19317
10	2.12692	2.09747	2.05720	1.99650	2.16386
11	2.10431	2.07556	2.03619	1.97674	2.14035
12	2.08578	2.05760	2.01897	1.96054	2.12108
13	2.07032	2.04262	2.00461	1.94702	2.10500
14	2.05723	2.02994	1.99244	1.93557	2.09137
15	2.04601	2.01906	1.98201	1.92575	2.07969
16	2.03627	2.00962	1.97296	1.91724	2.06955
17	2.02774	2.00136	1.96504	1.90978	2.06067
18	2.02022	1.99407	1.95805	1.90320	2.05284
19	2.01353	1.98759	1.95184	1.89734	2.04587
20	2.00754	1.98179	1.94628	1.89211	2.03964
21	2.00216	1.97657	1.94127	1.88739	2.03402
22	1.99728	1.97185	1.93674	1.88312	2.02894
23	1.99284	1.96755	1.93262	1.87924	2.02432
24	1.98880	1.96363	1.92886	1.87570	2.02010
25	1.98508	1.96004	1.92541	1.87245	2.01623
26	1.98167	1.95673	1.92223	1.86946	2.01268
27	1.97852	1.95367	1.91930	1.86670	2.00939
28	1.97560	1.95084	1.91659	1.86414	2.00635
29	1.97288	1.94822	1.91407	1.86176	2.00352
30	1.97036	1.94577	1.91173	1.85955	2.00089
35	1.95996	1.93570	1.90207	1.85045	1.99005
40	1.95223	1.92822	1.89488	1.84368	1.98199
45	1.94626	1.92243	1.88933	1.83844	1.97576
50	1.94150	1.91783	1.88491	1.83428	1.97080
60	1.93441	1.91096	1.87832	1.82806	1.96340
80	1.92561	1.90243	1.87014	1.82035	1.95422
100	1.92037	1.89736	1.86527	1.81575	1.94875
120	1.91689	1.89398	1.86204	1.81270	1.94512
200	1.90996	1.88728	1.85560	1.80662	1.93789
∞	1.89966	1.87730	1.84602	1.79759	1.92714

TABLE B.2

Two-sided Percentage Point (h) for Equal
Correlations $\left(\text{Case I with Correlation } \rho\right)$

$p = 3, \quad 1-\alpha = 0.90$

$\nu \downarrow$	ρ					
	0.0	0.1	0.2	0.3	0.4	0.5
2	4.37869	4.37072	4.34735	4.30874	4.25414	4.18171
3	3.36854	3.36303	3.34687	3.32010	3.28218	3.23183
4	2.97548	2.97098	2.95773	2.93574	2.90455	2.86308
5	2.76863	2.76467	2.75301	2.73362	2.70608	2.66943
6	2.64146	2.63785	2.62719	2.60945	2.58422	2.55060
7	2.55550	2.55213	2.54217	2.52557	2.50194	2.47042
8	2.49356	2.49036	2.48092	2.46516	2.44271	2.41273
9	2.44682	2.44376	2.43471	2.41961	2.39806	2.36926
10	2.41031	2.40736	2.39863	2.38404	2.36320	2.33534
11	2.38101	2.37815	2.36967	2.35550	2.33525	2.30814
12	2.35697	2.35419	2.34592	2.33210	2.31233	2.28585
13	2.33690	2.33418	2.32610	2.31256	2.29320	2.26726
14	2.31989	2.31722	2.30929	2.29601	2.27700	2.25151
15	2.30529	2.30267	2.29487	2.28181	2.26310	2.23800
16	2.29262	2.29004	2.28236	2.26949	2.25105	2.22628
17	2.28153	2.27898	2.27141	2.25870	2.24049	2.21603
18	2.27173	2.26922	2.26173	2.24918	2.23118	2.20698
19	2.26302	2.26053	2.25313	2.24071	2.22289	2.19893
20	2.25521	2.25275	2.24543	2.23313	2.21548	2.19173
21	2.24819	2.24575	2.23849	2.22630	2.20880	2.18525
22	2.24183	2.23941	2.23221	2.22012	2.20276	2.17938
23	2.23604	2.23365	2.22650	2.21450	2.19726	2.17405
24	2.23076	2.22838	2.22129	2.20937	2.19225	2.16918
25	2.22591	2.22355	2.21651	2.20466	2.18765	2.16471
26	2.22145	2.21911	2.21211	2.20033	2.18341	2.16061
27	2.21734	2.21500	2.20804	2.19634	2.17951	2.15681
28	2.21352	2.21120	2.20428	2.19264	2.17589	2.15330
29	2.20998	2.20767	2.20079	2.18920	2.17253	2.15004
30	2.20668	2.20439	2.19753	2.18600	2.16940	2.14701
35	2.19309	2.19084	2.18412	2.17281	2.15652	2.13451
40	2.18298	2.18077	2.17415	2.16300	2.14693	2.12521
45	2.17516	2.17297	2.16644	2.15542	2.13953	2.11803
50	2.16893	2.16677	2.16030	2.14938	2.13363	2.11231
60	2.15964	2.15751	2.15114	2.14038	2.12484	2.10379
80	2.14811	2.14602	2.13977	2.12920	2.11393	2.09321
100	2.14124	2.13917	2.13299	2.12254	2.10743	2.08691
120	2.13667	2.13463	2.12849	2.11812	2.10311	2.08273
200	2.12758	2.12557	2.11953	2.10932	2.09452	2.07441
∞	2.11405	2.11209	2.10620	2.09622	2.08175	2.06204

TABLE B.2

Two-sided Percentage Point (h) for Equal Correlations $\left(\text{Case I with Correlation } \rho\right)$

p = 3, 1-α = 0.90

ν ↓	ρ 0.6	0.7	0.8	0.9	1/(1+√3)
2	4.08804	3.96688	3.80553	3.57090	4.27457
3	3.16669	3.08243	2.97029	2.80737	3.29638
4	2.80937	2.73984	2.64719	2.51228	2.91623
5	2.62190	2.56029	2.47804	2.35787	2.71640
6	2.50697	2.45031	2.37451	2.26338	2.59368
7	2.42946	2.37620	2.30479	2.19973	2.51080
8	2.37372	2.32293	2.25470	2.15400	2.45113
9	2.33175	2.28284	2.21701	2.11957	2.40614
10	2.29901	2.25158	2.18762	2.09272	2.37102
11	2.27277	2.22653	2.16408	2.07121	2.34285
12	2.25127	2.20601	2.14480	2.05359	2.31975
13	2.23334	2.18890	2.12872	2.03888	2.30047
14	2.21816	2.17442	2.11511	2.02644	2.28414
15	2.20513	2.16200	2.10344	2.01576	2.27013
16	2.19385	2.15124	2.09332	2.00650	2.25798
17	2.18397	2.14181	2.08447	1.99840	2.24733
18	2.17525	2.13350	2.07665	1.99125	2.23794
19	2.16749	2.12611	2.06971	1.98489	2.22959
20	2.16056	2.11949	2.06349	1.97920	2.22211
21	2.15432	2.11354	2.05790	1.97408	2.21538
22	2.14867	2.10816	2.05284	1.96944	2.20928
23	2.14353	2.10326	2.04823	1.96523	2.20374
24	2.13884	2.09879	2.04403	1.96138	2.19868
25	2.13454	2.09469	2.04018	1.95785	2.19405
26	2.13059	2.09093	2.03664	1.95461	2.18978
27	2.12694	2.08745	2.03337	1.95161	2.18584
28	2.12356	2.08422	2.03034	1.94884	2.18219
29	2.12042	2.08123	2.02753	1.94626	2.17880
30	2.11750	2.07845	2.02491	1.94386	2.17564
35	2.10547	2.06698	2.01413	1.93398	2.16265
40	2.09652	2.05846	2.00611	1.92663	2.15298
45	2.08961	2.05187	1.99992	1.92096	2.14551
50	2.08411	2.04663	1.99499	1.91644	2.13956
60	2.07591	2.03882	1.98764	1.90970	2.13069
80	2.06574	2.02912	1.97852	1.90133	2.11968
100	2.05968	2.02335	1.97309	1.89635	2.11312
120	2.05566	2.01951	1.96949	1.89304	2.10877
200	2.04766	2.01189	1.96231	1.88646	2.10010
∞	2.03576	2.00055	1.95164	1.87666	2.08721

TABLE B.2

Two-sided Percentage Point (h) for Equal
Correlations $\left(\text{Case } I \text{ with Correlation } \rho\right)$

$p = 4, \quad 1-\alpha = 0.90$

$\nu \downarrow$	0.0	0.1	0.2	0.3	0.4	0.5
			ρ			
2	4.76587	4.75438	4.72162	4.66877	4.59547	4.49981
3	3.63708	3.62921	3.60667	3.57016	3.51939	3.45300
4	3.19694	3.19055	3.17216	3.14230	3.10067	3.04612
5	2.96487	2.95929	2.94319	2.91697	2.88033	2.83224
6	2.82198	2.81692	2.80227	2.77837	2.74490	2.70091
7	2.72527	2.72057	2.70694	2.68466	2.65340	2.61224
8	2.65551	2.65108	2.63821	2.61711	2.58749	2.54842
9	2.60283	2.59860	2.58631	2.56615	2.53778	2.50032
10	2.56165	2.55759	2.54576	2.52633	2.49896	2.46278
11	2.52857	2.52465	2.51321	2.49437	2.46782	2.43267
12	2.50143	2.49762	2.48649	2.46816	2.44229	2.40800
13	2.47876	2.47504	2.46418	2.44628	2.42097	2.38742
14	2.45953	2.45590	2.44527	2.42773	2.40292	2.36998
15	2.44302	2.43946	2.42904	2.41181	2.38743	2.35503
16	2.42870	2.42520	2.41495	2.39800	2.37399	2.34206
17	2.41615	2.41270	2.40261	2.38590	2.36222	2.33071
18	2.40506	2.40167	2.39171	2.37522	2.35183	2.32069
19	2.39519	2.39185	2.38201	2.36572	2.34260	2.31178
20	2.38636	2.38305	2.37333	2.35722	2.33433	2.30381
21	2.37841	2.37513	2.36552	2.34956	2.32688	2.29663
22	2.37120	2.36796	2.35844	2.34263	2.32015	2.29014
23	2.36465	2.36144	2.35200	2.33632	2.31402	2.28423
24	2.35867	2.35548	2.34612	2.33056	2.30843	2.27884
25	2.35318	2.35002	2.34073	2.32529	2.30330	2.27390
26	2.34812	2.34499	2.33577	2.32043	2.29858	2.26936
27	2.34346	2.34035	2.33119	2.31594	2.29422	2.26516
28	2.33914	2.33605	2.32694	2.31179	2.29018	2.26127
29	2.33512	2.33205	2.32300	2.30793	2.28644	2.25766
30	2.33138	2.32833	2.31933	2.30434	2.28295	2.25430
35	2.31598	2.31300	2.30420	2.28954	2.26858	2.24047
40	2.30451	2.30159	2.29295	2.27852	2.25789	2.23018
45	2.29564	2.29276	2.28425	2.27001	2.24963	2.22223
50	2.28858	2.28574	2.27732	2.26324	2.24306	2.21591
60	2.27804	2.27525	2.26698	2.25312	2.23325	2.20648
80	2.26495	2.26223	2.25414	2.24058	2.22109	2.19478
100	2.25714	2.25446	2.24649	2.23310	2.21384	2.18781
120	2.25196	2.24930	2.24141	2.22813	2.20903	2.18318
200	2.24163	2.23903	2.23129	2.21825	2.19945	2.17398
∞	2.22627	2.22375	2.21623	2.20355	2.18521	2.16029

TABLE B.2

Two-sided Percentage Point (h) for Equal
Correlations (*Case I with Correlation* ρ)

p = 4, 1-α = 0.90

ν ↓	ρ				
	0.6	0.7	0.8	0.9	$1/(1+\sqrt{4})$
2	4.37783	4.22203	4.01699	3.72232	4.64667
3	3.36825	3.25995	3.11743	2.91278	3.55486
4	2.97638	2.88713	2.76951	2.60021	3.12977
5	2.77065	2.69170	2.58742	2.43679	2.90595
6	2.64447	2.57198	2.47601	2.33686	2.76831
7	2.55936	2.49131	2.40101	2.26959	2.67527
8	2.49815	2.43334	2.34715	2.22128	2.60822
9	2.45204	2.38971	2.30663	2.18492	2.55763
10	2.41608	2.35570	2.27505	2.15659	2.51812
11	2.38727	2.32846	2.24976	2.13389	2.48641
12	2.36366	2.30614	2.22905	2.11529	2.46041
13	2.34396	2.28754	2.21179	2.09979	2.43869
14	2.32729	2.27179	2.19717	2.08666	2.42030
15	2.31300	2.25829	2.18464	2.07540	2.40451
16	2.30060	2.24659	2.17378	2.06564	2.39081
17	2.28975	2.23635	2.16428	2.05710	2.37881
18	2.28018	2.22731	2.15590	2.04956	2.36822
19	2.27167	2.21928	2.14844	2.04286	2.35880
20	2.26406	2.21210	2.14178	2.03686	2.35037
21	2.25721	2.20563	2.13578	2.03146	2.34278
22	2.25101	2.19978	2.13035	2.02658	2.33590
23	2.24537	2.19446	2.12541	2.02214	2.32966
24	2.24023	2.18961	2.12090	2.01808	2.32395
25	2.23551	2.18516	2.11677	2.01437	2.31871
26	2.23117	2.18106	2.11297	2.01095	2.31390
27	2.22716	2.17728	2.10947	2.00779	2.30945
28	2.22346	2.17378	2.10622	2.00487	2.30533
29	2.22001	2.17054	2.10321	2.00215	2.30151
30	2.21681	2.16751	2.10040	1.99963	2.29795
35	2.20361	2.15506	2.08884	1.98922	2.28328
40	2.19380	2.14581	2.08025	1.98148	2.27237
45	2.18622	2.13866	2.07361	1.97550	2.26393
50	2.18019	2.13297	2.06833	1.97075	2.25722
60	2.17120	2.12449	2.06046	1.96365	2.24720
80	2.16005	2.11397	2.05069	1.95484	2.23477
100	2.15340	2.10771	2.04487	1.94960	2.22737
120	2.14899	2.10355	2.04100	1.94611	2.22245
200	2.14022	2.09528	2.03332	1.93918	2.21266
∞	2.12719	2.08298	2.02189	1.92888	2.19810

BECHHOFER and DUNNETT

TABLE B.2

Two-sided Percentage Point (h) for Equal Correlations (*Case I with Correlation* ρ)

$$p = 5, \quad 1-\alpha = 0.90$$

$\nu \downarrow$	ρ					
	0.0	0.1	0.2	0.3	0.4	0.5
2	5.06299	5.04835	5.00753	4.94283	4.85433	4.74016
3	3.84387	3.83387	3.80583	3.76118	3.69990	3.62065
4	3.36758	3.35949	3.33667	3.30021	3.25001	3.18495
5	3.11601	3.10896	3.08904	3.05708	3.01297	2.95568
6	2.96087	2.95451	2.93643	2.90736	2.86713	2.81478
7	2.85575	2.84986	2.83308	2.80602	2.76851	2.71959
8	2.77984	2.77430	2.75849	2.73293	2.69743	2.65105
9	2.72246	2.71720	2.70214	2.67774	2.64379	2.59937
10	2.67758	2.67254	2.65807	2.63460	2.60189	2.55903
11	2.64151	2.63665	2.62268	2.59997	2.56827	2.52667
12	2.61189	2.60718	2.59362	2.57154	2.54069	2.50015
13	2.58714	2.58256	2.56934	2.54781	2.51767	2.47802
14	2.56614	2.56167	2.54876	2.52768	2.49816	2.45927
15	2.54811	2.54373	2.53108	2.51041	2.48142	2.44319
16	2.53245	2.52816	2.51574	2.49542	2.46689	2.42925
17	2.51873	2.51451	2.50229	2.48229	2.45417	2.41704
18	2.50661	2.50246	2.49042	2.47069	2.44295	2.40627
19	2.49582	2.49173	2.47985	2.46038	2.43296	2.39669
20	2.48615	2.48212	2.47039	2.45114	2.42402	2.38811
21	2.47745	2.47346	2.46187	2.44282	2.41597	2.38039
22	2.46957	2.46562	2.45415	2.43529	2.40869	2.37341
23	2.46240	2.45849	2.44713	2.42844	2.40206	2.36706
24	2.45584	2.45198	2.44072	2.42219	2.39601	2.36126
25	2.44983	2.44600	2.43484	2.41645	2.39047	2.35595
26	2.44430	2.44050	2.42943	2.41117	2.38536	2.35106
27	2.43919	2.43542	2.42443	2.40630	2.38065	2.34654
28	2.43446	2.43072	2.41980	2.40179	2.37629	2.34236
29	2.43006	2.42635	2.41550	2.39759	2.37224	2.33848
30	2.42597	2.42228	2.41150	2.39369	2.36846	2.33487
35	2.40908	2.40550	2.39499	2.37760	2.35292	2.31999
40	2.39651	2.39300	2.38271	2.36563	2.34136	2.30893
45	2.38679	2.38334	2.37320	2.35638	2.33243	2.30038
50	2.37905	2.37564	2.36564	2.34901	2.32532	2.29358
60	2.36748	2.36415	2.35435	2.33802	2.31471	2.28344
80	2.35312	2.34989	2.34033	2.32438	2.30155	2.27086
100	2.34456	2.34137	2.33197	2.31624	2.29371	2.26336
120	2.33887	2.33572	2.32641	2.31084	2.28851	2.25839
200	2.32754	2.32446	2.31536	2.30009	2.27815	2.24850
∞	2.31066	2.30770	2.29891	2.28410	2.26274	2.23379

TABLE B.2

*Two-sided Percentage Point (h) for Equal
Correlations* $\left(Case\ I\ with\ Correlation\ \rho\right)$

p = 5, 1-α = 0.90

ν ↓	ρ				
	0.6	0.7	0.8	0.9	1/(1+√5)
2	4.59596	4.41333	4.17484	3.83475	4.93584
3	3.52043	3.39339	3.22750	2.99113	3.75634
4	3.10252	2.99787	2.86100	2.66552	3.29625
5	2.88295	2.79044	2.66917	2.49536	3.05361
6	2.74819	2.66333	2.55181	2.39135	2.90420
7	2.65727	2.57767	2.47281	2.32136	2.80308
8	2.59185	2.51611	2.41609	2.27111	2.73014
9	2.54258	2.46978	2.37342	2.23331	2.67508
10	2.50414	2.43367	2.34018	2.20385	2.63204
11	2.47334	2.40474	2.31355	2.18026	2.59748
12	2.44810	2.38105	2.29176	2.16094	2.56913
13	2.42705	2.36129	2.27359	2.14482	2.54545
14	2.40922	2.34458	2.25821	2.13118	2.52538
15	2.39394	2.33025	2.24503	2.11949	2.50814
16	2.38069	2.31782	2.23361	2.10935	2.49319
17	2.36910	2.30696	2.22362	2.10048	2.48009
18	2.35887	2.29737	2.21480	2.09265	2.46853
19	2.34977	2.28884	2.20696	2.08569	2.45824
20	2.34163	2.28122	2.19995	2.07946	2.44903
21	2.33431	2.27436	2.19364	2.07386	2.44073
22	2.32768	2.26815	2.18793	2.06879	2.43322
23	2.32166	2.26251	2.18274	2.06418	2.42639
24	2.31616	2.25736	2.17800	2.05997	2.42015
25	2.31112	2.25264	2.17366	2.05611	2.41443
26	2.30648	2.24829	2.16966	2.05256	2.40917
27	2.30220	2.24428	2.16598	2.04928	2.40431
28	2.29824	2.24057	2.16256	2.04625	2.39980
29	2.29456	2.23713	2.15939	2.04343	2.39562
30	2.29113	2.23392	2.15644	2.04080	2.39173
35	2.27703	2.22072	2.14430	2.03000	2.37569
40	2.26655	2.21090	2.13527	2.02197	2.36375
45	2.25845	2.20332	2.12829	2.01577	2.35452
50	2.25201	2.19729	2.12274	2.01083	2.34718
60	2.24240	2.18830	2.11446	2.00346	2.33621
80	2.23049	2.17715	2.10420	1.99433	2.32261
100	2.22339	2.17050	2.09809	1.98888	2.31450
120	2.21868	2.16610	2.09403	1.98527	2.30912
200	2.20932	2.15733	2.08596	1.97808	2.29840
∞	2.19539	2.14429	2.07395	1.96738	2.28246

BECHHOFER and DUNNETT

TABLE B.2

*Two-sided Percentage Point (h) for Equal
Correlations* $\left(\text{Case } I \text{ with Correlation } \rho\right)$

$p = 6, \ 1-\alpha = 0.90$

	ρ					
ν ↓	0.0	0.1	0.2	0.3	0.4	0.5
2	5.30279	5.28532	5.23747	5.16270	5.06150	4.93203
3	4.01132	3.99941	3.96655	3.91493	3.84481	3.75489
4	3.50593	3.49630	3.46959	3.42746	3.37004	3.29622
5	3.23856	3.23020	3.20691	3.17002	3.11960	3.05462
6	3.07347	3.06593	3.04483	3.01131	2.96537	2.90603
7	2.96147	2.95451	2.93496	2.90379	2.86099	2.80559
8	2.88052	2.87399	2.85559	2.82619	2.78571	2.73322
9	2.81929	2.81309	2.79559	2.76755	2.72888	2.67864
10	2.77135	2.76542	2.74864	2.72170	2.68447	2.63603
11	2.73280	2.72710	2.71091	2.68487	2.64882	2.60184
12	2.70114	2.69562	2.67992	2.65463	2.61957	2.57381
13	2.67466	2.66930	2.65402	2.62937	2.59515	2.55042
14	2.65219	2.64696	2.63205	2.60795	2.57445	2.53060
15	2.63288	2.62778	2.61318	2.58956	2.55668	2.51360
16	2.61611	2.61111	2.59680	2.57360	2.54126	2.49886
17	2.60142	2.59651	2.58244	2.55961	2.52777	2.48595
18	2.58843	2.58360	2.56976	2.54726	2.51584	2.47456
19	2.57686	2.57211	2.55847	2.53627	2.50524	2.46443
20	2.56650	2.56182	2.54836	2.52643	2.49575	2.45536
21	2.55717	2.55255	2.53925	2.51757	2.48721	2.44720
22	2.54872	2.54415	2.53100	2.50954	2.47947	2.43982
23	2.54103	2.53651	2.52350	2.50224	2.47244	2.43310
24	2.53400	2.52953	2.51664	2.49558	2.46601	2.42697
25	2.52755	2.52313	2.51035	2.48946	2.46012	2.42135
26	2.52162	2.51723	2.50457	2.48384	2.45470	2.41618
27	2.51613	2.51179	2.49922	2.47864	2.44970	2.41140
28	2.51105	2.50674	2.49427	2.47383	2.44506	2.40698
29	2.50633	2.50206	2.48967	2.46936	2.44076	2.40288
30	2.50194	2.49769	2.48538	2.46519	2.43675	2.39906
35	2.48381	2.47970	2.46772	2.44804	2.42024	2.38333
40	2.47031	2.46629	2.45458	2.43527	2.40796	2.37163
45	2.45987	2.45592	2.44441	2.42540	2.39847	2.36259
50	2.45155	2.44766	2.43631	2.41754	2.39092	2.35540
60	2.43912	2.43532	2.42422	2.40581	2.37965	2.34467
80	2.42368	2.42000	2.40920	2.39126	2.36567	2.33137
100	2.41447	2.41086	2.40025	2.38258	2.35734	2.32344
120	2.40835	2.40479	2.39430	2.37682	2.35181	2.31818
200	2.39616	2.39269	2.38246	2.36534	2.34080	2.30772
∞	2.37800	2.37468	2.36483	2.34828	2.32444	2.29217

TABLE B.2

Two-sided Percentage Point (h) for Equal
Correlations (*Case I with Correlation* ρ)

$$p = 6, \quad 1-\alpha = 0.90$$

ν ↓	ρ				
	0.6	0.7	0.8	0.9	$1/(1+\sqrt{6})$
2	4.76968	4.56530	4.29990	3.92352	5.17144
3	3.64194	3.49965	3.31487	3.05307	3.92098
4	3.20333	3.08610	2.93364	2.71714	3.43240
5	2.97269	2.86909	2.73405	2.54163	3.17435
6	2.83106	2.73607	2.61193	2.43437	3.01525
7	2.73545	2.64640	2.52973	2.36221	2.90746
8	2.66665	2.58196	2.47072	2.31042	2.82965
9	2.61481	2.53345	2.42633	2.27147	2.77086
10	2.57437	2.49564	2.39175	2.24111	2.72488
11	2.54195	2.46535	2.36406	2.21681	2.68794
12	2.51539	2.44055	2.34139	2.19690	2.65762
13	2.49324	2.41987	2.32250	2.18031	2.63229
14	2.47448	2.40237	2.30651	2.16626	2.61081
15	2.45840	2.38736	2.29280	2.15422	2.59236
16	2.44445	2.37436	2.28093	2.14378	2.57635
17	2.43225	2.36299	2.27054	2.13464	2.56232
18	2.42148	2.35295	2.26137	2.12658	2.54993
19	2.41191	2.34403	2.25322	2.11942	2.53890
20	2.40334	2.33605	2.24593	2.11300	2.52903
21	2.39563	2.32887	2.23937	2.10723	2.52014
22	2.38866	2.32237	2.23344	2.10201	2.51209
23	2.38232	2.31646	2.22804	2.09727	2.50477
24	2.37653	2.31107	2.22312	2.09293	2.49808
25	2.37123	2.30613	2.21861	2.08896	2.49195
26	2.36635	2.30159	2.21446	2.08531	2.48630
27	2.36184	2.29739	2.21062	2.08193	2.48109
28	2.35767	2.29351	2.20708	2.07881	2.47626
29	2.35380	2.28991	2.20378	2.07591	2.47177
30	2.35020	2.28655	2.20072	2.07321	2.46760
35	2.33536	2.27273	2.18810	2.06209	2.45038
40	2.32433	2.26247	2.17871	2.05383	2.43757
45	2.31581	2.25453	2.17147	2.04744	2.42767
50	2.30903	2.24822	2.16570	2.04236	2.41978
60	2.29892	2.23882	2.15710	2.03478	2.40801
80	2.28639	2.22715	2.14644	2.02538	2.39341
100	2.27893	2.22021	2.14009	2.01978	2.38469
120	2.27397	2.21560	2.13587	2.01606	2.37891
200	2.26412	2.20643	2.12749	2.00866	2.36740
∞	2.24948	2.19280	2.11502	1.99766	2.35027

TABLE B.2

Two-sided Percentage Point (h) for Equal
Correlations $\left(Case\ I\ with\ Correlation\ \rho\right)$

p = 7, 1-α = 0.90

ν ↓	ρ					
	0.0	0.1	0.2	0.3	0.4	0.5
2	5.50307	5.48304	5.42899	5.34549	5.23342	5.09099
3	4.15160	4.13795	4.10081	4.04312	3.96541	3.86638
4	3.62197	3.61094	3.58078	3.53369	3.47004	3.38874
5	3.34141	3.33185	3.30555	3.26434	3.20847	3.13691
6	3.16797	3.15935	3.13556	3.09813	3.04724	2.98192
7	3.05018	3.04224	3.02021	2.98544	2.93805	2.87709
8	2.96498	2.95754	2.93683	2.90405	2.85926	2.80153
9	2.90048	2.89343	2.87375	2.84252	2.79975	2.74452
10	2.84995	2.84322	2.82437	2.79438	2.75323	2.69999
11	2.80930	2.80283	2.78466	2.75569	2.71587	2.66427
12	2.77588	2.76963	2.75203	2.72392	2.68521	2.63497
13	2.74793	2.74187	2.72476	2.69737	2.65961	2.61052
14	2.72420	2.71830	2.70161	2.67485	2.63790	2.58980
15	2.70381	2.69805	2.68172	2.65552	2.61927	2.57202
16	2.68609	2.68045	2.66446	2.63873	2.60310	2.55661
17	2.67056	2.66503	2.64932	2.62402	2.58894	2.54311
18	2.65682	2.65139	2.63594	2.61102	2.57643	2.53120
19	2.64460	2.63925	2.62403	2.59946	2.56531	2.52060
20	2.63364	2.62838	2.61337	2.58910	2.55535	2.51112
21	2.62377	2.61858	2.60376	2.57977	2.54638	2.50258
22	2.61482	2.60970	2.59505	2.57133	2.53826	2.49486
23	2.60668	2.60162	2.58714	2.56365	2.53088	2.48784
24	2.59925	2.59424	2.57990	2.55663	2.52414	2.48142
25	2.59242	2.58747	2.57326	2.55019	2.51795	2.47554
26	2.58614	2.58123	2.56715	2.54426	2.51226	2.47014
27	2.58033	2.57547	2.56151	2.53879	2.50701	2.46514
28	2.57495	2.57013	2.55628	2.53373	2.50215	2.46052
29	2.56995	2.56517	2.55142	2.52902	2.49763	2.45623
30	2.56530	2.56055	2.54690	2.52463	2.49342	2.45223
35	2.54610	2.54151	2.52825	2.50657	2.47609	2.43577
40	2.53179	2.52732	2.51436	2.49312	2.46320	2.42353
45	2.52072	2.51633	2.50361	2.48272	2.45323	2.41408
50	2.51190	2.50758	2.49506	2.47444	2.44530	2.40655
60	2.49872	2.49452	2.48228	2.46208	2.43347	2.39533
80	2.48235	2.47828	2.46641	2.44674	2.41879	2.38142
100	2.47257	2.46859	2.45694	2.43759	2.41004	2.37313
120	2.46608	2.46216	2.45065	2.43152	2.40423	2.36763
200	2.45314	2.44933	2.43812	2.41943	2.39268	2.35668
∞	2.43386	2.43023	2.41947	2.40144	2.37549	2.34042

TABLE B.2

Two-sided Percentage Point (h) for Equal
Correlations $\left[\text{Case I with Correlation } \rho\right]$

$p = 7, \ 1-\alpha = 0.90$

$\nu \downarrow$	ρ				
	0.6	0.7	0.8	0.9	$1/(1+\sqrt{7})$
2	4.91335	4.69077	4.40296	3.99650	5.36967
3	3.74268	3.58756	3.38700	3.10405	4.05986
4	3.28697	3.15915	2.99362	2.75963	3.54737
5	3.04716	2.93421	2.78762	2.57969	3.27633
6	2.89982	2.79627	2.66156	2.46974	3.10903
7	2.80031	2.70327	2.57670	2.39578	2.99557
8	2.72868	2.63642	2.51578	2.34273	2.91361
9	2.67470	2.58610	2.46996	2.30282	2.85163
10	2.63258	2.54687	2.43426	2.27172	2.80314
11	2.59881	2.51545	2.40568	2.24682	2.76415
12	2.57114	2.48971	2.38229	2.22644	2.73214
13	2.54806	2.46826	2.36279	2.20944	2.70539
14	2.52852	2.45010	2.34629	2.19506	2.68269
15	2.51176	2.43454	2.33215	2.18273	2.66319
16	2.49723	2.42105	2.31989	2.17204	2.64627
17	2.48452	2.40925	2.30917	2.16268	2.63144
18	2.47330	2.39883	2.29971	2.15443	2.61833
19	2.46332	2.38958	2.29131	2.14710	2.60667
20	2.45440	2.38130	2.28379	2.14053	2.59622
21	2.44637	2.37385	2.27702	2.13462	2.58682
22	2.43910	2.36711	2.27090	2.12928	2.57830
23	2.43249	2.36099	2.26534	2.12442	2.57055
24	2.42646	2.35540	2.26026	2.11998	2.56347
25	2.42094	2.35027	2.25560	2.11592	2.55697
26	2.41585	2.34556	2.25132	2.11218	2.55100
27	2.41116	2.34121	2.24737	2.10872	2.54548
28	2.40681	2.33718	2.24371	2.10553	2.54036
29	2.40278	2.33344	2.24031	2.10256	2.53561
30	2.39902	2.32996	2.23715	2.09980	2.53119
35	2.38356	2.31563	2.22413	2.08842	2.51296
40	2.37207	2.30498	2.21445	2.07996	2.49938
45	2.36319	2.29676	2.20698	2.07342	2.48889
50	2.35613	2.29022	2.20103	2.06822	2.48053
60	2.34560	2.28046	2.19217	2.06046	2.46805
80	2.33255	2.26837	2.18117	2.05084	2.45256
100	2.32477	2.26116	2.17462	2.04511	2.44332
120	2.31961	2.25638	2.17028	2.04131	2.43719
200	2.30935	2.24688	2.16163	2.03374	2.42497
∞	2.29410	2.23274	2.14878	2.02248	2.40680

TABLE B.2

Two-sided Percentage Point (h) for Equal
Correlations   (Case I with Correlation ρ)

p = 8, 1-α = 0.90

ν ↓	0.0	0.1	0.2	0.3	0.4	0.5
2	5.67455	5.65219	5.59261	5.50142	5.37988	5.22623
3	4.27203	4.25677	4.21581	4.15276	4.06839	3.96146
4	3.72172	3.70941	3.67613	3.62465	3.55553	3.46770
5	3.42986	3.41920	3.39020	3.34515	3.28447	3.20717
6	3.24925	3.23965	3.21343	3.17253	3.11727	3.04672
7	3.12649	3.11765	3.09338	3.05541	3.00397	2.93814
8	3.03762	3.02934	3.00654	2.97076	2.92216	2.85984
9	2.97029	2.96246	2.94081	2.90673	2.86035	2.80075
10	2.91752	2.91005	2.88933	2.85662	2.81201	2.75458
11	2.87504	2.86787	2.84791	2.81633	2.77318	2.71753
12	2.84011	2.83318	2.81387	2.78324	2.74131	2.68714
13	2.81088	2.80416	2.78540	2.75557	2.71468	2.66177
14	2.78605	2.77952	2.76123	2.73210	2.69210	2.64028
15	2.76471	2.75834	2.74046	2.71194	2.67272	2.62184
16	2.74616	2.73993	2.72242	2.69444	2.65590	2.60584
17	2.72989	2.72379	2.70660	2.67910	2.64117	2.59183
18	2.71551	2.70952	2.69262	2.66555	2.62815	2.57947
19	2.70270	2.69681	2.68018	2.65348	2.61658	2.56847
20	2.69122	2.68543	2.66903	2.64268	2.60621	2.55863
21	2.68087	2.67516	2.65898	2.63295	2.59688	2.54977
22	2.67150	2.66587	2.64988	2.62413	2.58843	2.54175
23	2.66297	2.65740	2.64160	2.61612	2.58074	2.53446
24	2.65517	2.64967	2.63403	2.60879	2.57372	2.52780
25	2.64801	2.64257	2.62709	2.60207	2.56729	2.52170
26	2.64142	2.63604	2.62069	2.59589	2.56136	2.51609
27	2.63533	2.63000	2.61479	2.59018	2.55590	2.51090
28	2.62969	2.62441	2.60932	2.58489	2.55083	2.50610
29	2.62445	2.61921	2.60424	2.57997	2.54613	2.50165
30	2.61956	2.61437	2.59950	2.57540	2.54175	2.49750
35	2.59941	2.59439	2.57998	2.55653	2.52370	2.48041
40	2.58440	2.57951	2.56544	2.54249	2.51028	2.46771
45	2.57277	2.56799	2.55419	2.53163	2.49990	2.45789
50	2.56351	2.55881	2.54523	2.52298	2.49164	2.45008
60	2.54966	2.54510	2.53185	2.51007	2.47931	2.43843
80	2.53246	2.52805	2.51522	2.49405	2.46402	2.42399
100	2.52219	2.51788	2.50530	2.48449	2.45491	2.41538
120	2.51536	2.51112	2.49871	2.47814	2.44886	2.40967
200	2.50176	2.49765	2.48559	2.46551	2.43682	2.39831
∞	2.48148	2.47758	2.46604	2.44671	2.41892	2.38142

TABLE B.2

*Two-sided Percentage Point (h) for Equal
Correlations* (*Case I with Correlation* ρ)

$$p = 8, \quad 1-\alpha = 0.90$$

$\nu \downarrow$	ρ				
	0.6	0.7	0.8	0.9	$1/(1+\sqrt{8})$
2	5.03543	4.79725	4.49030	4.05824	5.54040
3	3.82846	3.66231	3.44822	3.14723	4.17974
4	3.35825	3.22129	3.04455	2.79561	3.64671
5	3.11064	2.98961	2.83310	2.61192	3.36447
6	2.95842	2.84749	2.70368	2.49969	3.19009
7	2.85559	2.75165	2.61656	2.42422	3.07172
8	2.78154	2.68274	2.55401	2.37006	2.98615
9	2.72572	2.63086	2.50697	2.32933	2.92140
10	2.68216	2.59042	2.47032	2.29761	2.87070
11	2.64723	2.55802	2.44098	2.27221	2.82994
12	2.61861	2.53149	2.41696	2.25141	2.79644
13	2.59473	2.50937	2.39695	2.23408	2.76844
14	2.57451	2.49065	2.38001	2.21940	2.74467
15	2.55717	2.47460	2.36549	2.20683	2.72426
16	2.54214	2.46070	2.35292	2.19592	2.70652
17	2.52899	2.44853	2.34191	2.18639	2.69098
18	2.51738	2.43779	2.33220	2.17797	2.67725
19	2.50706	2.42825	2.32358	2.17049	2.66503
20	2.49782	2.41972	2.31586	2.16379	2.65408
21	2.48951	2.41204	2.30892	2.15777	2.64421
22	2.48199	2.40509	2.30263	2.15232	2.63528
23	2.47516	2.39878	2.29692	2.14736	2.62715
24	2.46892	2.39301	2.29171	2.14284	2.61972
25	2.46320	2.38773	2.28693	2.13869	2.61291
26	2.45794	2.38287	2.28254	2.13488	2.60664
27	2.45308	2.37839	2.27848	2.13136	2.60084
28	2.44859	2.37424	2.27473	2.12810	2.59548
29	2.44441	2.37038	2.27124	2.12507	2.59049
30	2.44053	2.36679	2.26800	2.12226	2.58585
35	2.42453	2.35202	2.25464	2.11065	2.56671
40	2.41264	2.34105	2.24471	2.10203	2.55246
45	2.40346	2.33257	2.23705	2.09537	2.54144
50	2.39615	2.32583	2.23094	2.09006	2.53266
60	2.38526	2.31577	2.22185	2.08216	2.51955
80	2.37175	2.30331	2.21057	2.07235	2.50327
100	2.36371	2.29589	2.20385	2.06651	2.49356
120	2.35837	2.29096	2.19939	2.06263	2.48711
200	2.34775	2.28116	2.19053	2.05491	2.47428
∞	2.33197	2.26660	2.17734	2.04344	2.45516

TABLE B.2

Two-sided Percentage Point (h) for Equal
Correlations   (*Case I with Correlation* ρ)

$$p = 9, \ 1-\alpha = 0.90$$

$\nu \downarrow$	ρ					
	0.0	0.1	0.2	0.3	0.4	0.5
2	5.82418	5.79967	5.73510	5.63707	5.50714	5.34362
3	4.37734	4.36061	4.31618	4.24833	4.15806	4.04415
4	3.80905	3.79555	3.75945	3.70403	3.63004	3.53644
5	3.50735	3.49567	3.46421	3.41571	3.35075	3.26836
6	3.32049	3.30997	3.28153	3.23750	3.17835	3.10316
7	3.19337	3.18369	3.15738	3.11651	3.06146	2.99131
8	3.10127	3.09222	3.06752	3.02902	2.97702	2.91063
9	3.03147	3.02291	2.99946	2.96281	2.91320	2.84971
10	2.97672	2.96856	2.94613	2.91097	2.86326	2.80211
11	2.93263	2.92480	2.90321	2.86927	2.82314	2.76390
12	2.89636	2.88881	2.86792	2.83502	2.79021	2.73255
13	2.86600	2.85868	2.83840	2.80637	2.76268	2.70638
14	2.84020	2.83309	2.81333	2.78207	2.73934	2.68421
15	2.81802	2.81109	2.79178	2.76118	2.71930	2.66517
16	2.79873	2.79196	2.77306	2.74305	2.70191	2.64867
17	2.78182	2.77518	2.75664	2.72715	2.68667	2.63421
18	2.76685	2.76035	2.74213	2.71310	2.67321	2.62145
19	2.75353	2.74714	2.72920	2.70060	2.66123	2.61010
20	2.74158	2.73530	2.71763	2.68940	2.65051	2.59994
21	2.73081	2.72462	2.70719	2.67931	2.64085	2.59080
22	2.72105	2.71495	2.69774	2.67017	2.63211	2.58252
23	2.71217	2.70615	2.68913	2.66186	2.62416	2.57500
24	2.70405	2.69810	2.68127	2.65426	2.61689	2.56813
25	2.69660	2.69071	2.67406	2.64730	2.61023	2.56183
26	2.68974	2.68391	2.66742	2.64088	2.60410	2.55603
27	2.68339	2.67763	2.66128	2.63496	2.59844	2.55068
28	2.67752	2.67181	2.65560	2.62947	2.59320	2.54572
29	2.67205	2.66639	2.65031	2.62437	2.58833	2.54112
30	2.66696	2.66135	2.64539	2.61963	2.58380	2.53684
35	2.64597	2.64056	2.62510	2.60006	2.56512	2.51920
40	2.63031	2.62505	2.60998	2.58549	2.55123	2.50608
45	2.61819	2.61305	2.59828	2.57422	2.54048	2.49595
50	2.60853	2.60348	2.58896	2.56524	2.53193	2.48789
60	2.59409	2.58919	2.57503	2.55184	2.51917	2.47586
80	2.57614	2.57142	2.55773	2.53521	2.50334	2.46095
100	2.56542	2.56082	2.54741	2.52528	2.49391	2.45206
120	2.55829	2.55377	2.54055	2.51870	2.48765	2.44617
200	2.54409	2.53972	2.52689	2.50558	2.47518	2.43444
∞	2.52292	2.51878	2.50654	2.48605	2.45665	2.41700

TABLE B.2

Two-sided Percentage Point (h) for Equal Correlations (*Case I with Correlation* ρ)

p = 9, 1-α = 0.90

ν ↓	ρ				
	0.6	0.7	0.8	0.9	$1/(1+\sqrt{9})$
2	5.14130	4.88950	4.56590	4.11161	5.69008
3	3.90299	3.72718	3.50128	3.18458	4.28506
4	3.42022	3.27525	3.08871	2.82674	3.73406
5	3.16585	3.03773	2.87253	2.63980	3.44201
6	3.00940	2.89197	2.74020	2.52559	3.26140
7	2.90366	2.79365	2.65111	2.44879	3.13871
8	2.82750	2.72295	2.58714	2.39369	3.04995
9	2.77008	2.66972	2.53903	2.35226	2.98274
10	2.72526	2.62822	2.50155	2.31998	2.93010
11	2.68931	2.59497	2.47155	2.29414	2.88775
12	2.65986	2.56774	2.44699	2.27299	2.85294
13	2.63528	2.54504	2.42652	2.25536	2.82382
14	2.61447	2.52582	2.40921	2.24044	2.79911
15	2.59662	2.50935	2.39436	2.22764	2.77787
16	2.58115	2.49508	2.38150	2.21656	2.75942
17	2.56761	2.48259	2.37025	2.20686	2.74325
18	2.55566	2.47158	2.36033	2.19830	2.72895
19	2.54503	2.46178	2.35151	2.19069	2.71622
20	2.53553	2.45303	2.34362	2.18388	2.70482
21	2.52697	2.44514	2.33652	2.17776	2.69455
22	2.51923	2.43801	2.33009	2.17222	2.68524
23	2.51220	2.43153	2.32426	2.16718	2.67677
24	2.50577	2.42562	2.31893	2.16258	2.66903
25	2.49988	2.42020	2.31405	2.15836	2.66193
26	2.49447	2.41521	2.30955	2.15448	2.65540
27	2.48947	2.41061	2.30541	2.15090	2.64936
28	2.48484	2.40635	2.30157	2.14759	2.64377
29	2.48054	2.40239	2.29801	2.14451	2.63857
30	2.47654	2.39871	2.29469	2.14165	2.63373
35	2.46008	2.38356	2.28104	2.12985	2.61378
40	2.44784	2.37229	2.27089	2.12108	2.59892
45	2.43838	2.36360	2.26305	2.11431	2.58742
50	2.43086	2.35668	2.25681	2.10892	2.57826
60	2.41965	2.34636	2.24752	2.10088	2.56458
80	2.40574	2.33357	2.23599	2.09091	2.54759
100	2.39747	2.32595	2.22912	2.08497	2.53746
120	2.39197	2.32090	2.22457	2.08103	2.53073
200	2.38104	2.31085	2.21550	2.07319	2.51732
∞	2.36480	2.29591	2.20203	2.06152	2.49736

TABLE B.2

Two-sided Percentage Point (h) for Equal
Correlations $\left(\text{Case I with Correlation } \rho\right)$

$$p = 10, \ 1-\alpha = 0.90$$

$\nu \downarrow$	0.0	0.1	0.2	0.3	0.4	0.5
2	5.95668	5.93019	5.86108	5.75688	5.61946	5.44715
3	4.47078	4.45269	4.40509	4.33290	4.23733	4.11719
4	3.88662	3.87202	3.83333	3.77434	3.69598	3.59721
5	3.57623	3.56359	3.52987	3.47824	3.40942	3.32247
6	3.38382	3.37245	3.34197	3.29510	3.23244	3.15308
7	3.25283	3.24237	3.21419	3.17069	3.11237	3.03834
8	3.15788	3.14810	3.12165	3.08067	3.02560	2.95554
9	3.08586	3.07662	3.05152	3.01253	2.95999	2.89301
10	3.02936	3.02055	2.99655	2.95915	2.90865	2.84414
11	2.98383	2.97539	2.95229	2.91621	2.86738	2.80489
12	2.94636	2.93822	2.91589	2.88091	2.83349	2.77270
13	2.91498	2.90710	2.88542	2.85139	2.80517	2.74582
14	2.88832	2.88067	2.85955	2.82634	2.78115	2.72303
15	2.86538	2.85792	2.83731	2.80481	2.76052	2.70348
16	2.84543	2.83815	2.81797	2.78611	2.74261	2.68651
17	2.82793	2.82081	2.80102	2.76972	2.72692	2.67166
18	2.81245	2.80547	2.78603	2.75523	2.71306	2.65854
19	2.79866	2.79180	2.77268	2.74233	2.70073	2.64688
20	2.78630	2.77955	2.76071	2.73078	2.68969	2.63644
21	2.77515	2.76851	2.74993	2.72037	2.67974	2.62704
22	2.76504	2.75850	2.74016	2.71094	2.67074	2.61854
23	2.75584	2.74939	2.73127	2.70236	2.66255	2.61080
24	2.74743	2.74106	2.72314	2.69452	2.65507	2.60374
25	2.73971	2.73341	2.71568	2.68733	2.64821	2.59726
26	2.73260	2.72637	2.70882	2.68071	2.64189	2.59130
27	2.72603	2.71987	2.70247	2.67460	2.63606	2.58580
28	2.71994	2.71384	2.69659	2.66893	2.63066	2.58071
29	2.71428	2.70823	2.69113	2.66367	2.62565	2.57598
30	2.70901	2.70301	2.68604	2.65877	2.62098	2.57158
35	2.68725	2.68147	2.66505	2.63856	2.60173	2.55344
40	2.67101	2.66541	2.64941	2.62352	2.58741	2.53996
45	2.65844	2.65297	2.63731	2.61188	2.57634	2.52954
50	2.64842	2.64305	2.62766	2.60261	2.56753	2.52125
60	2.63344	2.62824	2.61325	2.58877	2.55438	2.50889
80	2.61481	2.60981	2.59534	2.57159	2.53807	2.49355
100	2.60368	2.59881	2.58465	2.56134	2.52835	2.48442
120	2.59629	2.59150	2.57755	2.55453	2.52189	2.47836
200	2.58154	2.57692	2.56340	2.54098	2.50905	2.46630
∞	2.55955	2.55519	2.54233	2.52081	2.48994	2.44837

TABLE B.2

Two-sided Percentage Point (h) for Equal
Correlations (*Case I with Correlation* ρ)

p = 10, 1-α = 0.90

ν ↓	ρ				
	0.6	0.7	0.8	0.9	$1/(1+\sqrt{10})$
2	5.23458	4.97073	4.63240	4.15852	5.82316
3	3.96875	3.78437	3.54801	3.21744	4.37886
4	3.47495	3.32285	3.12762	2.85413	3.81193
5	3.21461	3.08018	2.90728	2.66432	3.51116
6	3.05443	2.93122	2.77238	2.54837	3.32500
7	2.94612	2.83071	2.68154	2.47040	3.19846
8	2.86810	2.75843	2.61632	2.41446	3.10684
9	2.80925	2.70399	2.56727	2.37240	3.03745
10	2.76332	2.66155	2.52906	2.33965	2.98306
11	2.72647	2.62754	2.49846	2.31342	2.93929
12	2.69628	2.59969	2.47342	2.29195	2.90329
13	2.67108	2.57648	2.45256	2.27406	2.87318
14	2.64974	2.55683	2.43490	2.25891	2.84761
15	2.63144	2.53998	2.41977	2.24593	2.82563
16	2.61558	2.52538	2.40666	2.23468	2.80653
17	2.60169	2.51261	2.39519	2.22484	2.78978
18	2.58944	2.50134	2.38507	2.21615	2.77498
19	2.57854	2.49133	2.37608	2.20843	2.76179
20	2.56880	2.48237	2.36804	2.20153	2.74998
21	2.56002	2.47431	2.36080	2.19531	2.73934
22	2.55208	2.46702	2.35425	2.18969	2.72969
23	2.54487	2.46039	2.34830	2.18458	2.72092
24	2.53828	2.45434	2.34287	2.17991	2.71290
25	2.53224	2.44880	2.33789	2.17564	2.70554
26	2.52669	2.44370	2.33332	2.17170	2.69876
27	2.52156	2.43899	2.32909	2.16807	2.69250
28	2.51682	2.43463	2.32518	2.16471	2.68671
29	2.51241	2.43059	2.32155	2.16158	2.68132
30	2.50831	2.42682	2.31816	2.15868	2.67630
35	2.49142	2.41132	2.30425	2.14671	2.65560
40	2.47887	2.39981	2.29391	2.13782	2.64018
45	2.46917	2.39091	2.28592	2.13094	2.62825
50	2.46146	2.38384	2.27956	2.12547	2.61874
60	2.44996	2.37329	2.27009	2.11732	2.60455
80	2.43571	2.36021	2.25834	2.10721	2.58691
100	2.42721	2.35242	2.25134	2.10118	2.57639
120	2.42158	2.34726	2.24670	2.09718	2.56939
200	2.41038	2.33698	2.23746	2.08923	2.55547
∞	2.39372	2.32170	2.22373	2.07740	2.53473

TABLE B.2

Two-sided Percentage Point (h) for Equal
Correlations [Case I with Correlation ρ]

p = 11, 1-α = 0.90

ν ↓	ρ					
	0.0	0.1	0.2	0.3	0.4	0.5
2	6.07542	6.04709	5.97382	5.86401	5.71981	5.53958
3	4.55466	4.53530	4.48479	4.40864	4.30827	4.18250
4	3.95633	3.94070	3.89962	3.83737	3.75503	3.65159
5	3.63815	3.62462	3.58882	3.53433	3.46200	3.37092
6	3.44077	3.42861	3.39625	3.34677	3.28092	3.19778
7	3.30632	3.29513	3.26521	3.21930	3.15801	3.08045
8	3.20880	3.19834	3.17026	3.12702	3.06915	2.99576
9	3.13479	3.12492	3.09828	3.05714	3.00193	2.93178
10	3.07670	3.06730	3.04183	3.00238	2.94932	2.88176
11	3.02988	3.02086	2.99637	2.95831	2.90702	2.84160
12	2.99133	2.98264	2.95897	2.92209	2.87228	2.80864
13	2.95904	2.95063	2.92765	2.89178	2.84324	2.78111
14	2.93158	2.92342	2.90105	2.86605	2.81861	2.75778
15	2.90796	2.90002	2.87818	2.84394	2.79745	2.73776
16	2.88742	2.87966	2.85830	2.82473	2.77908	2.72038
17	2.86939	2.86180	2.84086	2.80789	2.76298	2.70517
18	2.85344	2.84600	2.82543	2.79300	2.74876	2.69173
19	2.83922	2.83192	2.81170	2.77975	2.73611	2.67979
20	2.82648	2.81930	2.79938	2.76787	2.72478	2.66909
21	2.81498	2.80792	2.78828	2.75717	2.71457	2.65946
22	2.80456	2.79761	2.77823	2.74748	2.70533	2.65075
23	2.79508	2.78822	2.76907	2.73866	2.69692	2.64283
24	2.78640	2.77963	2.76071	2.73060	2.68925	2.63559
25	2.77844	2.77175	2.75303	2.72321	2.68220	2.62896
26	2.77111	2.76449	2.74596	2.71641	2.67572	2.62285
27	2.76433	2.75778	2.73942	2.71012	2.66974	2.61722
28	2.75804	2.75157	2.73337	2.70429	2.66420	2.61200
29	2.75220	2.74579	2.72774	2.69888	2.65905	2.60715
30	2.74676	2.74040	2.72250	2.69384	2.65425	2.60264
35	2.72430	2.71818	2.70088	2.67306	2.63450	2.58406
40	2.70754	2.70161	2.68476	2.65759	2.61980	2.57024
45	2.69456	2.68877	2.67229	2.64561	2.60843	2.55957
50	2.68420	2.67853	2.66234	2.63607	2.59938	2.55107
60	2.66873	2.66324	2.64749	2.62184	2.58588	2.53840
80	2.64948	2.64421	2.62903	2.60415	2.56913	2.52269
100	2.63798	2.63285	2.61801	2.59361	2.55914	2.51333
120	2.63033	2.62530	2.61068	2.58660	2.55251	2.50712
200	2.61508	2.61024	2.59609	2.57265	2.53932	2.49477
∞	2.59234	2.58778	2.57435	2.55189	2.51970	2.47640

TABLE B.2

Two-sided Percentage Point (h) for Equal
Correlations $\left(\text{Case I with Correlation } \rho\right)$

$p = 11, \ 1-\alpha = 0.90$

$\nu \downarrow$	ρ				
	0.6	0.7	0.8	0.9	$1/(1+\sqrt{11})$
2	5.31783	5.04317	4.69168	4.20030	5.94281
3	4.02752	3.83544	3.58971	3.24673	4.46333
4	3.52388	3.36538	3.16235	2.87854	3.88210
5	3.25824	3.11812	2.93830	2.68618	3.57351
6	3.09471	2.96629	2.80109	2.56866	3.38236
7	2.98411	2.86382	2.70870	2.48965	3.25234
8	2.90441	2.79012	2.64236	2.43296	3.15816
9	2.84429	2.73461	2.59246	2.39035	3.08678
10	2.79735	2.69132	2.55359	2.35716	3.03081
11	2.75970	2.65664	2.52247	2.33059	2.98575
12	2.72884	2.62823	2.49700	2.30884	2.94868
13	2.70309	2.60455	2.47577	2.29071	2.91766
14	2.68128	2.58451	2.45782	2.27537	2.89131
15	2.66257	2.56733	2.44242	2.26222	2.86865
16	2.64635	2.55244	2.42909	2.25082	2.84896
17	2.63216	2.53941	2.41742	2.24085	2.83169
18	2.61963	2.52792	2.40713	2.23205	2.81642
19	2.60849	2.51770	2.39799	2.22423	2.80283
20	2.59852	2.50857	2.38981	2.21723	2.79064
21	2.58955	2.50034	2.38245	2.21094	2.77966
22	2.58144	2.49291	2.37579	2.20524	2.76970
23	2.57406	2.48615	2.36974	2.20007	2.76065
24	2.56733	2.47998	2.36421	2.19534	2.75237
25	2.56115	2.47432	2.35915	2.19101	2.74477
26	2.55548	2.46912	2.35449	2.18702	2.73778
27	2.55023	2.46432	2.35020	2.18334	2.73132
28	2.54538	2.45987	2.34622	2.17994	2.72533
29	2.54088	2.45575	2.34252	2.17678	2.71976
30	2.53668	2.45191	2.33909	2.17383	2.71458
35	2.51942	2.43610	2.32493	2.16171	2.69320
40	2.50658	2.42435	2.31442	2.15270	2.67727
45	2.49667	2.41528	2.30629	2.14574	2.66494
50	2.48878	2.40807	2.29983	2.14021	2.65512
60	2.47703	2.39731	2.29020	2.13195	2.64044
80	2.46245	2.38397	2.27825	2.12171	2.62221
100	2.45377	2.37603	2.27114	2.11560	2.61132
120	2.44801	2.37076	2.26642	2.11156	2.60409
200	2.43656	2.36028	2.25702	2.10350	2.58969
∞	2.41953	2.34470	2.24307	2.09152	2.56824

TABLE B.2

Two-sided Percentage Point (h) for Equal
Correlations   (*Case I with Correlation* ρ)

p = 12, 1-α = 0.90

ν ↓	ρ					
	0.0	0.1	0.2	0.3	0.4	0.5
2	6.18288	6.15283	6.07572	5.96078	5.81041	5.62299
3	4.63069	4.61014	4.55694	4.47716	4.37240	4.24150
4	4.01956	4.00297	3.95968	3.89444	3.80845	3.70075
5	3.69436	3.67999	3.64226	3.58512	3.50958	3.41472
6	3.49249	3.47957	3.44546	3.39359	3.32480	3.23820
7	3.35490	3.34302	3.31148	3.26334	3.19932	3.11855
8	3.25504	3.24395	3.21436	3.16902	3.10857	3.03214
9	3.17924	3.16876	3.14070	3.09757	3.03991	2.96685
10	3.11970	3.10973	3.08291	3.04155	2.98614	2.91580
11	3.07170	3.06214	3.03635	2.99647	2.94291	2.87479
12	3.03217	3.02296	2.99803	2.95939	2.90739	2.84114
13	2.99904	2.99013	2.96595	2.92837	2.87770	2.81303
14	2.97087	2.96223	2.93869	2.90203	2.85251	2.78920
15	2.94662	2.93821	2.91524	2.87938	2.83087	2.76875
16	2.92553	2.91733	2.89486	2.85971	2.81208	2.75100
17	2.90702	2.89899	2.87697	2.84246	2.79561	2.73546
18	2.89063	2.88277	2.86115	2.82721	2.78106	2.72174
19	2.87603	2.86832	2.84706	2.81363	2.76812	2.70953
20	2.86294	2.85536	2.83443	2.80146	2.75652	2.69860
21	2.85113	2.84367	2.82304	2.79050	2.74608	2.68877
22	2.84042	2.83308	2.81273	2.78057	2.73662	2.67986
23	2.83067	2.82343	2.80333	2.77153	2.72802	2.67177
24	2.82176	2.81461	2.79474	2.76327	2.72016	2.66437
25	2.81357	2.80651	2.78686	2.75569	2.71296	2.65759
26	2.80603	2.79905	2.77961	2.74872	2.70632	2.65136
27	2.79906	2.79216	2.77290	2.74227	2.70020	2.64560
28	2.79260	2.78577	2.76668	2.73630	2.69452	2.64027
29	2.78659	2.77983	2.76091	2.73075	2.68925	2.63532
30	2.78099	2.77430	2.75553	2.72559	2.68434	2.63070
35	2.75789	2.75145	2.73332	2.70428	2.66412	2.61172
40	2.74064	2.73441	2.71677	2.68841	2.64907	2.59760
45	2.72728	2.72121	2.70396	2.67613	2.63743	2.58669
50	2.71662	2.71068	2.69374	2.66635	2.62817	2.57801
60	2.70069	2.69494	2.67848	2.65174	2.61435	2.56506
80	2.68087	2.67536	2.65951	2.63360	2.59719	2.54900
100	2.66902	2.66366	2.64818	2.62278	2.58697	2.53944
120	2.66114	2.65589	2.64065	2.61559	2.58018	2.53309
200	2.64544	2.64039	2.62565	2.60128	2.56667	2.52046
∞	2.62200	2.61726	2.60330	2.57997	2.54658	2.50169

TABLE B.2

*Two-sided Percentage Point (h) for Equal
Correlations* $\left(Case\ I\ with\ Correlation\ \rho\right)$

$p = 12,\ 1-\alpha = 0.90$

$\nu \downarrow$	0.6	0.7	0.8	0.9	$1/(1+\sqrt{12})$
2	5.39290	5.10846	4.74507	4.23790	6.05141
3	4.08058	3.88151	3.62730	3.27311	4.54010
4	3.56809	3.40377	3.19367	2.90054	3.94593
5	3.29765	3.15237	2.96627	2.70587	3.63024
6	3.13111	2.99796	2.82699	2.58694	3.43457
7	3.01844	2.89372	2.73320	2.50699	3.30138
8	2.93723	2.81873	2.66584	2.44962	3.20486
9	2.87596	2.76224	2.61517	2.40650	3.13167
10	2.82811	2.71819	2.57571	2.37292	3.07426
11	2.78973	2.68290	2.54411	2.34604	3.02802
12	2.75826	2.65399	2.51825	2.32403	2.98997
13	2.73200	2.62989	2.49670	2.30569	2.95811
14	2.70976	2.60949	2.47847	2.29017	2.93105
15	2.69069	2.59200	2.46284	2.27687	2.90778
16	2.67415	2.57685	2.44930	2.26534	2.88755
17	2.65967	2.56359	2.43746	2.25525	2.86980
18	2.64689	2.55189	2.42701	2.24636	2.85410
19	2.63553	2.54149	2.41773	2.23845	2.84012
20	2.62537	2.53219	2.40942	2.23137	2.82759
21	2.61622	2.52383	2.40195	2.22500	2.81630
22	2.60794	2.51626	2.39519	2.21924	2.80606
23	2.60042	2.50938	2.38905	2.21400	2.79675
24	2.59355	2.50310	2.38344	2.20922	2.78823
25	2.58725	2.49734	2.37830	2.20484	2.78042
26	2.58146	2.49205	2.37357	2.20081	2.77322
27	2.57612	2.48716	2.36921	2.19709	2.76657
28	2.57117	2.48264	2.36517	2.19364	2.76041
29	2.56657	2.47844	2.36142	2.19045	2.75468
30	2.56229	2.47453	2.35793	2.18747	2.74935
35	2.54468	2.45844	2.34357	2.17521	2.72734
40	2.53159	2.44649	2.33289	2.16610	2.71094
45	2.52148	2.43725	2.32465	2.15906	2.69824
50	2.51344	2.42991	2.31809	2.15346	2.68812
60	2.50145	2.41896	2.30831	2.14511	2.67300
80	2.48659	2.40539	2.29618	2.13475	2.65421
100	2.47773	2.39731	2.28896	2.12858	2.64299
120	2.47186	2.39194	2.28417	2.12448	2.63554
200	2.46017	2.38128	2.27464	2.11634	2.62069
∞	2.44281	2.36542	2.26047	2.10422	2.59857

BECHHOFER and DUNNETT

TABLE B.2

Two-sided Percentage Point (h) for Equal
Correlations $\left(\text{Case I with Correlation } \rho\right)$

p = 13, 1-α = 0.90

ν ↓	ρ					
	0.0	0.1	0.2	0.3	0.4	0.5
2	6.28094	6.24927	6.16860	6.04893	5.89289	5.69888
3	4.70015	4.67848	4.62278	4.53965	4.43086	4.29526
4	4.07739	4.05988	4.01454	3.94652	3.85718	3.74557
5	3.74579	3.73062	3.69109	3.63151	3.55300	3.45467
6	3.53982	3.52620	3.49045	3.43635	3.36485	3.27508
7	3.39936	3.38683	3.35379	3.30358	3.23703	3.15329
8	3.29738	3.28568	3.25467	3.20740	3.14456	3.06532
9	3.21992	3.20888	3.17948	3.13450	3.07457	2.99884
10	3.15907	3.14856	3.12046	3.07734	3.01975	2.94684
11	3.10999	3.09992	3.07290	3.03132	2.97567	2.90506
12	3.06955	3.05985	3.03375	2.99347	2.93944	2.87077
13	3.03565	3.02628	3.00096	2.96179	2.90915	2.84213
14	3.00683	2.99773	2.97309	2.93489	2.88345	2.81785
15	2.98201	2.97316	2.94912	2.91176	2.86137	2.79701
16	2.96041	2.95178	2.92827	2.89166	2.84219	2.77892
17	2.94145	2.93302	2.90998	2.87403	2.82539	2.76308
18	2.92467	2.91641	2.89380	2.85845	2.81054	2.74909
19	2.90972	2.90161	2.87939	2.84457	2.79732	2.73665
20	2.89630	2.88834	2.86646	2.83214	2.78549	2.72551
21	2.88420	2.87637	2.85481	2.82093	2.77483	2.71548
22	2.87323	2.86552	2.84425	2.81078	2.76517	2.70640
23	2.86323	2.85563	2.83464	2.80154	2.75639	2.69815
24	2.85409	2.84660	2.82585	2.79310	2.74837	2.69061
25	2.84570	2.83830	2.81778	2.78535	2.74101	2.68370
26	2.83797	2.83065	2.81035	2.77822	2.73423	2.67734
27	2.83082	2.82359	2.80348	2.77163	2.72798	2.67147
28	2.82420	2.81704	2.79712	2.76552	2.72218	2.66603
29	2.81804	2.81095	2.79120	2.75985	2.71680	2.66098
30	2.81230	2.80528	2.78569	2.75456	2.71179	2.65628
35	2.78859	2.78186	2.76296	2.73277	2.69114	2.63692
40	2.77090	2.76438	2.74600	2.71654	2.67576	2.62252
45	2.75719	2.75084	2.73287	2.70397	2.66388	2.61140
50	2.74625	2.74004	2.72241	2.69397	2.65442	2.60255
60	2.72989	2.72389	2.70677	2.67902	2.64030	2.58934
80	2.70953	2.70380	2.68732	2.66046	2.62277	2.57297
100	2.69736	2.69179	2.67571	2.64939	2.61233	2.56321
120	2.68927	2.68381	2.66800	2.64203	2.60539	2.55674
200	2.67313	2.66789	2.65262	2.62739	2.59159	2.54386
∞	2.64905	2.64414	2.62970	2.60558	2.57106	2.52472

TABLE B.2

Two-sided Percentage Point (h) for Equal
Correlations $\left(Case\ I\ with\ Correlation\ \rho\right)$

p = 13, 1-α = 0.90

ν ↓	ρ 0.6	0.7	0.8	0.9	1/(1+√13)
2	5.46118	5.16782	4.79360	4.27206	6.15073
3	4.12889	3.92345	3.66150	3.29709	4.61040
4	3.60837	3.43872	3.22217	2.92053	4.00443
5	3.33357	3.18356	2.99173	2.72376	3.68225
6	3.16429	3.02679	2.85056	2.60356	3.48243
7	3.04973	2.92095	2.75548	2.52274	3.34636
8	2.96714	2.84478	2.68720	2.46476	3.24769
9	2.90481	2.78741	2.63584	2.42118	3.17284
10	2.85614	2.74266	2.59583	2.38724	3.11411
11	2.81708	2.70681	2.56380	2.36007	3.06678
12	2.78507	2.67744	2.53758	2.33784	3.02783
13	2.75835	2.65295	2.51573	2.31930	2.99521
14	2.73571	2.63223	2.49725	2.30362	2.96749
15	2.71630	2.61446	2.48141	2.29018	2.94364
16	2.69946	2.59906	2.46768	2.27853	2.92291
17	2.68473	2.58559	2.45568	2.26834	2.90472
18	2.67172	2.57371	2.44508	2.25935	2.88863
19	2.66016	2.56314	2.43567	2.25135	2.87430
20	2.64982	2.55370	2.42726	2.24420	2.86145
21	2.64051	2.54519	2.41968	2.23777	2.84986
22	2.63208	2.53750	2.41283	2.23195	2.83937
23	2.62442	2.53051	2.40660	2.22666	2.82981
24	2.61743	2.52413	2.40092	2.22183	2.82107
25	2.61102	2.51828	2.39571	2.21740	2.81306
26	2.60513	2.51291	2.39092	2.21333	2.80567
27	2.59969	2.50794	2.38650	2.20957	2.79885
28	2.59465	2.50335	2.38240	2.20609	2.79252
29	2.58997	2.49908	2.37860	2.20286	2.78665
30	2.58561	2.49511	2.37506	2.19985	2.78117
35	2.56769	2.47876	2.36050	2.18746	2.75858
40	2.55436	2.46662	2.34968	2.17826	2.74174
45	2.54407	2.45724	2.34133	2.17115	2.72869
50	2.53589	2.44978	2.33468	2.16549	2.71830
60	2.52368	2.43865	2.32476	2.15705	2.70277
80	2.50855	2.42487	2.31248	2.14659	2.68346
100	2.49954	2.41666	2.30516	2.14035	2.67194
120	2.49356	2.41121	2.30030	2.13622	2.66428
200	2.48166	2.40037	2.29064	2.12799	2.64901
∞	2.46399	2.38427	2.27628	2.11575	2.62627

TABLE B.2

Two-sided Percentage Point (h) for Equal
Correlations   (*Case I with Correlation* ρ)

p = 14, 1-α = 0.90

ν ↓	0.0	0.1	0.2	0.3	0.4	0.5
			ρ			
2	6.37103	6.33784	6.25386	6.12980	5.96852	5.76845
3	4.76406	4.74133	4.68329	4.59705	4.48452	4.34459
4	4.13063	4.11226	4.06499	3.99439	3.90195	3.78671
5	3.79315	3.77724	3.73602	3.67417	3.59290	3.49137
6	3.58343	3.56913	3.53186	3.47568	3.40166	3.30895
7	3.44034	3.42719	3.39273	3.34059	3.27170	3.18521
8	3.33640	3.32412	3.29179	3.24270	3.17765	3.09581
9	3.25742	3.24584	3.21518	3.16848	3.10644	3.02822
10	3.19535	3.18433	3.15504	3.11027	3.05065	2.97535
11	3.14527	3.13472	3.10655	3.06339	3.00578	2.93287
12	3.10400	3.09384	3.06664	3.02482	2.96890	2.89800
13	3.06940	3.05958	3.03319	2.99254	2.93806	2.86887
14	3.03996	3.03044	3.00477	2.96512	2.91189	2.84416
15	3.01461	3.00535	2.98031	2.94154	2.88940	2.82296
16	2.99256	2.98352	2.95904	2.92105	2.86987	2.80456
17	2.97318	2.96435	2.94037	2.90308	2.85275	2.78844
18	2.95603	2.94739	2.92385	2.88718	2.83762	2.77421
19	2.94075	2.93227	2.90913	2.87303	2.82416	2.76154
20	2.92703	2.91871	2.89594	2.86035	2.81210	2.75021
21	2.91466	2.90648	2.88404	2.84892	2.80124	2.74000
22	2.90344	2.89538	2.87326	2.83856	2.79140	2.73077
23	2.89322	2.88528	2.86344	2.82913	2.78245	2.72237
24	2.88387	2.87604	2.85446	2.82052	2.77428	2.71470
25	2.87529	2.86756	2.84622	2.81261	2.76678	2.70766
26	2.86738	2.85975	2.83863	2.80533	2.75987	2.70119
27	2.86007	2.85252	2.83162	2.79861	2.75350	2.69521
28	2.85329	2.84582	2.82512	2.79238	2.74759	2.68968
29	2.84699	2.83960	2.81908	2.78659	2.74211	2.68454
30	2.84112	2.83380	2.81345	2.78120	2.73700	2.67976
35	2.81686	2.80984	2.79021	2.75896	2.71595	2.66005
40	2.79875	2.79196	2.77288	2.74239	2.70028	2.64540
45	2.78471	2.77810	2.75946	2.72956	2.68816	2.63407
50	2.77350	2.76704	2.74876	2.71934	2.67852	2.62506
60	2.75675	2.75051	2.73277	2.70409	2.66412	2.61162
80	2.73590	2.72994	2.71289	2.68514	2.64626	2.59495
100	2.72343	2.71765	2.70102	2.67383	2.63561	2.58503
120	2.71513	2.70947	2.69312	2.66632	2.62854	2.57843
200	2.69859	2.69317	2.67739	2.65136	2.61446	2.56532
∞	2.67390	2.66884	2.65394	2.62908	2.59353	2.54584

TABLE B.2

Two-sided Percentage Point (h) for Equal Correlations (*Case I with Correlation* ρ)

$p = 14, \ 1-\alpha = 0.90$

$\nu \downarrow$	ρ				
	0.6	0.7	0.8	0.9	$1/(1+\sqrt{14})$
2	5.52375	5.22220	4.83804	4.30332	6.24219
3	4.17320	3.96190	3.69283	3.31902	4.67520
4	3.64533	3.47078	3.24829	2.93883	4.05838
5	3.36654	3.21218	3.01506	2.74015	3.73024
6	3.19475	3.05325	2.87216	2.61877	3.52661
7	3.07846	2.94592	2.77591	2.53716	3.38787
8	2.99460	2.86868	2.70678	2.47862	3.28722
9	2.93131	2.81049	2.65477	2.43462	3.21084
10	2.88187	2.76511	2.61427	2.40035	3.15088
11	2.84220	2.72873	2.58183	2.37292	3.10255
12	2.80967	2.69894	2.55529	2.35047	3.06277
13	2.78252	2.67410	2.53317	2.33176	3.02944
14	2.75953	2.65308	2.51445	2.31593	3.00111
15	2.73980	2.63505	2.49841	2.30236	2.97673
16	2.72270	2.61943	2.48452	2.29060	2.95553
17	2.70773	2.60577	2.47236	2.28031	2.93693
18	2.69451	2.59371	2.46164	2.27123	2.92048
19	2.68276	2.58299	2.45211	2.26317	2.90582
20	2.67225	2.57341	2.44359	2.25595	2.89267
21	2.66279	2.56478	2.43592	2.24945	2.88082
22	2.65423	2.55698	2.42898	2.24358	2.87008
23	2.64644	2.54989	2.42268	2.23824	2.86030
24	2.63934	2.54342	2.41693	2.23336	2.85135
25	2.63283	2.53748	2.41165	2.22889	2.84315
26	2.62683	2.53203	2.40680	2.22478	2.83559
27	2.62130	2.52699	2.40233	2.22099	2.82860
28	2.61618	2.52233	2.39818	2.21747	2.82213
29	2.61143	2.51800	2.39433	2.21421	2.81611
30	2.60701	2.51397	2.39075	2.21118	2.81050
35	2.58879	2.49739	2.37601	2.19868	2.78736
40	2.57525	2.48507	2.36506	2.18938	2.77011
45	2.56479	2.47555	2.35660	2.18221	2.75675
50	2.55647	2.46798	2.34987	2.17650	2.74609
60	2.54406	2.45670	2.33983	2.16798	2.73018
80	2.52869	2.44271	2.32739	2.15742	2.71039
100	2.51953	2.43438	2.31999	2.15113	2.69857
120	2.51345	2.42886	2.31507	2.14695	2.69071
200	2.50136	2.41786	2.30529	2.13865	2.67506
∞	2.48340	2.40153	2.29076	2.12629	2.65173

BECHHOFER and DUNNETT

TABLE B.2

*Two-sided Percentage Point (h) for Equal
Correlations* $\left(\text{Case } I \text{ with Correlation } \rho\right)$

$p = 15,\ 1-\alpha = 0.90$

$\nu\downarrow$	ρ					
	0.0	0.1	0.2	0.3	0.4	0.5
2	6.45431	6.41969	6.33260	6.20446	6.03832	5.83263
3	4.82320	4.79947	4.73924	4.65010	4.53409	4.39013
4	4.17993	4.16075	4.11167	4.03866	3.94332	3.82472
5	3.83703	3.82042	3.77761	3.71362	3.62979	3.52527
6	3.62384	3.60891	3.57019	3.51207	3.43570	3.34025
7	3.47832	3.46458	3.42879	3.37484	3.30376	3.21471
8	3.37256	3.35974	3.32616	3.27536	3.20825	3.12398
9	3.29218	3.28009	3.24824	3.19992	3.13591	3.05538
10	3.22899	3.21748	3.18706	3.14074	3.07923	3.00170
11	3.17799	3.16697	3.13772	3.09306	3.03363	2.95857
12	3.13594	3.12534	3.09709	3.05383	2.99615	2.92316
13	3.10068	3.09043	3.06304	3.02099	2.96480	2.89357
14	3.07068	3.06074	3.03410	2.99309	2.93819	2.86848
15	3.04484	3.03517	3.00919	2.96910	2.91532	2.84694
16	3.02235	3.01293	2.98752	2.94824	2.89546	2.82825
17	3.00259	2.99339	2.96850	2.92995	2.87805	2.81187
18	2.98510	2.97609	2.95167	2.91377	2.86267	2.79741
19	2.96951	2.96067	2.93668	2.89936	2.84898	2.78455
20	2.95551	2.94684	2.92323	2.88645	2.83671	2.77303
21	2.94289	2.93436	2.91110	2.87480	2.82566	2.76266
22	2.93144	2.92305	2.90011	2.86426	2.81565	2.75328
23	2.92101	2.91275	2.89010	2.85466	2.80655	2.74474
24	2.91147	2.90332	2.88095	2.84589	2.79823	2.73695
25	2.90271	2.89466	2.87255	2.83783	2.79060	2.72980
26	2.89463	2.88669	2.86481	2.83042	2.78358	2.72322
27	2.88717	2.87932	2.85766	2.82357	2.77709	2.71715
28	2.88025	2.87248	2.85103	2.81722	2.77108	2.71152
29	2.87381	2.86613	2.84487	2.81133	2.76550	2.70630
30	2.86781	2.86020	2.83913	2.80583	2.76030	2.70144
35	2.84303	2.83574	2.81543	2.78317	2.73888	2.68141
40	2.82453	2.81748	2.79775	2.76629	2.72294	2.66652
45	2.81018	2.80333	2.78406	2.75322	2.71060	2.65501
50	2.79874	2.79204	2.77314	2.74281	2.70079	2.64585
60	2.78161	2.77515	2.75683	2.72726	2.68613	2.63219
80	2.76029	2.75413	2.73653	2.70794	2.66795	2.61525
100	2.74754	2.74157	2.72441	2.69641	2.65711	2.60516
120	2.73906	2.73321	2.71635	2.68876	2.64991	2.59846
200	2.72214	2.71654	2.70029	2.67351	2.63559	2.58514
∞	2.69687	2.69167	2.67635	2.65079	2.61428	2.56533

TABLE B.2

*Two-sided Percentage Point (h) for Equal
Correlations* $\left(\text{Case I with Correlation } \rho\right)$

$$p = 15, \quad 1-\alpha = 0.90$$

| $\nu \downarrow$ | \multicolumn{5}{c}{ρ} |
	0.6	0.7	0.8	0.9	$1/(1+\sqrt{15})$
2	5.58145	5.27232	4.87898	4.33212	6.32688
3	4.21410	3.99736	3.72173	3.33926	4.73527
4	3.67946	3.50037	3.27239	2.95571	4.10843
5	3.39700	3.23859	3.03659	2.75525	3.77477
6	3.22288	3.07767	2.89209	2.63279	3.56762
7	3.10499	2.96898	2.79475	2.55045	3.42641
8	3.01996	2.89075	2.72483	2.49139	3.32392
9	2.95578	2.83180	2.67224	2.44700	3.24612
10	2.90563	2.78582	2.63127	2.41243	3.18502
11	2.86539	2.74897	2.59847	2.38476	3.13576
12	2.83239	2.71879	2.57162	2.36211	3.09519
13	2.80485	2.69362	2.54925	2.34324	3.06120
14	2.78152	2.67232	2.53032	2.32727	3.03230
15	2.76151	2.65406	2.51410	2.31358	3.00743
16	2.74415	2.63823	2.50004	2.30171	2.98580
17	2.72896	2.62438	2.48775	2.29134	2.96682
18	2.71555	2.61216	2.47690	2.28218	2.95002
19	2.70363	2.60131	2.46727	2.27404	2.93505
20	2.69296	2.59159	2.45865	2.26676	2.92163
21	2.68336	2.58285	2.45089	2.26021	2.90953
22	2.67467	2.57495	2.44388	2.25429	2.89856
23	2.66677	2.56776	2.43750	2.24890	2.88857
24	2.65956	2.56120	2.43168	2.24398	2.87943
25	2.65295	2.55519	2.42635	2.23947	2.87105
26	2.64687	2.54966	2.42144	2.23533	2.86332
27	2.64126	2.54456	2.41692	2.23150	2.85619
28	2.63606	2.53984	2.41272	2.22796	2.84957
29	2.63124	2.53545	2.40883	2.22467	2.84342
30	2.62675	2.53137	2.40521	2.22161	2.83769
35	2.60826	2.51457	2.39030	2.20900	2.81404
40	2.59452	2.50208	2.37923	2.19963	2.79640
45	2.58391	2.49244	2.37067	2.19239	2.78274
50	2.57546	2.48477	2.36387	2.18663	2.77184
60	2.56287	2.47334	2.35372	2.17804	2.75556
80	2.54727	2.45917	2.34114	2.16739	2.73531
100	2.53797	2.45073	2.33365	2.16104	2.72322
120	2.53180	2.44513	2.32868	2.15683	2.71518
200	2.51954	2.43399	2.31879	2.14846	2.69916
∞	2.50131	2.41744	2.30409	2.13600	2.67527

BECHHOFER and DUNNETT

TABLE B.2

Two-sided Percentage Point (h) for Equal
Correlations (*Case I with Correlation* ρ)

$p = 16, \; 1-\alpha = 0.90$

$\nu \downarrow$	0.0	0.1	0.2	0.3	0.4	0.5
			ρ			
2	6.53169	6.49571	6.40570	6.27374	6.10307	5.89215
3	4.87820	4.85353	4.79123	4.69937	4.58012	4.43240
4	4.22581	4.20585	4.15507	4.07980	3.98176	3.86002
5	3.87789	3.86060	3.81629	3.75030	3.66408	3.55676
6	3.66148	3.64593	3.60585	3.54591	3.46734	3.36934
7	3.51369	3.49940	3.46234	3.40670	3.33356	3.24212
8	3.40626	3.39292	3.35815	3.30575	3.23669	3.15016
9	3.32457	3.31198	3.27902	3.22917	3.16331	3.08062
10	3.26033	3.24836	3.21686	3.16908	3.10580	3.02619
11	3.20847	3.19700	3.16673	3.12067	3.05952	2.98245
12	3.16570	3.15467	3.12544	3.08082	3.02148	2.94654
13	3.12983	3.11917	3.09083	3.04746	2.98965	2.91652
14	3.09930	3.08897	3.06140	3.01911	2.96264	2.89107
15	3.07300	3.06295	3.03607	2.99473	2.93942	2.86922
16	3.05010	3.04031	3.01403	2.97353	2.91925	2.85026
17	3.02999	3.02043	2.99468	2.95494	2.90157	2.83364
18	3.01218	3.00282	2.97757	2.93849	2.88595	2.81897
19	2.99629	2.98712	2.96231	2.92385	2.87204	2.80592
20	2.98204	2.97304	2.94863	2.91072	2.85958	2.79423
21	2.96918	2.96033	2.93629	2.89888	2.84835	2.78371
22	2.95752	2.94881	2.92510	2.88816	2.83819	2.77418
23	2.94689	2.93831	2.91492	2.87840	2.82894	2.76552
24	2.93717	2.92871	2.90560	2.86947	2.82049	2.75761
25	2.92824	2.91990	2.89705	2.86129	2.81274	2.75036
26	2.92001	2.91177	2.88917	2.85375	2.80560	2.74368
27	2.91240	2.90426	2.88189	2.84678	2.79901	2.73752
28	2.90534	2.89729	2.87514	2.84032	2.79291	2.73181
29	2.89878	2.89082	2.86887	2.83433	2.78724	2.72651
30	2.89267	2.88478	2.86302	2.82874	2.78195	2.72158
35	2.86740	2.85985	2.83889	2.80569	2.76019	2.70125
40	2.84853	2.84124	2.82088	2.78851	2.74398	2.68613
45	2.83390	2.82681	2.80694	2.77521	2.73145	2.67445
50	2.82222	2.81529	2.79581	2.76462	2.72147	2.66516
60	2.80474	2.79807	2.77919	2.74879	2.70658	2.65129
80	2.78298	2.77663	2.75851	2.72913	2.68810	2.63410
100	2.76996	2.76381	2.74616	2.71740	2.67708	2.62385
120	2.76130	2.75528	2.73795	2.70960	2.66976	2.61705
200	2.74403	2.73827	2.72158	2.69408	2.65520	2.60353
∞	2.71822	2.71288	2.69717	2.67096	2.63354	2.58342

TABLE B.2

Two-sided Percentage Point (h) for Equal
Correlations $\left(\text{Case I with Correlation } \rho\right)$

$p = 16, \quad 1-\alpha = 0.90$

$\nu \downarrow$	ρ				
	0.6	0.7	0.8	0.9	$1/(1+\sqrt{16})$
2	5.63495	5.31879	4.91693	4.35879	6.40570
3	4.25205	4.03027	3.74852	3.35801	4.79123
4	3.71114	3.52782	3.29474	2.97135	4.15507
5	3.42527	3.26310	3.05655	2.76925	3.81629
6	3.24901	3.10034	2.91058	2.64578	3.60585
7	3.12963	2.99038	2.81223	2.56277	3.46234
8	3.04352	2.91122	2.74158	2.50323	3.35815
9	2.97850	2.85158	2.68844	2.45847	3.27902
10	2.92770	2.80505	2.64704	2.42362	3.21686
11	2.88693	2.76775	2.61389	2.39572	3.16673
12	2.85349	2.73721	2.58676	2.37289	3.12544
13	2.82559	2.71173	2.56415	2.35387	3.09083
14	2.80194	2.69017	2.54503	2.33777	3.06140
15	2.78166	2.67169	2.52863	2.32397	3.03607
16	2.76407	2.65567	2.51443	2.31201	3.01403
17	2.74867	2.64165	2.50201	2.30155	2.99468
18	2.73508	2.62928	2.49105	2.29232	2.97757
19	2.72300	2.61829	2.48131	2.28412	2.96231
20	2.71219	2.60846	2.47260	2.27678	2.94863
21	2.70245	2.59962	2.46477	2.27018	2.93629
22	2.69365	2.59161	2.45768	2.26421	2.92510
23	2.68564	2.58434	2.45124	2.25878	2.91492
24	2.67833	2.57770	2.44536	2.25382	2.90560
25	2.67163	2.57162	2.43997	2.24928	2.89705
26	2.66547	2.56602	2.43501	2.24510	2.88917
27	2.65978	2.56086	2.43044	2.24124	2.88189
28	2.65452	2.55607	2.42620	2.23767	2.87514
29	2.64962	2.55164	2.42227	2.23435	2.86887
30	2.64507	2.54750	2.41861	2.23127	2.86302
35	2.62633	2.53050	2.40355	2.21856	2.83889
40	2.61240	2.51786	2.39236	2.20911	2.82088
45	2.60164	2.50810	2.38371	2.20182	2.80694
50	2.59308	2.50034	2.37684	2.19601	2.79581
60	2.58032	2.48877	2.36659	2.18736	2.77919
80	2.56450	2.47443	2.35388	2.17662	2.75851
100	2.55508	2.46588	2.34631	2.17023	2.74616
120	2.54883	2.46022	2.34129	2.16598	2.73795
200	2.53640	2.44894	2.33130	2.15755	2.72158
∞	2.51792	2.43219	2.31645	2.14499	2.69717

TABLE B.2

Two-sided Percentage Point (h) for Equal
Correlations (*Case I with Correlation* ρ)

p = 18, 1-α = 0.90

ν ↓	ρ					
	0.0	0.1	0.2	0.3	0.4	0.5
2	6.67161	6.63311	6.53775	6.39883	6.21992	5.99951
3	4.97779	4.95135	4.88526	4.78843	4.66327	4.50875
4	4.30896	4.28756	4.23364	4.15423	4.05125	3.92381
5	3.95197	3.93342	3.88635	3.81670	3.72609	3.61370
6	3.72975	3.71307	3.67047	3.60717	3.52459	3.42192
7	3.57788	3.56254	3.52315	3.46438	3.38749	3.29169
8	3.46740	3.45309	3.41612	3.36078	3.28817	3.19751
9	3.38335	3.36984	3.33480	3.28215	3.21290	3.12626
10	3.31721	3.30437	3.27089	3.22042	3.15389	3.07048
11	3.26379	3.25149	3.21932	3.17067	3.10638	3.02564
12	3.21972	3.20789	3.17682	3.12971	3.06732	2.98881
13	3.18273	3.17131	3.14119	3.09540	3.03463	2.95803
14	3.15124	3.14017	3.11088	3.06624	3.00688	2.93192
15	3.12410	3.11335	3.08479	3.04115	2.98302	2.90951
16	3.10047	3.08999	3.06208	3.01934	2.96230	2.89005
17	3.07970	3.06947	3.04214	3.00020	2.94413	2.87300
18	3.06131	3.05130	3.02449	2.98327	2.92807	2.85794
19	3.04490	3.03509	3.00876	2.96819	2.91377	2.84455
20	3.03018	3.02055	2.99465	2.95466	2.90096	2.83255
21	3.01688	3.00743	2.98192	2.94247	2.88941	2.82175
22	3.00483	2.99553	2.97038	2.93143	2.87896	2.81197
23	2.99384	2.98468	2.95987	2.92137	2.86945	2.80308
24	2.98378	2.97476	2.95026	2.91218	2.86075	2.79496
25	2.97455	2.96564	2.94143	2.90374	2.85278	2.78751
26	2.96604	2.95725	2.93330	2.89597	2.84544	2.78066
27	2.95816	2.94948	2.92579	2.88879	2.83866	2.77433
28	2.95086	2.94228	2.91882	2.88214	2.83238	2.76847
29	2.94407	2.93558	2.91234	2.87595	2.82655	2.76303
30	2.93774	2.92934	2.90630	2.87019	2.82111	2.75796
35	2.91159	2.90355	2.88138	2.84643	2.79871	2.73709
40	2.89204	2.88429	2.86277	2.82871	2.78204	2.72157
45	2.87687	2.86935	2.84836	2.81500	2.76914	2.70957
50	2.86477	2.85742	2.83686	2.80407	2.75887	2.70002
60	2.84665	2.83958	2.81968	2.78774	2.74354	2.68579
80	2.82407	2.81736	2.79829	2.76746	2.72451	2.66813
100	2.81057	2.80407	2.78551	2.75535	2.71317	2.65760
120	2.80158	2.79523	2.77701	2.74731	2.70564	2.65062
200	2.78364	2.77759	2.76008	2.73128	2.69064	2.63673
∞	2.75684	2.75125	2.73481	2.70741	2.66834	2.61608

TABLE B.2

Two-sided Percentage Point (h) for Equal
Correlations (Case I with Correlation ρ)

p = 18, 1-α = 0.90

ν ↓	ρ				
	0.6	0.7	0.8	0.9	$1/(1+\sqrt{18})$
2	5.73141	5.40254	4.98530	4.40684	6.54851
3	4.32056	4.08963	3.79684	3.39180	4.89274
4	3.76836	3.57739	3.33506	2.99954	4.23976
5	3.47635	3.30736	3.09258	2.79449	3.89170
6	3.29621	3.14126	2.94393	2.66920	3.67533
7	3.17416	3.02902	2.84376	2.58497	3.52765
8	3.08608	2.94820	2.77180	2.52455	3.42035
9	3.01956	2.88728	2.71766	2.47914	3.33882
10	2.96757	2.83975	2.67548	2.44378	3.27473
11	2.92584	2.80166	2.64171	2.41548	3.22301
12	2.89162	2.77045	2.61407	2.39232	3.18040
13	2.86304	2.74442	2.59104	2.37301	3.14466
14	2.83883	2.72239	2.57155	2.35668	3.11426
15	2.81806	2.70351	2.55485	2.34269	3.08809
16	2.80005	2.68714	2.54038	2.33056	3.06531
17	2.78428	2.67281	2.52772	2.31995	3.04531
18	2.77036	2.66018	2.51656	2.31058	3.02760
19	2.75798	2.64895	2.50664	2.30227	3.01182
20	2.74691	2.63890	2.49777	2.29482	2.99766
21	2.73694	2.62986	2.48979	2.28813	2.98489
22	2.72792	2.62169	2.48256	2.28207	2.97331
23	2.71972	2.61425	2.47600	2.27656	2.96276
24	2.71223	2.60747	2.47001	2.27153	2.95311
25	2.70537	2.60125	2.46452	2.26692	2.94426
26	2.69905	2.59553	2.45947	2.26269	2.93609
27	2.69323	2.59026	2.45481	2.25877	2.92855
28	2.68783	2.58537	2.45050	2.25515	2.92156
29	2.68282	2.58084	2.44650	2.25179	2.91506
30	2.67816	2.57661	2.44277	2.24866	2.90900
35	2.65896	2.55924	2.42742	2.23577	2.88398
40	2.64469	2.54632	2.41602	2.22619	2.86530
45	2.63366	2.53635	2.40722	2.21880	2.85083
50	2.62490	2.52842	2.40022	2.21291	2.83929
60	2.61182	2.51660	2.38977	2.20413	2.82203
80	2.59561	2.50194	2.37683	2.19324	2.80055
100	2.58596	2.49321	2.36912	2.18676	2.78772
120	2.57956	2.48742	2.36400	2.18246	2.77918
200	2.56682	2.47590	2.35383	2.17390	2.76217
∞	2.54789	2.45879	2.33871	2.16117	2.73679

TABLE B.2

Two-sided Percentage Point (h) for Equal
Correlations $\left(Case\ I\ with\ Correlation\ \rho\right)$

p = 20, 1-α = 0.90

ν ↓	ρ					
	0.0	0.1	0.2	0.3	0.4	0.5
2	6.79536	6.75456	6.65439	6.50926	6.32302	6.09419
3	5.06600	5.03796	4.96845	4.86717	4.73675	4.57616
4	4.38268	4.35996	4.30320	4.22008	4.11270	3.98018
5	4.01771	3.99800	3.94842	3.87548	3.78095	3.66404
6	3.79035	3.77263	3.72775	3.66144	3.57526	3.46843
7	3.63488	3.61858	3.57707	3.51549	3.43524	3.33553
8	3.52171	3.50650	3.46754	3.40954	3.33375	3.23939
9	3.43557	3.42121	3.38428	3.32910	3.25681	3.16663
10	3.36775	3.35410	3.31881	3.26592	3.19647	3.10965
11	3.31294	3.29988	3.26597	3.21498	3.14788	3.06384
12	3.26771	3.25515	3.22240	3.17303	3.10791	3.02620
13	3.22973	3.21760	3.18587	3.13788	3.07446	2.99474
14	3.19739	3.18564	3.15478	3.10801	3.04605	2.96806
15	3.16951	3.15809	3.12801	3.08229	3.02163	2.94513
16	3.14522	3.13410	3.10471	3.05993	3.00041	2.92524
17	3.12387	3.11301	3.08424	3.04031	2.98180	2.90780
18	3.10496	3.09434	3.06612	3.02294	2.96535	2.89240
19	3.08808	3.07768	3.04997	3.00747	2.95070	2.87870
20	3.07293	3.06273	3.03547	2.99360	2.93758	2.86643
21	3.05925	3.04923	3.02240	2.98110	2.92575	2.85538
22	3.04684	3.03699	3.01054	2.96976	2.91504	2.84538
23	3.03553	3.02583	2.99974	2.95944	2.90529	2.83629
24	3.02518	3.01563	2.98986	2.95001	2.89639	2.82798
25	3.01567	3.00625	2.98079	2.94135	2.88822	2.82036
26	3.00690	2.99760	2.97243	2.93337	2.88069	2.81335
27	2.99880	2.98961	2.96470	2.92600	2.87375	2.80687
28	2.99128	2.98220	2.95754	2.91917	2.86731	2.80088
29	2.98428	2.97530	2.95088	2.91282	2.86133	2.79531
30	2.97776	2.96888	2.94467	2.90691	2.85576	2.79012
35	2.95080	2.94231	2.91904	2.88250	2.83279	2.76877
40	2.93064	2.92246	2.89990	2.86431	2.81570	2.75288
45	2.91500	2.90707	2.88506	2.85022	2.80247	2.74060
50	2.90251	2.89477	2.87323	2.83899	2.79194	2.73083
60	2.88381	2.87638	2.85554	2.82222	2.77622	2.71626
80	2.86050	2.85346	2.83351	2.80137	2.75671	2.69819
100	2.84655	2.83975	2.82035	2.78893	2.74507	2.68742
120	2.83726	2.83062	2.81160	2.78066	2.73735	2.68027
200	2.81873	2.81241	2.79415	2.76418	2.72197	2.66605
∞	2.79102	2.78521	2.76811	2.73965	2.69909	2.64491

TABLE B.2

Two-sided Percentage Point (h) for Equal
Correlations $\left(\text{Case I with Correlation } \rho\right)$

p = 20, 1-α = 0.90

ν ↓	ρ				
	0.6	0.7	0.8	0.9	$1/(1+\sqrt{20})$
2	5.81645	5.47634	5.04552	4.44914	6.67510
3	4.38102	4.14200	3.83945	3.42158	4.98286
4	3.81890	3.62114	3.37063	3.02439	4.31500
5	3.52149	3.34644	3.12436	2.81672	3.95875
6	3.33793	3.17741	2.97336	2.68984	3.73711
7	3.21352	3.06314	2.87158	2.60453	3.58574
8	3.12370	2.98085	2.79845	2.54334	3.47570
9	3.05585	2.91881	2.74343	2.49735	3.39202
10	3.00282	2.87040	2.70056	2.46154	3.32622
11	2.96024	2.83159	2.66624	2.43288	3.27310
12	2.92531	2.79980	2.63815	2.40942	3.22930
13	2.89614	2.77328	2.61474	2.38988	3.19256
14	2.87143	2.75084	2.59493	2.37334	3.16129
15	2.85023	2.73159	2.57796	2.35917	3.13436
16	2.83184	2.71491	2.56326	2.34689	3.11092
17	2.81574	2.70032	2.55039	2.33614	3.09033
18	2.80152	2.68744	2.53905	2.32666	3.07210
19	2.78889	2.67600	2.52897	2.31824	3.05584
20	2.77758	2.66576	2.51995	2.31070	3.04126
21	2.76740	2.65655	2.51184	2.30392	3.02810
22	2.75819	2.64822	2.50450	2.29779	3.01616
23	2.74981	2.64065	2.49783	2.29221	3.00529
24	2.74217	2.63374	2.49174	2.28712	2.99534
25	2.73516	2.62740	2.48616	2.28246	2.98621
26	2.72871	2.62157	2.48103	2.27817	2.97780
27	2.72276	2.61619	2.47630	2.27421	2.97002
28	2.71725	2.61122	2.47191	2.27054	2.96280
29	2.71213	2.60659	2.46784	2.26714	2.95609
30	2.70737	2.60229	2.46405	2.26397	2.94984
35	2.68776	2.58459	2.44846	2.25092	2.92402
40	2.67319	2.57143	2.43688	2.24123	2.90474
45	2.66193	2.56127	2.42793	2.23374	2.88979
50	2.65298	2.55319	2.42081	2.22778	2.87787
60	2.63962	2.54114	2.41020	2.21889	2.86003
80	2.62307	2.52620	2.39705	2.20787	2.83783
100	2.61321	2.51731	2.38922	2.20131	2.82456
120	2.60667	2.51141	2.38402	2.19695	2.81573
200	2.59366	2.49967	2.37368	2.18829	2.79813
∞	2.57433	2.48223	2.35831	2.17540	2.77186

BECHHOFER and DUNNETT

TABLE B.3

Two-sided Percentage Point (h) for Equal
Correlations $\left(Case\ I\ with\ Correlation\ \rho\right)$

$p = 2,\ 1-\alpha = 0.95$

ν ↓	0.0	0.1	0.2	0.3	0.4	0.5
			ρ			
1	17.98317	17.96059	17.89197	17.77457	17.60326	17.36945
2	5.57139	5.56572	5.54850	5.51907	5.47620	5.41785
3	3.96016	3.95670	3.94619	3.92823	3.90208	3.86651
4	3.38199	3.37935	3.37131	3.35758	3.33758	3.31035
5	3.09052	3.08830	3.08155	3.07001	3.05318	3.03024
6	2.91611	2.91414	2.90818	2.89797	2.88308	2.86275
7	2.80038	2.79859	2.79316	2.78386	2.77028	2.75170
8	2.71812	2.71646	2.71141	2.70277	2.69012	2.67281
9	2.65669	2.65513	2.65037	2.64222	2.63029	2.61393
10	2.60910	2.60762	2.60309	2.59533	2.58396	2.56834
11	2.57116	2.56974	2.56540	2.55795	2.54702	2.53200
12	2.54021	2.53884	2.53465	2.52746	2.51690	2.50237
13	2.51449	2.51315	2.50909	2.50212	2.49187	2.47774
14	2.49277	2.49147	2.48752	2.48073	2.47074	2.45696
15	2.47419	2.47293	2.46907	2.46244	2.45267	2.43919
16	2.45812	2.45688	2.45311	2.44661	2.43705	2.42382
17	2.44409	2.44287	2.43917	2.43279	2.42340	2.41039
18	2.43172	2.43052	2.42688	2.42061	2.41137	2.39857
19	2.42074	2.41956	2.41598	2.40980	2.40070	2.38807
20	2.41093	2.40977	2.40623	2.40015	2.39116	2.37869
21	2.40211	2.40096	2.39748	2.39146	2.38258	2.37026
22	2.39413	2.39300	2.38956	2.38362	2.37484	2.36264
23	2.38689	2.38578	2.38237	2.37649	2.36780	2.35572
24	2.38029	2.37918	2.37581	2.36999	2.36138	2.34941
25	2.37424	2.37314	2.36980	2.36403	2.35550	2.34363
26	2.36868	2.36759	2.36428	2.35856	2.35010	2.33831
27	2.36355	2.36247	2.35918	2.35351	2.34511	2.33341
28	2.35880	2.35773	2.35447	2.34884	2.34050	2.32888
29	2.35439	2.35333	2.35010	2.34450	2.33622	2.32467
30	2.35029	2.34924	2.34603	2.34047	2.33224	2.32075
35	2.33344	2.33241	2.32929	2.32389	2.31586	2.30465
40	2.32093	2.31993	2.31687	2.31158	2.30372	2.29271
45	2.31128	2.31029	2.30729	2.30209	2.29434	2.28350
50	2.30360	2.30264	2.29967	2.29454	2.28690	2.27617
60	2.29218	2.29123	2.28833	2.28330	2.27580	2.26527
80	2.27803	2.27711	2.27429	2.26939	2.26207	2.25177
100	2.26961	2.26871	2.26593	2.26111	2.25390	2.24374
120	2.26403	2.26314	2.26039	2.25562	2.24848	2.23841
200	2.25294	2.25206	2.24938	2.24471	2.23772	2.22783
∞	2.23648	2.23563	2.23304	2.22853	2.22175	2.21213

TABLE B.3

Two-sided Percentage Point (h) for Equal
Correlations $\left(Case\ I\ with\ Correlation\ \rho \right)$

p = 2, 1-α = 0.95

ν $\downarrow$	ρ				
	0.6	0.7	0.8	0.9	$1/(1+\sqrt{2})$
1	17.05887	16.64635	16.08153	15.23281	17.57408
2	5.34062	5.23856	5.09986	4.89391	5.46892
3	3.81945	3.75728	3.67280	3.54724	3.89764
4	3.27429	3.22656	3.16151	3.06428	3.33418
5	2.99980	2.95941	2.90413	2.82096	3.05031
6	2.83572	2.79975	2.75030	2.67541	2.88054
7	2.72695	2.69392	2.64832	2.57887	2.76796
8	2.64970	2.61875	2.57588	2.51024	2.68797
9	2.59205	2.56266	2.52181	2.45900	2.62826
10	2.54740	2.51923	2.47994	2.41930	2.58201
11	2.51183	2.48462	2.44657	2.38765	2.54516
12	2.48282	2.45640	2.41935	2.36183	2.51510
13	2.45871	2.43294	2.39673	2.34036	2.49012
14	2.43837	2.41315	2.37764	2.32224	2.46903
15	2.42097	2.39622	2.36131	2.30674	2.45100
16	2.40593	2.38158	2.34719	2.29332	2.43541
17	2.39279	2.36880	2.33485	2.28161	2.42178
18	2.38121	2.35753	2.32398	2.27128	2.40978
19	2.37094	2.34753	2.31433	2.26212	2.39913
20	2.36175	2.33860	2.30571	2.25392	2.38962
21	2.35350	2.33057	2.29796	2.24656	2.38106
22	2.34604	2.32331	2.29096	2.23990	2.37333
23	2.33927	2.31672	2.28460	2.23386	2.36631
24	2.33309	2.31071	2.27879	2.22834	2.35990
25	2.32744	2.30520	2.27348	2.22329	2.35403
26	2.32223	2.30014	2.26859	2.21864	2.34864
27	2.31744	2.29547	2.26408	2.21436	2.34367
28	2.31300	2.29115	2.25991	2.21039	2.33907
29	2.30888	2.28714	2.25604	2.20671	2.33479
30	2.30505	2.28341	2.25244	2.20329	2.33082
35	2.28929	2.26807	2.23763	2.18920	2.31448
40	2.27760	2.25669	2.22664	2.17875	2.30236
45	2.26858	2.24791	2.21816	2.17069	2.29301
50	2.26141	2.24093	2.21142	2.16428	2.28558
60	2.25074	2.23054	2.20138	2.15473	2.27451
80	2.23752	2.21767	2.18895	2.14290	2.26080
100	2.22966	2.21002	2.18155	2.13586	2.25265
120	2.22445	2.20494	2.17665	2.13119	2.24725
200	2.21409	2.19486	2.16690	2.12192	2.23651
∞	2.19872	2.17988	2.15244	2.10814	2.22057

TABLE B.3

Two-sided Percentage Point (h) for Equal
Correlations $\left(\text{Case I with Correlation } \rho\right)$

$p = 3, \ 1-\alpha = 0.95$

$\nu \downarrow$	ρ					
	0.0	0.1	0.2	0.3	0.4	0.5
2	6.34048	6.32932	6.29663	6.24262	6.16620	6.06485
3	4.42967	4.42296	4.40323	4.37054	4.32419	4.26262
4	3.74458	3.73952	3.72460	3.69980	3.66457	3.61766
5	3.39924	3.39504	3.38262	3.36194	3.33248	3.29315
6	3.19259	3.18892	3.17805	3.15990	3.13398	3.09929
7	3.05547	3.05216	3.04235	3.02592	3.00241	2.97085
8	2.95801	2.95496	2.94591	2.93073	2.90896	2.87966
9	2.88523	2.88239	2.87391	2.85968	2.83923	2.81164
10	2.82886	2.82617	2.81815	2.80467	2.78525	2.75900
11	2.78391	2.78135	2.77370	2.76082	2.74223	2.71706
12	2.74724	2.74479	2.73745	2.72506	2.70716	2.68287
13	2.71677	2.71440	2.70732	2.69535	2.67803	2.65447
14	2.69105	2.68876	2.68189	2.67028	2.65344	2.63052
15	2.66905	2.66682	2.66014	2.64884	2.63242	2.61003
16	2.65002	2.64785	2.64133	2.63029	2.61425	2.59232
17	2.63339	2.63127	2.62490	2.61410	2.59837	2.57686
18	2.61875	2.61667	2.61043	2.59983	2.58439	2.56323
19	2.60575	2.60371	2.59758	2.58717	2.57198	2.55115
20	2.59413	2.59213	2.58611	2.57586	2.56090	2.54035
21	2.58369	2.58172	2.57579	2.56569	2.55093	2.53065
22	2.57425	2.57231	2.56646	2.55650	2.54193	2.52188
23	2.56568	2.56376	2.55799	2.54816	2.53375	2.51391
24	2.55786	2.55597	2.55027	2.54055	2.52630	2.50665
25	2.55070	2.54883	2.54319	2.53357	2.51947	2.50000
26	2.54411	2.54226	2.53669	2.52717	2.51319	2.49389
27	2.53804	2.53621	2.53069	2.52126	2.50740	2.48825
28	2.53242	2.53061	2.52514	2.51579	2.50205	2.48304
29	2.52721	2.52541	2.51999	2.51072	2.49708	2.47820
30	2.52235	2.52057	2.51520	2.50600	2.49245	2.47369
35	2.50241	2.50069	2.49550	2.48659	2.47345	2.45519
40	2.48760	2.48594	2.48088	2.47219	2.45935	2.44146
45	2.47619	2.47455	2.46961	2.46109	2.44848	2.43088
50	2.46711	2.46551	2.46065	2.45227	2.43984	2.42247
60	2.45359	2.45204	2.44730	2.43913	2.42697	2.40994
80	2.43686	2.43536	2.43079	2.42286	2.41105	2.39444
100	2.42691	2.42544	2.42096	2.41319	2.40158	2.38523
120	2.42031	2.41886	2.41445	2.40678	2.39531	2.37912
200	2.40720	2.40579	2.40150	2.39404	2.38284	2.36698
∞	2.38774	2.38640	2.38230	2.37514	2.36434	2.34897

TABLE B.3

Two-sided Percentage Point (h) for Equal
Correlations   (Case I with Correlation ρ)

p = 3, 1-α = 0.95

ν ↓	0.6	0.7	0.8	0.9	$1/(1+\sqrt{3})$
2	5.93379	5.76432	5.53875	5.21098	6.19480
3	4.18291	4.07974	3.94234	3.74244	4.34155
4	3.55678	3.47778	3.37217	3.21766	3.67777
5	3.24197	3.17532	3.08581	2.95391	3.34353
6	3.05401	2.99482	2.91494	2.79642	3.14371
7	2.92954	2.87533	2.80183	2.69209	3.01124
8	2.84120	2.79055	2.72157	2.61801	2.91714
9	2.77532	2.72733	2.66172	2.56273	2.84692
10	2.72434	2.67842	2.61541	2.51993	2.79255
11	2.68374	2.63946	2.57851	2.48582	2.74923
12	2.65065	2.60771	2.54844	2.45801	2.71391
13	2.62316	2.58135	2.52346	2.43490	2.68456
14	2.59998	2.55910	2.50238	2.41539	2.65979
15	2.58015	2.54008	2.48436	2.39871	2.63862
16	2.56302	2.52364	2.46878	2.38428	2.62030
17	2.54805	2.50928	2.45517	2.37167	2.60431
18	2.53487	2.49664	2.44318	2.36057	2.59022
19	2.52318	2.48541	2.43255	2.35071	2.57772
20	2.51273	2.47539	2.42304	2.34190	2.56655
21	2.50334	2.46638	2.41450	2.33398	2.55651
22	2.49486	2.45824	2.40678	2.32683	2.54744
23	2.48715	2.45085	2.39977	2.32033	2.53920
24	2.48013	2.44410	2.39337	2.31440	2.53169
25	2.47369	2.43793	2.38752	2.30897	2.52481
26	2.46778	2.43225	2.38213	2.30398	2.51848
27	2.46232	2.42702	2.37717	2.29937	2.51265
28	2.45728	2.42218	2.37258	2.29511	2.50725
29	2.45260	2.41768	2.36831	2.29116	2.50225
30	2.44824	2.41350	2.36435	2.28748	2.49759
35	2.43034	2.39632	2.34804	2.27235	2.47843
40	2.41706	2.38357	2.33595	2.26113	2.46423
45	2.40682	2.37374	2.32662	2.25247	2.45327
50	2.39868	2.36592	2.31920	2.24558	2.44456
60	2.38656	2.35429	2.30816	2.23533	2.43159
80	2.37157	2.33989	2.29448	2.22264	2.41555
100	2.36265	2.33133	2.28635	2.21508	2.40600
120	2.35674	2.32565	2.28096	2.21008	2.39968
200	2.34499	2.31437	2.27025	2.20012	2.38711
∞	2.32756	2.29763	2.25435	2.18535	2.36846

BECHHOFER and DUNNETT

TABLE B.3

Two-sided Percentage Point (h) for Equal Correlations (*Case I with Correlation* ρ)

$p = 4, \ 1-\alpha = 0.95$

$\nu \downarrow$	ρ					
	0.0	0.1	0.2	0.3	0.4	0.5
2	6.88627	6.87016	6.82422	6.75008	6.64722	6.51298
3	4.76439	4.75475	4.72710	4.68227	4.61987	4.53824
4	4.00304	3.99581	3.97497	3.94105	3.89368	3.83152
5	3.61890	3.61293	3.59566	3.56745	3.52792	3.47589
6	3.38884	3.38365	3.36859	3.34392	3.30922	3.26342
7	3.23609	3.23144	3.21790	3.19563	3.16423	3.12264
8	3.12746	3.12320	3.11076	3.09024	3.06124	3.02270
9	3.04632	3.04235	3.03075	3.01158	2.98439	2.94817
10	2.98343	2.97970	2.96877	2.95065	2.92489	2.89048
11	2.93329	2.92975	2.91935	2.90208	2.87748	2.84453
12	2.89237	2.88899	2.87904	2.86248	2.83883	2.80708
13	2.85836	2.85512	2.84554	2.82957	2.80672	2.77598
14	2.82965	2.82652	2.81726	2.80180	2.77963	2.74974
15	2.80509	2.80205	2.79307	2.77805	2.75647	2.72731
16	2.78384	2.78089	2.77215	2.75751	2.73644	2.70792
17	2.76528	2.76240	2.75388	2.73957	2.71895	2.69099
18	2.74893	2.74612	2.73778	2.72377	2.70354	2.67608
19	2.73441	2.73166	2.72349	2.70974	2.68987	2.66285
20	2.72144	2.71874	2.71072	2.69721	2.67766	2.65103
21	2.70977	2.70713	2.69925	2.68595	2.66669	2.64041
22	2.69923	2.69663	2.68887	2.67577	2.65677	2.63081
23	2.68966	2.68710	2.67945	2.66653	2.64776	2.62210
24	2.68093	2.67840	2.67086	2.65810	2.63955	2.61415
25	2.67293	2.67043	2.66299	2.65038	2.63203	2.60688
26	2.66557	2.66311	2.65576	2.64329	2.62512	2.60019
27	2.65879	2.65636	2.64908	2.63674	2.61874	2.59402
28	2.65251	2.65011	2.64291	2.63069	2.61284	2.58832
29	2.64669	2.64431	2.63718	2.62507	2.60737	2.58302
30	2.64127	2.63891	2.63185	2.61984	2.60228	2.57810
35	2.61899	2.61673	2.60994	2.59835	2.58135	2.55786
40	2.60246	2.60027	2.59368	2.58241	2.56583	2.54285
45	2.58970	2.58757	2.58114	2.57012	2.55387	2.53128
50	2.57956	2.57748	2.57117	2.56035	2.54435	2.52208
60	2.56447	2.56245	2.55633	2.54581	2.53020	2.50839
80	2.54578	2.54384	2.53796	2.52781	2.51268	2.49145
100	2.53466	2.53277	2.52704	2.51710	2.50226	2.48138
120	2.52729	2.52544	2.51980	2.51000	2.49535	2.47470
200	2.51265	2.51086	2.50541	2.49591	2.48164	2.46144
∞	2.49092	2.48923	2.48406	2.47500	2.46129	2.44177

TABLE B.3

Two-sided Percentage Point (h) for Equal
Correlations $\left(Case\ I\ with\ Correlation\ \rho\right)$

$p = 4, \ 1-\alpha = 0.95$

$\nu \downarrow$	\multicolumn{5}{c}{ρ}				
	0.6	0.7	0.8	0.9	$1/(1+\sqrt{4})$
2	6.34184	6.12333	5.83596	5.42342	6.71907
3	4.43400	4.30074	4.12536	3.87339	4.66348
4	3.75193	3.64989	3.51511	3.32041	3.92680
5	3.40905	3.32303	3.20888	3.04284	3.55557
6	3.20436	3.12807	3.02632	2.87727	3.33350
7	3.06884	2.99906	2.90554	2.76765	3.18621
8	2.97269	2.90756	2.81988	2.68987	3.08156
9	2.90100	2.83937	2.75605	2.63186	3.00344
10	2.84554	2.78663	2.70666	2.58695	2.94295
11	2.80138	2.74463	2.66734	2.55118	2.89473
12	2.76540	2.71042	2.63530	2.52202	2.85542
13	2.73552	2.68200	2.60869	2.49779	2.82276
14	2.71031	2.65804	2.58624	2.47734	2.79519
15	2.68877	2.63756	2.56705	2.45986	2.77162
16	2.67014	2.61985	2.55046	2.44473	2.75124
17	2.65388	2.60439	2.53597	2.43153	2.73344
18	2.63957	2.59078	2.52321	2.41989	2.71776
19	2.62686	2.57870	2.51189	2.40957	2.70384
20	2.61552	2.56791	2.50178	2.40034	2.69141
21	2.60532	2.55821	2.49269	2.39205	2.68023
22	2.59611	2.54945	2.48447	2.38456	2.67014
23	2.58774	2.54150	2.47701	2.37775	2.66097
24	2.58011	2.53424	2.47021	2.37154	2.65260
25	2.57313	2.52760	2.46398	2.36586	2.64494
26	2.56671	2.52149	2.45826	2.36063	2.63790
27	2.56079	2.51586	2.45298	2.35581	2.63141
28	2.55531	2.51066	2.44809	2.35135	2.62541
29	2.55023	2.50582	2.44356	2.34721	2.61983
30	2.54550	2.50133	2.43934	2.34336	2.61465
35	2.52607	2.48285	2.42200	2.32752	2.59333
40	2.51166	2.46914	2.40914	2.31577	2.57752
45	2.50055	2.45857	2.39922	2.30671	2.56533
50	2.49172	2.45017	2.39134	2.29951	2.55564
60	2.47858	2.43767	2.37960	2.28878	2.54121
80	2.46232	2.42219	2.36507	2.27549	2.52336
100	2.45265	2.41299	2.35643	2.26759	2.51274
120	2.44624	2.40689	2.35070	2.26235	2.50571
200	2.43350	2.39477	2.33932	2.25194	2.49173
∞	2.41462	2.37679	2.32243	2.23649	2.47099

TABLE B.3

Two-sided Percentage Point (h) for Equal
Correlations (Case I with Correlation ρ)

$p = 5, \ 1-\alpha = 0.95$

ν ↓	ρ					
	0.0	0.1	0.2	0.3	0.4	0.5
2	7.30606	7.28551	7.22817	7.13724	7.01283	6.85231
3	5.02319	5.01091	4.97638	4.92135	4.84575	4.74795
4	4.20321	4.19401	4.16802	4.12639	4.06898	3.99447
5	3.78902	3.78145	3.75995	3.72535	3.67748	3.61512
6	3.54072	3.53416	3.51545	3.48523	3.44325	3.38839
7	3.37573	3.36987	3.35308	3.32585	3.28790	3.23813
8	3.25831	3.25296	3.23757	3.21253	3.17752	3.13145
9	3.17056	3.16560	3.15128	3.12791	3.09514	3.05188
10	3.10253	3.09787	3.08440	3.06236	3.03134	2.99029
11	3.04825	3.04385	3.03107	3.01010	2.98051	2.94124
12	3.00396	2.99976	2.98756	2.96747	2.93906	2.90126
13	2.96713	2.96311	2.95139	2.93205	2.90463	2.86805
14	2.93604	2.93216	2.92085	2.90215	2.87558	2.84004
15	2.90943	2.90568	2.89473	2.87658	2.85074	2.81610
16	2.88641	2.88277	2.87213	2.85446	2.82926	2.79540
17	2.86629	2.86276	2.85239	2.83515	2.81050	2.77733
18	2.84857	2.84512	2.83500	2.81813	2.79398	2.76141
19	2.83284	2.82947	2.81956	2.80303	2.77932	2.74729
20	2.81877	2.81548	2.80577	2.78954	2.76622	2.73468
21	2.80613	2.80290	2.79336	2.77741	2.75445	2.72334
22	2.79470	2.79153	2.78216	2.76645	2.74382	2.71310
23	2.78432	2.78120	2.77198	2.75650	2.73416	2.70380
24	2.77486	2.77178	2.76269	2.74742	2.72535	2.69532
25	2.76618	2.76315	2.75419	2.73911	2.71729	2.68756
26	2.75821	2.75522	2.74637	2.73146	2.70987	2.68042
27	2.75085	2.74791	2.73916	2.72442	2.70304	2.67384
28	2.74405	2.74114	2.73249	2.71789	2.69671	2.66776
29	2.73773	2.73485	2.72630	2.71184	2.69085	2.66211
30	2.73185	2.72901	2.72053	2.70621	2.68539	2.65685
35	2.70769	2.70497	2.69685	2.68307	2.66295	2.63526
40	2.68976	2.68713	2.67928	2.66591	2.64631	2.61925
45	2.67592	2.67337	2.66573	2.65267	2.63348	2.60691
50	2.66493	2.66244	2.65496	2.64215	2.62328	2.59710
60	2.64855	2.64615	2.63892	2.62649	2.60810	2.58250
80	2.62828	2.62599	2.61906	2.60710	2.58932	2.56443
100	2.61622	2.61400	2.60725	2.59558	2.57816	2.55369
120	2.60822	2.60604	2.59943	2.58794	2.57076	2.54657
200	2.59234	2.59025	2.58387	2.57276	2.55605	2.53243
∞	2.56876	2.56681	2.56080	2.55025	2.53425	2.51146

TABLE B.3

*Two-sided Percentage Point (h) for Equal
Correlations* $\left(\text{Case I with Correlation } \rho\right)$

$$p = 5, \quad 1-\alpha = 0.95$$

$\nu \downarrow$	ρ				
	0.6	0.7	0.8	0.9	$1/(1+\sqrt{5})$
2	6.64964	6.39305	6.05823	5.58138	7.12742
3	4.62422	4.46739	4.26263	3.97091	4.91539
4	3.89994	3.77976	3.62233	3.39690	4.12187
5	3.53574	3.43446	3.30116	3.10899	3.72159
6	3.31830	3.22850	3.10974	2.93735	3.48194
7	3.17432	3.09225	2.98316	2.82378	3.32288
8	3.07218	2.99563	2.89342	2.74322	3.20979
9	2.99603	2.92364	2.82656	2.68314	3.12535
10	2.93712	2.86797	2.77485	2.63667	3.05994
11	2.89022	2.82365	2.73369	2.59964	3.00779
12	2.85200	2.78754	2.70015	2.56947	2.96526
13	2.82027	2.75756	2.67230	2.54440	2.92992
14	2.79351	2.73228	2.64881	2.52325	2.90009
15	2.77064	2.71068	2.62873	2.50516	2.87458
16	2.75087	2.69200	2.61138	2.48952	2.85251
17	2.73361	2.67570	2.59622	2.47586	2.83324
18	2.71841	2.66135	2.58288	2.46383	2.81627
19	2.70493	2.64861	2.57104	2.45316	2.80120
20	2.69289	2.63723	2.56046	2.44362	2.78774
21	2.68207	2.62701	2.55096	2.43504	2.77564
22	2.67229	2.61778	2.54237	2.42730	2.76471
23	2.66341	2.60939	2.53457	2.42026	2.75478
24	2.65532	2.60175	2.52746	2.41384	2.74572
25	2.64791	2.59474	2.52095	2.40796	2.73743
26	2.64109	2.58831	2.51496	2.40256	2.72980
27	2.63481	2.58237	2.50944	2.39758	2.72277
28	2.62900	2.57688	2.50433	2.39297	2.71627
29	2.62361	2.57179	2.49959	2.38869	2.71023
30	2.61860	2.56705	2.49518	2.38471	2.70462
35	2.59798	2.54758	2.47706	2.36834	2.68153
40	2.58270	2.53314	2.46362	2.35620	2.66441
45	2.57092	2.52200	2.45326	2.34683	2.65121
50	2.56156	2.51315	2.44502	2.33939	2.64071
60	2.54762	2.49998	2.43275	2.32830	2.62509
80	2.53038	2.48368	2.41757	2.31458	2.60576
100	2.52013	2.47399	2.40855	2.30641	2.59426
120	2.51333	2.46757	2.40256	2.30100	2.58664
200	2.49983	2.45480	2.39067	2.29024	2.57150
∞	2.47981	2.43587	2.37303	2.27428	2.54905

TABLE B.3

Two-sided Percentage Point (h) for Equal
Correlations $\left(\text{Case I with Correlation } \rho\right)$

p = 6, 1-α = 0.95

ν ↓	ρ 0.0	0.1	0.2	0.3	0.4	0.5
2	7.64539	7.62085	7.55355	7.44832	7.30587	7.12361
3	5.23336	5.21869	5.17813	5.11432	5.02759	4.91630
4	4.36610	4.35512	4.32458	4.27628	4.21036	4.12549
5	3.92754	3.91852	3.89327	3.85314	3.79815	3.72711
6	3.66438	3.65658	3.63463	3.59960	3.55140	3.48890
7	3.48938	3.48242	3.46274	3.43120	3.38765	3.33099
8	3.36476	3.35841	3.34040	3.31143	3.27127	3.21884
9	3.27157	3.26569	3.24896	3.22195	3.18439	3.13519
10	3.19928	3.19378	3.17807	3.15261	3.11710	3.07043
11	3.14159	3.13640	3.12151	3.09731	3.06346	3.01885
12	3.09450	3.08956	3.07535	3.05221	3.01973	2.97681
13	3.05533	3.05061	3.03698	3.01472	2.98340	2.94189
14	3.02225	3.01771	3.00458	2.98307	2.95274	2.91244
15	2.99394	2.98955	2.97685	2.95600	2.92652	2.88726
16	2.96944	2.96519	2.95287	2.93258	2.90385	2.86549
17	2.94803	2.94391	2.93191	2.91213	2.88405	2.84649
18	2.92917	2.92515	2.91345	2.89412	2.86661	2.82975
19	2.91241	2.90849	2.89706	2.87812	2.85114	2.81490
20	2.89744	2.89361	2.88241	2.86383	2.83732	2.80164
21	2.88398	2.88022	2.86924	2.85099	2.82489	2.78972
22	2.87181	2.86813	2.85734	2.83938	2.81367	2.77895
23	2.86075	2.85714	2.84653	2.82884	2.80347	2.76918
24	2.85067	2.84711	2.83667	2.81923	2.79418	2.76026
25	2.84143	2.83793	2.82763	2.81042	2.78566	2.75210
26	2.83294	2.82949	2.81933	2.80233	2.77784	2.74460
27	2.82510	2.82170	2.81167	2.79486	2.77062	2.73768
28	2.81785	2.81449	2.80458	2.78796	2.76395	2.73128
29	2.81112	2.80781	2.79801	2.78155	2.75775	2.72534
30	2.80486	2.80158	2.79189	2.77559	2.75199	2.71982
35	2.77911	2.77599	2.76673	2.75107	2.72831	2.69711
40	2.76001	2.75700	2.74806	2.73289	2.71074	2.68028
45	2.74526	2.74235	2.73366	2.71887	2.69720	2.66731
50	2.73354	2.73071	2.72222	2.70773	2.68644	2.65700
60	2.71609	2.71336	2.70517	2.69114	2.67042	2.64165
80	2.69448	2.69189	2.68408	2.67060	2.65060	2.62266
100	2.68163	2.67912	2.67153	2.65840	2.63882	2.61137
120	2.67310	2.67065	2.66321	2.65030	2.63101	2.60389
200	2.65617	2.65383	2.64669	2.63423	2.61549	2.58903
∞	2.63104	2.62886	2.62217	2.61038	2.59249	2.56700

TABLE B.3

Two-sided Percentage Point (h) for Equal
Correlations   (*Case I with Correlation* ρ)

p = 6, 1-α = 0.95

ν ↓	ρ				
	0.6	0.7	0.8	0.9	$1/(1+\sqrt{6})$
2	6.89512	6.60760	6.23455	5.70624	7.46063
3	4.77650	4.60040	4.37181	4.04812	5.12180
4	4.01857	3.88351	3.70765	3.45745	4.28195
5	3.63731	3.52346	3.37455	3.16132	3.85786
6	3.40962	3.30870	3.17608	2.98485	3.60372
7	3.25884	3.16663	3.04485	2.86814	3.43493
8	3.15185	3.06590	2.95184	2.78537	3.31485
9	3.07210	2.99084	2.88255	2.72365	3.22514
10	3.01040	2.93280	2.82898	2.67592	3.15562
11	2.96128	2.88661	2.78633	2.63790	3.10018
12	2.92125	2.84898	2.75159	2.60691	3.05496
13	2.88802	2.81774	2.72275	2.58118	3.01736
14	2.86000	2.79140	2.69843	2.55947	2.98563
15	2.83605	2.76889	2.67764	2.54090	2.95848
16	2.81534	2.74943	2.65967	2.52485	2.93500
17	2.79727	2.73244	2.64398	2.51083	2.91449
18	2.78136	2.71749	2.63017	2.49849	2.89643
19	2.76724	2.70422	2.61791	2.48753	2.88039
20	2.75464	2.69237	2.60696	2.47774	2.86606
21	2.74331	2.68173	2.59712	2.46894	2.85318
22	2.73307	2.67211	2.58823	2.46099	2.84154
23	2.72378	2.66337	2.58016	2.45377	2.83096
24	2.71530	2.65541	2.57280	2.44719	2.82132
25	2.70755	2.64812	2.56606	2.44116	2.81249
26	2.70042	2.64142	2.55986	2.43561	2.80437
27	2.69384	2.63524	2.55415	2.43050	2.79688
28	2.68776	2.62952	2.54886	2.42577	2.78996
29	2.68212	2.62422	2.54396	2.42138	2.78353
30	2.67687	2.61928	2.53940	2.41730	2.77755
35	2.65530	2.59901	2.52065	2.40051	2.75296
40	2.63930	2.58397	2.50674	2.38805	2.73473
45	2.62697	2.57238	2.49601	2.37844	2.72066
50	2.61717	2.56317	2.48749	2.37081	2.70948
60	2.60259	2.54945	2.47480	2.35944	2.69284
80	2.58455	2.53249	2.45910	2.34536	2.67224
100	2.57382	2.52240	2.44976	2.33699	2.66000
120	2.56672	2.51571	2.44357	2.33144	2.65188
200	2.55259	2.50243	2.43128	2.32040	2.63575
∞	2.53165	2.48273	2.41303	2.30404	2.61183

BECHHOFER and DUNNETT

TABLE B.3

Two-sided Percentage Point (h) for Equal
Correlations $\left(\text{Case I with Correlation } \rho\right)$

$p = 7, \ 1-\alpha = 0.95$

$\nu \downarrow$	ρ					
	0.0	0.1	0.2	0.3	0.4	0.5
2	7.92913	7.90096	7.82488	7.70724	7.54932	7.34861
3	5.40981	5.39295	5.34702	5.27556	5.17921	5.05640
4	4.50311	4.49051	4.45590	4.40176	4.32845	4.23469
5	4.04417	4.03381	4.00520	3.96020	3.89902	3.82050
6	3.76853	3.75958	3.73472	3.69543	3.64181	3.57273
7	3.58509	3.57711	3.55484	3.51949	3.47105	3.40842
8	3.45437	3.44711	3.42674	3.39428	3.34963	3.29171
9	3.35657	3.34986	3.33095	3.30071	3.25896	3.20462
10	3.28067	3.27440	3.25666	3.22817	3.18872	3.13721
11	3.22008	3.21416	3.19737	3.17032	3.13274	3.08350
12	3.17060	3.16497	3.14897	3.12311	3.08707	3.03972
13	3.12943	3.12406	3.10872	3.08387	3.04913	3.00337
14	3.09465	3.08950	3.07473	3.05073	3.01711	2.97269
15	3.06489	3.05991	3.04565	3.02239	2.98973	2.94647
16	3.03912	3.03431	3.02048	2.99787	2.96605	2.92381
17	3.01660	3.01194	2.99848	2.97646	2.94537	2.90402
18	2.99675	2.99222	2.97910	2.95759	2.92716	2.88659
19	2.97913	2.97470	2.96190	2.94084	2.91100	2.87113
20	2.96337	2.95905	2.94652	2.92587	2.89656	2.85731
21	2.94920	2.94498	2.93270	2.91242	2.88358	2.84490
22	2.93640	2.93225	2.92020	2.90026	2.87185	2.83369
23	2.92476	2.92070	2.90885	2.88922	2.86120	2.82351
24	2.91415	2.91015	2.89849	2.87915	2.85149	2.81422
25	2.90442	2.90049	2.88901	2.86992	2.84260	2.80572
26	2.89548	2.89161	2.88029	2.86144	2.83442	2.79791
27	2.88723	2.88342	2.87224	2.85362	2.82689	2.79071
28	2.87960	2.87583	2.86480	2.84639	2.81991	2.78404
29	2.87251	2.86880	2.85789	2.83967	2.81344	2.77786
30	2.86592	2.86225	2.85147	2.83342	2.80742	2.77211
35	2.83880	2.83532	2.82504	2.80774	2.78268	2.74847
40	2.81868	2.81534	2.80543	2.78869	2.76433	2.73094
45	2.80315	2.79992	2.79030	2.77400	2.75018	2.71743
50	2.79080	2.78766	2.77828	2.76232	2.73894	2.70670
60	2.77242	2.76940	2.76037	2.74494	2.72221	2.69072
80	2.74965	2.74680	2.73821	2.72342	2.70150	2.67095
100	2.73610	2.73335	2.72503	2.71063	2.68920	2.65920
120	2.72712	2.72444	2.71629	2.70215	2.68104	2.65141
200	2.70928	2.70672	2.69893	2.68531	2.66483	2.63594
∞	2.68280	2.68044	2.67317	2.66032	2.64080	2.61300

TABLE B.3

Two-sided Percentage Point (h) for Equal
Correlations (*Case I with Correlation* ρ)

p = 7, 1-α = 0.95

ν ↓	ρ				
	0.6	0.7	0.8	0.9	1/(1+√7)
2	7.09835	6.78492	6.37998	5.80897	7.74131
3	4.90297	4.71063	4.46208	4.11174	5.29629
4	4.11722	3.96958	3.77823	3.50734	4.41749
5	3.72180	3.59730	3.43526	3.20443	3.97330
6	3.48559	3.37523	3.23092	3.02397	3.70689
7	3.32913	3.22832	3.09584	2.90465	3.52981
8	3.21810	3.12415	3.00011	2.82006	3.40377
9	3.13533	3.04654	2.92880	2.75696	3.30956
10	3.07129	2.98653	2.87367	2.70820	3.23652
11	3.02031	2.93876	2.82980	2.66936	3.17825
12	2.97877	2.89986	2.79406	2.63770	3.13070
13	2.94428	2.86756	2.76439	2.61142	3.09117
14	2.91519	2.84033	2.73937	2.58924	3.05779
15	2.89033	2.81706	2.71799	2.57028	3.02924
16	2.86885	2.79695	2.69951	2.55388	3.00454
17	2.85009	2.77940	2.68337	2.53957	2.98295
18	2.83358	2.76394	2.66917	2.52696	2.96394
19	2.81893	2.75023	2.65656	2.51577	2.94706
20	2.80585	2.73798	2.64530	2.50578	2.93198
21	2.79409	2.72698	2.63519	2.49679	2.91842
22	2.78347	2.71704	2.62605	2.48868	2.90616
23	2.77383	2.70801	2.61775	2.48130	2.89503
24	2.76504	2.69979	2.61018	2.47458	2.88488
25	2.75699	2.69225	2.60325	2.46842	2.87558
26	2.74959	2.68533	2.59688	2.46276	2.86703
27	2.74277	2.67894	2.59101	2.45754	2.85915
28	2.73646	2.67303	2.58557	2.45271	2.85185
29	2.73060	2.66755	2.58053	2.44823	2.84508
30	2.72515	2.66246	2.57584	2.44406	2.83879
35	2.70277	2.64151	2.55656	2.42692	2.81289
40	2.68618	2.62597	2.54227	2.41421	2.79368
45	2.67339	2.61400	2.53124	2.40440	2.77887
50	2.66323	2.60448	2.52249	2.39661	2.76709
60	2.64810	2.59031	2.50944	2.38500	2.74956
80	2.62939	2.57279	2.49331	2.37063	2.72786
100	2.61826	2.56237	2.48371	2.36209	2.71496
120	2.61089	2.55546	2.47735	2.35642	2.70640
200	2.59624	2.54174	2.46471	2.34516	2.68941
∞	2.57452	2.52139	2.44596	2.32846	2.66421

TABLE B.3

Two-sided Percentage Point (h) for Equal
Correlations   (*Case I with Correlation* ρ)

p = 8, 1-α = 0.95

ν ↓	ρ					
	0.0	0.1	0.2	0.3	0.4	0.5
2	8.17229	8.14081	8.05687	7.92831	7.75690	7.54021
3	5.56153	5.54268	5.49192	5.41367	5.30889	5.17605
4	4.62115	4.60704	4.56877	4.50942	4.42961	4.32808
5	4.14473	4.13314	4.10149	4.05214	3.98550	3.90042
6	3.85836	3.84836	3.82086	3.77777	3.71934	3.64448
7	3.66765	3.65874	3.63411	3.59534	3.54256	3.47470
8	3.53167	3.52356	3.50105	3.46546	3.41682	3.35406
9	3.42988	3.42239	3.40151	3.36836	3.32290	3.26403
10	3.35085	3.34386	3.32427	3.29307	3.25012	3.19433
11	3.28773	3.28114	3.26262	3.23300	3.19210	3.13879
12	3.23617	3.22992	3.21228	3.18397	3.14477	3.09352
13	3.19327	3.18730	3.17040	3.14321	3.10544	3.05592
14	3.15701	3.15129	3.13503	3.10880	3.07225	3.02420
15	3.12597	3.12046	3.10476	3.07935	3.04386	2.99708
16	3.09910	3.09377	3.07856	3.05387	3.01931	2.97363
17	3.07561	3.07044	3.05566	3.03162	2.99787	2.95316
18	3.05490	3.04988	3.03549	3.01201	2.97898	2.93514
19	3.03651	3.03162	3.01757	2.99460	2.96222	2.91914
20	3.02007	3.01530	3.00156	2.97905	2.94724	2.90486
21	3.00528	3.00062	2.98716	2.96506	2.93379	2.89202
22	2.99191	2.98735	2.97414	2.95242	2.92162	2.88042
23	2.97977	2.97529	2.96232	2.94095	2.91058	2.86989
24	2.96868	2.96429	2.95153	2.93048	2.90051	2.86029
25	2.95853	2.95421	2.94165	2.92089	2.89129	2.85150
26	2.94920	2.94494	2.93257	2.91207	2.88281	2.84342
27	2.94058	2.93639	2.92419	2.90394	2.87499	2.83596
28	2.93261	2.92848	2.91643	2.89642	2.86776	2.82907
29	2.92521	2.92114	2.90924	2.88944	2.86105	2.82268
30	2.91833	2.91431	2.90254	2.88294	2.85480	2.81673
35	2.89001	2.88620	2.87500	2.85624	2.82914	2.79228
40	2.86899	2.86534	2.85457	2.83643	2.81011	2.77415
45	2.85277	2.84924	2.83880	2.82115	2.79544	2.76018
50	2.83987	2.83645	2.82627	2.80901	2.78378	2.74907
60	2.82066	2.81739	2.80761	2.79093	2.76642	2.73255
80	2.79687	2.79378	2.78451	2.76856	2.74495	2.71210
100	2.78271	2.77974	2.77077	2.75525	2.73218	2.69995
120	2.77333	2.77043	2.76166	2.74643	2.72372	2.69189
200	2.75468	2.75193	2.74356	2.72892	2.70691	2.67590
∞	2.72701	2.72448	2.71670	2.70293	2.68199	2.65218

TABLE B.3

Two-sided Percentage Point (h) for Equal
Correlations $\Big(Case\ I\ with\ Correlation\ \rho\Big)$

p = 8, 1-α = 0.95

ν ↓	ρ				
	0.6	0.7	0.8	0.9	1/(1+√8)
2	7.27118	6.93552	6.50334	5.89594	7.98327
3	5.01082	4.80448	4.53879	4.16567	5.44717
4	4.20145	4.04292	3.83824	3.54963	4.53487
5	3.79397	3.66024	3.48688	3.24096	4.07333
6	3.55048	3.43194	3.27755	3.05711	3.79629
7	3.38916	3.28088	3.13917	2.93557	3.61202
8	3.27467	3.17377	3.04112	2.84944	3.48079
9	3.18931	3.09397	2.96809	2.78516	3.38266
10	3.12327	3.03227	2.91164	2.73553	3.30654
11	3.07068	2.98316	2.86671	2.69598	3.24580
12	3.02784	2.94317	2.83011	2.66376	3.19622
13	2.99226	2.90997	2.79974	2.63700	3.15498
14	2.96227	2.88197	2.77412	2.61443	3.12016
15	2.93663	2.85805	2.75223	2.59513	3.09037
16	2.91447	2.83738	2.73331	2.57845	3.06459
17	2.89513	2.81933	2.71680	2.56388	3.04206
18	2.87809	2.80345	2.70226	2.55105	3.02221
19	2.86299	2.78935	2.68935	2.53966	3.00459
20	2.84949	2.77677	2.67783	2.52949	2.98884
21	2.83737	2.76546	2.66748	2.52035	2.97468
22	2.82642	2.75524	2.65812	2.51209	2.96188
23	2.81647	2.74597	2.64963	2.50459	2.95026
24	2.80741	2.73751	2.64188	2.49775	2.93965
25	2.79910	2.72977	2.63479	2.49148	2.92994
26	2.79147	2.72265	2.62827	2.48572	2.92101
27	2.78444	2.71609	2.62226	2.48041	2.91278
28	2.77793	2.71002	2.61670	2.47550	2.90516
29	2.77189	2.70439	2.61154	2.47094	2.89809
30	2.76628	2.69915	2.60674	2.46670	2.89151
35	2.74320	2.67762	2.58702	2.44926	2.86445
40	2.72609	2.66166	2.57239	2.43632	2.84438
45	2.71290	2.64935	2.56111	2.42634	2.82890
50	2.70243	2.63958	2.55215	2.41842	2.81659
60	2.68683	2.62502	2.53880	2.40661	2.79827
80	2.66753	2.60702	2.52229	2.39199	2.77559
100	2.65606	2.59631	2.51247	2.38330	2.76210
120	2.64846	2.58921	2.50597	2.37754	2.75316
200	2.63336	2.57512	2.49304	2.36609	2.73540
∞	2.61097	2.55422	2.47386	2.34909	2.70905

TABLE B.3

Two-sided Percentage Point (h) for Equal
Correlations $\left(Case\ I\ with\ Correlation\ \rho\right)$

$p = 9,\ 1-\alpha = 0.95$

ν ↓	0.0	0.1	0.2	0.3	0.4	0.5
			ρ			
2	8.38461	8.35009	8.25906	8.12074	7.93740	7.70663
3	5.69440	5.67370	5.61856	5.53423	5.42196	5.28024
4	4.72469	4.70919	4.66758	4.60354	4.51793	4.40950
5	4.23302	4.22029	4.18586	4.13258	4.06105	3.97015
6	3.93728	3.92628	3.89636	3.84984	3.78710	3.70709
7	3.74019	3.73040	3.70361	3.66174	3.60507	3.53253
8	3.59958	3.59069	3.56620	3.52777	3.47555	3.40847
9	3.49428	3.48607	3.46337	3.42758	3.37878	3.31586
10	3.41249	3.40483	3.38355	3.34987	3.30378	3.24416
11	3.34715	3.33993	3.31981	3.28785	3.24397	3.18702
12	3.29375	3.28691	3.26776	3.23723	3.19518	3.14044
13	3.24930	3.24278	3.22445	3.19514	3.15463	3.10174
14	3.21173	3.20548	3.18786	3.15958	3.12040	3.06910
15	3.17956	3.17355	3.15654	3.12916	3.09112	3.04119
16	3.15171	3.14590	3.12942	3.10284	3.06580	3.01706
17	3.12736	3.12173	3.10573	3.07985	3.04369	2.99600
18	3.10588	3.10042	3.08485	3.05958	3.02421	2.97745
19	3.08681	3.08149	3.06630	3.04160	3.00692	2.96098
20	3.06976	3.06457	3.04972	3.02552	2.99148	2.94628
21	3.05442	3.04936	3.03481	3.01107	2.97759	2.93307
22	3.04056	3.03560	3.02134	2.99800	2.96505	2.92113
23	3.02796	3.02310	3.00910	2.98614	2.95366	2.91029
24	3.01646	3.01169	2.99793	2.97532	2.94327	2.90041
25	3.00592	3.00124	2.98769	2.96540	2.93375	2.89136
26	2.99624	2.99163	2.97829	2.95629	2.92501	2.88304
27	2.98730	2.98276	2.96961	2.94788	2.91694	2.87537
28	2.97903	2.97456	2.96158	2.94011	2.90948	2.86828
29	2.97135	2.96694	2.95412	2.93289	2.90256	2.86170
30	2.96420	2.95985	2.94719	2.92617	2.89612	2.85557
35	2.93481	2.93070	2.91866	2.89857	2.86964	2.83041
40	2.91298	2.90906	2.89749	2.87808	2.85001	2.81175
45	2.89614	2.89235	2.88116	2.86229	2.83487	2.79737
50	2.88275	2.87907	2.86818	2.84973	2.82284	2.78594
60	2.86280	2.85929	2.84884	2.83104	2.80493	2.76894
80	2.83809	2.83480	2.82490	2.80790	2.78278	2.74790
100	2.82339	2.82022	2.81066	2.79414	2.76961	2.73539
120	2.81364	2.81056	2.80122	2.78502	2.76088	2.72710
200	2.79427	2.79136	2.78246	2.76691	2.74354	2.71063
∞	2.76553	2.76286	2.75463	2.74004	2.71782	2.68622

TABLE B.3

Two-sided Percentage Point (h) for Equal Correlations $\left(Case\ I\ with\ Correlation\ \rho\right)$

$p = 9, \ 1-\alpha = 0.95$

$\nu \downarrow$	0.6	0.7	0.8	0.9	$1/(1+\sqrt{9})$
			ρ		
2	7.42117	7.06609	6.61017	5.97117	8.19555
3	5.10462	4.88601	4.60533	4.21236	5.57990
4	4.27479	4.10669	3.89032	3.58626	4.63826
5	3.85684	3.71498	3.53168	3.27259	4.16150
6	3.60701	3.48125	3.31801	3.08580	3.87511
7	3.44146	3.32659	3.17678	2.96233	3.68451
8	3.32395	3.21692	3.07670	2.87488	3.54869
9	3.23632	3.13521	3.00217	2.80955	3.44708
10	3.16853	3.07203	2.94456	2.75918	3.36823
11	3.11454	3.02175	2.89871	2.71901	3.30529
12	3.07055	2.98080	2.86137	2.68630	3.25390
13	3.03403	2.94680	2.83038	2.65913	3.21116
14	3.00323	2.91814	2.80424	2.63621	3.17505
15	2.97691	2.89365	2.78191	2.61662	3.14414
16	2.95416	2.87249	2.76261	2.59968	3.11740
17	2.93430	2.85402	2.74576	2.58490	3.09403
18	2.91682	2.83775	2.73093	2.57187	3.07343
19	2.90131	2.82332	2.71777	2.56031	3.05515
20	2.88745	2.81044	2.70601	2.54999	3.03880
21	2.87501	2.79886	2.69545	2.54071	3.02411
22	2.86376	2.78841	2.68591	2.53233	3.01082
23	2.85355	2.77891	2.67725	2.52471	2.99875
24	2.84425	2.77025	2.66935	2.51777	2.98775
25	2.83572	2.76233	2.66211	2.51141	2.97766
26	2.82789	2.75504	2.65547	2.50557	2.96839
27	2.82067	2.74833	2.64933	2.50018	2.95984
28	2.81399	2.74211	2.64366	2.49519	2.95193
29	2.80779	2.73635	2.63840	2.49056	2.94458
30	2.80203	2.73099	2.63351	2.48625	2.93775
35	2.77834	2.70895	2.61339	2.46856	2.90965
40	2.76078	2.69262	2.59847	2.45543	2.88881
45	2.74724	2.68002	2.58697	2.44530	2.87272
50	2.73648	2.67002	2.57783	2.43726	2.85994
60	2.72047	2.65512	2.56422	2.42528	2.84090
80	2.70067	2.63670	2.54738	2.41045	2.81734
100	2.68889	2.62574	2.53737	2.40162	2.80333
120	2.68109	2.61848	2.53074	2.39578	2.79404
200	2.66559	2.60406	2.51755	2.38416	2.77558
∞	2.64261	2.58267	2.49800	2.36692	2.74820

TABLE B.3

Two-sided Percentage Point (h) for Equal
Correlations $\left(\text{Case I with Correlation } \rho\right)$

p = 10, 1-α = 0.95

ν ↓	ρ					
	0.0	0.1	0.2	0.3	0.4	0.5
2	8.57273	8.53540	8.43791	8.29082	8.09678	7.85347
3	5.81243	5.79001	5.73087	5.64103	5.52202	5.37237
4	4.81680	4.80001	4.75533	4.68704	4.59619	4.48158
5	4.31164	4.29784	4.26085	4.20400	4.12805	4.03190
6	4.00757	3.99566	3.96351	3.91384	3.84721	3.76256
7	3.80482	3.79421	3.76543	3.72073	3.66053	3.58377
8	3.66010	3.65046	3.62416	3.58313	3.52765	3.45667
9	3.55166	3.54278	3.51839	3.48019	3.42835	3.36178
10	3.46741	3.45912	3.43627	3.40033	3.35137	3.28830
11	3.40008	3.39227	3.37068	3.33658	3.28997	3.22973
12	3.34503	3.33764	3.31709	3.28453	3.23988	3.18199
13	3.29921	3.29216	3.27250	3.24124	3.19824	3.14232
14	3.26046	3.25371	3.23482	3.20468	3.16309	3.10886
15	3.22728	3.22079	3.20256	3.17339	3.13303	3.08024
16	3.19854	3.19228	3.17463	3.14631	3.10702	3.05551
17	3.17341	3.16735	3.15022	3.12265	3.08431	3.03391
18	3.15126	3.14537	3.12870	3.10181	3.06430	3.01489
19	3.13157	3.12585	3.10959	3.08330	3.04654	2.99801
20	3.11397	3.10839	3.09251	3.06675	3.03067	2.98294
21	3.09814	3.09269	3.07714	3.05188	3.01641	2.96939
22	3.08382	3.07849	3.06325	3.03844	3.00352	2.95715
23	3.07081	3.06559	3.05063	3.02622	2.99182	2.94604
24	3.05893	3.05382	3.03912	3.01509	2.98114	2.93590
25	3.04806	3.04303	3.02857	3.00488	2.97137	2.92662
26	3.03805	3.03311	3.01887	2.99550	2.96238	2.91810
27	3.02882	3.02396	3.00993	2.98685	2.95410	2.91023
28	3.02028	3.01549	3.00165	2.97885	2.94643	2.90296
29	3.01234	3.00762	2.99396	2.97142	2.93932	2.89621
30	3.00496	3.00031	2.98681	2.96450	2.93270	2.88993
35	2.97460	2.97021	2.95740	2.93608	2.90550	2.86413
40	2.95205	2.94786	2.93556	2.91500	2.88532	2.84500
45	2.93464	2.93061	2.91872	2.89873	2.86977	2.83025
50	2.92080	2.91689	2.90533	2.88581	2.85741	2.81853
60	2.90018	2.89646	2.88538	2.86656	2.83901	2.80110
80	2.87464	2.87115	2.86069	2.84274	2.81624	2.77952
100	2.85945	2.85610	2.84600	2.82857	2.80271	2.76670
120	2.84937	2.84611	2.83626	2.81918	2.79374	2.75820
200	2.82934	2.82627	2.81690	2.80052	2.77592	2.74132
∞	2.79963	2.79683	2.78819	2.77285	2.74950	2.71629

TABLE B.3

Two-sided Percentage Point (h) for Equal
Correlations $\left(Case\ I\ with\ Correlation\ \rho\right)$

$p = 10,\ 1-\alpha = 0.95$

$\nu\ \downarrow$	0.6	0.7	0.8	0.9	$1/(1+\sqrt{10})$
			ρ		
2	7.55340	7.18111	6.70420	6.03731	8.38439
3	5.18749	4.95796	4.66399	4.25346	5.69824
4	4.33963	4.16300	3.93626	3.61850	4.73057
5	3.91245	3.76334	3.57120	3.30043	4.24027
6	3.65703	3.52481	3.35370	3.11105	3.94555
7	3.48774	3.36696	3.20993	2.98587	3.74929
8	3.36754	3.25502	3.10807	2.89726	3.60937
9	3.27791	3.17162	3.03221	2.83099	3.50464
10	3.20855	3.10714	2.97357	2.77997	3.42334
11	3.15333	3.05581	2.92691	2.73926	3.35843
12	3.10832	3.01401	2.88891	2.70611	3.30541
13	3.07096	2.97932	2.85738	2.67858	3.26130
14	3.03945	2.95007	2.83078	2.65536	3.22403
15	3.01252	2.92507	2.80806	2.63551	3.19213
16	2.98924	2.90347	2.78842	2.61835	3.16451
17	2.96893	2.88462	2.77128	2.60337	3.14038
18	2.95104	2.86802	2.75618	2.59017	3.11911
19	2.93517	2.85329	2.74279	2.57846	3.10022
20	2.92099	2.84015	2.73083	2.56801	3.08333
21	2.90826	2.82833	2.72009	2.55860	3.06815
22	2.89676	2.81766	2.71038	2.55011	3.05443
23	2.88631	2.80797	2.70157	2.54240	3.04196
24	2.87679	2.79914	2.69353	2.53536	3.03058
25	2.86807	2.79105	2.68617	2.52892	3.02016
26	2.86006	2.78361	2.67941	2.52300	3.01058
27	2.85267	2.77676	2.67317	2.51754	3.00174
28	2.84584	2.77042	2.66740	2.51249	2.99357
29	2.83949	2.76454	2.66205	2.50780	2.98598
30	2.83360	2.75907	2.65707	2.50344	2.97891
35	2.80936	2.73658	2.63660	2.48551	2.94987
40	2.79140	2.71991	2.62143	2.47222	2.92831
45	2.77755	2.70706	2.60973	2.46196	2.91168
50	2.76654	2.69685	2.60043	2.45381	2.89846
60	2.75017	2.68166	2.58659	2.44167	2.87878
80	2.72991	2.66286	2.56947	2.42665	2.85441
100	2.71786	2.65168	2.55928	2.41772	2.83991
120	2.70988	2.64427	2.55253	2.41180	2.83030
200	2.69403	2.62955	2.53912	2.40003	2.81121
∞	2.67052	2.60773	2.51924	2.38257	2.78289

TABLE B.3

*Two-sided Percentage Point (h) for Equal
Correlations   (Case I with Correlation ρ)*

p = 11,   1-α = 0.95

ν ↓	0.0	0.1	0.2	0.3	0.4	0.5
				ρ		
2	8.74140	8.70145	8.59805	8.44296	8.23926	7.98465
3	5.91848	5.89446	5.83164	5.73678	5.61165	5.45482
4	4.89968	4.88167	4.83416	4.76199	4.66637	4.54616
5	4.38243	4.36763	4.32827	4.26815	4.18817	4.08726
6	4.07091	4.05812	4.02391	3.97136	3.90116	3.81229
7	3.86307	3.85169	3.82105	3.77375	3.71031	3.62972
8	3.71464	3.70431	3.67631	3.63289	3.57443	3.49989
9	3.60339	3.59386	3.56791	3.52749	3.47286	3.40296
10	3.51691	3.50803	3.48371	3.44568	3.39410	3.32787
11	3.44778	3.43942	3.41645	3.38037	3.33127	3.26803
12	3.39125	3.38333	3.36148	3.32704	3.28001	3.21923
13	3.34417	3.33663	3.31573	3.28268	3.23739	3.17869
14	3.30436	3.29715	3.27707	3.24520	3.20141	3.14449
15	3.27026	3.26332	3.24396	3.21312	3.17063	3.11524
16	3.24072	3.23403	3.21529	3.18536	3.14400	3.08996
17	3.21489	3.20842	3.19023	3.16111	3.12075	3.06788
18	3.19211	3.18583	3.16814	3.13973	3.10026	3.04844
19	3.17187	3.16576	3.14852	3.12075	3.08208	3.03118
20	3.15376	3.14782	3.13097	3.10378	3.06583	3.01577
21	3.13748	3.13167	3.11519	3.08853	3.05122	3.00192
22	3.12275	3.11708	3.10093	3.07474	3.03803	2.98941
23	3.10937	3.10381	3.08797	3.06222	3.02604	2.97805
24	3.09715	3.09171	3.07614	3.05079	3.01511	2.96769
25	3.08596	3.08061	3.06531	3.04032	3.00510	2.95820
26	3.07567	3.07041	3.05534	3.03070	2.99590	2.94949
27	3.06617	3.06100	3.04615	3.02183	2.98741	2.94145
28	3.05738	3.05229	3.03765	3.01361	2.97956	2.93401
29	3.04921	3.04420	3.02975	3.00599	2.97227	2.92711
30	3.04162	3.03667	3.02240	2.99890	2.96549	2.92069
35	3.01036	3.00571	2.99218	2.96974	2.93763	2.89431
40	2.98715	2.98271	2.96975	2.94811	2.91697	2.87476
45	2.96923	2.96496	2.95243	2.93142	2.90104	2.85968
50	2.95498	2.95085	2.93867	2.91816	2.88838	2.84770
60	2.93375	2.92982	2.91817	2.89841	2.86953	2.82987
80	2.90745	2.90378	2.89279	2.87396	2.84621	2.80782
100	2.89180	2.88828	2.87769	2.85942	2.83235	2.79471
120	2.88142	2.87800	2.86768	2.84978	2.82316	2.78602
200	2.86079	2.85758	2.84778	2.83064	2.80491	2.76876
∞	2.83018	2.82727	2.81826	2.80224	2.77784	2.74317

TABLE B.3

*Two-sided Percentage Point (h) for Equal
Correlations* (*Case I with Correlation* ρ)

p = 11, 1-α = 0.95

ν ↓	ρ				
	0.6	0.7	0.8	0.9	$1/(1+\sqrt{11})$
2	7.67146	7.28374	6.78804	6.09623	8.55427
3	5.26160	5.02225	4.71637	4.29011	5.80491
4	4.39768	4.21337	3.97729	3.64725	4.81387
5	3.96225	3.80659	3.60651	3.32526	4.31140
6	3.70182	3.56378	3.38558	3.13355	4.00918
7	3.52918	3.40308	3.23955	3.00685	3.80782
8	3.40659	3.28911	3.13608	2.91721	3.66418
9	3.31515	3.20419	3.05903	2.85010	3.55663
10	3.24440	3.13853	2.99948	2.79850	3.47312
11	3.18805	3.08627	2.95209	2.75730	3.40641
12	3.14214	3.04371	2.91350	2.72376	3.35192
13	3.10402	3.00838	2.88147	2.69591	3.30656
14	3.07187	2.97860	2.85447	2.67242	3.26824
15	3.04439	2.95315	2.83139	2.65234	3.23542
16	3.02064	2.93116	2.81145	2.63498	3.20702
17	2.99991	2.91196	2.79404	2.61982	3.18219
18	2.98166	2.89506	2.77872	2.60647	3.16030
19	2.96547	2.88007	2.76512	2.59463	3.14086
20	2.95100	2.86669	2.75298	2.58405	3.12349
21	2.93801	2.85466	2.74207	2.57454	3.10786
22	2.92627	2.84379	2.73221	2.56595	3.09373
23	2.91562	2.83393	2.72326	2.55815	3.08089
24	2.90590	2.82494	2.71510	2.55103	3.06918
25	2.89700	2.81670	2.70763	2.54452	3.05845
26	2.88883	2.80913	2.70076	2.53853	3.04859
27	2.88129	2.80216	2.69443	2.53301	3.03949
28	2.87432	2.79570	2.68858	2.52790	3.03107
29	2.86785	2.78971	2.68314	2.52315	3.02325
30	2.86183	2.78414	2.67808	2.51874	3.01598
35	2.83710	2.76126	2.65731	2.50061	2.98606
40	2.81877	2.74429	2.64190	2.48716	2.96386
45	2.80464	2.73121	2.63003	2.47679	2.94673
50	2.79342	2.72081	2.62059	2.46855	2.93311
60	2.77671	2.70535	2.60654	2.45627	2.91283
80	2.75604	2.68621	2.58915	2.44108	2.88771
100	2.74376	2.67483	2.57882	2.43205	2.87278
120	2.73561	2.66729	2.57197	2.42606	2.86287
200	2.71944	2.65231	2.55836	2.41416	2.84320
∞	2.69546	2.63010	2.53817	2.39650	2.81401

TABLE B.3

Two-sided Percentage Point (h) for Equal Correlations $\left(\text{Case I with Correlation } \rho\right)$

$p = 12,\ 1\text{-}\alpha = 0.95$

ν ↓	ρ					
	0.0	0.1	0.2	0.3	0.4	0.5
2	8.89411	8.85172	8.74285	8.58045	8.36793	8.10305
3	6.01469	5.98916	5.92293	5.82346	5.69274	5.52937
4	4.97496	4.95581	4.90567	4.82991	4.72993	4.60460
5	4.44678	4.43103	4.38948	4.32633	4.24264	4.13738
6	4.12850	4.11490	4.07876	4.02354	3.95007	3.85733
7	3.91605	3.90394	3.87157	3.82186	3.75545	3.67133
8	3.76426	3.75327	3.72369	3.67805	3.61684	3.53904
9	3.65044	3.64031	3.61289	3.57041	3.51321	3.44025
10	3.56195	3.55250	3.52681	3.48684	3.43283	3.36371
11	3.49117	3.48229	3.45803	3.42012	3.36871	3.30271
12	3.43329	3.42488	3.40181	3.36562	3.31638	3.25296
13	3.38507	3.37706	3.35500	3.32027	3.27287	3.21162
14	3.34429	3.33663	3.31544	3.28196	3.23613	3.17675
15	3.30935	3.30199	3.28155	3.24917	3.20471	3.14693
16	3.27908	3.27198	3.25221	3.22079	3.17752	3.12114
17	3.25260	3.24574	3.22656	3.19599	3.15377	3.09863
18	3.22925	3.22259	3.20394	3.17413	3.13285	3.07880
19	3.20849	3.20203	3.18385	3.15471	3.11427	3.06120
20	3.18993	3.18364	3.16589	3.13736	3.09768	3.04549
21	3.17323	3.16709	3.14973	3.12176	3.08276	3.03136
22	3.15813	3.15213	3.13513	3.10766	3.06928	3.01860
23	3.14440	3.13853	3.12185	3.09485	3.05704	3.00702
24	3.13188	3.12612	3.10974	3.08316	3.04587	2.99645
25	3.12039	3.11475	3.09864	3.07246	3.03565	2.98678
26	3.10984	3.10429	3.08844	3.06261	3.02625	2.97789
27	3.10009	3.09463	3.07902	3.05353	3.01758	2.96969
28	3.09107	3.08570	3.07031	3.04513	3.00956	2.96210
29	3.08270	3.07741	3.06222	3.03733	3.00212	2.95507
30	3.07490	3.06969	3.05469	3.03008	2.99519	2.94852
35	3.04283	3.03793	3.02373	3.00025	2.96673	2.92162
40	3.01901	3.01434	3.00074	2.97811	2.94562	2.90167
45	3.00061	2.99613	2.98300	2.96103	2.92934	2.88629
50	2.98599	2.98165	2.96890	2.94746	2.91641	2.87408
60	2.96419	2.96008	2.94789	2.92725	2.89715	2.85589
80	2.93719	2.93335	2.92187	2.90223	2.87333	2.83340
100	2.92112	2.91745	2.90639	2.88735	2.85916	2.82003
120	2.91046	2.90690	2.89613	2.87749	2.84978	2.81117
200	2.88928	2.88594	2.87574	2.85790	2.83113	2.79357
∞	2.85784	2.85483	2.84548	2.82883	2.80347	2.76747

TABLE B.3

Two-sided Percentage Point (h) for Equal
Correlations $\left(\text{Case } I \text{ with Correlation } \rho\right)$

$p = 12, \ 1-\alpha = 0.95$

$\nu \downarrow$			ρ		
	0.6	0.7	0.8	0.9	$1/(1+\sqrt{12})$
2	7.77795	7.37627	6.86360	6.14929	8.70850
3	5.32856	5.08031	4.76362	4.32314	5.90194
4	4.45017	4.25887	4.01433	3.67317	4.88972
5	4.00730	3.84568	3.63837	3.34763	4.37621
6	3.74235	3.59900	3.41436	3.15384	4.06718
7	3.56667	3.43572	3.26628	3.02575	3.86116
8	3.44191	3.31991	3.16136	2.93502	3.71415
9	3.34885	3.23362	3.08324	2.86730	3.60403
10	3.27682	3.16689	3.02285	2.81520	3.51849
11	3.21946	3.11379	2.97481	2.77355	3.45014
12	3.17272	3.07054	2.93568	2.73966	3.39429
13	3.13391	3.03464	2.90321	2.71152	3.34780
14	3.10118	3.00437	2.87583	2.68778	3.30850
15	3.07321	2.97851	2.85244	2.66749	3.27486
16	3.04903	2.95616	2.83222	2.64995	3.24572
17	3.02792	2.93666	2.81457	2.63464	3.22025
18	3.00934	2.91948	2.79904	2.62115	3.19780
19	2.99285	2.90425	2.78525	2.60918	3.17786
20	2.97813	2.89065	2.77295	2.59850	3.16002
21	2.96490	2.87843	2.76189	2.58889	3.14399
22	2.95295	2.86739	2.75189	2.58021	3.12949
23	2.94210	2.85737	2.74282	2.57233	3.11631
24	2.93221	2.84823	2.73455	2.56514	3.10429
25	2.92315	2.83986	2.72698	2.55856	3.09328
26	2.91483	2.83217	2.72002	2.55251	3.08315
27	2.90716	2.82508	2.71360	2.54693	3.07381
28	2.90006	2.81852	2.70766	2.54176	3.06517
29	2.89347	2.81244	2.70216	2.53697	3.05714
30	2.88734	2.80678	2.69703	2.53252	3.04967
35	2.86217	2.78353	2.67597	2.51420	3.01895
40	2.84351	2.76629	2.66036	2.50062	2.99615
45	2.82912	2.75299	2.64832	2.49014	2.97856
50	2.81770	2.74244	2.63875	2.48181	2.96457
60	2.80069	2.72672	2.62451	2.46941	2.94373
80	2.77964	2.70728	2.60690	2.45407	2.91793
100	2.76714	2.69572	2.59642	2.44494	2.90259
120	2.75885	2.68806	2.58948	2.43889	2.89241
200	2.74238	2.67284	2.57568	2.42688	2.87219
∞	2.71797	2.65028	2.55522	2.40904	2.84219

BECHHOFER and DUNNETT

TABLE B.3

Two-sided Percentage Point (h) for Equal
Correlations ⎛Case I with Correlation ρ⎞

p = 13, 1-α = 0.95

ν ↓	ρ					
	0.0	0.1	0.2	0.3	0.4	0.5
2	9.03350	8.98882	8.87488	8.70573	8.48512	8.21083
3	6.10265	6.07572	6.00631	5.90257	5.76670	5.59733
4	5.04387	5.02364	4.97105	4.89197	4.78796	4.65792
5	4.50573	4.48909	4.44547	4.37951	4.29240	4.18313
6	4.18128	4.16691	4.12896	4.07126	3.99476	3.89845
7	3.96461	3.95182	3.91782	3.86587	3.79670	3.70933
8	3.80976	3.79814	3.76707	3.71937	3.65561	3.57479
9	3.69359	3.68288	3.65409	3.60967	3.55009	3.47430
10	3.60324	3.59326	3.56628	3.52450	3.46824	3.39644
11	3.53096	3.52158	3.49610	3.45648	3.40293	3.33437
12	3.47183	3.46295	3.43873	3.40091	3.34962	3.28375
13	3.42257	3.41411	3.39096	3.35467	3.30530	3.24169
14	3.38089	3.37281	3.35057	3.31559	3.26787	3.20620
15	3.34518	3.33741	3.31597	3.28214	3.23584	3.17585
16	3.31423	3.30675	3.28602	3.25319	3.20814	3.14961
17	3.28716	3.27992	3.25982	3.22789	3.18394	3.12669
18	3.26328	3.25627	3.23672	3.20558	3.16261	3.10651
19	3.24206	3.23524	3.21620	3.18578	3.14369	3.08860
20	3.22307	3.21644	3.19785	3.16807	3.12677	3.07261
21	3.20599	3.19952	3.18134	3.15215	3.11157	3.05823
22	3.19054	3.18422	3.16642	3.13776	3.09783	3.04524
23	3.17650	3.17032	3.15286	3.12468	3.08535	3.03345
24	3.16368	3.15763	3.14048	3.11275	3.07397	3.02270
25	3.15193	3.14599	3.12914	3.10183	3.06355	3.01285
26	3.14113	3.13530	3.11871	3.09178	3.05397	3.00380
27	3.13115	3.12542	3.10909	3.08251	3.04513	2.99545
28	3.12192	3.11628	3.10018	3.07394	3.03695	2.98773
29	3.11335	3.10780	3.09192	3.06598	3.02937	2.98057
30	3.10538	3.09990	3.08423	3.05857	3.02231	2.97390
35	3.07255	3.06741	3.05258	3.02812	2.99329	2.94652
40	3.04816	3.04327	3.02908	3.00552	2.97177	2.92622
45	3.02933	3.02464	3.01094	2.98809	2.95518	2.91057
50	3.01435	3.00982	2.99653	2.97423	2.94199	2.89813
60	2.99203	2.98774	2.97504	2.95359	2.92236	2.87962
80	2.96437	2.96038	2.94844	2.92805	2.89807	2.85672
100	2.94791	2.94410	2.93262	2.91286	2.88363	2.84311
120	2.93700	2.93330	2.92212	2.90278	2.87406	2.83409
200	2.91530	2.91184	2.90127	2.88277	2.85504	2.81617
∞	2.88310	2.87998	2.87032	2.85310	2.82685	2.78961

TABLE B.3

Two-sided Percentage Point (h) for Equal
Correlations $\left(Case\ I\ with\ Correlation\ \rho\right)$

p = 13, 1-α = 0.95

ν ↓	ρ				
	0.6	0.7	0.8	0.9	1/(1+√13)
2	7.87486	7.46043	6.93229	6.19750	8.84963
3	5.38957	5.13317	4.80662	4.35318	5.99086
4	4.49802	4.30033	4.04805	3.69674	4.95930
5	4.04839	3.88131	3.66739	3.36798	4.43569
6	3.77932	3.63111	3.44056	3.17228	4.12042
7	3.60088	3.46546	3.29061	3.04296	3.91015
8	3.47414	3.34798	3.18437	2.95134	3.76004
9	3.37958	3.26043	3.10527	2.88309	3.64755
10	3.30640	3.19274	3.04413	2.83037	3.56015
11	3.24812	3.13886	2.99548	2.78832	3.49029
12	3.20062	3.09498	2.95586	2.75411	3.43320
13	3.16118	3.05856	2.92299	2.72570	3.38566
14	3.12792	3.02785	2.89527	2.70174	3.34547
15	3.09949	3.00161	2.87158	2.68126	3.31105
16	3.07491	2.97894	2.85112	2.66355	3.28124
17	3.05346	2.95915	2.83325	2.64810	3.25518
18	3.03458	2.94173	2.81752	2.63448	3.23220
19	3.01783	2.92627	2.80357	2.62241	3.21179
20	3.00286	2.91247	2.79111	2.61162	3.19354
21	2.98942	2.90007	2.77991	2.60192	3.17712
22	2.97727	2.88887	2.76980	2.59316	3.16228
23	2.96625	2.87871	2.76062	2.58521	3.14879
24	2.95619	2.86944	2.75224	2.57795	3.13648
25	2.94699	2.86095	2.74458	2.57131	3.12521
26	2.93853	2.85315	2.73753	2.56520	3.11484
27	2.93073	2.84596	2.73103	2.55957	3.10527
28	2.92352	2.83930	2.72502	2.55436	3.09641
29	2.91682	2.83313	2.71945	2.54953	3.08820
30	2.91060	2.82739	2.71426	2.54503	3.08055
35	2.88501	2.80380	2.69295	2.52655	3.04908
40	2.86605	2.78631	2.67714	2.51283	3.02572
45	2.85143	2.77283	2.66496	2.50226	3.00770
50	2.83981	2.76212	2.65527	2.49386	2.99336
60	2.82253	2.74617	2.64086	2.48135	2.97201
80	2.80115	2.72645	2.62303	2.46586	2.94557
100	2.78843	2.71473	2.61242	2.45665	2.92985
120	2.78001	2.70696	2.60540	2.45055	2.91942
200	2.76328	2.69152	2.59144	2.43842	2.89869
∞	2.73847	2.66863	2.57073	2.42042	2.86794

TABLE B.3

*Two-sided Percentage Point (h) for Equal
Correlations* (*Case I with Correlation* ρ)

p = 14, 1-α = 0.95

ν ↓	ρ					
	0.0	0.1	0.2	0.3	0.4	0.5
2	9.16162	9.11478	8.99611	8.82071	8.59262	8.30967
3	6.18363	6.15536	6.08298	5.97528	5.83464	5.65973
4	5.10736	5.08612	5.03123	4.94906	4.84131	4.70691
5	4.56008	4.54259	4.49705	4.42847	4.33818	4.22519
6	4.22997	4.21486	4.17522	4.11520	4.03588	3.93627
7	4.00943	3.99598	3.96045	3.90640	3.83466	3.74427
8	3.85174	3.83953	3.80706	3.75742	3.69129	3.60767
9	3.73341	3.72215	3.69206	3.64584	3.58403	3.50562
10	3.64135	3.63086	3.60267	3.55919	3.50083	3.42654
11	3.56768	3.55783	3.53120	3.48997	3.43442	3.36349
12	3.50740	3.49808	3.47277	3.43341	3.38022	3.31206
13	3.45717	3.44830	3.42410	3.38634	3.33514	3.26933
14	3.41467	3.40618	3.38295	3.34657	3.29707	3.23327
15	3.37824	3.37009	3.34770	3.31251	3.26450	3.20244
16	3.34667	3.33881	3.31717	3.28303	3.23632	3.17578
17	3.31904	3.31146	3.29047	3.25726	3.21170	3.15249
18	3.29468	3.28732	3.26692	3.23455	3.19000	3.13199
19	3.27302	3.26588	3.24600	3.21438	3.17075	3.11379
20	3.25364	3.24669	3.22730	3.19634	3.15354	3.09753
21	3.23620	3.22943	3.21047	3.18013	3.13807	3.08292
22	3.22043	3.21382	3.19525	3.16547	3.12409	3.06973
23	3.20609	3.19963	3.18142	3.15215	3.11140	3.05774
24	3.19301	3.18667	3.16880	3.14000	3.09981	3.04681
25	3.18101	3.17480	3.15723	3.12886	3.08921	3.03681
26	3.16998	3.16388	3.14660	3.11863	3.07946	3.02761
27	3.15979	3.15380	3.13679	3.10919	3.07047	3.01913
28	3.15037	3.14447	3.12770	3.10045	3.06215	3.01128
29	3.14161	3.13581	3.11927	3.09234	3.05443	3.00400
30	3.13347	3.12775	3.11143	3.08480	3.04724	2.99723
35	3.09993	3.09457	3.07914	3.05377	3.01772	2.96940
40	3.07502	3.06993	3.05517	3.03074	2.99582	2.94877
45	3.05577	3.05089	3.03667	3.01297	2.97893	2.93286
50	3.04047	3.03576	3.02196	2.99885	2.96551	2.92022
60	3.01766	3.01320	3.00004	2.97782	2.94553	2.90141
80	2.98940	2.98525	2.97289	2.95179	2.92081	2.87814
100	2.97258	2.96862	2.95674	2.93630	2.90611	2.86431
120	2.96142	2.95759	2.94603	2.92604	2.89637	2.85514
200	2.93924	2.93566	2.92474	2.90564	2.87702	2.83693
∞	2.90632	2.90311	2.89316	2.87539	2.84832	2.80993

TABLE B.3

Two-sided Percentage Point (h) for Equal Correlations (*Case I with Correlation* ρ)

p = 14, 1-α = 0.95

ν ↓	ρ				
	0.6	0.7	0.8	0.9	1/(1+√14)
2	7.96369	7.53755	6.99520	6.24164	8.97962
3	5.44556	5.18167	4.84605	4.38067	6.07288
4	4.54197	4.33838	4.07898	3.71834	5.02355
5	4.08614	3.91402	3.69401	3.38662	4.49064
6	3.81330	3.66058	3.46459	3.18917	4.16963
7	3.63232	3.49278	3.31293	3.05871	3.95543
8	3.50375	3.37375	3.20548	2.96628	3.80245
9	3.40782	3.28505	3.12547	2.89744	3.68778
10	3.33358	3.21646	3.06363	2.84427	3.59865
11	3.27444	3.16187	3.01443	2.80184	3.52740
12	3.22625	3.11741	2.97437	2.76734	3.46915
13	3.18623	3.08051	2.94112	2.73868	3.42063
14	3.15247	3.04939	2.91309	2.71452	3.37962
15	3.12363	3.02281	2.88913	2.69386	3.34448
16	3.09869	2.99984	2.86844	2.67601	3.31405
17	3.07692	2.97979	2.85037	2.66042	3.28744
18	3.05776	2.96213	2.83447	2.64669	3.26397
19	3.04076	2.94648	2.82036	2.63451	3.24312
20	3.02557	2.93250	2.80776	2.62363	3.22448
21	3.01193	2.91993	2.79643	2.61385	3.20771
22	2.99960	2.90859	2.78621	2.60502	3.19255
23	2.98842	2.89828	2.77692	2.59700	3.17877
24	2.97821	2.88889	2.76846	2.58968	3.16619
25	2.96887	2.88029	2.76070	2.58298	3.15467
26	2.96029	2.87239	2.75358	2.57682	3.14407
27	2.95237	2.86510	2.74701	2.57115	3.13430
28	2.94505	2.85836	2.74093	2.56589	3.12525
29	2.93826	2.85211	2.73529	2.56102	3.11685
30	2.93194	2.84629	2.73005	2.55648	3.10903
35	2.90598	2.82239	2.70850	2.53784	3.07687
40	2.88674	2.80467	2.69252	2.52402	3.05299
45	2.87190	2.79102	2.68020	2.51335	3.03456
50	2.86011	2.78016	2.67041	2.50488	3.01990
60	2.84257	2.76401	2.65583	2.49227	2.99807
80	2.82087	2.74403	2.63780	2.47665	2.97103
100	2.80798	2.73216	2.62708	2.46737	2.95494
120	2.79943	2.72428	2.61998	2.46121	2.94428
200	2.78245	2.70865	2.60586	2.44899	2.92308
∞	2.75728	2.68546	2.58493	2.43084	2.89162

TABLE B.3

Two-sided Percentage Point (h) for Equal
Correlations (*Case I with Correlation* ρ)

$p = 15, \ 1-\alpha = 0.95$

ν ↓	ρ					
	0.0	0.1	0.2	0.3	0.4	0.5
2	9.28007	9.23120	9.10810	8.92688	8.69185	8.40086
3	6.25861	6.22908	6.15391	6.04251	5.89743	5.71737
4	5.16621	5.14400	5.08695	5.00188	4.89065	4.75220
5	4.61048	4.59219	4.54482	4.47379	4.38053	4.26408
6	4.27514	4.25933	4.21809	4.15590	4.07393	3.97124
7	4.05101	4.03694	3.99997	3.94394	3.86980	3.77660
8	3.89071	3.87792	3.84413	3.79267	3.72432	3.63809
9	3.77037	3.75859	3.72727	3.67935	3.61546	3.53459
10	3.67672	3.66575	3.63641	3.59133	3.53100	3.45438
11	3.60177	3.59146	3.56375	3.52100	3.46358	3.39042
12	3.54042	3.53067	3.50433	3.46353	3.40854	3.33826
13	3.48929	3.48001	3.45484	3.41569	3.36276	3.29491
14	3.44602	3.43714	3.41298	3.37526	3.32410	3.25832
15	3.40892	3.40040	3.37712	3.34064	3.29102	3.22703
16	3.37677	3.36856	3.34605	3.31067	3.26240	3.19998
17	3.34863	3.34071	3.31888	3.28447	3.23739	3.17635
18	3.32381	3.31613	3.29492	3.26138	3.21535	3.15554
19	3.30174	3.29429	3.27363	3.24087	3.19579	3.13708
20	3.28200	3.27475	3.25459	3.22253	3.17831	3.12058
21	3.26423	3.25717	3.23746	3.20604	3.16260	3.10576
22	3.24816	3.24126	3.22197	3.19113	3.14839	3.09236
23	3.23355	3.22680	3.20789	3.17758	3.13549	3.08020
24	3.22021	3.21360	3.19504	3.16522	3.12373	3.06911
25	3.20798	3.20151	3.18327	3.15390	3.11295	3.05896
26	3.19674	3.19038	3.17244	3.14349	3.10305	3.04962
27	3.18635	3.18011	3.16245	3.13389	3.09391	3.04101
28	3.17674	3.17061	3.15320	3.12500	3.08545	3.03305
29	3.16782	3.16178	3.14462	3.11675	3.07761	3.02566
30	3.15951	3.15356	3.13663	3.10908	3.07031	3.01879
35	3.12532	3.11975	3.10375	3.07751	3.04031	2.99055
40	3.09991	3.09462	3.07934	3.05408	3.01806	2.96961
45	3.08029	3.07522	3.06049	3.03601	3.00090	2.95346
50	3.06467	3.05979	3.04550	3.02164	2.98726	2.94064
60	3.04141	3.03679	3.02317	3.00024	2.96695	2.92155
80	3.01257	3.00829	2.99552	2.97375	2.94183	2.89793
100	2.99541	2.99133	2.97906	2.95799	2.92689	2.88389
120	2.98402	2.98007	2.96815	2.94755	2.91699	2.87458
200	2.96139	2.95771	2.94646	2.92679	2.89733	2.85610
∞	2.92780	2.92451	2.91428	2.89601	2.86816	2.82870

TABLE B.3

Two-sided Percentage Point (h) for Equal
Correlations   (Case I with Correlation ρ)

p = 15, 1-α = 0.95

ν ↓	ρ				
	0.6	0.7	0.8	0.9	$1/(1+\sqrt{15})$
2	8.04561	7.60865	7.05319	6.28230	9.10002
3	5.49726	5.22643	4.88242	4.40604	6.14896
4	4.58258	4.37352	4.10752	3.73825	5.08318
5	4.12102	3.94423	3.71858	3.40381	4.54168
6	3.84470	3.68780	3.48676	3.20475	4.21534
7	3.66137	3.51801	3.33353	3.07323	3.99750
8	3.53112	3.39755	3.22496	2.98005	3.84187
9	3.43393	3.30779	3.14412	2.91066	3.72517
10	3.35870	3.23837	3.08163	2.85707	3.63444
11	3.29877	3.18312	3.03192	2.81430	3.56188
12	3.24993	3.13813	2.99144	2.77953	3.50255
13	3.20938	3.10078	2.95785	2.75065	3.45313
14	3.17517	3.06929	2.92952	2.72629	3.41134
15	3.14593	3.04238	2.90533	2.70548	3.37554
16	3.12066	3.01913	2.88441	2.68748	3.34452
17	3.09860	2.99884	2.86616	2.67177	3.31740
18	3.07918	2.98097	2.85009	2.65794	3.29347
19	3.06195	2.96513	2.83584	2.64566	3.27222
20	3.04656	2.95098	2.82311	2.63470	3.25321
21	3.03273	2.93826	2.81167	2.62484	3.23611
22	3.02024	2.92678	2.80134	2.61594	3.22065
23	3.00890	2.91635	2.79196	2.60786	3.20659
24	2.99856	2.90685	2.78341	2.60048	3.19377
25	2.98909	2.89815	2.77558	2.59373	3.18201
26	2.98039	2.89015	2.76838	2.58753	3.17120
27	2.97237	2.88277	2.76174	2.58181	3.16123
28	2.96495	2.87595	2.75560	2.57651	3.15200
29	2.95806	2.86962	2.74991	2.57160	3.14343
30	2.95166	2.86374	2.74461	2.56703	3.13545
35	2.92535	2.83955	2.72284	2.54825	3.10264
40	2.90584	2.82162	2.70669	2.53432	3.07827
45	2.89081	2.80780	2.69425	2.52357	3.05946
50	2.87886	2.79682	2.68436	2.51504	3.04450
60	2.86108	2.78047	2.66964	2.50232	3.02221
80	2.83909	2.76025	2.65143	2.48659	2.99461
100	2.82602	2.74824	2.64060	2.47724	2.97819
120	2.81735	2.74027	2.63342	2.47103	2.96730
200	2.80015	2.72445	2.61916	2.45872	2.94565
∞	2.77463	2.70098	2.59802	2.44043	2.91354

BECHHOFER and DUNNETT

TABLE B.3

Two-sided Percentage Point (h) for Equal
Correlations (*Case I with Correlation* ρ)

p = 16, 1-α = 0.95

ν ↓	ρ					
	0.0	0.1	0.2	0.3	0.4	0.5
2	9.39017	9.33936	9.21211	9.02543	8.78392	8.45545
3	6.32838	6.29765	6.21986	6.10499	5.95577	5.77090
4	5.22101	5.19789	5.13880	5.05102	4.93653	4.79429
5	4.65745	4.63840	4.58931	4.51596	4.41992	4.30023
6	4.31724	4.30077	4.25801	4.19378	4.10934	4.00376
7	4.08979	4.07512	4.03678	3.97890	3.90250	3.80666
8	3.92705	3.91372	3.87867	3.82550	3.75505	3.66637
9	3.80485	3.79257	3.76008	3.71056	3.64471	3.56153
10	3.70972	3.69828	3.66785	3.62126	3.55908	3.48027
11	3.63357	3.62282	3.59408	3.54990	3.49072	3.41547
12	3.57123	3.56106	3.53374	3.49157	3.43490	3.36261
13	3.51925	3.50958	3.48348	3.44302	3.38847	3.31868
14	3.47526	3.46602	3.44096	3.40198	3.34926	3.28161
15	3.43754	3.42867	3.40453	3.36684	3.31570	3.24990
16	3.40484	3.39630	3.37296	3.33641	3.28666	3.22248
17	3.37623	3.36798	3.34536	3.30981	3.26129	3.19853
18	3.35098	3.34299	3.32101	3.28636	3.23894	3.17744
19	3.32853	3.32078	3.29937	3.26553	3.21909	3.15873
20	3.30845	3.30090	3.28002	3.24691	3.20136	3.14201
21	3.29037	3.28302	3.26261	3.23016	3.18541	3.12698
22	3.27401	3.26684	3.24686	3.21501	3.17100	3.11341
23	3.25914	3.25213	3.23255	3.20126	3.15791	3.10108
24	3.24557	3.23871	3.21949	3.18870	3.14597	3.08984
25	3.23313	3.22640	3.20752	3.17720	3.13504	3.07954
26	3.22168	3.21508	3.19651	3.16663	3.12498	3.07008
27	3.21111	3.20463	3.18635	3.15688	3.11571	3.06136
28	3.20133	3.19496	3.17695	3.14785	3.10713	3.05329
29	3.19224	3.18597	3.16822	3.13947	3.09917	3.04580
30	3.18379	3.17761	3.16009	3.13167	3.09177	3.03883
35	3.14898	3.14320	3.12666	3.09960	3.06132	3.01020
40	3.12310	3.11762	3.10183	3.07580	3.03874	2.98897
45	3.10312	3.09787	3.08266	3.05743	3.02132	2.97261
50	3.08722	3.08216	3.06742	3.04283	3.00748	2.95961
60	3.06352	3.05875	3.04471	3.02109	2.98687	2.94025
80	3.03415	3.02973	3.01657	2.99417	2.96137	2.91631
100	3.01666	3.01245	2.99983	2.97816	2.94621	2.90208
120	3.00506	3.00100	2.98873	2.96755	2.93616	2.89265
200	2.98200	2.97822	2.96666	2.94646	2.91620	2.87391
∞	2.94778	2.94441	2.93392	2.91517	2.88660	2.84613

TABLE B.3

*Two-sided Percentage Point (h) for Equal
Correlations* $\left[$ *Case I with Correlation* $\rho\right]$

$p = 16, \quad 1-\alpha = 0.95$

ν ↓	0.6	0.7	0.8	0.9	$1/(1+\sqrt{16})$
2	8.12160	7.67458	7.10694	6.31998	9.21211
3	5.54525	5.26796	4.91616	4.42955	6.21986
4	4.62030	4.40614	4.13400	3.75671	5.13880
5	4.15344	3.97228	3.74137	3.41975	4.58931
6	3.87388	3.71309	3.50735	3.21919	4.25801
7	3.68838	3.54144	3.35265	3.08669	4.03678
8	3.55656	3.41966	3.24303	2.99282	3.87867
9	3.45819	3.32891	3.16141	2.92291	3.76008
10	3.38204	3.25872	3.09833	2.86895	3.66785
11	3.32138	3.20286	3.04814	2.82585	3.59408
12	3.27194	3.15736	3.00728	2.79082	3.53374
13	3.23089	3.11960	2.97336	2.76173	3.48348
14	3.19625	3.08776	2.94477	2.73720	3.44096
15	3.16666	3.06055	2.92034	2.71624	3.40453
16	3.14107	3.03704	2.89923	2.69811	3.37296
17	3.11874	3.01652	2.88081	2.68229	3.34536
18	3.09907	2.99846	2.86459	2.66836	3.32101
19	3.08163	2.98244	2.85020	2.65599	3.29937
20	3.06605	2.96813	2.83735	2.64495	3.28002
21	3.05205	2.95528	2.82580	2.63503	3.26261
22	3.03940	2.94367	2.81537	2.62606	3.24686
23	3.02792	2.93313	2.80590	2.61792	3.23255
24	3.01745	2.92352	2.79727	2.61049	3.21949
25	3.00787	2.91472	2.78937	2.60370	3.20752
26	2.99906	2.90663	2.78210	2.59745	3.19651
27	2.99094	2.89917	2.77540	2.59168	3.18635
28	2.98342	2.89228	2.76921	2.58635	3.17695
29	2.97645	2.88588	2.76346	2.58141	3.16822
30	2.96997	2.87993	2.75811	2.57680	3.16009
35	2.94333	2.85547	2.73613	2.55789	3.12666
40	2.92358	2.83735	2.71984	2.54386	3.10183
45	2.90836	2.82337	2.70728	2.53304	3.08266
50	2.89627	2.81227	2.69729	2.52444	3.06742
60	2.87827	2.79574	2.68244	2.51164	3.04471
80	2.85600	2.77530	2.66406	2.49580	3.01657
100	2.84277	2.76315	2.65313	2.48638	2.99983
120	2.83400	2.75510	2.64588	2.48013	2.98873
200	2.81657	2.73910	2.63149	2.46773	2.96666
∞	2.79074	2.71538	2.61015	2.44931	2.93392

TABLE B.3

Two-sided Percentage Point (h) for Equal
Correlations   (Case I with Correlation ρ)

p = 18, 1-α = 0.95

ν ↓	ρ					
	0.0	0.1	0.2	0.3	0.4	0.5
2	9.58931	9.53492	9.40003	9.20342	8.95013	8.63809
3	6.45479	6.42183	6.33922	6.21800	6.06122	5.86763
4	5.32042	5.29559	5.23274	5.13998	5.01953	4.87040
5	4.74271	4.72223	4.66996	4.59238	4.49125	4.36565
6	4.39373	4.37600	4.33044	4.26245	4.17347	4.06262
7	4.16025	4.14446	4.10358	4.04228	3.96174	3.86108
8	3.99309	3.97875	3.94137	3.88503	3.81074	3.71758
9	3.86751	3.85429	3.81964	3.76716	3.69771	3.61031
10	3.76970	3.75739	3.72492	3.67555	3.60995	3.52715
11	3.69137	3.67980	3.64914	3.60231	3.53988	3.46082
12	3.62722	3.61628	3.58714	3.54244	3.48266	3.40670
13	3.57372	3.56331	3.53547	3.49259	3.43505	3.36173
14	3.52841	3.51847	3.49175	3.45045	3.39484	3.32376
15	3.48956	3.48003	3.45428	3.41435	3.36042	3.29129
16	3.45587	3.44670	3.42181	3.38309	3.33063	3.26321
17	3.42639	3.41752	3.39341	3.35576	3.30460	3.23868
18	3.40036	3.39178	3.36835	3.33166	3.28166	3.21708
19	3.37721	3.36889	3.34609	3.31025	3.26130	3.19791
20	3.35650	3.34841	3.32617	3.29111	3.24310	3.18078
21	3.33786	3.32998	3.30824	3.27389	3.22673	3.16539
22	3.32099	3.31330	3.29203	3.25832	3.21194	3.15148
23	3.30565	3.29813	3.27729	3.24418	3.19851	3.13885
24	3.29164	3.28429	3.26384	3.23127	3.18625	3.12734
25	3.27880	3.27160	3.25151	3.21945	3.17503	3.11679
26	3.26698	3.25992	3.24018	3.20857	3.16471	3.10710
27	3.25608	3.24914	3.22971	3.19854	3.15520	3.09816
28	3.24598	3.23916	3.22003	3.18926	3.14639	3.08989
29	3.23660	3.22990	3.21103	3.18064	3.13822	3.08222
30	3.22787	3.22127	3.20266	3.17262	3.13061	3.07508
35	3.19192	3.18576	3.16821	3.13963	3.09936	3.04575
40	3.16519	3.15936	3.14262	3.11515	3.07617	3.02400
45	3.14455	3.13897	3.12287	3.09625	3.05828	3.00723
50	3.12812	3.12275	3.10715	3.08123	3.04407	2.99391
60	3.10362	3.09857	3.08373	3.05886	3.02291	2.97408
80	3.07326	3.06859	3.05472	3.03116	2.99673	2.94955
100	3.05518	3.05075	3.03746	3.01468	2.98116	2.93496
120	3.04319	3.03891	3.02601	3.00375	2.97084	2.92530
200	3.01935	3.01537	3.00325	2.98205	2.95034	2.90610
∞	2.98395	2.98043	2.96947	2.94984	2.91994	2.87764

TABLE B.3

Two-sided Percentage Point (h) for Equal
Correlations $\left(\text{Case I with Correlation } \rho\right)$

$p = 18, \ 1-\alpha = 0.95$

$\nu \downarrow$	0.6	0.7	0.8	0.9	$1/(1+\sqrt{18})$
			ρ		
2	8.25865	7.79345	7.20383	6.38786	9.41526
3	5.63194	5.34295	4.97704	4.47195	6.34858
4	4.68848	4.46508	4.18181	3.79001	5.23988
5	4.21206	4.02297	3.78254	3.44849	4.67592
6	3.92666	3.75878	3.54452	3.24523	4.33565
7	3.73723	3.58378	3.38716	3.11095	4.10827
8	3.60258	3.45961	3.27566	3.01583	3.94566
9	3.50208	3.36706	3.19264	2.94500	3.82362
10	3.42427	3.29549	3.12847	2.89035	3.72867
11	3.36228	3.23851	3.07742	2.84666	3.65268
12	3.31175	3.19211	3.03586	2.81118	3.59051
13	3.26978	3.15359	3.00136	2.78171	3.53869
14	3.23439	3.12112	2.97228	2.75686	3.49485
15	3.20413	3.09337	2.94744	2.73563	3.45728
16	3.17798	3.06940	2.92597	2.71727	3.42471
17	3.15515	3.04847	2.90723	2.70124	3.39622
18	3.13504	3.03005	2.89073	2.68713	3.37109
19	3.11721	3.01371	2.87610	2.67461	3.34875
20	3.10128	2.99912	2.86303	2.66342	3.32877
21	3.08697	2.98600	2.85129	2.65337	3.31079
22	3.07404	2.97416	2.84069	2.64429	3.29453
23	3.06231	2.96341	2.83106	2.63605	3.27974
24	3.05160	2.95361	2.82228	2.62853	3.26625
25	3.04180	2.94464	2.81424	2.62164	3.25388
26	3.03280	2.93639	2.80685	2.61531	3.24250
27	3.02450	2.92879	2.80004	2.60948	3.23201
28	3.01682	2.92175	2.79374	2.60408	3.22229
29	3.00969	2.91523	2.78789	2.59907	3.21326
30	3.00306	2.90916	2.78246	2.59441	3.20486
35	2.97583	2.88422	2.76011	2.57525	3.17030
40	2.95564	2.86573	2.74354	2.56104	3.14462
45	2.94008	2.85148	2.73077	2.55009	3.12479
50	2.92771	2.84016	2.72062	2.54138	3.10902
60	2.90931	2.82331	2.70551	2.52842	3.08552
80	2.88655	2.80246	2.68682	2.51238	3.05640
100	2.87302	2.79007	2.67571	2.50284	3.03907
120	2.86405	2.78186	2.66835	2.49651	3.02758
200	2.84624	2.76554	2.65372	2.48395	3.00473
∞	2.81984	2.74136	2.63202	2.46531	2.97082

BECHHOFER and DUNNETT

TABLE B.3

Two-sided Percentage Point (h) for Equal
Correlations ⎛Case I with Correlation ρ⎞

p = 20, 1-α = 0.95

ν ↓	ρ					
	0.0	0.1	0.2	0.3	0.4	0.5
2	9.76551	9.70785	9.56609	9.36060	9.09684	8.77276
3	6.56686	6.53186	6.44488	6.31798	6.15446	5.95311
4	5.40867	5.38226	5.31601	5.21878	5.09301	4.93773
5	4.81847	4.79667	4.74152	4.66012	4.55443	4.42356
6	4.46172	4.44285	4.39474	4.32335	4.23031	4.11475
7	4.22292	4.20610	4.16291	4.09851	4.01426	3.90928
8	4.05186	4.03657	3.99707	3.93786	3.86012	3.76294
9	3.92328	3.90919	3.87255	3.81740	3.74470	3.65352
10	3.82309	3.80997	3.77564	3.72374	3.65507	3.56867
11	3.74282	3.73050	3.69808	3.64884	3.58349	3.50099
12	3.67706	3.66540	3.63459	3.58760	3.52501	3.44576
13	3.62220	3.61111	3.58167	3.53660	3.47636	3.39985
14	3.57573	3.56514	3.53689	3.49347	3.43525	3.36109
15	3.53587	3.52571	3.49850	3.45652	3.40006	3.32795
16	3.50130	3.49152	3.46523	3.42452	3.36961	3.29927
17	3.47102	3.46159	3.43611	3.39654	3.34299	3.27423
18	3.44430	3.43517	3.41042	3.37186	3.31954	3.25218
19	3.42053	3.41168	3.38759	3.34994	3.29871	3.23260
20	3.39926	3.39066	3.36717	3.33033	3.28010	3.21511
21	3.38011	3.37173	3.34878	3.31270	3.26336	3.19939
22	3.36277	3.35461	3.33215	3.29675	3.24823	3.18519
23	3.34701	3.33903	3.31704	3.28226	3.23449	3.17229
24	3.33262	3.32481	3.30324	3.26904	3.22196	3.16053
25	3.31942	3.31178	3.29059	3.25693	3.21047	3.14976
26	3.30728	3.29978	3.27896	3.24578	3.19992	3.13986
27	3.29606	3.28871	3.26822	3.23550	3.19018	3.13073
28	3.28568	3.27846	3.25828	3.22599	3.18117	3.12229
29	3.27604	3.26894	3.24905	3.21716	3.17281	3.11445
30	3.26706	3.26007	3.24046	3.20894	3.16503	3.10716
35	3.23009	3.22357	3.20510	3.17514	3.13305	3.07720
40	3.20260	3.19643	3.17883	3.15004	3.10932	3.05499
45	3.18135	3.17547	3.15854	3.13067	3.09102	3.03786
50	3.16444	3.15879	3.14240	3.11527	3.07648	3.02425
60	3.13923	3.13391	3.11835	3.09233	3.05483	3.00400
80	3.10798	3.10308	3.08856	3.06393	3.02803	2.97894
100	3.08936	3.08472	3.07082	3.04703	3.01210	2.96404
120	3.07701	3.07253	3.05905	3.03583	3.00153	2.95417
200	3.05246	3.04831	3.03567	3.01357	2.98055	2.93456
∞	3.01599	3.01235	3.00096	2.98054	2.94943	2.90548

TABLE B.3

Two-sided Percentage Point (h) for Equal
Correlations (Case I with Correlation ρ)

p = 20, 1-α = 0.95

ν ↓	0.6	0.7	0.8	0.9	1/(1+√20)
			ρ		
2	8.37951	7.89824	7.28920	6.44764	9.59541
3	5.70851	5.40916	5.03075	4.50933	6.46292
4	4.74876	4.51715	4.22402	3.81936	5.32979
5	4.26391	4.06778	3.81888	3.47383	4.75302
6	3.97336	3.79917	3.57734	3.26818	4.40480
7	3.78045	3.62121	3.41763	3.13234	4.17196
8	3.64330	3.49493	3.30446	3.03612	4.00536
9	3.54092	3.40079	3.22020	2.96447	3.88026
10	3.46163	3.32798	3.15508	2.90920	3.78288
11	3.39846	3.27002	3.10326	2.86500	3.70492
12	3.34697	3.22282	3.06108	2.82912	3.64111
13	3.30420	3.18364	3.02607	2.79931	3.58791
14	3.26812	3.15060	2.99656	2.77418	3.54288
15	3.23728	3.12237	2.97134	2.75271	3.50428
16	3.21062	3.09798	2.94956	2.73414	3.47082
17	3.18735	3.07669	2.93054	2.71794	3.44154
18	3.16686	3.05795	2.91380	2.70366	3.41570
19	3.14868	3.04132	2.89895	2.69100	3.39274
20	3.13244	3.02648	2.88569	2.67969	3.37219
21	3.11785	3.01314	2.87378	2.66953	3.35370
22	3.10467	3.00109	2.86301	2.66035	3.33697
23	3.09271	2.99016	2.85324	2.65201	3.32176
24	3.08180	2.98019	2.84433	2.64441	3.30787
25	3.07181	2.97106	2.83618	2.63744	3.29515
26	3.06263	2.96267	2.82868	2.63105	3.28344
27	3.05417	2.95493	2.82177	2.62514	3.27263
28	3.04634	2.94778	2.81538	2.61969	3.26263
29	3.03907	2.94114	2.80944	2.61462	3.25334
30	3.03232	2.93496	2.80393	2.60991	3.24470
35	3.00455	2.90960	2.78125	2.59054	3.20911
40	2.98397	2.89079	2.76444	2.57617	3.18266
45	2.96811	2.87629	2.75148	2.56510	3.16223
50	2.95550	2.86477	2.74118	2.55630	3.14598
60	2.93674	2.84763	2.72586	2.54319	3.12177
80	2.91354	2.82643	2.70690	2.52697	3.09176
100	2.89975	2.81382	2.69562	2.51733	3.07390
120	2.89060	2.80547	2.68815	2.51093	3.06205
200	2.87245	2.78888	2.67331	2.49824	3.03849
∞	2.84553	2.76427	2.65129	2.47939	3.00353

BECHHOFER and DUNNETT

TABLE B.4

Two-sided Percentage Point (h) for Equal
Correlations $\begin{pmatrix} Case\ I\ with\ Correlation\ \rho \end{pmatrix}$

$$p = 2, \quad 1-\alpha = 0.99$$

			ρ			
$\nu \downarrow$	0.0	0.1	0.2	0.3	0.4	0.5
1	90.02699	89.91411	89.57109	88.98422	88.12788	86.95921
2	12.72666	12.71415	12.67613	12.61116	12.51653	12.38774
3	7.12671	7.12106	7.10393	7.07465	7.03199	6.97389
4	5.46191	5.45831	5.44737	5.42864	5.40130	5.36395
5	4.70031	4.69765	4.68956	4.67568	4.65536	4.62746
6	4.27114	4.26901	4.26254	4.25140	4.23503	4.21244
7	3.99782	3.99604	3.99060	3.98121	3.96736	3.94814
8	3.80925	3.80770	3.80298	3.79480	3.78268	3.76577
9	3.67161	3.67023	3.66603	3.65874	3.64787	3.63263
10	3.56686	3.56561	3.56181	3.55519	3.54528	3.53130
11	3.48454	3.48340	3.47990	3.47381	3.46465	3.45166
12	3.41818	3.41713	3.41388	3.40821	3.39965	3.38745
13	3.36358	3.36259	3.35955	3.35422	3.34615	3.33460
14	3.31787	3.31694	3.31407	3.30903	3.30137	3.29036
15	3.27906	3.27818	3.27546	3.27066	3.26334	3.25279
16	3.24569	3.24486	3.24226	3.23767	3.23065	3.22048
17	3.21671	3.21591	3.21342	3.20902	3.20225	3.19242
18	3.19131	3.19053	3.18814	3.18389	3.17735	3.16781
19	3.16885	3.16811	3.16580	3.16169	3.15535	3.14606
20	3.14886	3.14814	3.14591	3.14192	3.13575	3.12670
21	3.13096	3.13026	3.12809	3.12422	3.11820	3.10935
22	3.11483	3.11415	3.11204	3.10826	3.10239	3.09373
23	3.10022	3.09956	3.09750	3.09382	3.08807	3.07957
24	3.08693	3.08629	3.08428	3.08068	3.07504	3.06669
25	3.07479	3.07416	3.07220	3.06867	3.06314	3.05492
26	3.06365	3.06304	3.06111	3.05765	3.05222	3.04413
27	3.05340	3.05280	3.05091	3.04751	3.04217	3.03419
28	3.04393	3.04334	3.04149	3.03815	3.03288	3.02501
29	3.03516	3.03458	3.03276	3.02947	3.02428	3.01651
30	3.02701	3.02644	3.02466	3.02141	3.01629	3.00861
35	2.99366	2.99313	2.99146	2.98842	2.98358	2.97626
40	2.96905	2.96855	2.96697	2.96407	2.95944	2.95239
45	2.95015	2.94967	2.94816	2.94537	2.94090	2.93406
50	2.93518	2.93472	2.93326	2.93056	2.92621	2.91953
60	2.91297	2.91254	2.91115	2.90859	2.90441	2.89797
80	2.88562	2.88522	2.88393	2.88152	2.87757	2.87142
100	2.86942	2.86904	2.86781	2.86549	2.86167	2.85569
120	2.85872	2.85834	2.85715	2.85489	2.85116	2.84529
200	2.83751	2.83716	2.83604	2.83390	2.83033	2.82468
∞	2.80623	2.80591	2.80489	2.80293	2.79960	2.79427

TABLE B.4

Two-sided Percentage Point (h) for Equal
Correlations [*Case I with Correlation* ρ]

$p = 2$, $1-\alpha = 0.99$

$\nu \downarrow$	0.6	0.7	0.8	0.9	$1/(1+\sqrt{2})$
1	85.40681	83.34491	80.52197	76.28043	87.98209
2	12.21728	11.99203	11.68591	11.23133	12.50045
3	6.89692	6.79507	6.65630	6.44920	7.02473
4	5.31427	5.24814	5.15730	5.02003	5.39664
5	4.59013	4.54006	4.47056	4.36405	4.65189
6	4.18201	4.14084	4.08309	3.99341	4.23223
7	3.92205	3.88647	3.83605	3.75685	3.96498
8	3.74266	3.71087	3.66543	3.59332	3.78059
9	3.61167	3.58262	3.54077	3.47375	3.64600
10	3.51196	3.48497	3.44581	3.38262	3.54356
11	3.43358	3.40819	3.37112	3.31092	3.46306
12	3.37038	3.34627	3.31087	3.25305	3.39816
13	3.31835	3.29530	3.26126	3.20539	3.34474
14	3.27480	3.25261	3.21970	3.16545	3.30003
15	3.23780	3.21635	3.18440	3.13152	3.26206
16	3.20599	3.18517	3.15403	3.10232	3.22942
17	3.17836	3.15807	3.12764	3.07695	3.20106
18	3.15412	3.13431	3.10450	3.05469	3.17620
19	3.13270	3.11330	3.08403	3.03500	3.15423
20	3.11363	3.09460	3.06581	3.01747	3.13466
21	3.09654	3.07784	3.04948	3.00176	3.11714
22	3.08114	3.06274	3.03477	2.98760	3.10135
23	3.06720	3.04906	3.02144	2.97477	3.08705
24	3.05451	3.03661	3.00931	2.96310	3.07404
25	3.04292	3.02523	2.99822	2.95243	3.06215
26	3.03228	3.01480	2.98805	2.94264	3.05125
27	3.02249	3.00519	2.97869	2.93363	3.04121
28	3.01344	2.99632	2.97004	2.92530	3.03194
29	3.00506	2.98810	2.96202	2.91758	3.02336
30	2.99728	2.98046	2.95458	2.91041	3.01538
35	2.96540	2.94917	2.92407	2.88104	2.98271
40	2.94187	2.92608	2.90156	2.85935	2.95860
45	2.92380	2.90834	2.88425	2.84269	2.94009
50	2.90947	2.89428	2.87054	2.82948	2.92542
60	2.88822	2.87342	2.85019	2.80987	2.90366
80	2.86204	2.84771	2.82511	2.78570	2.87685
100	2.84652	2.83247	2.81025	2.77137	2.86097
120	2.83627	2.82240	2.80042	2.76190	2.85047
200	2.81595	2.80244	2.78094	2.74312	2.82968
∞	2.78595	2.77298	2.75218	2.71539	2.79898

TABLE B.4

Two-sided Percentage Point (h) for Equal
Correlations $\left(Case\ I\ with\ Correlation\ \rho\right)$

p = 3, 1-α = 0.99

ν ↓	ρ 0.0	0.1	0.2	0.3	0.4	0.5
2	14.43661	14.41190	14.33948	14.21976	14.05040	13.82575
3	7.91419	7.90314	7.87065	7.81676	7.74029	7.63862
4	5.98543	5.97844	5.95778	5.92338	5.87437	5.80893
5	5.10563	5.10051	5.08532	5.05990	5.02353	4.97470
6	4.61086	4.60680	4.59472	4.57441	4.54520	4.50573
7	4.29627	4.29289	4.28280	4.26578	4.24114	4.20763
8	4.07952	4.07661	4.06790	4.05314	4.03166	4.00226
9	3.92149	3.91892	3.91123	3.89812	3.87895	3.85252
10	3.80134	3.79904	3.79211	3.78027	3.76284	3.73868
11	3.70700	3.70491	3.69859	3.68773	3.67168	3.64928
12	3.63102	3.62909	3.62326	3.61320	3.59825	3.57727
13	3.56854	3.56675	3.56131	3.55190	3.53785	3.51803
14	3.51627	3.51459	3.50949	3.50063	3.48733	3.46848
15	3.47191	3.47033	3.46551	3.45711	3.44445	3.42641
16	3.43380	3.43230	3.42773	3.41972	3.40761	3.39026
17	3.40071	3.39929	3.39492	3.38725	3.37561	3.35887
18	3.37171	3.37035	3.36617	3.35880	3.34757	3.33136
19	3.34610	3.34479	3.34077	3.33366	3.32279	3.30704
20	3.32330	3.32205	3.31817	3.31130	3.30075	3.28541
21	3.30289	3.30168	3.29793	3.29127	3.28100	3.26603
22	3.28451	3.28333	3.27970	3.27322	3.26322	3.24857
23	3.26787	3.26673	3.26319	3.25689	3.24711	3.23277
24	3.25273	3.25162	3.24818	3.24203	3.23247	3.21839
25	3.23891	3.23783	3.23447	3.22846	3.21909	3.20525
26	3.22623	3.22518	3.22190	3.21601	3.20682	3.19321
27	3.21456	3.21353	3.21033	3.20456	3.19552	3.18212
28	3.20379	3.20278	3.19964	3.19398	3.18509	3.17188
29	3.19381	3.19282	3.18975	3.18419	3.17544	3.16240
30	3.18454	3.18358	3.18056	3.17509	3.16646	3.15359
35	3.14662	3.14573	3.14294	3.13785	3.12974	3.11752
40	3.11866	3.11783	3.11521	3.11039	3.10266	3.09092
45	3.09720	3.09642	3.09392	3.08931	3.08186	3.07049
50	3.08021	3.07946	3.07706	3.07261	3.06539	3.05432
60	3.05502	3.05431	3.05207	3.04786	3.04097	3.03032
80	3.02401	3.02337	3.02130	3.01738	3.01090	3.00076
100	3.00567	3.00506	3.00309	2.99934	2.99310	2.98327
120	2.99354	2.99295	2.99106	2.98742	2.98133	2.97170
200	2.96954	2.96900	2.96723	2.96382	2.95803	2.94879
∞	2.93416	2.93368	2.93211	2.92901	2.92366	2.91500

TABLE B.4

Two-sided Percentage Point (h) for Equal
Correlations $\left(\text{Case } I \text{ with Correlation } \rho\right)$

$p = 3, \quad 1-\alpha = 0.99$

ν ↓	0.6	0.7	0.8	0.9	$1/(1+\sqrt{3})$
			ρ		
2	13.53531	13.15980	12.66017	11.93455	14.11378
3	7.50690	7.33624	7.10865	6.77684	7.76893
4	5.72374	5.61273	5.46353	5.24351	5.89275
5	4.91070	4.82664	4.71252	4.54195	5.03719
6	4.45362	4.38458	4.28986	4.14640	4.55619
7	4.16305	4.10346	4.02088	3.89431	4.25043
8	3.96284	3.90970	3.83537	3.72024	4.03978
9	3.81685	3.76837	3.69997	3.59307	3.88620
10	3.70583	3.66085	3.59693	3.49622	3.76945
11	3.61863	3.57639	3.51594	3.42005	3.67777
12	3.54838	3.50832	3.45065	3.35860	3.60393
13	3.49059	3.45232	3.39691	3.30801	3.54321
14	3.44223	3.40544	3.35192	3.26564	3.49241
15	3.40118	3.36564	3.31371	3.22964	3.44929
16	3.36590	3.33143	3.28085	3.19867	3.41224
17	3.33525	3.30171	3.25231	3.17177	3.38007
18	3.30839	3.27565	3.22728	3.14817	3.35188
19	3.28465	3.25262	3.20515	3.12730	3.32697
20	3.26352	3.23212	3.18546	3.10872	3.30480
21	3.24460	3.21376	3.16781	3.09208	3.28495
22	3.22755	3.19722	3.15191	3.07707	3.26707
23	3.21211	3.18223	3.13751	3.06348	3.25088
24	3.19807	3.16860	3.12440	3.05111	3.23616
25	3.18524	3.15615	3.11243	3.03981	3.22271
26	3.17347	3.14472	3.10144	3.02944	3.21037
27	3.16263	3.13420	3.09133	3.01990	3.19902
28	3.15263	3.12449	3.08199	3.01108	3.18853
29	3.14336	3.11549	3.07334	3.00291	3.17882
30	3.13475	3.10713	3.06530	2.99532	3.16981
35	3.09951	3.07291	3.03237	2.96422	3.13289
40	3.07351	3.04765	3.00807	2.94126	3.10567
45	3.05354	3.02825	2.98941	2.92362	3.08477
50	3.03772	3.01288	2.97462	2.90965	3.06822
60	3.01425	2.99008	2.95267	2.88890	3.04367
80	2.98535	2.96199	2.92562	2.86333	3.01345
100	2.96823	2.94535	2.90960	2.84818	2.99556
120	2.95692	2.93435	2.89901	2.83816	2.98373
200	2.93451	2.91256	2.87802	2.81831	2.96032
∞	2.90143	2.88040	2.84704	2.78899	2.92579

TABLE B.4

Two-sided Percentage Point (h) for Equal
Correlations $\left(\text{Case } I \text{ with Correlation } \rho\right)$

$p = 4,\ 1-\alpha = 0.99$

$\nu \downarrow$	ρ					
	0.0	0.1	0.2	0.3	0.4	0.5
2	15.65380	15.61804	15.51604	15.35136	15.12285	14.82463
3	8.47959	8.46363	8.41781	8.34342	8.23980	8.10414
4	6.36237	6.35228	6.32313	6.27556	6.20896	6.12137
5	5.39743	5.39006	5.36864	5.33348	5.28400	5.21853
6	4.85507	4.84924	4.83223	4.80417	4.76442	4.71150
7	4.51037	4.50554	4.49137	4.46787	4.43437	4.38946
8	4.27297	4.26883	4.25664	4.23630	4.20713	4.16775
9	4.09994	4.09631	4.08557	4.06755	4.04155	4.00620
10	3.96844	3.96519	3.95556	3.93931	3.91573	3.88344
11	3.86523	3.86229	3.85352	3.83866	3.81697	3.78709
12	3.78212	3.77943	3.77136	3.75762	3.73746	3.70949
13	3.71380	3.71131	3.70382	3.69100	3.67208	3.64570
14	3.65667	3.65434	3.64733	3.63529	3.61741	3.59234
15	3.60820	3.60601	3.59941	3.58802	3.57102	3.54707
16	3.56656	3.56450	3.55825	3.54741	3.53117	3.50817
17	3.53042	3.52846	3.52251	3.51216	3.49658	3.47439
18	3.49876	3.49689	3.49120	3.48127	3.46626	3.44480
19	3.47079	3.46900	3.46355	3.45399	3.43948	3.41865
20	3.44591	3.44420	3.43895	3.42972	3.41566	3.39538
21	3.42364	3.42199	3.41692	3.40799	3.39432	3.37455
22	3.40358	3.40199	3.39709	3.38842	3.37511	3.35579
23	3.38543	3.38388	3.37913	3.37070	3.35772	3.33880
24	3.36892	3.36742	3.36280	3.35459	3.34190	3.32335
25	3.35384	3.35238	3.34789	3.33987	3.32745	3.30923
26	3.34001	3.33860	3.33422	3.32638	3.31420	3.29629
27	3.32729	3.32591	3.32163	3.31396	3.30200	3.28437
28	3.31555	3.31420	3.31002	3.30250	3.29075	3.27338
29	3.30467	3.30335	3.29926	3.29188	3.28032	3.26319
30	3.29457	3.29328	3.28927	3.28202	3.27063	3.25373
35	3.25325	3.25207	3.24839	3.24167	3.23100	3.21499
40	3.22280	3.22171	3.21827	3.21193	3.20179	3.18644
45	3.19944	3.19841	3.19515	3.18911	3.17936	3.16451
50	3.18094	3.17996	3.17684	3.17104	3.16160	3.14715
60	3.15353	3.15262	3.14971	3.14424	3.13527	3.12139
80	3.11981	3.11899	3.11633	3.11127	3.10285	3.08969
100	3.09987	3.09909	3.09658	3.09176	3.08367	3.07093
120	3.08669	3.08595	3.08354	3.07887	3.07099	3.05852
200	3.06061	3.05994	3.05771	3.05335	3.04590	3.03396
∞	3.02220	3.02162	3.01966	3.01574	3.00889	2.99774

TABLE B.4

Two-sided Percentage Point (h) for Equal
Correlations $\left(Case\ I\ with\ Correlation\ \rho\right)$

$p = 4, \ 1-\alpha = 0.99$

$\nu \downarrow$	ρ 0.6	0.7	0.8	0.9	$1/(1+\sqrt{4})$
2	14.44451	13.95929	13.32146	12.40657	15.28247
3	7.93082	7.70915	7.41721	6.99735	8.31222
4	6.00892	5.86431	5.67253	5.39385	6.25554
5	5.13394	5.02431	4.87754	4.66151	5.31864
6	4.64260	4.55254	4.43074	4.24912	4.79227
7	4.33053	4.25284	4.14669	3.98653	4.45786
8	4.11568	4.04643	3.95094	3.80534	4.22761
9	3.95911	3.89598	3.80817	3.67304	4.05982
10	3.84012	3.78160	3.69956	3.57231	3.93232
11	3.74670	3.69178	3.61424	3.49312	3.83225
12	3.67147	3.61942	3.54548	3.42926	3.75167
13	3.60961	3.55990	3.48890	3.37669	3.68543
14	3.55786	3.51011	3.44155	3.33266	3.63003
15	3.51393	3.46783	3.40134	3.29527	3.58303
16	3.47619	3.43150	3.36677	3.26311	3.54265
17	3.44342	3.39995	3.33674	3.23517	3.50760
18	3.41470	3.37229	3.31042	3.21066	3.47689
19	3.38932	3.34785	3.28715	3.18900	3.44976
20	3.36674	3.32610	3.26644	3.16971	3.42563
21	3.34652	3.30662	3.24788	3.15243	3.40402
22	3.32830	3.28906	3.23116	3.13686	3.38455
23	3.31181	3.27317	3.21602	3.12275	3.36694
24	3.29680	3.25871	3.20225	3.10992	3.35091
25	3.28309	3.24550	3.18966	3.09819	3.33628
26	3.27052	3.23338	3.17812	3.08743	3.32286
27	3.25895	3.22223	3.16749	3.07752	3.31051
28	3.24827	3.21193	3.15767	3.06837	3.29911
29	3.23837	3.20239	3.14858	3.05989	3.28855
30	3.22918	3.19353	3.14013	3.05201	3.27874
35	3.19155	3.15724	3.10554	3.01975	3.23861
40	3.16380	3.13048	3.08002	2.99594	3.20903
45	3.14249	3.10992	3.06041	2.97764	3.18633
50	3.12561	3.09364	3.04488	2.96314	3.16835
60	3.10057	3.06948	3.02183	2.94163	3.14170
80	3.06974	3.03972	2.99343	2.91511	3.10890
100	3.05149	3.02210	2.97662	2.89940	3.08949
120	3.03942	3.01045	2.96549	2.88902	3.07666
200	3.01553	2.98738	2.94347	2.86843	3.05127
∞	2.98028	2.95333	2.91095	2.83805	3.01384

BECHHOFER and DUNNETT

TABLE B.4

Two-sided Percentage Point (h) for Equal
Correlations ⟨*Case I with Correlation* ρ⟩

p = 5, 1-α = 0.99

ν ↓	ρ					
	0.0	0.1	0.2	0.3	0.4	0.5
2	16.59167	16.54601	16.41850	16.21625	15.93945	15.58232
3	8.91869	8.89827	8.84081	8.74909	8.62299	8.45975
4	6.65619	6.64328	6.60665	6.54782	6.46652	6.36077
5	5.62520	5.61576	5.58882	5.54528	5.48475	5.40556
6	5.04569	5.03825	5.01686	4.98208	4.93342	4.86933
7	4.67739	4.67122	4.65343	4.62430	4.58328	4.52888
8	4.42372	4.41846	4.40316	4.37797	4.34226	4.29457
9	4.23886	4.23425	4.22079	4.19849	4.16669	4.12390
10	4.09837	4.09426	4.08221	4.06213	4.03330	3.99424
11	3.98812	3.98440	3.97346	3.95512	3.92863	3.89249
12	3.89935	3.89596	3.88591	3.86898	3.84436	3.81058
13	3.82639	3.82326	3.81394	3.79817	3.77511	3.74325
14	3.76538	3.76247	3.75377	3.73896	3.71719	3.68694
15	3.71363	3.71090	3.70272	3.68873	3.66806	3.63917
16	3.66918	3.66661	3.65888	3.64559	3.62585	3.59813
17	3.63060	3.62817	3.62082	3.60815	3.58922	3.56251
18	3.59681	3.59449	3.58748	3.57534	3.55713	3.53129
19	3.56696	3.56475	3.55804	3.54637	3.52878	3.50372
20	3.54042	3.53830	3.53185	3.52060	3.50356	3.47919
21	3.51665	3.51461	3.50840	3.49752	3.48098	3.45722
22	3.49525	3.49329	3.48729	3.47674	3.46065	3.43744
23	3.47589	3.47399	3.46818	3.45794	3.44224	3.41953
24	3.45828	3.45644	3.45081	3.44083	3.42550	3.40325
25	3.44220	3.44042	3.43494	3.42521	3.41022	3.38837
26	3.42745	3.42572	3.42039	3.41089	3.39620	3.37473
27	3.41389	3.41220	3.40700	3.39771	3.38330	3.36217
28	3.40137	3.39972	3.39465	3.38555	3.37139	3.35058
29	3.38977	3.38817	3.38320	3.37428	3.36036	3.33985
30	3.37900	3.37744	3.37258	3.36382	3.35012	3.32988
35	3.33496	3.33354	3.32910	3.32101	3.30820	3.28907
40	3.30252	3.30121	3.29707	3.28947	3.27731	3.25899
45	3.27763	3.27640	3.27249	3.26526	3.25360	3.23589
50	3.25793	3.25676	3.25304	3.24610	3.23483	3.21761
60	3.22874	3.22767	3.22421	3.21770	3.20699	3.19049
80	3.19285	3.19188	3.18875	3.18275	3.17274	3.15711
100	3.17163	3.17072	3.16778	3.16208	3.15248	3.13736
120	3.15761	3.15675	3.15392	3.14842	3.13908	3.12430
200	3.12988	3.12909	3.12651	3.12139	3.11257	3.09845
∞	3.08904	3.08837	3.08612	3.08155	3.07350	3.06033

TABLE B.4

Two-sided Percentage Point (h) for Equal Correlations $\left(Case\ I\ with\ Correlation\ \rho\right)$

p = 5, 1-α = 0.99

ν ↓	0.6	0.7	0.8	0.9	1/(1+√5)
			ρ		
2	15.13146	14.56086	13.81672	12.75795	16.19438
3	8.25319	7.99130	7.64932	7.16194	8.73915
4	6.22633	6.05502	5.82991	5.50607	6.54143
5	5.30424	5.17420	5.00179	4.75071	5.54053
6	4.78675	4.67987	4.53676	4.32571	4.97827
7	4.45823	4.36602	4.24132	4.05527	4.62110
8	4.23215	4.14998	4.03783	3.86874	4.37519
9	4.06747	3.99257	3.88947	3.73259	4.19603
10	3.94234	3.87294	3.77666	3.62896	4.05991
11	3.84414	3.77903	3.68806	3.54752	3.95309
12	3.76508	3.70340	3.61667	3.48184	3.86709
13	3.70008	3.64120	3.55795	3.42779	3.79641
14	3.64571	3.58917	3.50880	3.38253	3.73730
15	3.59958	3.54501	3.46707	3.34409	3.68716
16	3.55995	3.50706	3.43121	3.31103	3.64410
17	3.52554	3.47411	3.40006	3.28232	3.60672
18	3.49538	3.44523	3.37275	3.25713	3.57397
19	3.46874	3.41971	3.34862	3.23487	3.54505
20	3.44504	3.39700	3.32713	3.21505	3.51932
21	3.42381	3.37666	3.30789	3.19730	3.49628
22	3.40469	3.35833	3.29056	3.18129	3.47554
23	3.38738	3.34174	3.27486	3.16680	3.45676
24	3.37164	3.32665	3.26057	3.15362	3.43969
25	3.35725	3.31286	3.24752	3.14157	3.42410
26	3.34406	3.30022	3.23555	3.13051	3.40980
27	3.33192	3.28858	3.22454	3.12034	3.39664
28	3.32072	3.27783	3.21436	3.11094	3.38450
29	3.31033	3.26788	3.20493	3.10223	3.37325
30	3.30069	3.25863	3.19618	3.09414	3.36281
35	3.26122	3.22077	3.16032	3.06100	3.32007
40	3.23212	3.19285	3.13387	3.03655	3.28858
45	3.20977	3.17140	3.11355	3.01776	3.26441
50	3.19207	3.15442	3.09745	3.00287	3.24528
60	3.16582	3.12922	3.07357	2.98078	3.21692
80	3.13350	3.09819	3.04415	2.95355	3.18203
100	3.11437	3.07982	3.02673	2.93743	3.16139
120	3.10172	3.06767	3.01521	2.92676	3.14775
200	3.07668	3.04362	2.99239	2.90563	3.12076
∞	3.03975	3.00812	2.95872	2.87444	3.08099

TABLE B.4

*Two-sided Percentage Point (h) for Equal
Correlations* $\left(Case\ I\ with\ Correlation\ \rho\right)$

$p = 6,\ 1-\alpha = 0.99$

$\nu \downarrow$	ρ					
	0.0	0.1	0.2	0.3	0.4	0.5
2	17.35072	17.29613	17.14635	16.91202	16.59473	16.18880
3	9.27639	9.25193	9.18420	9.07752	8.93236	8.74603
4	6.89641	6.88091	6.83764	6.76901	6.67513	6.55405
5	5.81173	5.80040	5.76852	5.71763	5.64759	5.55673
6	5.20190	5.19297	5.16764	5.12694	5.07057	4.99696
7	4.81425	4.80686	4.78578	4.75168	4.70413	4.64161
8	4.54722	4.54090	4.52280	4.49331	4.45192	4.39709
9	4.35259	4.34707	4.33115	4.30506	4.26819	4.21901
10	4.20468	4.19977	4.18553	4.16204	4.12863	4.08374
11	4.08859	4.08416	4.07124	4.04981	4.01912	3.97761
12	3.99513	3.99109	3.97924	3.95947	3.93097	3.89218
13	3.91831	3.91458	3.90362	3.88522	3.85853	3.82196
14	3.85408	3.85062	3.84039	3.82313	3.79796	3.76325
15	3.79959	3.79636	3.78675	3.77047	3.74657	3.71344
16	3.75280	3.74976	3.74069	3.72524	3.70244	3.67067
17	3.71219	3.70932	3.70072	3.68598	3.66414	3.63353
18	3.67662	3.67389	3.66570	3.65160	3.63059	3.60100
19	3.64521	3.64261	3.63477	3.62123	3.60095	3.57226
20	3.61727	3.61478	3.60726	3.59421	3.57459	3.54669
21	3.59226	3.58987	3.58263	3.57003	3.55098	3.52380
22	3.56974	3.56744	3.56046	3.54825	3.52973	3.50319
23	3.54936	3.54715	3.54040	3.52854	3.51049	3.48453
24	3.53083	3.52869	3.52215	3.51062	3.49300	3.46756
25	3.51392	3.51184	3.50549	3.49425	3.47702	3.45206
26	3.49840	3.49639	3.49021	3.47925	3.46237	3.43785
27	3.48413	3.48218	3.47615	3.46544	3.44889	3.42477
28	3.47096	3.46905	3.46318	3.45269	3.43645	3.41269
29	3.45876	3.45691	3.45117	3.44089	3.42492	3.40151
30	3.44744	3.44563	3.44001	3.42993	3.41422	3.39112
35	3.40112	3.39948	3.39438	3.38509	3.37043	3.34862
40	3.36701	3.36550	3.36076	3.35205	3.33815	3.31729
45	3.34084	3.33943	3.33497	3.32670	3.31339	3.29324
50	3.32014	3.31881	3.31456	3.30664	3.29378	3.27420
60	3.28946	3.28824	3.28432	3.27690	3.26471	3.24597
80	3.25175	3.25066	3.24712	3.24032	3.22895	3.21121
100	3.22946	3.22845	3.22513	3.21868	3.20779	3.19065
120	3.21473	3.21377	3.21060	3.20438	3.19381	3.17706
200	3.18562	3.18475	3.18186	3.17609	3.16613	3.15015
∞	3.14276	3.14202	3.13953	3.13442	3.12534	3.11048

TABLE B.4

Two-sided Percentage Point (h) for Equal Correlations $\left(\text{Case I with Correlation } \rho\right)$

p = 6, 1-α = 0.99

ν ↓	ρ				
	0.6	0.7	0.8	0.9	1/(1+√6)
2	15.67995	15.03991	14.20998	13.03591	16.93942
3	8.51192	8.21703	7.83431	7.29243	9.09002
4	6.40125	6.20787	5.95548	5.59506	6.77708
5	5.44138	5.29440	5.10093	4.82142	5.72363
6	4.90285	4.78196	4.62133	4.38641	5.13175
7	4.56107	4.45675	4.31678	4.10972	4.75572
8	4.32592	4.23296	4.10710	3.91895	4.49681
9	4.15467	4.06995	3.95427	3.77974	4.30817
10	4.02458	3.94610	3.83809	3.67380	4.16485
11	3.92251	3.84889	3.74686	3.59057	4.05238
12	3.84034	3.77062	3.67337	3.52344	3.96184
13	3.77280	3.70626	3.61292	3.46822	3.88743
14	3.71631	3.65243	3.56234	3.42197	3.82522
15	3.66838	3.60675	3.51940	3.38270	3.77244
16	3.62722	3.56750	3.48249	3.34894	3.72711
17	3.59148	3.53341	3.45044	3.31960	3.68777
18	3.56016	3.50354	3.42235	3.29388	3.65331
19	3.53249	3.47715	3.39752	3.27115	3.62288
20	3.50788	3.45367	3.37542	3.25090	3.59581
21	3.48583	3.43263	3.35563	3.23277	3.57157
22	3.46598	3.41369	3.33779	3.21643	3.54975
23	3.44801	3.39654	3.32165	3.20163	3.53000
24	3.43167	3.38094	3.30696	3.18816	3.51204
25	3.41673	3.36668	3.29354	3.17586	3.49564
26	3.40304	3.35361	3.28123	3.16457	3.48060
27	3.39044	3.34158	3.26990	3.15418	3.46676
28	3.37881	3.33047	3.25943	3.14458	3.45399
29	3.36803	3.32018	3.24974	3.13569	3.44216
30	3.35802	3.31062	3.24073	3.12743	3.43118
35	3.31705	3.27149	3.20386	3.09360	3.38625
40	3.28685	3.24263	3.17666	3.06863	3.35314
45	3.26366	3.22047	3.15577	3.04945	3.32774
50	3.24530	3.20292	3.13923	3.03426	3.30764
60	3.21806	3.17688	3.11467	3.01170	3.27784
80	3.18453	3.14482	3.08443	2.98392	3.24118
100	3.16469	3.12584	3.06653	2.96746	3.21951
120	3.15157	3.11330	3.05469	2.95658	3.20518
200	3.12559	3.08845	3.03124	2.93501	3.17684
∞	3.08728	3.05179	2.99663	2.90318	3.13509

TABLE B.4

Two-sided Percentage Point (h) for Equal
Correlations $\left(\text{Case I with Correlation } \rho\right)$

p = 7, 1−α = 0.99

ν ↓	ρ					
	0.0	0.1	0.2	0.3	0.4	0.5
2	17.98599	17.92328	17.75378	17.49162	17.13960	16.69220
3	9.57738	9.54922	9.47234	9.35254	9.19090	8.98477
4	7.09920	7.08133	7.03208	6.95480	6.84995	6.71563
5	5.96947	5.95639	5.92006	5.86263	5.78425	5.68328
6	5.33412	5.32380	5.29490	5.24892	5.18575	5.10385
7	4.93012	4.92158	4.89752	4.85897	4.80565	4.73604
8	4.65176	4.64447	4.62380	4.59046	4.54402	4.48296
9	4.44884	4.44248	4.42431	4.39481	4.35344	4.29865
10	4.29461	4.28895	4.27271	4.24615	4.20867	4.15867
11	4.17355	4.16845	4.15373	4.12951	4.09508	4.04885
12	4.07608	4.07143	4.05794	4.03561	4.00366	3.96046
13	3.99596	3.99169	3.97921	3.95843	3.92852	3.88781
14	3.92897	3.92501	3.91338	3.89390	3.86570	3.82708
15	3.87215	3.86844	3.85754	3.83917	3.81241	3.77555
16	3.82335	3.81987	3.80958	3.79217	3.76665	3.73130
17	3.78100	3.77771	3.76796	3.75137	3.72693	3.69290
18	3.74390	3.74079	3.73150	3.71564	3.69214	3.65925
19	3.71114	3.70818	3.69931	3.68408	3.66141	3.62953
20	3.68200	3.67917	3.67067	3.65601	3.63408	3.60309
21	3.65592	3.65321	3.64504	3.63088	3.60961	3.57942
22	3.63244	3.62984	3.62196	3.60825	3.58757	3.55811
23	3.61120	3.60868	3.60107	3.58777	3.56763	3.53881
24	3.59188	3.58945	3.58208	3.56915	3.54949	3.52126
25	3.57424	3.57189	3.56474	3.55215	3.53293	3.50524
26	3.55806	3.55579	3.54884	3.53656	3.51775	3.49054
27	3.54319	3.54098	3.53421	3.52222	3.50377	3.47702
28	3.52945	3.52731	3.52071	3.50898	3.49087	3.46454
29	3.51674	3.51465	3.50820	3.49671	3.47893	3.45298
30	3.50493	3.50289	3.49660	3.48533	3.46784	3.44224
35	3.45665	3.45482	3.44911	3.43875	3.42245	3.39830
40	3.42110	3.41942	3.41413	3.40444	3.38901	3.36592
45	3.39384	3.39227	3.38731	3.37811	3.36334	3.34106
50	3.37227	3.37079	3.36608	3.35728	3.34303	3.32138
60	3.34031	3.33896	3.33462	3.32640	3.31291	3.29220
80	3.30103	3.29983	3.29593	3.28842	3.27586	3.25629
100	3.27782	3.27671	3.27307	3.26596	3.25394	3.23504
120	3.26249	3.26144	3.25796	3.25111	3.23945	3.22099
200	3.23218	3.23124	3.22808	3.22175	3.21079	3.19319
∞	3.18757	3.18679	3.18409	3.17850	3.16854	3.15221

TABLE B.4

Two-sided Percentage Point (h) for Equal
Correlations (*Case I with Correlation* ρ)

$p = 7, \quad 1-\alpha = 0.99$

$\nu \downarrow$	ρ				
	0.6	0.7	0.8	0.9	$1/(1+\sqrt{7})$
2	16.13441	15.43614	14.53460	13.26476	17.56757
3	8.72724	8.40445	7.98749	7.40008	9.38731
4	6.54712	6.33499	6.05956	5.66850	6.97728
5	5.55586	5.39442	5.18312	4.87976	5.87937
6	4.99979	4.86693	4.69144	4.43647	5.26235
7	4.64694	4.53224	4.37932	4.15462	4.87026
8	4.40421	4.30200	4.16449	3.96034	4.60024
9	4.22745	4.13431	4.00794	3.81860	4.40349
10	4.09321	4.00693	3.88896	3.71075	4.25398
11	3.98789	3.90697	3.79555	3.62605	4.13666
12	3.90312	3.82649	3.72030	3.55771	4.04222
13	3.83344	3.76033	3.65842	3.50153	3.96460
14	3.77518	3.70500	3.60665	3.45446	3.89970
15	3.72574	3.65804	3.56270	3.41451	3.84464
16	3.68329	3.61770	3.52493	3.38016	3.79737
17	3.64643	3.58267	3.49213	3.35031	3.75633
18	3.61413	3.55197	3.46338	3.32414	3.72039
19	3.58561	3.52485	3.43797	3.30102	3.68865
20	3.56023	3.50072	3.41536	3.28043	3.66042
21	3.53750	3.47911	3.39511	3.26198	3.63514
22	3.51703	3.45965	3.37687	3.24536	3.61239
23	3.49850	3.44202	3.36035	3.23030	3.59179
24	3.48165	3.42599	3.34532	3.21661	3.57306
25	3.46626	3.41135	3.33159	3.20409	3.55596
26	3.45215	3.39792	3.31899	3.19261	3.54028
27	3.43916	3.38556	3.30740	3.18204	3.52586
28	3.42716	3.37415	3.29670	3.17228	3.51254
29	3.41605	3.36358	3.28678	3.16324	3.50021
30	3.40574	3.35376	3.27757	3.15484	3.48876
35	3.36351	3.31356	3.23986	3.12043	3.44192
40	3.33238	3.28392	3.21204	3.09504	3.40742
45	3.30849	3.26116	3.19067	3.07554	3.38095
50	3.28956	3.24314	3.17375	3.06009	3.36000
60	3.26150	3.21640	3.14864	3.03715	3.32895
80	3.22695	3.18348	3.11772	3.00890	3.29077
100	3.20650	3.16399	3.09940	2.99217	3.26819
120	3.19298	3.15111	3.08730	2.98110	3.25327
200	3.16622	3.12559	3.06332	2.95918	3.22376
∞	3.12676	3.08796	3.02794	2.92682	3.18029

TABLE B.4

Two-sided Percentage Point (h) for Equal Correlations (*Case I with Correlation* ρ)

$$p = 8, \quad 1-\alpha = 0.99$$

$\nu \downarrow$	ρ 0.0	0.1	0.2	0.3	0.4	0.5
2	18.53078	18.46066	18.27354	17.98685	17.60451	17.12114
3	9.83668	9.80512	9.71998	9.58855	9.41241	9.18900
4	7.27440	7.25433	7.19966	7.11465	7.00010	6.85416
5	6.10599	6.09128	6.05087	5.98758	5.90179	5.79188
6	5.44864	5.43703	5.40486	5.35411	5.28488	5.19564
7	5.03052	5.02092	4.99411	4.95153	4.89304	4.81714
8	4.74236	4.73416	4.71113	4.67428	4.62332	4.55670
9	4.53224	4.52509	4.50485	4.47224	4.42683	4.36705
10	4.37252	4.36616	4.34806	4.31872	4.27757	4.22301
11	4.24713	4.24140	4.22501	4.19825	4.16046	4.11001
12	4.14617	4.14095	4.12593	4.10126	4.06620	4.01907
13	4.06317	4.05838	4.04450	4.02155	3.98873	3.94433
14	3.99377	3.98933	3.97640	3.95491	3.92397	3.88185
15	3.93490	3.93076	3.91864	3.89838	3.86903	3.82884
16	3.88434	3.88045	3.86904	3.84983	3.82186	3.78332
17	3.84046	3.83680	3.82598	3.80770	3.78091	3.74382
18	3.80203	3.79856	3.78827	3.77079	3.74504	3.70921
19	3.76809	3.76479	3.75497	3.73820	3.71337	3.67864
20	3.73791	3.73475	3.72535	3.70921	3.68519	3.65145
21	3.71089	3.70787	3.69883	3.68325	3.65997	3.62710
22	3.68656	3.68367	3.67496	3.65989	3.63726	3.60518
23	3.66455	3.66176	3.65336	3.63874	3.61670	3.58534
24	3.64454	3.64185	3.63371	3.61951	3.59801	3.56729
25	3.62626	3.62366	3.61578	3.60195	3.58094	3.55081
26	3.60951	3.60699	3.59933	3.58586	3.56529	3.53570
27	3.59410	3.59166	3.58420	3.57104	3.55089	3.52179
28	3.57987	3.57750	3.57024	3.55737	3.53760	3.50896
29	3.56670	3.56439	3.55731	3.54471	3.52529	3.49707
30	3.55448	3.55222	3.54530	3.53296	3.51386	3.48603
35	3.50447	3.50246	3.49620	3.48487	3.46708	3.44085
40	3.46765	3.46581	3.46003	3.44944	3.43262	3.40755
45	3.43942	3.43771	3.43229	3.42226	3.40618	3.38200
50	3.41709	3.41547	3.41034	3.40075	3.38525	3.36176
60	3.38400	3.38254	3.37782	3.36888	3.35422	3.33176
80	3.34335	3.34205	3.33783	3.32968	3.31604	3.29485
100	3.31932	3.31813	3.31420	3.30649	3.29346	3.27300
120	3.30346	3.30233	3.29858	3.29118	3.27854	3.25857
200	3.27210	3.27109	3.26771	3.26088	3.24901	3.22999
∞	3.22596	3.22513	3.22225	3.21625	3.20550	3.18787

TABLE B.4

Two-sided Percentage Point (h) for Equal
Correlations $\left(\text{Case I with Correlation } \rho\right)$

$p = 8, \ 1-\alpha = 0.99$

$\nu \downarrow$	ρ				
	0.6	0.7	0.8	0.9	$1/(1+\sqrt{8})$
2	16.52116	15.77289	14.81010	13.45862	18.10942
3	8.91113	8.56424	8.11783	7.49143	9.64484
4	6.67193	6.44353	6.14821	5.73083	7.15112
5	5.65389	5.47986	5.25314	4.92928	6.01478
6	5.08283	4.93952	4.75116	4.47896	5.37597
7	4.72050	4.59674	4.43259	4.19272	4.96990
8	4.47127	4.36097	4.21336	3.99545	4.69021
9	4.28980	4.18929	4.05364	3.85156	4.48636
10	4.15199	4.05888	3.93227	3.74209	4.33145
11	4.04388	3.95657	3.83699	3.65613	4.20988
12	3.95687	3.87420	3.76025	3.58677	4.11201
13	3.88535	3.80649	3.69714	3.52977	4.03156
14	3.82556	3.74986	3.64435	3.48200	3.96430
15	3.77483	3.70181	3.59954	3.44147	3.90725
16	3.73126	3.66053	3.56103	3.40661	3.85825
17	3.69344	3.62469	3.52759	3.37634	3.81573
18	3.66031	3.59329	3.49828	3.34979	3.77848
19	3.63104	3.56555	3.47238	3.32633	3.74558
20	3.60500	3.54086	3.44933	3.30544	3.71633
21	3.58168	3.51875	3.42869	3.28673	3.69013
22	3.56069	3.49884	3.41009	3.26987	3.66655
23	3.54168	3.48082	3.39325	3.25460	3.64521
24	3.52439	3.46442	3.37794	3.24070	3.62581
25	3.50860	3.44944	3.36394	3.22801	3.60809
26	3.49413	3.43571	3.35110	3.21636	3.59184
27	3.48080	3.42307	3.33929	3.20565	3.57690
28	3.46850	3.41139	3.32838	3.19574	3.56310
29	3.45711	3.40058	3.31827	3.18657	3.55032
30	3.44652	3.39054	3.30889	3.17805	3.53846
35	3.40322	3.34943	3.27045	3.14316	3.48994
40	3.37129	3.31912	3.24210	3.11741	3.45420
45	3.34678	3.29584	3.22033	3.09763	3.42678
50	3.32737	3.27741	3.20308	3.08196	3.40508
60	3.29859	3.25007	3.17750	3.05870	3.37293
80	3.26317	3.21641	3.14599	3.03006	3.33339
100	3.24220	3.19649	3.12734	3.01309	3.31002
120	3.22834	3.18331	3.11500	3.00187	3.29457
200	3.20090	3.15723	3.09057	2.97964	3.26403
∞	3.16044	3.11875	3.05452	2.94683	3.21904

TABLE B.4

Two-sided Percentage Point (h) for Equal
Correlations (*Case I with Correlation ρ*)

p = 9, 1-α = 0.99

ν ↓	ρ					
	0.0	0.1	0.2	0.3	0.4	0.5
2	19.00671	18.92979	18.72678	18.41815	18.00897	17.49396
3	10.06409	10.02938	9.93674	9.79485	9.60579	9.36708
4	7.42844	7.40632	7.34670	7.25471	7.13146	6.97518
5	6.22620	6.20997	6.16583	6.09721	6.00475	5.88687
6	5.54958	5.53675	5.50157	5.44649	5.37178	5.27595
7	5.11906	5.10845	5.07911	5.03285	4.96968	4.88812
8	4.82226	4.81321	4.78799	4.74793	4.69286	4.62125
9	4.60580	4.59790	4.57573	4.54028	4.49119	4.42692
10	4.44122	4.43421	4.41439	4.38248	4.33799	4.27932
11	4.31201	4.30569	4.28774	4.25863	4.21778	4.16354
12	4.20795	4.20220	4.18576	4.15893	4.12103	4.07035
13	4.12240	4.11712	4.10193	4.07699	4.04152	3.99377
14	4.05086	4.04597	4.03184	4.00848	3.97504	3.92975
15	3.99017	3.98561	3.97237	3.95036	3.91865	3.87545
16	3.93805	3.93378	3.92130	3.90045	3.87023	3.82882
17	3.89282	3.88879	3.87698	3.85713	3.82820	3.78834
18	3.85319	3.84938	3.83816	3.81919	3.79139	3.75289
19	3.81820	3.81458	3.80387	3.78568	3.75888	3.72157
20	3.78708	3.78363	3.77338	3.75587	3.72996	3.69372
21	3.75923	3.75592	3.74608	3.72919	3.70408	3.66878
22	3.73415	3.73098	3.72150	3.70517	3.68077	3.64632
23	3.71145	3.70841	3.69926	3.68343	3.65967	3.62600
24	3.69082	3.68789	3.67904	3.66367	3.64049	3.60751
25	3.67198	3.66915	3.66058	3.64562	3.62297	3.59063
26	3.65471	3.65197	3.64365	3.62907	3.60691	3.57515
27	3.63883	3.63617	3.62807	3.61384	3.59213	3.56091
28	3.62416	3.62158	3.61370	3.59979	3.57849	3.54776
29	3.61059	3.60807	3.60039	3.58677	3.56585	3.53559
30	3.59798	3.59553	3.58803	3.57469	3.55412	3.52428
35	3.54644	3.54425	3.53748	3.52526	3.50613	3.47801
40	3.50849	3.50650	3.50026	3.48884	3.47077	3.44391
45	3.47939	3.47754	3.47170	3.46091	3.44363	3.41773
50	3.45638	3.45464	3.44911	3.43881	3.42216	3.39701
60	3.42228	3.42071	3.41564	3.40605	3.39032	3.36629
80	3.38040	3.37902	3.37450	3.36576	3.35116	3.32849
100	3.35565	3.35438	3.35018	3.34194	3.32799	3.30613
120	3.33931	3.33811	3.33411	3.32620	3.31268	3.29134
200	3.30701	3.30595	3.30235	3.29507	3.28239	3.26209
∞	3.25950	3.25863	3.25559	3.24922	3.23776	3.21895

TABLE B.4

Two-sided Percentage Point (h) for Equal Correlations (*Case I with Correlation* ρ)

$$p = 9, \quad 1-\alpha = 0.99$$

$\nu \downarrow$	ρ				
	0.6	0.7	0.8	0.9	$1/(1+\sqrt{9})$
2	16.85696	16.06499	15.04881	13.62635	18.58508
3	9.07129	8.70322	8.23102	7.57059	9.87171
4	6.78081	6.53805	6.22526	5.78487	7.30461
5	5.73948	5.55432	5.31401	4.97220	6.13449
6	5.15537	5.00279	4.80308	4.51577	5.47646
7	4.78476	4.65296	4.47889	4.22572	5.05806
8	4.52985	4.41236	4.25583	4.02586	4.76979
9	4.34425	4.23719	4.09335	3.88011	4.55966
10	4.20332	4.10414	3.96989	3.76923	4.39995
11	4.09277	3.99977	3.87299	3.68218	4.27460
12	4.00379	3.91575	3.79495	3.61193	4.17367
13	3.93067	3.84668	3.73078	3.55422	4.09071
14	3.86953	3.78893	3.67710	3.50585	4.02134
15	3.81767	3.73992	3.63153	3.46482	3.96250
16	3.77313	3.69783	3.59238	3.42951	3.91197
17	3.73447	3.66128	3.55838	3.39887	3.86811
18	3.70060	3.62926	3.52859	3.37199	3.82969
19	3.67068	3.60097	3.50226	3.34823	3.79576
20	3.64406	3.57580	3.47883	3.32709	3.76559
21	3.62023	3.55326	3.45784	3.30815	3.73858
22	3.59877	3.53296	3.43894	3.29108	3.71426
23	3.57934	3.51458	3.42182	3.27562	3.69225
24	3.56167	3.49786	3.40625	3.26156	3.67224
25	3.54553	3.48259	3.39202	3.24871	3.65396
26	3.53073	3.46859	3.37898	3.23692	3.63721
27	3.51712	3.45570	3.36697	3.22607	3.62179
28	3.50454	3.44380	3.35588	3.21605	3.60757
29	3.49290	3.43277	3.34561	3.20677	3.59439
30	3.48208	3.42254	3.33607	3.19814	3.58216
35	3.43782	3.38063	3.29700	3.16282	3.53213
40	3.40520	3.34973	3.26819	3.13676	3.49528
45	3.38015	3.32601	3.24606	3.11674	3.46701
50	3.36032	3.30722	3.22854	3.10088	3.44464
60	3.33091	3.27935	3.20254	3.07735	3.41150
80	3.29471	3.24504	3.17052	3.04836	3.37075
100	3.27329	3.22473	3.15156	3.03118	3.34665
120	3.25913	3.21131	3.13903	3.01983	3.33074
200	3.23110	3.18472	3.11421	2.99734	3.29927
∞	3.18976	3.14551	3.07758	2.96414	3.25292

TABLE B.4

Two-sided Percentage Point (h) for Equal Correlations $\left(Case\ I\ with\ Correlation\ \rho \right)$

$p = 10,\ 1-\alpha = 0.99$

$\nu \downarrow$	ρ					
	0.0	0.1	0.2	0.3	0.4	0.5
2	19.42863	19.34540	19.12788	18.79951	18.36626	17.82301
3	10.26634	10.22870	10.12920	9.97783	9.77714	9.52471
4	7.56574	7.54172	7.47754	7.37918	7.24808	7.08249
5	6.33350	6.31585	6.26826	6.19478	6.09625	5.97117
6	5.63976	5.62580	5.58782	5.52875	5.44906	5.34727
7	5.19820	5.18664	5.15495	5.10530	5.03786	4.95116
8	4.89370	4.88384	4.85658	4.81356	4.75474	4.67859
9	4.67158	4.66297	4.63900	4.60092	4.54846	4.48009
10	4.50266	4.49501	4.47359	4.43930	4.39175	4.32933
11	4.37001	4.36313	4.34372	4.31245	4.26879	4.21107
12	4.26317	4.25692	4.23915	4.21033	4.16981	4.11589
13	4.17534	4.16959	4.15318	4.12638	4.08847	4.03768
14	4.10187	4.09656	4.08129	4.05620	4.02046	3.97229
15	4.03955	4.03460	4.02030	3.99666	3.96278	3.91683
16	3.98602	3.98138	3.96792	3.94553	3.91324	3.86920
17	3.93957	3.93520	3.92246	3.90115	3.87025	3.82786
18	3.89887	3.89474	3.88263	3.86228	3.83259	3.79165
19	3.86293	3.85901	3.84747	3.82795	3.79934	3.75967
20	3.83097	3.82723	3.81619	3.79742	3.76976	3.73123
21	3.80236	3.79879	3.78819	3.77008	3.74327	3.70576
22	3.77660	3.77318	3.76298	3.74548	3.71943	3.68283
23	3.75329	3.75000	3.74016	3.72320	3.69785	3.66207
24	3.73210	3.72893	3.71942	3.70295	3.67823	3.64320
25	3.71275	3.70969	3.70048	3.68446	3.66031	3.62596
26	3.69501	3.69205	3.68312	3.66751	3.64388	3.61015
27	3.67870	3.67583	3.66714	3.65191	3.62877	3.59561
28	3.66364	3.66085	3.65239	3.63751	3.61481	3.58218
29	3.64969	3.64699	3.63874	3.62418	3.60189	3.56975
30	3.63675	3.63411	3.62607	3.61180	3.58989	3.55820
35	3.58381	3.58147	3.57422	3.56117	3.54080	3.51096
40	3.54484	3.54271	3.53604	3.52387	3.50463	3.47614
45	3.51496	3.51299	3.50676	3.49526	3.47688	3.44941
50	3.49133	3.48948	3.48359	3.47261	3.45491	3.42826
60	3.45633	3.45465	3.44926	3.43906	3.42236	3.39690
80	3.41332	3.41186	3.40707	3.39780	3.38230	3.35830
100	3.38792	3.38658	3.38213	3.37340	3.35861	3.33547
120	3.37115	3.36989	3.36567	3.35728	3.34296	3.32038
200	3.33800	3.33689	3.33310	3.32540	3.31198	3.29051
∞	3.28926	3.28835	3.28517	3.27845	3.26634	3.24648

TABLE B.4

Two-sided Percentage Point (h) for Equal Correlations (Case I with Correlation ρ)

p = 10, 1-α = 0.99

ν ↓	0.6	0.7	0.8	0.9	1/(1+√10)
			ρ		
2	17.15314	16.32240	15.25900	13.77388	19.00841
3	9.21291	8.82599	8.33088	7.64031	10.07424
4	6.87724	6.62165	6.29329	5.83247	7.44190
5	5.81533	5.62020	5.36778	5.01000	6.24169
6	5.21967	5.05878	4.84893	4.54820	5.56651
7	4.84173	4.70271	4.51978	4.25478	5.13707
8	4.58178	4.45784	4.29334	4.05265	4.84113
9	4.39253	4.27957	4.12842	3.90525	4.62536
10	4.24882	4.14419	4.00311	3.79312	4.46133
11	4.13610	4.03799	3.90477	3.70511	4.33257
12	4.04538	3.95250	3.82558	3.63408	4.22889
13	3.97084	3.88224	3.76046	3.57574	4.14366
14	3.90851	3.82348	3.70599	3.52683	4.07239
15	3.85564	3.77363	3.65977	3.48536	4.01194
16	3.81023	3.73081	3.62005	3.44967	3.96001
17	3.77082	3.69364	3.58555	3.41869	3.91495
18	3.73629	3.66106	3.55532	3.39152	3.87548
19	3.70579	3.63229	3.52861	3.36751	3.84062
20	3.67866	3.60669	3.50484	3.34614	3.80961
21	3.65437	3.58376	3.48356	3.32699	3.78185
22	3.63249	3.56311	3.46438	3.30974	3.75686
23	3.61269	3.54442	3.44702	3.29412	3.73425
24	3.59468	3.52742	3.43122	3.27991	3.71369
25	3.57823	3.51189	3.41679	3.26692	3.69491
26	3.56315	3.49765	3.40356	3.25501	3.67770
27	3.54927	3.48454	3.39138	3.24404	3.66186
28	3.53646	3.47244	3.38013	3.23391	3.64724
29	3.52459	3.46123	3.36971	3.22453	3.63371
30	3.51357	3.45082	3.36004	3.21581	3.62114
35	3.46846	3.40820	3.32042	3.18012	3.56973
40	3.43521	3.37678	3.29120	3.15378	3.53188
45	3.40969	3.35266	3.26875	3.13355	3.50284
50	3.38948	3.33355	3.25098	3.11753	3.47986
60	3.35951	3.30522	3.22461	3.09375	3.44582
80	3.32263	3.27033	3.19214	3.06445	3.40396
100	3.30080	3.24968	3.17292	3.04710	3.37922
120	3.28637	3.23603	3.16021	3.03562	3.36288
200	3.25780	3.20900	3.13504	3.01290	3.33056
∞	3.21568	3.16914	3.09790	2.97935	3.28298

TABLE B.4

Two-sided Percentage Point (h) for Equal Correlations (Case I with Correlation ρ)

p = 11, 1-α = 0.99

ν ↓	ρ					
	0.0	0.1	0.2	0.3	0.4	0.5
2	19.80705	19.71796	19.48714	19.14079	18.68581	18.11710
3	10.44827	10.40789	10.30206	10.14203	9.93077	9.66593
4	7.68949	7.66368	7.59528	7.49109	7.35282	7.17878
5	6.43034	6.41135	6.36055	6.28259	6.17852	6.04688
6	5.72121	5.70618	5.66559	5.60285	5.51858	5.41135
7	5.26972	5.25726	5.22337	5.17058	5.09921	5.00782
8	4.95828	4.94765	4.91848	4.87272	4.81043	4.73013
9	4.73104	4.72176	4.69611	4.65557	4.60002	4.52789
10	4.55819	4.54996	4.52702	4.49052	4.44014	4.37429
11	4.42244	4.41503	4.39425	4.36096	4.31470	4.25379
12	4.31309	4.30635	4.28733	4.25665	4.21372	4.15682
13	4.22318	4.21699	4.19942	4.17090	4.13073	4.07713
14	4.14797	4.14225	4.12591	4.09920	4.06134	4.01051
15	4.08416	4.07883	4.06354	4.03838	4.00249	3.95400
16	4.02936	4.02437	4.00997	3.98614	3.95195	3.90548
17	3.98179	3.97709	3.96347	3.94081	3.90808	3.86337
18	3.94012	3.93568	3.92275	3.90109	3.86966	3.82647
19	3.90332	3.89911	3.88678	3.86602	3.83573	3.79389
20	3.87059	3.86658	3.85478	3.83483	3.80555	3.76491
21	3.84129	3.83746	3.82615	3.80691	3.77853	3.73897
22	3.81491	3.81124	3.80036	3.78177	3.75420	3.71561
23	3.79104	3.78752	3.77703	3.75901	3.73219	3.69446
24	3.76934	3.76595	3.75581	3.73832	3.71217	3.67523
25	3.74953	3.74625	3.73644	3.71943	3.69388	3.65767
26	3.73136	3.72820	3.71868	3.70211	3.67712	3.64157
27	3.71465	3.71159	3.70234	3.68618	3.66170	3.62676
28	3.69923	3.69626	3.68726	3.67147	3.64746	3.61308
29	3.68495	3.68207	3.67329	3.65785	3.63428	3.60041
30	3.67170	3.66889	3.66033	3.64520	3.62204	3.58865
35	3.61749	3.61500	3.60730	3.59347	3.57196	3.54053
40	3.57759	3.57532	3.56825	3.55537	3.53506	3.50506
45	3.54699	3.54490	3.53830	3.52614	3.50675	3.47784
50	3.52280	3.52084	3.51461	3.50301	3.48434	3.45629
60	3.48696	3.48519	3.47951	3.46873	3.45113	3.42435
80	3.44294	3.44140	3.43636	3.42659	3.41027	3.38504
100	3.41694	3.41553	3.41086	3.40167	3.38610	3.36178
120	3.39977	3.39845	3.39402	3.38520	3.37014	3.34641
200	3.36585	3.36469	3.36072	3.35264	3.33854	3.31599
∞	3.31597	3.31503	3.31172	3.30469	3.29199	3.27115

TABLE B.4

Two-sided Percentage Point (h) for Equal
Correlations (*Case I with Correlation* ρ)

$p = 11$, $1-\alpha = 0.99$

$\nu \downarrow$	0.6	0.7	0.8	0.9	$1/(1+\sqrt{11})$
			ρ		
2	17.41766	16.55219	15.44648	13.90535	19.38937
3	9.33968	8.93579	8.42011	7.70250	10.25699
4	6.96366	6.69650	6.35412	5.87495	7.56601
5	5.88337	5.67921	5.41586	5.04375	6.33870
6	5.27736	5.10895	4.88994	4.57713	5.64805
7	4.89286	4.74728	4.55635	4.28072	5.20865
8	4.62839	4.49858	4.32688	4.07654	4.90575
9	4.43586	4.31754	4.15977	3.92767	4.68486
10	4.28966	4.18006	4.03281	3.81444	4.51692
11	4.17499	4.07222	3.93318	3.72556	4.38507
12	4.08270	3.98542	3.85296	3.65383	4.27888
13	4.00687	3.91408	3.78700	3.59493	4.19159
14	3.94347	3.85442	3.73182	3.54554	4.11859
15	3.88969	3.80381	3.68500	3.50368	4.05666
16	3.84350	3.76034	3.64477	3.46764	4.00347
17	3.80342	3.72260	3.60983	3.43637	3.95731
18	3.76830	3.68954	3.57922	3.40894	3.91687
19	3.73728	3.66033	3.55217	3.38470	3.88116
20	3.70969	3.63434	3.52809	3.36312	3.84939
21	3.68498	3.61107	3.50653	3.34379	3.82095
22	3.66273	3.59011	3.48711	3.32638	3.79535
23	3.64259	3.57113	3.46953	3.31061	3.77218
24	3.62428	3.55388	3.45354	3.29626	3.75112
25	3.60755	3.53811	3.43892	3.28315	3.73188
26	3.59221	3.52366	3.42552	3.27113	3.71425
27	3.57810	3.51035	3.41319	3.26006	3.69802
28	3.56507	3.49807	3.40180	3.24983	3.68305
29	3.55300	3.48669	3.39125	3.24036	3.66918
30	3.54179	3.47612	3.38145	3.23157	3.65630
35	3.49592	3.43287	3.34133	3.19554	3.60365
40	3.46211	3.40098	3.31174	3.16896	3.56486
45	3.43616	3.37650	3.28902	3.14854	3.53512
50	3.41561	3.35711	3.27102	3.13236	3.51159
60	3.38513	3.32835	3.24433	3.10836	3.47672
80	3.34763	3.29295	3.21145	3.07879	3.43385
100	3.32543	3.27200	3.19199	3.06128	3.40852
120	3.31076	3.25814	3.17912	3.04970	3.39178
200	3.28172	3.23072	3.15364	3.02676	3.35869
∞	3.23890	3.19026	3.11604	2.99291	3.30998

TABLE B.4

Two-sided Percentage Point (h) for Equal
Correlations (*Case I with Correlation* ρ)

$p = 12, \; 1-\alpha = 0.99$

	ρ					
$\nu \downarrow$	0.0	0.1	0.2	0.3	0.4	0.5
2	20.14977	20.05521	19.81210	19.44931	18.97446	18.38260
3	10.61344	10.57050	10.45880	10.29081	10.06987	9.79370
4	7.80205	7.77455	7.70223	7.59265	7.44779	7.26601
5	6.51853	6.49827	6.44448	6.36236	6.25319	6.11553
6	5.79545	5.77940	5.73637	5.67021	5.58172	5.46948
7	5.33493	5.32163	5.28566	5.22996	5.15495	5.05923
8	5.01719	5.00583	4.97486	4.92654	4.86104	4.77689
9	4.78529	4.77537	4.74813	4.70530	4.64686	4.57126
10	4.60886	4.60006	4.57569	4.53712	4.48412	4.41508
11	4.47028	4.46236	4.44028	4.40510	4.35641	4.29256
12	4.35863	4.35142	4.33122	4.29879	4.25361	4.19396
13	4.26681	4.26020	4.24155	4.21140	4.16912	4.11293
14	4.19001	4.18390	4.16655	4.13832	4.09848	4.04519
15	4.12484	4.11916	4.10291	4.07633	4.03856	3.98773
16	4.06887	4.06354	4.04826	4.02309	3.98710	3.93839
17	4.02028	4.01527	4.00082	3.97687	3.94244	3.89557
18	3.97771	3.97298	3.95926	3.93639	3.90332	3.85806
19	3.94012	3.93564	3.92256	3.90065	3.86878	3.82493
20	3.90669	3.90242	3.88991	3.86885	3.83805	3.79547
21	3.87676	3.87268	3.86069	3.84038	3.81054	3.76909
22	3.84981	3.84591	3.83438	3.81476	3.78577	3.74534
23	3.82543	3.82168	3.81057	3.79156	3.76336	3.72384
24	3.80326	3.79966	3.78892	3.77047	3.74298	3.70429
25	3.78302	3.77954	3.76915	3.75122	3.72436	3.68643
26	3.76446	3.76110	3.75103	3.73356	3.70730	3.67006
27	3.74739	3.74414	3.73435	3.71732	3.69160	3.65500
28	3.73164	3.72848	3.71896	3.70232	3.67710	3.64109
29	3.71705	3.71399	3.70471	3.68844	3.66368	3.62822
30	3.70351	3.70053	3.69148	3.67555	3.65122	3.61626
35	3.64813	3.64549	3.63737	3.62282	3.60023	3.56733
40	3.60737	3.60498	3.59753	3.58398	3.56267	3.53127
45	3.57612	3.57391	3.56697	3.55418	3.53385	3.50360
50	3.55140	3.54934	3.54279	3.53061	3.51103	3.48170
60	3.51480	3.51294	3.50697	3.49567	3.47722	3.44922
80	3.46984	3.46823	3.46295	3.45271	3.43563	3.40926
100	3.44328	3.44182	3.43694	3.42731	3.41103	3.38562
120	3.42576	3.42438	3.41976	3.41053	3.39477	3.36999
200	3.39112	3.38991	3.38579	3.37735	3.36261	3.33907
∞	3.34020	3.33923	3.33580	3.32848	3.31522	3.29348

TABLE B.4

Two-sided Percentage Point (h) for Equal
Correlations (Case I with Correlation ρ)

p = 12, 1-α = 0.99

$\nu \downarrow$	\multicolumn{5}{c}{ρ}
	0.6	0.7	0.8	0.9	$1/(1+\sqrt{12})$
2	17.65636	16.75941	15.61548	14.02378	19.73540
3	9.45431	9.03501	8.50066	7.75859	10.42337
4	7.04190	6.76419	6.40908	5.91327	7.67918
5	5.94500	5.73261	5.45931	5.07418	6.42725
6	5.32964	5.15434	4.92700	4.60323	5.72253
7	4.93919	4.78762	4.58939	4.30410	5.27404
8	4.67064	4.53546	4.35718	4.09809	4.96480
9	4.47512	4.35190	4.18809	3.94788	4.73924
10	4.32666	4.21252	4.05964	3.83365	4.56771
11	4.21022	4.10319	3.95884	3.74399	4.43303
12	4.11652	4.01520	3.87768	3.67164	4.32455
13	4.03952	3.94288	3.81096	3.61222	4.23536
14	3.97514	3.88241	3.75515	3.56241	4.16077
15	3.92054	3.83111	3.70778	3.52019	4.09749
16	3.87365	3.78705	3.66709	3.48383	4.04314
17	3.83295	3.74880	3.63176	3.45230	3.99596
18	3.79729	3.71529	3.60079	3.42463	3.95463
19	3.76580	3.68568	3.57343	3.40019	3.91813
20	3.73779	3.65934	3.54908	3.37843	3.88567
21	3.71270	3.63576	3.52728	3.35894	3.85661
22	3.69012	3.61452	3.50764	3.34137	3.83044
23	3.66967	3.59529	3.48986	3.32547	3.80676
24	3.65108	3.57780	3.47368	3.31100	3.78523
25	3.63410	3.56182	3.45890	3.29778	3.76557
26	3.61853	3.54717	3.44535	3.28565	3.74755
27	3.60420	3.53369	3.43288	3.27449	3.73097
28	3.59097	3.52124	3.42136	3.26418	3.71566
29	3.57872	3.50971	3.41069	3.25463	3.70149
30	3.56734	3.49900	3.40078	3.24576	3.68833
35	3.52078	3.45518	3.36021	3.20943	3.63451
40	3.48646	3.42286	3.33029	3.18263	3.59488
45	3.46011	3.39805	3.30731	3.16204	3.56449
50	3.43925	3.37840	3.28911	3.14572	3.54044
60	3.40832	3.34927	3.26212	3.12152	3.50481
80	3.37025	3.31339	3.22888	3.09171	3.46101
100	3.34772	3.29216	3.20920	3.07405	3.43512
120	3.33283	3.27813	3.19618	3.06238	3.41803
200	3.30336	3.25034	3.17042	3.03925	3.38422
∞	3.25989	3.20935	3.13240	3.00512	3.33447

TABLE B.4

Two-sided Percentage Point (h) for Equal
Correlations [*Case I with Correlation* ρ]

p = 13, 1-α = 0.99

ν ↓	ρ					
	0.0	0.1	0.2	0.3	0.4	0.5
2	20.46269	20.36298	20.10849	19.73051	19.23743	18.62437
3	10.76459	10.71924	10.60207	10.42670	10.19685	9.91027
4	7.90521	7.87612	7.80013	7.68555	7.53460	7.34569
5	6.59945	6.57799	6.52139	6.43541	6.32150	6.17828
6	5.86362	5.84660	5.80128	5.73193	5.63951	5.52264
7	5.39484	5.38073	5.34282	5.28439	5.20599	5.10626
8	5.07132	5.05926	5.02659	4.97587	4.90739	4.81968
9	4.83515	4.82462	4.79587	4.75090	4.68977	4.61095
10	4.65543	4.64608	4.62037	4.57986	4.52440	4.45240
11	4.51424	4.50583	4.48254	4.44557	4.39462	4.32803
12	4.40048	4.39284	4.37151	4.33743	4.29015	4.22793
13	4.30692	4.29990	4.28021	4.24853	4.20428	4.14567
14	4.22864	4.22216	4.20385	4.17419	4.13249	4.07691
15	4.16222	4.15619	4.13905	4.11112	4.07159	4.01858
16	4.10517	4.09952	4.08340	4.05695	4.01929	3.96849
17	4.05564	4.05033	4.03509	4.00994	3.97391	3.92502
18	4.01225	4.00724	3.99277	3.96875	3.93415	3.88695
19	3.97393	3.96918	3.95539	3.93238	3.89904	3.85332
20	3.93984	3.93532	3.92214	3.90003	3.86780	3.82341
21	3.90933	3.90501	3.89238	3.87106	3.83985	3.79663
22	3.88186	3.87773	3.86558	3.84499	3.81467	3.77252
23	3.85700	3.85304	3.84133	3.82139	3.79189	3.75069
24	3.83439	3.83058	3.81928	3.79993	3.77118	3.73085
25	3.81375	3.81008	3.79915	3.78034	3.75226	3.71272
26	3.79484	3.79129	3.78069	3.76237	3.73492	3.69611
27	3.77743	3.77400	3.76371	3.74584	3.71896	3.68082
28	3.76137	3.75804	3.74803	3.73059	3.70423	3.66670
29	3.74650	3.74327	3.73352	3.71646	3.69059	3.65363
30	3.73269	3.72955	3.72004	3.70334	3.67792	3.64149
35	3.67623	3.67346	3.66493	3.64969	3.62610	3.59183
40	3.63467	3.63216	3.62435	3.61017	3.58792	3.55523
45	3.60281	3.60050	3.59322	3.57986	3.55863	3.52714
50	3.57761	3.57545	3.56860	3.55587	3.53545	3.50491
60	3.54030	3.53836	3.53212	3.52032	3.50108	3.47194
80	3.49447	3.49280	3.48729	3.47661	3.45881	3.43138
100	3.46740	3.46588	3.46080	3.45077	3.43381	3.40739
120	3.44954	3.44811	3.44330	3.43370	3.41729	3.39153
200	3.41424	3.41299	3.40871	3.39994	3.38461	3.36014
∞	3.36236	3.36136	3.35782	3.35022	3.33645	3.31387

TABLE B.4

Two-sided Percentage Point (h) for Equal
Correlations ⎡*Case I with Correlation* ρ⎤

p = 13, 1-α = 0.99

ν ↓	ρ				
	0.6	0.7	0.8	0.9	1/(1+√13)
2	17.87361	16.94794	15.76915	14.13140	20.05208
3	9.55883	9.12541	8.57401	7.80961	10.57597
4	7.11331	6.82593	6.45915	5.94814	7.78313
5	6.00129	5.78133	5.49890	5.10187	6.50867
6	5.37740	5.19577	4.96077	4.62697	5.79105
7	4.98153	4.82441	4.61950	4.32538	5.33422
8	4.70924	4.56911	4.38480	4.11768	5.01916
9	4.51100	4.38326	4.21390	3.96627	4.78929
10	4.36048	4.24214	4.08409	3.85113	4.61446
11	4.24242	4.13146	3.98222	3.76075	4.47717
12	4.14741	4.04237	3.90021	3.68784	4.36657
13	4.06934	3.96916	3.83279	3.62794	4.27564
14	4.00408	3.90795	3.77640	3.57775	4.19958
15	3.94872	3.85602	3.72854	3.53520	4.13504
16	3.90118	3.81142	3.68743	3.49856	4.07961
17	3.85992	3.77270	3.65172	3.46679	4.03150
18	3.82377	3.73878	3.62044	3.43890	3.98935
19	3.79185	3.70881	3.59280	3.41427	3.95212
20	3.76345	3.68215	3.56820	3.39234	3.91901
21	3.73802	3.65828	3.54617	3.37270	3.88937
22	3.71513	3.63678	3.52633	3.35500	3.86268
23	3.69440	3.61732	3.50836	3.33898	3.83853
24	3.67555	3.59961	3.49202	3.32440	3.81657
25	3.65834	3.58345	3.47709	3.31108	3.79651
26	3.64256	3.56862	3.46340	3.29886	3.77813
27	3.62803	3.55497	3.45080	3.28761	3.76122
28	3.61462	3.54237	3.43917	3.27722	3.74560
29	3.60220	3.53070	3.42839	3.26760	3.73115
30	3.59067	3.51987	3.41838	3.25866	3.71773
35	3.54347	3.47551	3.37740	3.22206	3.66283
40	3.50868	3.44280	3.34717	3.19505	3.62241
45	3.48198	3.41770	3.32396	3.17430	3.59141
50	3.46083	3.39781	3.30558	3.15787	3.56688
60	3.42948	3.36833	3.27831	3.13349	3.53054
80	3.39090	3.33203	3.24474	3.10345	3.48588
100	3.36806	3.31054	3.22486	3.08566	3.45948
120	3.35297	3.29633	3.21172	3.07390	3.44205
200	3.32309	3.26821	3.18569	3.05060	3.40758
∞	3.27904	3.22674	3.14730	3.01622	3.35686

TABLE B.4

*Two-sided Percentage Point (h) for Equal
Correlations* (*Case I with Correlation* ρ)

p = 14, 1-α = 0.99

ν ↓	ρ					
	0.0	0.1	0.2	0.3	0.4	0.5
2	20.75037	20.64581	20.38071	19.98865	19.47870	18.84612
3	10.90383	10.85619	10.73390	10.55168	10.31356	10.01737
4	8.00038	7.96978	7.89035	7.77110	7.61449	7.41898
5	6.67417	6.65158	6.59233	6.50274	6.38442	6.23603
6	5.92661	5.90867	5.86119	5.78885	5.69276	5.57159
7	5.45023	5.43534	5.39559	5.33459	5.25304	5.14957
8	5.12138	5.10865	5.07438	5.02141	4.95012	4.85908
9	4.88126	4.87015	4.83997	4.79299	4.72933	4.64750
10	4.69851	4.68864	4.66165	4.61930	4.56154	4.48678
11	4.55492	4.54604	4.52157	4.48293	4.42985	4.36070
12	4.43920	4.43113	4.40873	4.37310	4.32384	4.25922
13	4.34401	4.33661	4.31593	4.28281	4.23671	4.17583
14	4.26438	4.25754	4.23830	4.20730	4.16385	4.10612
15	4.19679	4.19043	4.17244	4.14323	4.10205	4.04699
16	4.13874	4.13278	4.11586	4.08821	4.04897	3.99621
17	4.08833	4.08274	4.06674	4.04045	4.00291	3.95215
18	4.04418	4.03890	4.02371	3.99861	3.96256	3.91354
19	4.00518	4.00017	3.98571	3.96166	3.92693	3.87945
20	3.97048	3.96573	3.95190	3.92879	3.89523	3.84913
21	3.93943	3.93489	3.92164	3.89937	3.86686	3.82198
22	3.91147	3.90713	3.89439	3.87288	3.84131	3.79754
23	3.88617	3.88200	3.86974	3.84890	3.81819	3.77542
24	3.86316	3.85916	3.84731	3.82710	3.79717	3.75530
25	3.84215	3.83830	3.82684	3.80720	3.77797	3.73693
26	3.82290	3.81917	3.80807	3.78895	3.76036	3.72008
27	3.80518	3.80158	3.79080	3.77215	3.74417	3.70458
28	3.78883	3.78534	3.77486	3.75665	3.72922	3.69027
29	3.77369	3.77031	3.76010	3.74230	3.71537	3.67702
30	3.75964	3.75635	3.74640	3.72898	3.70252	3.66472
35	3.70217	3.69927	3.69036	3.67447	3.64993	3.61437
40	3.65987	3.65725	3.64909	3.63432	3.61118	3.57727
45	3.62744	3.62503	3.61744	3.60352	3.58146	3.54880
50	3.60180	3.59954	3.59240	3.57915	3.55793	3.52626
60	3.56382	3.56180	3.55531	3.54304	3.52305	3.49285
80	3.51718	3.51544	3.50973	3.49863	3.48016	3.45173
100	3.48964	3.48806	3.48279	3.47238	3.45479	3.42741
120	3.47146	3.46998	3.46500	3.45504	3.43802	3.41133
200	3.43554	3.43425	3.42983	3.42074	3.40485	3.37952
∞	3.38276	3.38174	3.37809	3.37024	3.35598	3.33262

TABLE B.4

Two-sided Percentage Point (h) for Equal
Correlations $\left(\text{Case I with Correlation } \rho\right)$

$p = 14, \ 1-\alpha = 0.99$

$\nu \downarrow$	ρ				
	0.6	0.7	0.8	0.9	$1/(1+\sqrt{14})$
2	18.07278	17.12073	15.90994	14.22995	20.34384
3	9.65480	9.20839	8.64130	7.85637	10.71682
4	7.17895	6.88264	6.50511	5.98010	7.87920
5	6.05305	5.82610	5.53525	5.12725	6.58398
6	5.42134	5.23385	4.99177	4.64873	5.85447
7	5.02048	4.85827	4.64714	4.34487	5.38994
8	4.74475	4.60004	4.41014	4.13564	5.06949
9	4.54401	4.41207	4.23759	3.98312	4.83565
10	4.39159	4.26936	4.10652	3.86715	4.65776
11	4.27204	4.15743	4.00368	3.77611	4.51804
12	4.17583	4.06734	3.92089	3.70268	4.40549
13	4.09678	3.99331	3.85282	3.64235	4.31292
14	4.03070	3.93141	3.79589	3.59180	4.23550
15	3.97464	3.87890	3.74758	3.54895	4.16980
16	3.92651	3.83380	3.70608	3.51205	4.11337
17	3.88473	3.79466	3.67005	3.48006	4.06438
18	3.84813	3.76035	3.63847	3.45197	4.02147
19	3.81580	3.73006	3.61056	3.42717	3.98356
20	3.78705	3.70310	3.58574	3.40509	3.94985
21	3.76131	3.67896	3.56350	3.38531	3.91967
22	3.73812	3.65723	3.54348	3.36749	3.89249
23	3.71714	3.63755	3.52534	3.35135	3.86790
24	3.69806	3.61965	3.50885	3.33667	3.84554
25	3.68063	3.60331	3.49378	3.32326	3.82512
26	3.66465	3.58831	3.47996	3.31095	3.80640
27	3.64994	3.57452	3.46725	3.29963	3.78917
28	3.63637	3.56178	3.45550	3.28917	3.77327
29	3.62379	3.54998	3.44462	3.27948	3.75855
30	3.61212	3.53903	3.43452	3.27048	3.74489
35	3.56433	3.49418	3.39316	3.23362	3.68899
40	3.52911	3.46112	3.36266	3.20643	3.64783
45	3.50208	3.43574	3.33923	3.18554	3.61626
50	3.48067	3.41564	3.32068	3.16899	3.59128
60	3.44893	3.38583	3.29316	3.14444	3.55428
80	3.40987	3.34913	3.25928	3.11420	3.50881
100	3.38675	3.32741	3.23922	3.09629	3.48194
120	3.37148	3.31305	3.22596	3.08445	3.46419
200	3.34123	3.28462	3.19969	3.06099	3.42910
∞	3.29663	3.24270	3.16095	3.02637	3.37747

TABLE B.4

Two-sided Percentage Point (h) for Equal
Correlations $\begin{pmatrix} Case\ I\ with\ Correlation\ \rho \end{pmatrix}$

p = 15, 1-α = 0.99

ν ↓	ρ					
	0.0	0.1	0.2	0.3	0.4	0.5
2	21.01642	20.90729	20.63222	20.22705	19.70146	19.05077
3	11.03282	10.98302	10.85593	10.66730	10.42149	10.11636
4	8.08867	8.05663	7.97396	7.85034	7.68845	7.48678
5	6.74356	6.71988	6.65814	6.56515	6.44271	6.28950
6	5.98514	5.96632	5.91680	5.84165	5.74212	5.61692
7	5.50172	5.48609	5.44460	5.38117	5.29665	5.18969
8	5.16793	5.15456	5.11877	5.06366	4.98974	4.89559
9	4.92415	4.91248	4.88095	4.83205	4.76602	4.68136
10	4.73858	4.72821	4.70000	4.65592	4.59598	4.51863
11	4.59275	4.58342	4.55785	4.51761	4.46253	4.39097
12	4.47521	4.46673	4.44331	4.40622	4.35509	4.28821
13	4.37852	4.37074	4.34912	4.31463	4.26677	4.20377
14	4.29761	4.29043	4.27032	4.23803	4.19293	4.13318
15	4.22895	4.22226	4.20345	4.17304	4.13029	4.07331
16	4.16995	4.16370	4.14601	4.11722	4.07649	4.02189
17	4.11874	4.11287	4.09614	4.06876	4.02980	3.97727
18	4.07386	4.06832	4.05245	4.02632	3.98891	3.93818
19	4.03423	4.02898	4.01387	3.98883	3.95279	3.90366
20	3.99897	3.99398	3.97954	3.95548	3.92066	3.87296
21	3.96741	3.96265	3.94881	3.92563	3.89190	3.84547
22	3.93899	3.93444	3.92115	3.89875	3.86600	3.82072
23	3.91328	3.90891	3.89611	3.87443	3.84257	3.79832
24	3.88989	3.88570	3.87334	3.85231	3.82126	3.77794
25	3.86854	3.86450	3.85255	3.83211	3.80179	3.75934
26	3.84897	3.84507	3.83349	3.81360	3.78395	3.74228
27	3.83096	3.82720	3.81595	3.79656	3.76754	3.72659
28	3.81434	3.81069	3.79977	3.78083	3.75238	3.71210
29	3.79896	3.79542	3.78478	3.76627	3.73835	3.69868
30	3.78467	3.78123	3.77087	3.75275	3.72532	3.68622
35	3.72626	3.72323	3.71395	3.69744	3.67201	3.63525
40	3.68326	3.68053	3.67205	3.65671	3.63274	3.59768
45	3.65030	3.64779	3.63991	3.62546	3.60260	3.56885
50	3.62424	3.62190	3.61448	3.60073	3.57876	3.54603
60	3.58564	3.58355	3.57681	3.56409	3.54341	3.51220
80	3.53824	3.53644	3.53053	3.51904	3.49993	3.47057
100	3.51025	3.50862	3.50317	3.49240	3.47421	3.44594
120	3.49178	3.49025	3.48511	3.47481	3.45722	3.42966
200	3.45528	3.45396	3.44940	3.44001	3.42360	3.39745
∞	3.40165	3.40061	3.39686	3.38877	3.37406	3.34997

TABLE B.4

*Two-sided Percentage Point (h) for Equal
Correlations* (*Case I with Correlation* ρ)

$$p = 15, \quad 1-\alpha = 0.99$$

$\nu \downarrow$	0.6	0.7	0.8	0.9	$1/(1+\sqrt{15})$
			ρ		
2	18.25655	17.28008	16.03973	14.32076	20.61415
3	9.74347	9.28502	8.70340	7.89949	10.84755
4	7.23965	6.93505	6.54755	6.00958	7.96848
5	6.10094	5.86749	5.56882	5.15067	6.65403
6	5.46200	5.26905	5.02041	4.66881	5.91349
7	5.05653	4.88956	4.67267	4.36286	5.44182
8	4.77762	4.62863	4.43355	4.15221	5.11636
9	4.57457	4.43872	4.25946	3.99866	4.87882
10	4.42038	4.29453	4.12724	3.88192	4.69808
11	4.29945	4.18144	4.02350	3.79027	4.55611
12	4.20214	4.09043	3.93998	3.71636	4.44172
13	4.12218	4.01564	3.87132	3.65563	4.34764
14	4.05533	3.95310	3.81390	3.60476	4.26894
15	3.99863	3.90006	3.76517	3.56163	4.20215
16	3.94994	3.85450	3.72331	3.52448	4.14478
17	3.90769	3.81495	3.68696	3.49230	4.09498
18	3.87067	3.78030	3.65511	3.46402	4.05135
19	3.83797	3.74969	3.62697	3.43906	4.01281
20	3.80889	3.72246	3.60193	3.41684	3.97853
21	3.78285	3.69808	3.57950	3.39693	3.94784
22	3.75940	3.67613	3.55930	3.37900	3.92021
23	3.73818	3.65625	3.54102	3.36276	3.89521
24	3.71888	3.63817	3.52438	3.34798	3.87247
25	3.70125	3.62166	3.50919	3.33449	3.85170
26	3.68509	3.60652	3.49525	3.32210	3.83267
27	3.67021	3.59258	3.48242	3.31071	3.81515
28	3.65648	3.57972	3.47058	3.30018	3.79899
29	3.64377	3.56780	3.45961	3.29043	3.78402
30	3.63196	3.55673	3.44942	3.28137	3.77012
35	3.58363	3.51144	3.40771	3.24428	3.71328
40	3.54801	3.47804	3.37695	3.21691	3.67143
45	3.52067	3.45240	3.35333	3.19590	3.63933
50	3.49902	3.43210	3.33462	3.17924	3.61393
60	3.46692	3.40199	3.30687	3.15454	3.57631
80	3.42741	3.36493	3.27270	3.12411	3.53008
100	3.40404	3.34299	3.25247	3.10609	3.50276
120	3.38858	3.32849	3.23910	3.09417	3.48472
200	3.35800	3.29978	3.21261	3.07057	3.44905
∞	3.31290	3.25744	3.17355	3.03574	3.39657

TABLE B.4

*Two-sided Percentage Point (h) for Equal
Correlations*   (*Case I with Correlation* ρ)

$$p = 16, \quad 1-\alpha = 0.99$$

ν ↓	ρ					
	0.0	0.1	0.2	0.3	0.4	0.5
2	21.26373	21.15027	20.86584	20.44839	19.90820	19.24063
3	11.15293	11.10107	10.96945	10.77481	10.52180	10.20833
4	8.17098	8.13757	8.05183	7.92410	7.75726	7.54984
5	6.80830	6.78358	6.71948	6.62329	6.49698	6.33925
6	6.03979	6.02013	5.96866	5.89086	5.78810	5.65912
7	5.54981	5.53348	5.49032	5.42461	5.33729	5.22704
8	5.21142	5.19744	5.16019	5.10307	5.02666	4.92959
9	4.96423	4.95202	4.91919	4.86849	4.80021	4.71290
10	4.77603	4.76518	4.73580	4.69007	4.62809	4.54830
11	4.62811	4.61835	4.59171	4.54996	4.49299	4.41915
12	4.50887	4.49999	4.47560	4.43710	4.38421	4.31521
13	4.41076	4.40263	4.38010	4.34431	4.29480	4.22979
14	4.32867	4.32115	4.30021	4.26669	4.22003	4.15838
15	4.25899	4.25200	4.23240	4.20084	4.15661	4.09781
16	4.19912	4.19259	4.17416	4.14428	4.10214	4.04580
17	4.14714	4.14100	4.12359	4.09517	4.05487	4.00066
18	4.10160	4.09581	4.07928	4.05216	4.01346	3.96112
19	4.06137	4.05588	4.04015	4.01417	3.97688	3.92620
20	4.02558	4.02037	4.00534	3.98038	3.94435	3.89514
21	3.99355	3.98858	3.97417	3.95012	3.91523	3.86733
22	3.96470	3.95995	3.94611	3.92289	3.88901	3.84229
23	3.93859	3.93404	3.92072	3.89823	3.86528	3.81963
24	3.91486	3.91048	3.89763	3.87582	3.84370	3.79902
25	3.89318	3.88897	3.87654	3.85535	3.82399	3.78020
26	3.87331	3.86925	3.85721	3.83658	3.80593	3.76295
27	3.85503	3.85111	3.83942	3.81931	3.78930	3.74707
28	3.83816	3.83436	3.82300	3.80338	3.77396	3.73242
29	3.82254	3.81885	3.80780	3.78862	3.75975	3.71884
30	3.80804	3.80446	3.79369	3.77491	3.74655	3.70624
35	3.74874	3.74559	3.73596	3.71886	3.69258	3.65467
40	3.70509	3.70225	3.69345	3.67757	3.65281	3.61667
45	3.67163	3.66902	3.66085	3.64590	3.62230	3.58751
50	3.64517	3.64274	3.63506	3.62084	3.59815	3.56442
60	3.60599	3.60382	3.59686	3.58371	3.56236	3.53020
80	3.55787	3.55601	3.54991	3.53805	3.51833	3.48809
100	3.52946	3.52778	3.52216	3.51105	3.49229	3.46318
120	3.51071	3.50914	3.50384	3.49322	3.47508	3.44671
200	3.47367	3.47231	3.46762	3.45795	3.44104	3.41413
∞	3.41925	3.41818	3.41434	3.40602	3.39088	3.36609

TABLE B.4

Two-sided Percentage Point (h) for Equal
Correlations (*Case I with Correlation* ρ)

p = 16, 1-α = 0.99

ν ↓	ρ 0.6	0.7	0.8	0.9	1/(1+√16)
2	18.42700	17.42786	16.16006	14.40492	20.86584
3	9.82583	9.35616	8.76103	7.93948	10.96945
4	7.29607	6.98374	6.58696	6.03693	8.05183
5	6.14548	5.90595	5.60000	5.17239	6.71948
6	5.49983	5.30178	5.04701	4.68743	5.96866
7	5.09007	4.91864	4.69639	4.37954	5.49032
8	4.80821	4.65522	4.45530	4.16757	5.16019
9	4.60300	4.46349	4.27978	4.01307	4.91919
10	4.44718	4.31793	4.14648	3.89561	4.73580
11	4.32496	4.20376	4.04190	3.80340	4.59171
12	4.22661	4.11188	3.95770	3.72906	4.47560
13	4.14580	4.03638	3.88849	3.66794	4.38010
14	4.07825	3.97326	3.83061	3.61678	4.30021
15	4.02095	3.91972	3.78149	3.57339	4.23240
16	3.97175	3.87373	3.73930	3.53601	4.17416
17	3.92904	3.83381	3.70266	3.50364	4.12359
18	3.89163	3.79883	3.67056	3.47519	4.07928
19	3.85859	3.76794	3.64219	3.45009	4.04015
20	3.82920	3.74046	3.61696	3.42773	4.00534
21	3.80289	3.71585	3.59435	3.40771	3.97417
22	3.77919	3.69368	3.57400	3.38967	3.94611
23	3.75775	3.67362	3.55556	3.37333	3.92072
24	3.73824	3.65538	3.53880	3.35847	3.89763
25	3.72043	3.63871	3.52348	3.34489	3.87654
26	3.70410	3.62343	3.50944	3.33244	3.85721
27	3.68907	3.60936	3.49651	3.32097	3.83942
28	3.67519	3.59638	3.48457	3.31038	3.82300
29	3.66234	3.58435	3.47352	3.30058	3.80780
30	3.65041	3.57318	3.46325	3.29147	3.79369
35	3.60157	3.52746	3.42121	3.25416	3.73596
40	3.56558	3.49376	3.39021	3.22663	3.69345
45	3.53795	3.46789	3.36640	3.20549	3.66085
50	3.51607	3.44740	3.34755	3.18875	3.63506
60	3.48364	3.41701	3.31959	3.16390	3.59686
80	3.44372	3.37961	3.28515	3.13329	3.54991
100	3.42010	3.35747	3.26477	3.11517	3.52216
120	3.40449	3.34283	3.25129	3.10318	3.50384
200	3.37358	3.31386	3.22460	3.07944	3.46762
∞	3.32801	3.27112	3.18523	3.04441	3.41434

BECHHOFER and DUNNETT

TABLE B.4

Two-sided Percentage Point (h) for Equal
Correlations $\quad$ (*Case I with Correlation* ρ)

$$p = 18, \quad 1-\alpha = 0.99$$

$\nu \downarrow$	ρ					
	0.0	0.1	0.2	0.3	0.4	0.5
2	21.71121	21.58969	21.28809	20.84824	20.28149	19.58333
3	11.37070	11.31500	11.17505	10.96940	10.70327	10.37462
4	8.32044	8.28448	8.19308	8.05780	7.88192	7.66400
5	6.92601	6.89935	6.83087	6.72879	6.59538	6.42940
6	6.13923	6.11799	6.06291	5.98021	5.87152	5.73561
7	5.63738	5.61971	5.57346	5.50353	5.41105	5.29478
8	5.29064	5.27551	5.23555	5.17468	5.09370	4.99124
9	5.03726	5.02402	4.98878	4.93471	4.86230	4.77010
10	4.84428	4.83252	4.80095	4.75216	4.68638	4.60210
11	4.69255	4.68197	4.65334	4.60877	4.54829	4.47028
12	4.57021	4.56059	4.53437	4.49326	4.43710	4.36418
13	4.46954	4.46071	4.43650	4.39827	4.34569	4.27698
14	4.38528	4.37713	4.35460	4.31881	4.26925	4.20409
15	4.31375	4.30617	4.28510	4.25138	4.20440	4.14226
16	4.25228	4.24520	4.22538	4.19346	4.14871	4.08916
17	4.19890	4.19225	4.17353	4.14318	4.10037	4.04308
18	4.15213	4.14586	4.12810	4.09913	4.05803	4.00271
19	4.11081	4.10488	4.08797	4.06023	4.02063	3.96706
20	4.07406	4.06842	4.05227	4.02562	3.98736	3.93535
21	4.04115	4.03578	4.02031	3.99463	3.95758	3.90696
22	4.01152	4.00638	3.99153	3.96674	3.93077	3.88141
23	3.98470	3.97978	3.96548	3.94149	3.90650	3.85827
24	3.96031	3.95559	3.94180	3.91853	3.88443	3.83724
25	3.93805	3.93350	3.92017	3.89756	3.86428	3.81802
26	3.91763	3.91325	3.90034	3.87834	3.84581	3.80041
27	3.89885	3.89462	3.88209	3.86065	3.82881	3.78420
28	3.88152	3.87742	3.86525	3.84433	3.81312	3.76924
29	3.86547	3.86150	3.84966	3.82921	3.79858	3.75538
30	3.85057	3.84671	3.83518	3.81517	3.78509	3.74252
35	3.78964	3.78625	3.77596	3.75775	3.72989	3.68988
40	3.74479	3.74174	3.73235	3.71546	3.68922	3.65108
45	3.71040	3.70762	3.69891	3.68302	3.65802	3.62131
50	3.68321	3.68063	3.67245	3.65735	3.63332	3.59775
60	3.64296	3.64066	3.63326	3.61931	3.59672	3.56281
80	3.59352	3.59156	3.58509	3.57253	3.55169	3.51983
100	3.56434	3.56257	3.55663	3.54488	3.52506	3.49439
120	3.54508	3.54343	3.53784	3.52662	3.50747	3.47759
200	3.50704	3.50562	3.50069	3.49049	3.47266	3.44432
∞	3.45115	3.45005	3.44604	3.43730	3.42136	3.39529

TABLE B.4

*Two-sided Percentage Point (h) for Equal
Correlations* (*Case I with Correlation* ρ)

p = 18, 1-α = 0.99

ν ↓	ρ				
	0.6	0.7	0.8	0.9	1/(1+√18)
2	18.73453	17.69437	16.37701	14.55657	21.32214
3	9.97467	9.48467	8.86508	8.01162	11.19091
4	7.39815	7.07177	6.65814	6.08628	8.20348
5	6.22612	5.97553	5.65634	5.21158	6.83869
6	5.56833	5.36098	5.09507	4.72102	6.06922
7	5.15083	4.97127	4.73923	4.40963	5.57878
8	4.86362	4.70332	4.49458	4.19528	5.24016
9	4.65450	4.50831	4.31649	4.03906	4.99286
10	4.49571	4.36025	4.18124	3.92031	4.80461
11	4.37117	4.24414	4.07513	3.82707	4.65667
12	4.27095	4.15070	3.98972	3.75196	4.53743
13	4.18860	4.07392	3.91951	3.69013	4.43933
14	4.11976	4.00973	3.86080	3.63846	4.35725
15	4.06137	3.95527	3.81098	3.59458	4.28758
16	4.01123	3.90851	3.76818	3.55680	4.22772
17	3.96771	3.86792	3.73102	3.52410	4.17575
18	3.92959	3.83235	3.69846	3.49533	4.13021
19	3.89593	3.80094	3.66969	3.46997	4.08998
20	3.86598	3.77299	3.64409	3.44737	4.05420
21	3.83917	3.74797	3.62117	3.42714	4.02216
22	3.81502	3.72544	3.60052	3.40890	3.99331
23	3.79317	3.70504	3.58183	3.39240	3.96720
24	3.77330	3.68649	3.56483	3.37738	3.94346
25	3.75515	3.66954	3.54930	3.36365	3.92177
26	3.73851	3.65400	3.53506	3.35107	3.90190
27	3.72319	3.63971	3.52195	3.33948	3.88361
28	3.70906	3.62650	3.50984	3.32878	3.86673
29	3.69596	3.61428	3.49863	3.31887	3.85110
30	3.68380	3.60292	3.48822	3.30967	3.83658
35	3.63405	3.55644	3.44559	3.27197	3.77722
40	3.59737	3.52218	3.41415	3.24416	3.73351
45	3.56923	3.49587	3.39001	3.22279	3.69999
50	3.54694	3.47504	3.37090	3.20587	3.67347
60	3.51390	3.44415	3.34254	3.18077	3.63419
80	3.47323	3.40613	3.30763	3.14984	3.58592
100	3.44916	3.38362	3.28696	3.13153	3.55739
120	3.43325	3.36874	3.27329	3.11942	3.53856
200	3.40177	3.33929	3.24624	3.09544	3.50133
∞	3.35534	3.29585	3.20632	3.06004	3.44657

TABLE B.4

Two-sided Percentage Point (h) for Equal
Correlations   ($Case\ I\ with\ Correlation\ \rho$)

$$p = 20,\ 1-\alpha = 0.99$$

$\nu \downarrow$	0.0	0.1	0.2	0.3	0.4	0.5
2	22.10722	21.97836	21.66132	21.20145	20.61108	19.88576
3	11.56390	11.50470	11.35721	11.14170	10.86385	10.52170
4	8.45330	8.41499	8.31846	8.17639	7.99240	7.76511
5	7.03079	7.00234	6.92987	6.82248	6.68269	6.50931
6	6.22785	6.20514	6.14677	6.05963	5.94559	5.80346
7	5.71547	5.69656	5.64749	5.57371	5.47658	5.35488
8	5.36132	5.34511	5.30267	5.23840	5.15327	5.04597
9	5.10244	5.08825	5.05079	4.99365	4.91748	4.82088
10	4.90520	4.89259	4.85901	4.80742	4.73821	4.64986
11	4.75009	4.73874	4.70827	4.66113	4.59746	4.51567
12	4.62499	4.61467	4.58675	4.54325	4.48411	4.40765
13	4.52202	4.51255	4.48676	4.44631	4.39093	4.31888
14	4.43582	4.42708	4.40309	4.36520	4.31300	4.24466
15	4.36263	4.35451	4.33207	4.29637	4.24689	4.18170
16	4.29973	4.29214	4.27104	4.23725	4.19010	4.12764
17	4.24511	4.23798	4.21804	4.18592	4.14082	4.08072
18	4.19724	4.19052	4.17161	4.14094	4.09764	4.03962
19	4.15495	4.14859	4.13058	4.10122	4.05951	4.00333
20	4.11732	4.11128	4.09409	4.06588	4.02559	3.97104
21	4.08363	4.07787	4.06141	4.03424	3.99522	3.94214
22	4.05329	4.04779	4.03199	4.00576	3.96788	3.91611
23	4.02583	4.02057	4.00536	3.97997	3.94313	3.89256
24	4.00086	3.99581	3.98114	3.95653	3.92063	3.87114
25	3.97806	3.97320	3.95903	3.93512	3.90008	3.85158
26	3.95716	3.95247	3.93875	3.91549	3.88124	3.83364
27	3.93792	3.93340	3.92009	3.89742	3.86390	3.81714
28	3.92017	3.91580	3.90287	3.88075	3.84790	3.80191
29	3.90374	3.89950	3.88693	3.86531	3.83308	3.78780
30	3.88848	3.88437	3.87212	3.85098	3.81932	3.77470
35	3.82608	3.82248	3.81156	3.79234	3.76303	3.72110
40	3.78014	3.77691	3.76697	3.74914	3.72155	3.68160
45	3.74493	3.74198	3.73277	3.71601	3.68973	3.65129
50	3.71708	3.71435	3.70571	3.68979	3.66455	3.62729
60	3.67586	3.67343	3.66563	3.65094	3.62722	3.59172
80	3.62523	3.62317	3.61637	3.60317	3.58130	3.54796
100	3.59535	3.59349	3.58726	3.57493	3.55414	3.52206
120	3.57563	3.57390	3.56804	3.55627	3.53620	3.50495
200	3.53668	3.53520	3.53005	3.51937	3.50070	3.47108
∞	3.47948	3.47834	3.47417	3.46505	3.44838	3.42116

TABLE B.4

Two-sided Percentage Point (h) for Equal
Correlations $\left(\text{Case } I \text{ with Correlation } \rho\right)$

$p = 20, \quad 1-\alpha = 0.99$

$\nu \downarrow$	ρ				
	0.6	0.7	0.8	0.9	$1/(1+\sqrt{20})$
2	19.00580	17.92937	16.56824	14.69017	21.72688
3	10.10623	9.59820	8.95694	8.07525	11.38782
4	7.48850	7.14963	6.72104	6.12981	8.33856
5	6.29754	6.03710	5.70613	5.24617	6.94501
6	5.62904	5.41339	5.13756	4.75066	6.15900
7	5.20469	5.01785	4.77711	4.43618	5.65780
8	4.91274	4.74591	4.52930	4.21973	5.31162
9	4.70016	4.54798	4.34893	4.06199	5.05871
10	4.53874	4.39772	4.21196	3.94210	4.86613
11	4.41214	4.27988	4.10451	3.84795	4.71475
12	4.31025	4.18506	4.01801	3.77215	4.59270
13	4.22654	4.10714	3.94692	3.70969	4.49228
14	4.15655	4.04200	3.88747	3.65758	4.40824
15	4.09720	3.98674	3.83702	3.61326	4.33689
16	4.04623	3.93929	3.79369	3.57513	4.27558
17	4.00199	3.89810	3.75607	3.54213	4.22235
18	3.96324	3.86201	3.72310	3.51308	4.17570
19	3.92901	3.83014	3.69398	3.48749	4.13449
20	3.89857	3.80178	3.66806	3.46468	4.09783
21	3.87131	3.77639	3.64486	3.44426	4.06500
22	3.84677	3.75353	3.62396	3.42586	4.03544
23	3.82456	3.73283	3.60503	3.40920	4.00869
24	3.80436	3.71401	3.58782	3.39404	3.98436
25	3.78591	3.69682	3.57210	3.38019	3.96214
26	3.76899	3.68105	3.55768	3.36749	3.94177
27	3.75343	3.66654	3.54441	3.35580	3.92303
28	3.73906	3.65315	3.53215	3.34500	3.90573
29	3.72575	3.64074	3.52080	3.33500	3.88971
30	3.71339	3.62922	3.51026	3.32571	3.87484
35	3.66281	3.58207	3.46711	3.28766	3.81400
40	3.62554	3.54730	3.43529	3.25960	3.76920
45	3.59692	3.52062	3.41085	3.23804	3.73485
50	3.57427	3.49949	3.39150	3.22096	3.70767
60	3.54068	3.46815	3.36280	3.19563	3.66741
80	3.49935	3.42957	3.32746	3.16442	3.61794
100	3.47489	3.40674	3.30654	3.14594	3.58871
120	3.45872	3.39165	3.29271	3.13373	3.56942
200	3.42671	3.36177	3.26533	3.10953	3.53127
∞	3.37953	3.31770	3.22493	3.07381	3.47518

TABLES C.1 to C.3

These tables give the one-sided upper equicoordinate $100(1-\alpha)$ percentage point $g_1 = g_1((p_1, p_2), v; \alpha)$ of a central p-variate Student t-distribution based on v d.f. with a correlation matrix $\underline{P} = \{\rho_{ij}\}$ where $p_1 + p_2 = p$ and

$$\rho_{ij} = \begin{cases} 1 & \text{if } i = j & (1 \le i, j \le p) \\ 1/2 & \text{if } i \ne j & (1 \le i, j \le p_1) \\ 1/2 & \text{if } i \ne j & (p_1+1 \le i, j \le p) \\ 0 & \text{if} & (1 \le i \le p_1, \ p_1+1 \le j \le p) \\ 0 & \text{if} & (p_1+1 \le i \le p, \ 1 \le j \le p_1). \end{cases}$$

That is, it is the solution of the equation

$$P\{ \max_{1 \le i \le p} T_i \le g_1 \} = \int_{-\infty}^{g_1} \cdots \int_{-\infty}^{g_1} f_v(t_1, \ldots, t_p; \underline{P}) dt_1 \cdots dt_p$$

$$= \int_0^\infty \prod_{i=1}^{2} \left[\int_{-\infty}^{+\infty} \left\{ \Phi(x_i + \sqrt{2}\, g_1 y) \right\}^{P_i} d\Phi(x_i) \right] q_v(y) dy = 1-\alpha$$

where $f_v(t_1, \ldots, t_p; \underline{P})$ is the p-variate Student t density function, $\Phi(\bullet)$ is the cdf of a standard normal r.v., and $q_v(\bullet)$ is the density function of $x_v/\sqrt{v}$, i.e.,

$$q_v(y) = \frac{2}{\Gamma(v/2)} (v/2)^{v/2} y^{v-1} \exp(-vy^2/2).$$

The constant g_1 is tabulated to 5 decimal places for

$(p_1, p_2) = (a, b)$ where $a = 1(1)4$, $b = a(1)6, 9$.

$1-\alpha = 0.80$ (Table C.1), 0.90 (Table C.2), 0.95 (Table C.3)

$v = 5(1)30(5)50, 60(20)120, 200, \infty$.

TABLE C.1

One-sided Percentage Point (g_1) for Block
Correlations $\left(\text{ Case II with }\; \rho_1=\rho_2=0.5,\; \rho_3=0.0\right)$

$1-\alpha = 0.80$

$\nu \downarrow$	(p_1,p_2)					
	$(1,1)$	$(1,2)$	$(1,3)$	$(1,4)$	$(1,5)$	$(1,6)$
5	1.41210	1.62586	1.76242	1.86227	1.94071	2.00513
6	1.38259	1.58823	1.71932	1.81508	1.89024	1.95193
7	1.36217	1.56226	1.68961	1.78256	1.85547	1.91530
8	1.34721	1.54326	1.66790	1.75880	1.83008	1.88855
9	1.33578	1.52877	1.65134	1.74069	1.81072	1.86816
10	1.32676	1.51734	1.63829	1.72642	1.79548	1.85210
11	1.31946	1.50811	1.62775	1.71489	1.78316	1.83913
12	1.31344	1.50049	1.61905	1.70539	1.77301	1.82844
13	1.30838	1.49409	1.61176	1.69741	1.76449	1.81946
14	1.30407	1.48865	1.60555	1.69063	1.75724	1.81183
15	1.30036	1.48396	1.60020	1.68479	1.75100	1.80526
16	1.29713	1.47988	1.59555	1.67970	1.74557	1.79954
17	1.29429	1.47630	1.59147	1.67524	1.74080	1.79452
18	1.29178	1.47313	1.58785	1.67128	1.73658	1.79007
19	1.28954	1.47030	1.58463	1.66776	1.73282	1.78611
20	1.28753	1.46777	1.58174	1.66460	1.72945	1.78256
21	1.28571	1.46548	1.57913	1.66175	1.72640	1.77936
22	1.28407	1.46340	1.57676	1.65917	1.72365	1.77645
23	1.28257	1.46151	1.57461	1.65682	1.72113	1.77380
24	1.28120	1.45979	1.57264	1.65467	1.71884	1.77139
25	1.27994	1.45820	1.57083	1.65269	1.71673	1.76916
26	1.27878	1.45674	1.56917	1.65087	1.71478	1.76712
27	1.27771	1.45539	1.56763	1.64919	1.71299	1.76523
28	1.27671	1.45413	1.56620	1.64763	1.71132	1.76347
29	1.27579	1.45297	1.56487	1.64618	1.70978	1.76184
30	1.27493	1.45188	1.56364	1.64483	1.70833	1.76033
35	1.27137	1.44740	1.55853	1.63926	1.70238	1.75406
40	1.26872	1.44406	1.55472	1.63510	1.69794	1.74938
45	1.26666	1.44147	1.55177	1.63187	1.69450	1.74575
50	1.26502	1.43940	1.54942	1.62930	1.69175	1.74286
60	1.26256	1.43631	1.54590	1.62546	1.68765	1.73854
80	1.25950	1.43247	1.54152	1.62068	1.68254	1.73317
100	1.25768	1.43017	1.53891	1.61782	1.67949	1.72995
120	1.25646	1.42864	1.53716	1.61592	1.67746	1.72782
200	1.25404	1.42560	1.53370	1.61213	1.67342	1.72356
∞	1.25042	1.42105	1.52852	1.60648	1.66738	1.71720

TABLE C.1

One-sided Percentage Point (g_1) for Block
Correlations $\left(\text{Case II with } \rho_1=\rho_2=0.5, \; \rho_3=0.0\right)$

$1-\alpha = 0.80$

$\nu \downarrow$	(p_1, p_2)					
	(1,9)	(2,2)	(2,3)	(2,4)	(2,5)	(2,6)
5	2.14853	1.79028	1.90103	1.98442	2.05118	2.10678
6	2.08919	1.74562	1.85157	1.93131	1.99514	2.04829
7	2.04834	1.71484	1.81750	1.89474	1.95656	2.00802
8	2.01851	1.69235	1.79261	1.86802	1.92837	1.97861
9	1.99578	1.67519	1.77363	1.84766	1.90689	1.95619
10	1.97788	1.66168	1.75868	1.83162	1.88997	1.93854
11	1.96342	1.65076	1.74661	1.81866	1.87631	1.92428
12	1.95150	1.64176	1.73665	1.80798	1.86503	1.91252
13	1.94150	1.63420	1.72829	1.79902	1.85558	1.90265
14	1.93299	1.62777	1.72118	1.79139	1.84754	1.89425
15	1.92567	1.62224	1.71506	1.78482	1.84061	1.88703
16	1.91929	1.61742	1.70974	1.77911	1.83458	1.88073
17	1.91369	1.61319	1.70506	1.77409	1.82929	1.87521
18	1.90874	1.60944	1.70092	1.76965	1.82460	1.87032
19	1.90433	1.60610	1.69723	1.76569	1.82043	1.86596
20	1.90037	1.60311	1.69392	1.76214	1.81668	1.86205
21	1.89680	1.60041	1.69093	1.75894	1.81330	1.85853
22	1.89356	1.59796	1.68823	1.75603	1.81024	1.85533
23	1.89061	1.59573	1.68576	1.75339	1.80745	1.85242
24	1.88791	1.59369	1.68351	1.75097	1.80490	1.84976
25	1.88544	1.59182	1.68144	1.74875	1.80256	1.84732
26	1.88316	1.59009	1.67953	1.74670	1.80040	1.84506
27	1.88105	1.58850	1.67777	1.74481	1.79841	1.84298
28	1.87910	1.58702	1.67614	1.74306	1.79656	1.84105
29	1.87728	1.58565	1.67462	1.74143	1.79484	1.83926
30	1.87559	1.58437	1.67320	1.73991	1.79324	1.83759
35	1.86860	1.57908	1.66736	1.73365	1.78663	1.83069
40	1.86339	1.57514	1.66300	1.72897	1.78169	1.82554
45	1.85935	1.57208	1.65962	1.72535	1.77787	1.82155
50	1.85613	1.56964	1.65693	1.72246	1.77482	1.81837
60	1.85131	1.56600	1.65290	1.71814	1.77026	1.81361
80	1.84532	1.56146	1.64789	1.71276	1.76459	1.80769
100	1.84174	1.55875	1.64490	1.70955	1.76120	1.80415
120	1.83936	1.55695	1.64290	1.70741	1.75895	1.80180
200	1.83461	1.55336	1.63893	1.70315	1.75446	1.79711
∞	1.82752	1.54800	1.63301	1.69679	1.74775	1.79011

BECHHOFER and DUNNETT

TABLE C.1

*One-sided Percentage Point (g₁) for Block
Correlations* $\left(\text{Case II with } \rho_1=\rho_2=0.5, \rho_3=0.0 \right)$

$1-\alpha = 0.80$

ν ↓	(2,9)	(3,3)	(3,4)	(3,5)	(3,6)
			(p_1,p_2)		
5	2.23278	1.99720	2.07086	2.13051	2.18062
6	2.16873	1.94334	2.01364	2.07058	2.11842
7	2.12463	1.90625	1.97424	2.02932	2.07558
8	2.09242	1.87916	1.94546	1.99918	2.04430
9	2.06787	1.85850	1.92352	1.97620	2.02045
10	2.04853	1.84224	1.90625	1.95810	2.00166
11	2.03291	1.82910	1.89229	1.94348	1.98648
12	2.02003	1.81826	1.88078	1.93142	1.97397
13	2.00922	1.80917	1.87112	1.92131	1.96347
14	2.00002	1.80143	1.86291	1.91270	1.95453
15	1.99210	1.79477	1.85583	1.90529	1.94684
16	1.98521	1.78898	1.84967	1.89884	1.94014
17	1.97916	1.78389	1.84427	1.89318	1.93426
18	1.97380	1.77938	1.83948	1.88817	1.92906
19	1.96903	1.77537	1.83522	1.88370	1.92442
20	1.96475	1.77176	1.83139	1.87969	1.92026
21	1.96088	1.76852	1.82794	1.87607	1.91650
22	1.95738	1.76557	1.82481	1.87280	1.91310
23	1.95419	1.76289	1.82196	1.86981	1.91000
24	1.95127	1.76044	1.81936	1.86708	1.90716
25	1.94860	1.75818	1.81696	1.86457	1.90456
26	1.94613	1.75611	1.81476	1.86226	1.90216
27	1.94385	1.75419	1.81272	1.86013	1.89995
28	1.94173	1.75241	1.81083	1.85815	1.89789
29	1.93977	1.75076	1.80908	1.85631	1.89598
30	1.93794	1.74922	1.80744	1.85460	1.89420
35	1.93037	1.74286	1.80069	1.84752	1.88685
40	1.92473	1.73812	1.79565	1.84224	1.88137
45	1.92036	1.73444	1.79174	1.83815	1.87712
50	1.91687	1.73151	1.78863	1.83488	1.87373
60	1.91165	1.72713	1.78397	1.83000	1.86866
80	1.90516	1.72168	1.77817	1.82393	1.86235
100	1.90129	1.71842	1.77471	1.82030	1.85858
120	1.89871	1.71625	1.77241	1.81789	1.85608
200	1.89356	1.71193	1.76781	1.81307	1.85108
∞	1.88588	1.70547	1.76096	1.80589	1.84361

TABLE C.1

*One-sided Percentage Point (g₁) for Block
Correlations* (*Case II with* $\rho_1=\rho_2=0.5,\ \rho_3=0.0$)

$1-\alpha = 0.80$

ν ↓	(p₁,p₂)				
	(3,9)	(4,4)	(4,5)	(4,6)	(4,9)
5	2.29547	2.13785	2.19255	2.23877	2.34557
6	2.22806	2.07747	2.12962	2.17369	2.27557
7	2.18165	2.03590	2.08628	2.12887	2.22735
8	2.14774	2.00554	2.05463	2.09614	2.19212
9	2.12189	1.98239	2.03050	2.07118	2.16525
10	2.10152	1.96416	2.01149	2.05152	2.14409
11	2.08507	1.94943	1.99614	2.03563	2.12698
12	2.07150	1.93728	1.98347	2.02253	2.11288
13	2.06011	1.92709	1.97285	2.01154	2.10104
14	2.05042	1.91842	1.96380	2.00218	2.09096
15	2.04208	1.91095	1.95602	1.99413	2.08229
16	2.03482	1.90446	1.94924	1.98712	2.07473
17	2.02844	1.89875	1.94329	1.98096	2.06810
18	2.02279	1.89370	1.93802	1.97551	2.06223
19	2.01776	1.88920	1.93333	1.97065	2.05699
20	2.01324	1.88516	1.92912	1.96629	2.05230
21	2.00917	1.88151	1.92532	1.96236	2.04806
22	2.00548	1.87821	1.92187	1.95880	2.04422
23	2.00212	1.87520	1.91873	1.95555	2.04072
24	1.99904	1.87245	1.91587	1.95258	2.03752
25	1.99622	1.86993	1.91323	1.94986	2.03458
26	1.99361	1.86760	1.91080	1.94734	2.03187
27	1.99121	1.86545	1.90856	1.94502	2.02937
28	1.98898	1.86346	1.90648	1.94287	2.02705
29	1.98691	1.86160	1.90455	1.94087	2.02490
30	1.98498	1.85988	1.90275	1.93900	2.02289
35	1.97700	1.85274	1.89531	1.93131	2.01459
40	1.97105	1.84742	1.88976	1.92556	2.00839
45	1.96643	1.84330	1.88545	1.92111	2.00359
50	1.96275	1.84001	1.88202	1.91755	1.99976
60	1.95725	1.83509	1.87689	1.91224	1.99403
80	1.95040	1.82897	1.87051	1.90563	1.98690
100	1.94631	1.82531	1.86669	1.90168	1.98264
120	1.94359	1.82288	1.86415	1.89906	1.97980
200	1.93816	1.81803	1.85909	1.89381	1.97415
∞	1.93005	1.81079	1.85153	1.88599	1.96571

BECHHOFER and DUNNETT

TABLE C.2

One-sided Percentage Point (g_1) for Block
Correlations $\left(\ \text{Case II with}\ \ \rho_1=\rho_2=0.5,\ \rho_3=0.0\right)$

$$1-\alpha = 0.90$$

$\nu \downarrow$	(p_1, p_2) (1,1)	(1,2)	(1,3)	(1,4)	(1,5)	(1,6)
5	1.96893	2.19849	2.34833	2.45915	2.54684	2.61922
6	1.90475	2.12001	2.26025	2.36388	2.44582	2.51343
7	1.86105	2.06671	2.20050	2.29928	2.37734	2.44174
8	1.82940	2.02816	2.15732	2.25262	2.32790	2.38998
9	1.80543	1.99901	2.12468	2.21736	2.29054	2.35088
10	1.78664	1.97619	2.09914	2.18978	2.26132	2.32029
11	1.77153	1.95784	2.07862	2.16761	2.23785	2.29572
12	1.75911	1.94277	2.06177	2.14942	2.21857	2.27555
13	1.74872	1.93017	2.04768	2.13421	2.20247	2.25870
14	1.73990	1.91948	2.03573	2.12131	2.18881	2.24441
15	1.73232	1.91030	2.02547	2.11024	2.17708	2.23213
16	1.72574	1.90233	2.01656	2.10062	2.16690	2.22148
17	1.71997	1.89534	2.00876	2.09220	2.15798	2.21214
18	1.71487	1.88917	2.00186	2.08475	2.15010	2.20390
19	1.71033	1.88368	1.99572	2.07813	2.14309	2.19656
20	1.70626	1.87876	1.99022	2.07220	2.13681	2.18999
21	1.70260	1.87432	1.98527	2.06686	2.13115	2.18407
22	1.69928	1.87031	1.98079	2.06202	2.12603	2.17871
23	1.69626	1.86666	1.97671	2.05762	2.12137	2.17383
24	1.69350	1.86332	1.97298	2.05360	2.11711	2.16938
25	1.69096	1.86026	1.96956	2.04991	2.11321	2.16529
26	1.68863	1.85744	1.96642	2.04651	2.10961	2.16153
27	1.68648	1.85484	1.96351	2.04338	2.10629	2.15806
28	1.68448	1.85243	1.96082	2.04047	2.10322	2.15484
29	1.68263	1.85019	1.95832	2.03778	2.10036	2.15186
30	1.68090	1.84810	1.95599	2.03526	2.09770	2.14907
35	1.67379	1.83950	1.94639	2.02491	2.08674	2.13760
40	1.66848	1.83310	1.93924	2.01720	2.07857	2.12906
45	1.66438	1.82814	1.93371	2.01123	2.07226	2.12246
50	1.66111	1.82420	1.92931	2.00648	2.06723	2.11719
60	1.65623	1.81830	1.92273	1.99939	2.05973	2.10934
80	1.65017	1.81099	1.91457	1.99058	2.05040	2.09959
100	1.64655	1.80662	1.90969	1.98533	2.04484	2.09377
120	1.64414	1.80372	1.90646	1.98184	2.04115	2.08991
200	1.63935	1.79794	1.90001	1.97489	2.03379	2.08221
∞	1.63222	1.78934	1.89042	1.96454	2.02284	2.07075

TABLE C.2

One-sided Percentage Point (g_1) for Block
Correlations $\left(\text{ Case II with } \rho_1=\rho_2=0.5, \rho_3=0.0\right)$

$1-\alpha = 0.90$

$\nu \downarrow$	(p_1,p_2)					
	(1,9)	(2,2)	(2,3)	(2,4)	(2,5)	(2,6)
5	2.78141	2.37521	2.49697	2.58979	2.66470	2.72743
6	2.66484	2.28461	2.39806	2.48456	2.55437	2.61285
7	2.58587	2.22317	2.33102	2.41325	2.47963	2.53522
8	2.52887	2.17878	2.28261	2.36177	2.42567	2.47919
9	2.48581	2.14523	2.24602	2.32287	2.38491	2.43686
10	2.45213	2.11898	2.21741	2.29245	2.35302	2.40376
11	2.42508	2.09788	2.19441	2.26801	2.32741	2.37716
12	2.40288	2.08056	2.17554	2.24795	2.30639	2.35533
13	2.38432	2.06609	2.15976	2.23118	2.28882	2.33709
14	2.36859	2.05381	2.14639	2.21696	2.27392	2.32162
15	2.35508	2.04326	2.13490	2.20475	2.26113	2.30834
16	2.34335	2.03411	2.12493	2.19415	2.25002	2.29681
17	2.33307	2.02609	2.11619	2.18487	2.24029	2.28670
18	2.32400	2.01900	2.10847	2.17666	2.23170	2.27777
19	2.31592	2.01269	2.10160	2.16936	2.22405	2.26983
20	2.30868	2.00705	2.09545	2.16282	2.21719	2.26272
21	2.30217	2.00196	2.08990	2.15694	2.21102	2.25631
22	2.29627	1.99735	2.08489	2.15160	2.20544	2.25051
23	2.29090	1.99316	2.08032	2.14675	2.20035	2.24523
24	2.28600	1.98933	2.07615	2.14232	2.19571	2.24041
25	2.28150	1.98582	2.07233	2.13825	2.19145	2.23599
26	2.27736	1.98258	2.06880	2.13451	2.18753	2.23192
27	2.27354	1.97960	2.06555	2.13106	2.18391	2.22816
28	2.27000	1.97683	2.06254	2.12786	2.18056	2.22467
29	2.26671	1.97426	2.05974	2.12488	2.17744	2.22144
30	2.26365	1.97187	2.05714	2.12211	2.17454	2.21843
35	2.25102	1.96201	2.04639	2.11070	2.16258	2.20601
40	2.24162	1.95466	2.03840	2.10220	2.15367	2.19676
45	2.23435	1.94898	2.03221	2.09563	2.14679	2.18961
50	2.22856	1.94446	2.02728	2.09039	2.14130	2.18391
60	2.21991	1.93770	2.01993	2.08258	2.13311	2.17541
80	2.20917	1.92931	2.01079	2.07287	2.12294	2.16485
100	2.20277	1.92431	2.00534	2.06708	2.11687	2.15855
120	2.19852	1.92098	2.00173	2.06323	2.11284	2.15436
200	2.19005	1.91436	1.99452	2.05557	2.10482	2.14603
∞	2.17743	1.90450	1.98378	2.04417	2.09287	2.13362

TABLE C.2

One-sided Percentage Point (g_1) for Block
Correlations (Case II with $\rho_1=\rho_2=0.5$, $\rho_3=0.0$)

$1-\alpha = 0.90$

			(p_1,p_2)		
$\nu \downarrow$	(2,9)	(3,3)	(3,4)	(3,5)	(3,6)
5	2.87062	2.60254	2.68446	2.75138	2.80793
6	2.74632	2.49609	2.57222	2.63446	2.68707
7	2.66213	2.42395	2.49618	2.55525	2.60520
8	2.60136	2.37188	2.44130	2.49808	2.54611
9	2.55545	2.33254	2.39983	2.45489	2.50146
10	2.51955	2.30177	2.36740	2.42111	2.46655
11	2.49071	2.27704	2.34134	2.39397	2.43850
12	2.46703	2.25675	2.31996	2.37169	2.41547
13	2.44725	2.23979	2.30208	2.35308	2.39623
14	2.43047	2.22541	2.28693	2.33729	2.37991
15	2.41606	2.21306	2.27391	2.32373	2.36590
16	2.40355	2.20234	2.26262	2.31197	2.35373
17	2.39259	2.19295	2.25272	2.30165	2.34307
18	2.38291	2.18465	2.24397	2.29254	2.33366
19	2.37429	2.17726	2.23619	2.28444	2.32528
20	2.36658	2.17065	2.22922	2.27718	2.31777
21	2.35963	2.16469	2.22294	2.27064	2.31101
22	2.35333	2.15930	2.21726	2.26472	2.30489
23	2.34761	2.15439	2.21208	2.25933	2.29932
24	2.34238	2.14991	2.20736	2.25441	2.29423
25	2.33758	2.14580	2.20303	2.24989	2.28957
26	2.33316	2.14201	2.19904	2.24574	2.28527
27	2.32908	2.13852	2.19535	2.24190	2.28130
28	2.32531	2.13528	2.19194	2.23835	2.27763
29	2.32180	2.13227	2.18877	2.23504	2.27422
30	2.31853	2.12947	2.18582	2.23197	2.27104
35	2.30505	2.11792	2.17365	2.21929	2.25793
40	2.29502	2.10933	2.16459	2.20985	2.24817
45	2.28726	2.10268	2.15758	2.20255	2.24062
50	2.28108	2.09738	2.15200	2.19674	2.23461
60	2.27185	2.08947	2.14367	2.18806	2.22564
80	2.26040	2.07965	2.13332	2.17727	2.21449
100	2.25356	2.07380	2.12714	2.17084	2.20784
120	2.24902	2.06991	2.12304	2.16657	2.20342
200	2.23997	2.06216	2.11487	2.15806	2.19462
∞	2.22650	2.05062	2.10271	2.14539	2.18152

TABLE C.2

*One-sided Percentage Point (g_1) for Block
Correlations* (*Case II with* $\rho_1=\rho_2=0.5$, $\rho_3=0.0$)

$1-\alpha = 0.90$

$\nu \downarrow$	\multicolumn{5}{c}{(p_1,p_2)}
	(3,9)	(4,4)	(4,5)	(4,6)	(4,9)
5	2.93856	2.75883	2.82011	2.87223	2.99367
6	2.80868	2.64119	2.69807	2.74649	2.85940
7	2.72069	2.56150	2.61541	2.66131	2.76843
8	2.65718	2.50398	2.55574	2.59983	2.70277
9	2.60920	2.46052	2.51065	2.55337	2.65315
10	2.57167	2.42653	2.47540	2.51704	2.61433
11	2.54152	2.39923	2.44707	2.48785	2.58315
12	2.51677	2.37681	2.42382	2.46389	2.55755
13	2.49608	2.35808	2.40439	2.44387	2.53615
14	2.47854	2.34220	2.38791	2.42688	2.51800
15	2.46347	2.32856	2.37375	2.41230	2.50242
16	2.45039	2.31672	2.36147	2.39964	2.48888
17	2.43893	2.30634	2.35070	2.38854	2.47703
18	2.42880	2.29717	2.34119	2.37874	2.46655
19	2.41979	2.28902	2.33273	2.37002	2.45723
20	2.41172	2.28171	2.32515	2.36221	2.44887
21	2.40445	2.27513	2.31832	2.35517	2.44135
22	2.39787	2.26917	2.31214	2.34880	2.43454
23	2.39188	2.26375	2.30651	2.34300	2.42834
24	2.38641	2.25880	2.30138	2.33770	2.42268
25	2.38139	2.25426	2.29666	2.33284	2.41748
26	2.37677	2.25008	2.29232	2.32837	2.41270
27	2.37250	2.24622	2.28832	2.32424	2.40829
28	2.36855	2.24264	2.28461	2.32042	2.40420
29	2.36488	2.23932	2.28116	2.31686	2.40040
30	2.36146	2.23622	2.27795	2.31355	2.39686
35	2.34736	2.22346	2.26471	2.29991	2.38226
40	2.33687	2.21397	2.25485	2.28974	2.37140
45	2.32874	2.20662	2.24723	2.28188	2.36299
50	2.32228	2.20077	2.24115	2.27562	2.35630
60	2.31262	2.19203	2.23209	2.26628	2.34630
80	2.30063	2.18118	2.22082	2.25467	2.33388
100	2.29347	2.17471	2.21411	2.24774	2.32647
120	2.28872	2.17041	2.20964	2.24314	2.32154
200	2.27925	2.16184	2.20075	2.23397	2.31174
∞	2.26514	2.14909	2.18752	2.22032	2.29713

BECHHOFER and DUNNETT

TABLE C.3

One-sided Percentage Point (g₁) for Block

Correlations $\left(\text{Case II with } \rho_1=\rho_2=0.5, \rho_3=0.0 \right)$

$1-\alpha = 0.95$

ν ↓	(p₁,p₂)					
	(1,1)	(1,2)	(1,3)	(1,4)	(1,5)	(1,6)
5	2.53197	2.78383	2.95060	3.07490	3.17373	3.25560
6	2.41699	2.64694	2.79902	2.91229	3.00231	3.07687
7	2.33998	2.55550	2.69788	2.80386	2.88807	2.95778
8	2.28487	2.49017	2.62568	2.72651	2.80658	2.87286
9	2.24349	2.44120	2.57160	2.66858	2.74557	2.80929
10	2.21130	2.40314	2.52959	2.62359	2.69820	2.75993
11	2.18555	2.37272	2.49603	2.58766	2.66037	2.72051
12	2.16448	2.34785	2.46860	2.55829	2.62946	2.68831
13	2.14693	2.32715	2.44576	2.53386	2.60373	2.66152
14	2.13208	2.30964	2.42646	2.51320	2.58199	2.63887
15	2.11936	2.29465	2.40993	2.49552	2.56338	2.61948
16	2.10833	2.28166	2.39562	2.48020	2.54726	2.60269
17	2.09869	2.27030	2.38311	2.46682	2.53317	2.58802
18	2.09018	2.26029	2.37207	2.45501	2.52075	2.57508
19	2.08262	2.25139	2.36227	2.44453	2.50971	2.56359
20	2.07586	2.24343	2.35350	2.43515	2.49985	2.55331
21	2.06978	2.23627	2.34562	2.42672	2.49097	2.54407
22	2.06427	2.22980	2.33849	2.41909	2.48295	2.53571
23	2.05927	2.22392	2.33201	2.41216	2.47566	2.52812
24	2.05470	2.21855	2.32610	2.40584	2.46901	2.52119
25	2.05052	2.21363	2.32068	2.40004	2.46291	2.51484
26	2.04667	2.20910	2.31570	2.39472	2.45730	2.50900
27	2.04312	2.20492	2.31110	2.38980	2.45213	2.50361
28	2.03983	2.20106	2.30684	2.38525	2.44734	2.49863
29	2.03677	2.19747	2.30289	2.38102	2.44290	2.49400
30	2.03393	2.19413	2.29921	2.37709	2.43876	2.48969
35	2.02223	2.18038	2.28408	2.36091	2.42174	2.47196
40	2.01353	2.17017	2.27284	2.34889	2.40910	2.45880
45	2.00681	2.16228	2.26416	2.33962	2.39934	2.44863
50	2.00147	2.15601	2.25726	2.33224	2.39157	2.44055
60	1.99350	2.14666	2.24697	2.32124	2.38000	2.42850
80	1.98362	2.13507	2.23422	2.30761	2.36567	2.41357
100	1.97773	2.12816	2.22663	2.29950	2.35713	2.40469
120	1.97383	2.12358	2.22159	2.29411	2.35147	2.39879
200	1.96606	2.11447	2.21158	2.28341	2.34021	2.38707
∞	1.95451	2.10094	2.19669	2.26750	2.32348	2.36965

TABLE C.3

One-sided Percentage Point (g_1) for Block
Correlations $\left(\text{ Case II with } \rho_1=\rho_2=0.5, \rho_3=0.0\right)$

$1-\alpha = 0.95$

$\nu \downarrow$	(p_1,p_2)					
	(1,9)	(2,2)	(2,3)	(2,4)	(2,5)	(2,6)
5	3.43983	2.97788	3.11362	3.21794	3.30258	3.37373
6	3.24459	2.82273	2.94586	3.04057	3.11746	3.18211
7	3.11456	2.71925	2.83405	2.92241	2.99416	3.05451
8	3.02186	2.64541	2.75431	2.83816	2.90627	2.96356
9	2.95248	2.59010	2.69461	2.77510	2.84048	2.89549
10	2.89862	2.54715	2.64826	2.72614	2.78942	2.84266
11	2.85561	2.51283	2.61124	2.68705	2.74864	2.80047
12	2.82047	2.48479	2.58100	2.65511	2.71534	2.76601
13	2.79124	2.46146	2.55583	2.62854	2.68762	2.73734
14	2.76653	2.44173	2.53455	2.60608	2.66420	2.71311
15	2.74538	2.42484	2.51634	2.58685	2.64415	2.69236
16	2.72707	2.41021	2.50057	2.57020	2.62679	2.67440
17	2.71106	2.39742	2.48679	2.55565	2.61161	2.65870
18	2.69694	2.38615	2.47463	2.54282	2.59824	2.64486
19	2.68441	2.37613	2.46384	2.53143	2.58635	2.63257
20	2.67320	2.36718	2.45418	2.52124	2.57573	2.62158
21	2.66312	2.35912	2.44550	2.51207	2.56618	2.61170
22	2.65400	2.35184	2.43765	2.50379	2.55754	2.60276
23	2.64572	2.34522	2.43052	2.49626	2.54968	2.59464
24	2.63816	2.33917	2.42401	2.48939	2.54252	2.58723
25	2.63124	2.33364	2.41804	2.48309	2.53596	2.58044
26	2.62487	2.32855	2.41256	2.47730	2.52992	2.57419
27	2.61899	2.32385	2.40750	2.47196	2.52435	2.56843
28	2.61355	2.31950	2.40281	2.46702	2.51920	2.56310
29	2.60851	2.31547	2.39846	2.46243	2.51441	2.55815
30	2.60381	2.31171	2.39441	2.45816	2.50996	2.55354
35	2.58447	2.29625	2.37776	2.44058	2.49164	2.53459
40	2.57012	2.28477	2.36539	2.42753	2.47803	2.52051
45	2.55903	2.27591	2.35584	2.41746	2.46752	2.50965
50	2.55021	2.26886	2.34825	2.40944	2.45917	2.50100
60	2.53707	2.25835	2.33693	2.39750	2.44672	2.48812
80	2.52079	2.24533	2.32291	2.38270	2.43129	2.47216
100	2.51110	2.23757	2.31456	2.37389	2.42210	2.46266
120	2.50467	2.23243	2.30902	2.36805	2.41601	2.45636
200	2.49188	2.22220	2.29800	2.35642	2.40389	2.44383
∞	2.47289	2.20700	2.28164	2.33916	2.38590	2.42521

TABLE C.3

One-sided Percentage Point (g_1) for Block
Correlations $\left(\text{Case II with } \rho_1=\rho_2=0.5, \ \rho_3=0.0 \right)$

$1-\alpha = 0.95$

$\nu \downarrow$	(p_1,p_2)				
	(2,9)	(3,3)	(3,4)	(3,5)	(3,6)
5	3.53690	3.23118	3.32321	3.39883	3.46299
6	3.33045	3.05207	3.13536	3.20388	3.26206
7	3.19302	2.93277	3.01027	3.07408	3.12829
8	3.09506	2.84771	2.92110	2.98156	3.03296
9	3.02176	2.78404	2.85437	2.91233	2.96161
10	2.96487	2.73462	2.80257	2.85859	2.90624
11	2.91944	2.69515	2.76121	2.81569	2.86203
12	2.88234	2.66291	2.72742	2.78064	2.82592
13	2.85146	2.63608	2.69931	2.75148	2.79587
14	2.82537	2.61341	2.67556	2.72684	2.77048
15	2.80304	2.59400	2.65522	2.70574	2.74875
16	2.78370	2.57720	2.63761	2.68748	2.72993
17	2.76680	2.56251	2.62222	2.67151	2.71348
18	2.75190	2.54956	2.60865	2.65744	2.69898
19	2.73866	2.53806	2.59660	2.64494	2.68610
20	2.72683	2.52777	2.58583	2.63376	2.67458
21	2.71618	2.51852	2.57614	2.62371	2.66422
22	2.70656	2.51016	2.56737	2.61462	2.65486
23	2.69782	2.50256	2.55941	2.60636	2.64635
24	2.68984	2.49563	2.55215	2.59883	2.63858
25	2.68253	2.48927	2.54549	2.59192	2.63147
26	2.67580	2.48343	2.53937	2.58557	2.62492
27	2.66960	2.47804	2.53372	2.57971	2.61889
28	2.66386	2.47305	2.52849	2.57429	2.61330
29	2.65853	2.46841	2.52364	2.56926	2.60811
30	2.65357	2.46410	2.51912	2.56457	2.60329
35	2.63316	2.44637	2.50054	2.54530	2.58343
40	2.61800	2.43319	2.48674	2.53098	2.56868
45	2.60630	2.42302	2.47608	2.51993	2.55729
50	2.59699	2.41493	2.46761	2.51114	2.54823
60	2.58312	2.40288	2.45498	2.49805	2.53474
80	2.56594	2.38794	2.43934	2.48182	2.51801
100	2.55571	2.37905	2.43002	2.47216	2.50806
120	2.54892	2.37315	2.42384	2.46575	2.50145
200	2.53542	2.36142	2.41155	2.45300	2.48832
∞	2.51537	2.34399	2.39330	2.43407	2.46881

TABLE C.3

One-sided Percentage Point (g_1) for Block
Correlations $\left(\text{ Case II with } \rho_1=\rho_2=0.5, \rho_3=0.0\right)$

$1-\alpha = 0.95$

$\nu \downarrow$	(p_1,p_2) (3,9)	(4,4)	(4,5)	(4,6)	(4,9)
5	3.61195	3.40667	3.47586	3.53496	3.67342
6	3.39729	3.21069	3.27324	3.32673	3.45224
7	3.25439	3.08021	3.13836	3.18812	3.30502
8	3.15256	2.98722	3.04223	3.08934	3.20010
9	3.07635	2.91762	2.97029	3.01542	3.12159
10	3.01721	2.86361	2.91446	2.95805	3.06065
11	2.96998	2.82048	2.86988	2.91225	3.01199
12	2.93141	2.78525	2.83347	2.87483	2.97225
13	2.89931	2.75594	2.80317	2.84370	2.93918
14	2.87219	2.73117	2.77757	2.81739	2.91123
15	2.84897	2.70997	2.75566	2.79487	2.88730
16	2.82887	2.69161	2.73668	2.77537	2.86659
17	2.81130	2.67556	2.72009	2.75833	2.84848
18	2.79581	2.66141	2.70547	2.74330	2.83251
19	2.78205	2.64885	2.69248	2.72996	2.81833
20	2.76974	2.63762	2.68087	2.71802	2.80565
21	2.75868	2.62751	2.67043	2.70729	2.79425
22	2.74867	2.61838	2.66099	2.69759	2.78394
23	2.73958	2.61007	2.65241	2.68877	2.77457
24	2.73128	2.60250	2.64458	2.68072	2.76602
25	2.72368	2.59556	2.63740	2.67335	2.75819
26	2.71669	2.58918	2.63080	2.66657	2.75098
27	2.71024	2.58329	2.62472	2.66032	2.74434
28	2.70427	2.57784	2.61908	2.65453	2.73818
29	2.69873	2.57278	2.61385	2.64915	2.73247
30	2.69358	2.56807	2.60899	2.64415	2.72716
35	2.67236	2.54869	2.58896	2.62357	2.70528
40	2.65660	2.53430	2.57409	2.60829	2.68904
45	2.64443	2.52320	2.56261	2.59649	2.67650
50	2.63475	2.51436	2.55347	2.58710	2.66653
60	2.62033	2.50120	2.53987	2.57312	2.65166
80	2.60246	2.48488	2.52300	2.55579	2.63324
100	2.59182	2.47517	2.51296	2.54547	2.62227
120	2.58476	2.46873	2.50630	2.53862	2.61499
200	2.57073	2.45592	2.49306	2.52501	2.60053
∞	2.54988	2.43689	2.47339	2.50479	2.57903

TABLES D.1 to D.3

These tables give a one-sided upper non-equicoordinate $100(1-\alpha)$ percentage point $g_2 = g_2((p_1, p_2), v; \alpha)$ of a central p-variate Student t-distribution based on v d.f. with a correlation matrix $\underline{P} = \{\rho_{ij}\}$ where $p_1 + p_2 = p$ and

$$\rho_{ij} = \begin{cases} 1 & \text{if } i = j & (1 \leq i, j \leq p) \\ 1/2 & \text{if } i \neq j & (1 \leq i, j \leq p_1) \\ 1/2 & \text{if } i \neq j & (p_1+1 \leq i, j \leq p) \\ 0 & \text{if } i \neq j & (1 \leq i \leq p_1, \ p_1+1 \leq j \leq p) \\ 0 & \text{if } i \neq j & (p_1+1 \leq i \leq p, \ 1 \leq j \leq p_1). \end{cases}$$

That is, it is the solution of the equation

$$P\{ \max_{1 \leq i \leq p_1} T_i \leq g_2 \sqrt{\frac{p_2+1}{p_1+1}} , \quad \max_{p_1+1 \leq i \leq p} T_i \leq g_2 \}$$

$$= \int_{-\infty}^{g_2 \sqrt{\frac{p_2+1}{p_1+1}}} \cdots \int_{-\infty}^{g_2 \sqrt{\frac{p_2+1}{p_1+1}}} \left[\int_{-\infty}^{g_2} \cdots \int_{-\infty}^{g_2} f_v(t_1, \ldots, t_p; \underline{P}) \prod_{i=p_1+1}^{p} dt_i \right] \prod_{j=1}^{p_1} dt_j$$

$$= \int_0^\infty \prod_{i=1}^2 \left[\int_{-\infty}^{+\infty} \left\{ \Phi(x_i + \sqrt{2(p_2+1)/(p_i+1)} \ g_2 y) \right\}^{P_i} d\Phi(x_i) \right] q_v(y) dy = 1-\alpha$$

where $f_v(t_1, \ldots, t_p; \underline{P})$ is the p-variate Student t density function, $\Phi(\bullet)$ is the cdf of a standard normal r.v., and $q_v(\bullet)$ is the density function of $x_v/\sqrt{v}$, i.e.,

$$q_v(y) = \frac{2}{\Gamma(v/2)} (v/2)^{v/2} y^{v-1} \exp(-vy^2/2).$$

The constant g_2 is tabulated to 5 decimal places for

$(p_1, p_2) = (a, b)$ where $a = 1(1)4, \ b = a(1)6, 9$

$1-\alpha = 0.80$ (Table D.1), 0.90 (Table D.2), 0.95 (Table D.3)

$v = 5(1)30(5)50, 60(20)120, 200, \infty.$

TABLE D.1

One-sided Non-equicoordinate Percentage Point (g_2) for Block Correlations $\left(\text{Case II with }\rho_1=\rho_2=0.5,\ \rho_3=0.0\right)$

$$1-\alpha = 0.80$$

$\nu\downarrow$	\(p_1,p_2\) (1,1)	(1,2)	(1,3)	(1,4)	(1,5)	(1,6)
5	1.41210	1.51804	1.61791	1.70845	1.78855	1.85874
6	1.38259	1.48278	1.57838	1.66573	1.74331	1.81136
7	1.36217	1.45843	1.55110	1.63629	1.71217	1.77879
8	1.34721	1.44062	1.53115	1.61477	1.68944	1.75503
9	1.33578	1.42702	1.51591	1.59835	1.67212	1.73694
10	1.32676	1.41630	1.50391	1.58541	1.65849	1.72272
11	1.31946	1.40762	1.49420	1.57497	1.64748	1.71124
12	1.31344	1.40047	1.48619	1.56634	1.63840	1.70178
13	1.30838	1.39446	1.47946	1.55911	1.63079	1.69385
14	1.30407	1.38935	1.47374	1.55296	1.62432	1.68711
15	1.30036	1.38494	1.46881	1.54765	1.61875	1.68132
16	1.29713	1.38111	1.46452	1.54304	1.61390	1.67627
17	1.29429	1.37774	1.46075	1.53899	1.60965	1.67185
18	1.29178	1.37476	1.45741	1.53540	1.60588	1.66793
19	1.28954	1.37210	1.45443	1.53221	1.60253	1.66445
20	1.28753	1.36972	1.45176	1.52934	1.59952	1.66132
21	1.28571	1.36757	1.44936	1.52675	1.59681	1.65850
22	1.28407	1.36562	1.44717	1.52441	1.59435	1.65595
23	1.28257	1.36384	1.44518	1.52227	1.59211	1.65363
24	1.28120	1.36222	1.44336	1.52032	1.59007	1.65150
25	1.27994	1.36072	1.44169	1.51853	1.58819	1.64955
26	1.27878	1.35935	1.44015	1.51688	1.58646	1.64775
27	1.27771	1.35808	1.43873	1.51535	1.58486	1.64609
28	1.27671	1.35690	1.43741	1.51393	1.58337	1.64455
29	1.27579	1.35581	1.43619	1.51262	1.58200	1.64313
30	1.27493	1.35478	1.43504	1.51139	1.58071	1.64179
35	1.27137	1.35057	1.43032	1.50633	1.57541	1.63630
40	1.26872	1.34742	1.42681	1.50255	1.57147	1.63220
45	1.26666	1.34499	1.42407	1.49962	1.56840	1.62903
50	1.26502	1.34304	1.42189	1.49729	1.56596	1.62650
60	1.26256	1.34013	1.41864	1.49380	1.56232	1.62273
80	1.25950	1.33651	1.41458	1.48946	1.55778	1.61803
100	1.25768	1.33435	1.41216	1.48686	1.55507	1.61523
120	1.25646	1.33291	1.41055	1.48514	1.55327	1.61337
200	1.25404	1.33004	1.40734	1.48169	1.54968	1.60965
∞	1.25042	1.32576	1.40254	1.47656	1.54433	1.60412

TABLE D.1

One-sided Non-equicoordinate Percentage Point (g₂) for
Block Correlations $\left(\; Case\; II\; with\quad \rho_1=\rho_2=0.5,\; \rho_3=0.0\right)$

$1-\alpha = 0.80$

ν ↓	(p₁,p₂)					
	(1,9)	(2,2)	(2,3)	(2,4)	(2,5)	(2,6)
5	2.02279	1.79028	1.79551	1.82698	1.86871	1.91372
6	1.97003	1.74562	1.74876	1.77815	1.81808	1.86156
7	1.93380	1.71484	1.71654	1.74450	1.78318	1.82563
8	1.90740	1.69235	1.69300	1.71990	1.75767	1.79938
9	1.88733	1.67519	1.67504	1.70113	1.73822	1.77936
10	1.87154	1.66168	1.66090	1.68634	1.72289	1.76360
11	1.85881	1.65076	1.64947	1.67439	1.71049	1.75086
12	1.84832	1.64176	1.64004	1.66453	1.70027	1.74036
13	1.83954	1.63420	1.63213	1.65625	1.69169	1.73154
14	1.83207	1.62777	1.62540	1.64921	1.68438	1.72405
15	1.82565	1.62224	1.61960	1.64314	1.67809	1.71759
16	1.82006	1.61742	1.61456	1.63786	1.67262	1.71197
17	1.81516	1.61319	1.61013	1.63322	1.66781	1.70703
18	1.81082	1.60944	1.60620	1.62912	1.66355	1.70266
19	1.80696	1.60610	1.60271	1.62546	1.65976	1.69877
20	1.80350	1.60311	1.59957	1.62218	1.65635	1.69528
21	1.80038	1.60041	1.59674	1.61921	1.65328	1.69213
22	1.79755	1.59796	1.59418	1.61652	1.65050	1.68927
23	1.79497	1.59573	1.59184	1.61408	1.64796	1.68667
24	1.79262	1.59369	1.58971	1.61184	1.64564	1.68429
25	1.79046	1.59182	1.58775	1.60979	1.64351	1.68211
26	1.78847	1.59009	1.58594	1.60789	1.64155	1.68010
27	1.78663	1.58850	1.58427	1.60614	1.63973	1.67824
28	1.78493	1.58702	1.58272	1.60452	1.63805	1.67651
29	1.78335	1.58565	1.58128	1.60301	1.63649	1.67491
30	1.78187	1.58437	1.57994	1.60161	1.63503	1.67342
35	1.77579	1.57908	1.57440	1.59580	1.62901	1.66725
40	1.77125	1.57514	1.57027	1.59147	1.62452	1.66265
45	1.76774	1.57208	1.56707	1.58811	1.62104	1.65908
50	1.76494	1.56964	1.56451	1.58544	1.61826	1.65624
60	1.76075	1.56600	1.56069	1.58143	1.61410	1.65199
80	1.75555	1.56146	1.55594	1.57645	1.60894	1.64670
100	1.75245	1.55875	1.55310	1.57347	1.60585	1.64354
120	1.75038	1.55695	1.55121	1.57149	1.60379	1.64143
200	1.74627	1.55336	1.54744	1.56753	1.59969	1.63724
∞	1.74013	1.54800	1.54182	1.56163	1.59357	1.63099

BECHHOFER and DUNNETT

TABLE D.1

One-sided Non-equicoordinate Percentage Point (g₂) for
Block Correlations $\left(\text{Case II with } \rho_1=\rho_2=0.5, \ \rho_3=0.0\right)$

$1-\alpha = 0.80$

ν ↓	(2,9)	(3,3)	(3,4)	(3,5)	(3,6)
			(p_1,p_2)		
5	2.04233	1.99721	1.97535	1.97931	1.99671
6	1.98653	1.94334	1.92078	1.92373	1.94008
7	1.94820	1.90625	1.88318	1.88542	1.90105
8	1.92026	1.87916	1.85572	1.85743	1.87252
9	1.89901	1.85850	1.83478	1.83607	1.85075
10	1.88230	1.84224	1.81829	1.81926	1.83361
11	1.86882	1.82910	1.80496	1.80566	1.81974
12	1.85772	1.81826	1.79397	1.79444	1.80830
13	1.84842	1.80917	1.78475	1.78503	1.79870
14	1.84052	1.80143	1.77690	1.77702	1.79053
15	1.83372	1.79477	1.77014	1.77012	1.78349
16	1.82781	1.78898	1.76426	1.76411	1.77737
17	1.82262	1.78389	1.75910	1.75884	1.77198
18	1.81804	1.77938	1.75452	1.75417	1.76722
19	1.81395	1.77537	1.75045	1.75000	1.76297
20	1.81029	1.77176	1.74679	1.74627	1.75916
21	1.80699	1.76852	1.74350	1.74290	1.75572
22	1.80400	1.76557	1.74051	1.73984	1.75260
23	1.80128	1.76289	1.73778	1.73706	1.74976
24	1.79879	1.76044	1.73529	1.73452	1.74716
25	1.79650	1.75818	1.73301	1.73218	1.74478
26	1.79440	1.75611	1.73090	1.73003	1.74258
27	1.79246	1.75419	1.72895	1.72804	1.74055
28	1.79066	1.75241	1.72715	1.72619	1.73866
29	1.78899	1.75076	1.72547	1.72448	1.73691
30	1.78743	1.74922	1.72391	1.72288	1.73528
35	1.78100	1.74286	1.71745	1.71627	1.72854
40	1.77621	1.73812	1.71263	1.71135	1.72351
45	1.77250	1.73444	1.70890	1.70753	1.71961
50	1.76955	1.73151	1.70592	1.70448	1.71649
60	1.76514	1.72713	1.70147	1.69992	1.71184
80	1.75965	1.72168	1.69593	1.69425	1.70605
100	1.75638	1.71842	1.69262	1.69086	1.70258
120	1.75420	1.71625	1.69041	1.68861	1.70028
200	1.74987	1.71193	1.68602	1.68411	1.69568
∞	1.74342	1.70547	1.67946	1.67740	1.68882

TABLE D.1

One-sided Non-equicoordinate Percentage Point (g₂) for
Block Correlations (Case II with $\rho_1=\rho_2=0.5$, $\rho_3=0.0$)

$1-\alpha = 0.80$

ν ↓	(3,9)	(4,4)	(4,5)	(4,6)	(4,9)
			(p_1,p_2)		
5	2.07862	2.13785	2.10667	2.09746	2.12977
6	2.01936	2.07747	2.04621	2.03658	2.06715
7	1.97856	2.03590	2.00458	1.99463	2.02400
8	1.94878	2.00554	1.97416	1.96398	1.99246
9	1.92608	1.98239	1.95097	1.94060	1.96840
10	1.90821	1.96416	1.93270	1.92218	1.94944
11	1.89378	1.94943	1.91794	1.90729	1.93412
12	1.88188	1.93728	1.90576	1.89500	1.92147
13	1.87191	1.92709	1.89555	1.88470	1.91086
14	1.86342	1.91842	1.88685	1.87592	1.90183
15	1.85611	1.91095	1.87937	1.86837	1.89405
16	1.84975	1.90446	1.87285	1.86179	1.88727
17	1.84417	1.89875	1.86713	1.85601	1.88132
18	1.83923	1.89370	1.86206	1.85090	1.87606
19	1.83483	1.88920	1.85755	1.84633	1.87136
20	1.83088	1.88516	1.85350	1.84225	1.86714
21	1.82732	1.88151	1.84984	1.83855	1.86334
22	1.82409	1.87821	1.84653	1.83521	1.85990
23	1.82115	1.87520	1.84351	1.83216	1.85676
24	1.81846	1.87245	1.84075	1.82937	1.85388
25	1.81600	1.86993	1.83822	1.82681	1.85125
26	1.81372	1.86760	1.83588	1.82445	1.84882
27	1.81162	1.86545	1.83373	1.82227	1.84657
28	1.80968	1.86346	1.83173	1.82025	1.84449
29	1.80787	1.86160	1.82987	1.81837	1.84255
30	1.80618	1.85988	1.82813	1.81662	1.84075
35	1.79922	1.85274	1.82098	1.80938	1.83329
40	1.79403	1.84742	1.81564	1.80398	1.82773
45	1.79001	1.84330	1.81149	1.79980	1.82342
50	1.78680	1.84001	1.80819	1.79646	1.81997
60	1.78201	1.83509	1.80326	1.79147	1.81483
80	1.77604	1.82897	1.79711	1.78525	1.80842
100	1.77248	1.82531	1.79344	1.78153	1.80459
120	1.77011	1.82288	1.79100	1.77906	1.80204
200	1.76539	1.81803	1.78612	1.77413	1.79696
∞	1.75835	1.81079	1.77885	1.76677	1.78937

TABLE D.2

One-sided Non-equicoordinate Percentage Point (g2) for
Block Correlations $\left(\text{ Case II with } \rho_1 = \rho_2 = 0.5, \ \rho_3 = 0.0\right)$

$1-\alpha = 0.90$

$\nu \downarrow$	(p_1, p_2)					
	$(1,1)$	$(1,2)$	$(1,3)$	$(1,4)$	$(1,5)$	$(1,6)$
5	1.96892	2.06281	2.17925	2.28781	2.38268	2.46455
6	1.90475	1.98960	2.09923	2.20233	2.29248	2.37008
7	1.86105	1.93986	2.04494	2.14443	2.23145	2.30619
8	1.82940	1.90388	2.00572	2.10266	2.18745	2.26014
9	1.80543	1.87666	1.97607	2.07112	2.15424	2.22540
10	1.78663	1.85534	1.95287	2.04647	2.12831	2.19826
11	1.77153	1.83821	1.93424	2.02667	2.10749	2.17648
12	1.75911	1.82413	1.91894	2.01043	2.09041	2.15862
13	1.74872	1.81236	1.90615	1.99687	2.07616	2.14371
14	1.73990	1.80238	1.89530	1.98537	2.06408	2.13107
15	1.73232	1.79380	1.88599	1.97550	2.05372	2.12023
16	1.72574	1.78635	1.87791	1.96694	2.04472	2.11082
17	1.71997	1.77982	1.87082	1.95944	2.03685	2.10258
18	1.71487	1.77405	1.86456	1.95282	2.02990	2.09531
19	1.71033	1.76891	1.85899	1.94692	2.02371	2.08884
20	1.70627	1.76431	1.85400	1.94166	2.01819	2.08304
21	1.70260	1.76017	1.84952	1.93690	2.01319	2.07783
22	1.69928	1.75642	1.84545	1.93260	2.00868	2.07311
23	1.69626	1.75300	1.84175	1.92869	2.00458	2.06882
24	1.69350	1.74988	1.83837	1.92512	2.00083	2.06490
25	1.69096	1.74702	1.83527	1.92184	1.99740	2.06131
26	1.68863	1.74439	1.83241	1.91883	1.99424	2.05800
27	1.68648	1.74195	1.82978	1.91604	1.99132	2.05495
28	1.68448	1.73970	1.82734	1.91347	1.98861	2.05212
29	1.68263	1.73760	1.82507	1.91107	1.98610	2.04949
30	1.68090	1.73565	1.82296	1.90884	1.98377	2.04704
35	1.67379	1.72761	1.81425	1.89966	1.97414	2.03697
40	1.66848	1.72164	1.80778	1.89282	1.96697	2.02948
45	1.66438	1.71699	1.80276	1.88754	1.96144	2.02367
50	1.66111	1.71330	1.79877	1.88334	1.95703	2.01906
60	1.65623	1.70779	1.79281	1.87706	1.95046	2.01218
80	1.65017	1.70094	1.78541	1.86927	1.94230	2.00363
100	1.64655	1.69686	1.78100	1.86463	1.93744	1.99854
120	1.64414	1.69414	1.77807	1.86154	1.93421	1.99516
200	1.63935	1.68874	1.77223	1.85541	1.92778	1.98843
∞	1.63223	1.68068	1.76353	1.84628	1.91823	1.97840

TABLE D.2

One-sided Non-equicoordinate Percentage Point (g₂) for
Block Correlations $\left(\right.$ *Case II with* $\rho_1=\rho_2=0.5, \ \rho_3=0.0\left.\right)$

$1-\alpha = 0.90$

ν ↓	(p₁,p₂)					
	(1,9)	(2,2)	(2,3)	(2,4)	(2,5)	(2,6)
5	2.65235	2.37520	2.36426	2.40060	2.45334	2.51010
6	2.54714	2.28461	2.27100	2.30444	2.35448	2.40869
7	2.47596	2.22317	2.20777	2.23929	2.28755	2.34010
8	2.42463	2.17878	2.16210	2.19225	2.23927	2.29066
9	2.38589	2.14523	2.12759	2.15671	2.20281	2.25335
10	2.35560	2.11899	2.10059	2.12893	2.17430	2.22423
11	2.33130	2.09788	2.07890	2.10659	2.15143	2.20083
12	2.31135	2.08056	2.06109	2.08826	2.13266	2.18165
13	2.29469	2.06609	2.04620	2.07294	2.11698	2.16564
14	2.28056	2.05381	2.03358	2.05995	2.10368	2.15207
15	2.26844	2.04326	2.02274	2.04880	2.09227	2.14043
16	2.25791	2.03411	2.01332	2.03912	2.08236	2.13033
17	2.24870	2.02609	2.00508	2.03063	2.07369	2.12148
18	2.24055	2.01900	1.99779	2.02314	2.06603	2.11367
19	2.23331	2.01269	1.99130	2.01647	2.05921	2.10673
20	2.22683	2.00705	1.98548	2.01051	2.05310	2.10051
21	2.22098	2.00196	1.98027	2.00512	2.04761	2.09491
22	2.21569	1.99735	1.97553	2.00025	2.04263	2.08985
23	2.21088	1.99316	1.97122	1.99582	2.03811	2.08524
24	2.20649	1.98933	1.96728	1.99177	2.03397	2.08103
25	2.20246	1.98582	1.96367	1.98805	2.03018	2.07717
26	2.19875	1.98258	1.96035	1.98464	2.02669	2.07362
27	2.19532	1.97960	1.95728	1.98148	2.02347	2.07034
28	2.19215	1.97683	1.95443	1.97856	2.02049	2.06731
29	2.18920	1.97426	1.95179	1.97584	2.01771	2.06449
30	2.18646	1.97187	1.94933	1.97331	2.01513	2.06186
35	2.17514	1.96201	1.93919	1.96289	2.00450	2.05105
40	2.16672	1.95466	1.93165	1.95512	1.99658	2.04301
45	2.16021	1.94898	1.92579	1.94912	1.99046	2.03679
50	2.15502	1.94446	1.92114	1.94434	1.98559	2.03184
60	2.14728	1.93770	1.91420	1.93720	1.97832	2.02446
80	2.13767	1.92931	1.90557	1.92834	1.96929	2.01530
100	2.13194	1.92431	1.90042	1.92305	1.96391	2.00984
120	2.12813	1.92098	1.89700	1.91954	1.96033	2.00622
200	2.12054	1.91436	1.89020	1.91254	1.95322	1.99901
∞	2.10926	1.90450	1.88004	1.90213	1.94263	1.98827

TABLE D.2

One-sided Non-equicoordinate Percentage Point (g2) for
Block Correlations (*Case II with* $\rho_1=\rho_2=0.5$, $\rho_3=0.0$)

$1-\alpha = 0.90$

$\nu \downarrow$	(p_1,p_2)				
	(2,9)	(3,3)	(3,4)	(3,5)	(3,6)
5	2.66676	2.60254	2.56443	2.56749	2.59055
6	2.55811	2.49609	2.45753	2.45948	2.48122
7	2.48472	2.42395	2.38510	2.38632	2.40720
8	2.43187	2.37188	2.33281	2.33350	2.35379
9	2.39202	2.33254	2.29330	2.29360	2.31345
10	2.36092	2.30176	2.26239	2.26239	2.28190
11	2.33598	2.27704	2.23758	2.23733	2.25659
12	2.31553	2.25675	2.21720	2.21675	2.23580
13	2.29846	2.23979	2.20017	2.19956	2.21844
14	2.28400	2.22541	2.18572	2.18497	2.20371
15	2.27160	2.21306	2.17332	2.17245	2.19107
16	2.26084	2.20234	2.16256	2.16158	2.18009
17	2.25143	2.19295	2.15312	2.15205	2.17048
18	2.24311	2.18465	2.14479	2.14364	2.16199
19	2.23572	2.17726	2.13737	2.13615	2.15443
20	2.22911	2.17064	2.13074	2.12946	2.14767
21	2.22315	2.16469	2.12474	2.12340	2.14157
22	2.21775	2.15930	2.11933	2.11793	2.13606
23	2.21285	2.15439	2.11440	2.11295	2.13104
24	2.20838	2.14991	2.10989	2.10841	2.12645
25	2.20427	2.14580	2.10576	2.10424	2.12225
26	2.20050	2.14201	2.10196	2.10040	2.11838
27	2.19701	2.13852	2.09845	2.09685	2.11480
28	2.19378	2.13528	2.09520	2.09357	2.11149
29	2.19078	2.13227	2.09218	2.09052	2.10842
30	2.18799	2.12947	2.08936	2.08768	2.10555
35	2.17649	2.11792	2.07776	2.07597	2.09375
40	2.16793	2.10933	2.06911	2.06724	2.08495
45	2.16133	2.10268	2.06245	2.06050	2.07817
50	2.15607	2.09738	2.05712	2.05513	2.07276
60	2.14822	2.08947	2.04918	2.04711	2.06468
80	2.13848	2.07965	2.03931	2.03715	2.05465
100	2.13268	2.07380	2.03343	2.03121	2.04867
120	2.12883	2.06991	2.02952	2.02726	2.04469
200	2.12116	2.06216	2.02173	2.01940	2.03678
∞	2.10976	2.05062	2.01014	2.00770	2.02502

TABLE D.2

One-sided Non-equicoordinate Percentage Point (g₂) for Block Correlations (*Case II with* $\rho_1 = \rho_2 = 0.5$, $\rho_3 = 0.0$)

$1 - \alpha = 0.90$

			(p_1, p_2)		
$\nu \downarrow$	(3,9)	(4,4)	(4,5)	(4,6)	(4,9)
5	2.69725	2.75882	2.71228	2.69925	2.74517
6	2.58360	2.64119	2.59514	2.58196	2.62587
7	2.50681	2.56150	2.51580	2.50251	2.54517
8	2.45150	2.50398	2.45853	2.44516	2.48699
9	2.40979	2.46052	2.41526	2.40183	2.44307
10	2.37721	2.42654	2.38140	2.36793	2.40876
11	2.35111	2.39923	2.35423	2.34072	2.38121
12	2.32970	2.37681	2.33191	2.31837	2.35861
13	2.31183	2.35808	2.31326	2.29969	2.33974
14	2.29669	2.34220	2.29744	2.28385	2.32375
15	2.28370	2.32856	2.28385	2.27024	2.31002
16	2.27244	2.31672	2.27206	2.25843	2.29811
17	2.26258	2.30634	2.26173	2.24808	2.28768
18	2.25387	2.29717	2.25259	2.23894	2.27846
19	2.24613	2.28902	2.24447	2.23081	2.27027
20	2.23921	2.28170	2.23720	2.22353	2.26292
21	2.23297	2.27513	2.23064	2.21695	2.25632
22	2.22733	2.26917	2.22470	2.21101	2.25034
23	2.22220	2.26375	2.21930	2.20560	2.24490
24	2.21751	2.25880	2.21437	2.20066	2.23994
25	2.21322	2.25426	2.20985	2.19613	2.23538
26	2.20927	2.25008	2.20568	2.19196	2.23119
27	2.20562	2.24622	2.20183	2.18811	2.22732
28	2.20224	2.24264	2.19827	2.18454	2.22373
29	2.19910	2.23932	2.19496	2.18122	2.22040
30	2.19618	2.23622	2.19188	2.17813	2.21730
35	2.18415	2.22346	2.17917	2.16540	2.20453
40	2.17520	2.21398	2.16969	2.15592	2.19503
45	2.16830	2.20662	2.16238	2.14859	2.18768
50	2.16280	2.20077	2.15655	2.14275	2.18183
60	2.15460	2.19203	2.14784	2.13403	2.17310
80	2.14443	2.18118	2.13703	2.12320	2.16227
100	2.13837	2.17471	2.13058	2.11674	2.15582
120	2.13435	2.17041	2.12629	2.11245	2.15153
200	2.12634	2.16184	2.11776	2.10390	2.14300
∞	2.11446	2.14909	2.10505	2.09116	2.13032

TABLE D.3

One-sided Non-equicoordinate Percentage Point (g₂) for
Block Correlations $\left(\ \textit{Case II with}\quad \rho_1 = \rho_2 = 0.5,\ \rho_3 = 0.0 \right)$

$1-\alpha = 0.95$

$\nu \downarrow$	(p_1, p_2)					
	$(1,1)$	$(1,2)$	$(1,3)$	$(1,4)$	$(1,5)$	$(1,6)$
5	2.53198	2.62136	2.75760	2.88489	2.99481	3.08875
6	2.41699	2.49388	2.62005	2.73863	2.84070	2.92755
7	2.33998	2.40875	2.52836	2.64125	2.73814	2.82026
8	2.28487	2.34793	2.46297	2.57187	2.66508	2.74382
9	2.24349	2.30235	2.41402	2.51996	2.61043	2.68664
10	2.21132	2.26691	2.37600	2.47968	2.56802	2.64227
11	2.18555	2.23861	2.34568	2.44756	2.53421	2.60685
12	2.16448	2.21547	2.32089	2.42133	2.50658	2.57793
13	2.14693	2.19620	2.30027	2.39951	2.48361	2.55387
14	2.13208	2.17990	2.28285	2.38108	2.46420	2.53354
15	2.11936	2.16595	2.26793	2.36530	2.44760	2.51614
16	2.10833	2.15386	2.25502	2.35165	2.43322	2.50108
17	2.09869	2.14329	2.24374	2.33972	2.42066	2.48792
18	2.09018	2.13398	2.23379	2.32921	2.40959	2.47632
19	2.08262	2.12570	2.22496	2.31988	2.39976	2.46601
20	2.07584	2.11831	2.21708	2.31152	2.39096	2.45679
21	2.06978	2.11163	2.20996	2.30403	2.38307	2.44851
22	2.06427	2.10561	2.20354	2.29725	2.37592	2.44102
23	2.05927	2.10013	2.19770	2.29109	2.36943	2.43421
24	2.05470	2.09514	2.19238	2.28547	2.36351	2.42800
25	2.05052	2.09056	2.18750	2.28032	2.35808	2.42231
26	2.04667	2.08635	2.18302	2.27558	2.35310	2.41708
27	2.04312	2.08246	2.17888	2.27122	2.34849	2.41225
28	2.03983	2.07887	2.17505	2.26717	2.34424	2.40779
29	2.03677	2.07553	2.17150	2.26342	2.34028	2.40364
30	2.03393	2.07242	2.16819	2.25993	2.33660	2.39978
35	2.02223	2.05963	2.15459	2.24558	2.32147	2.38390
40	2.01355	2.05013	2.14449	2.23493	2.31025	2.37209
45	2.00681	2.04279	2.13669	2.22669	2.30157	2.36300
50	2.00147	2.03696	2.13049	2.22016	2.29467	2.35576
60	1.99350	2.02826	2.12125	2.21042	2.28440	2.34497
80	1.98362	2.01748	2.10981	2.19835	2.27167	2.33160
100	1.97773	2.01106	2.10300	2.19118	2.26410	2.32364
120	1.97383	2.00680	2.09849	2.18642	2.25908	2.31836
200	1.96606	1.99833	2.08951	2.17695	2.24909	2.30786
∞	1.95450	1.98572	2.07616	2.16290	2.23427	2.29225

TABLE D.3

One-sided Non-equicoordinate Percentage Point (g_2) for Block Correlations $\Big($ *Case II with* $\rho_1=\rho_2=0.5,\ \rho_3=0.0$ $\Big)$

$1-\alpha = 0.95$

$\nu \downarrow$	(p_1,p_2)					
	(1,9)	(2,2)	(2,3)	(2,4)	(2,5)	(2,6)
5	3.30251	2.97787	2.95354	2.99690	3.06155	3.13023
6	3.12384	2.82273	2.79550	2.83510	2.89574	2.96024
7	3.00485	2.71925	2.69019	2.72739	2.78547	2.84728
8	2.92001	2.64541	2.61509	2.65065	2.70697	2.76691
9	2.85651	2.59010	2.55886	2.59323	2.64828	2.70685
10	2.80722	2.54715	2.51523	2.54869	2.60279	2.66030
11	2.76784	2.51283	2.48036	2.51313	2.56648	2.62319
12	2.73568	2.48479	2.45188	2.48409	2.53686	2.59290
13	2.70891	2.46146	2.42818	2.45994	2.51222	2.56772
14	2.68628	2.44173	2.40815	2.43954	2.49142	2.54647
15	2.66691	2.42484	2.39101	2.42207	2.47363	2.52829
16	2.65014	2.41021	2.37616	2.40696	2.45823	2.51256
17	2.63547	2.39742	2.36318	2.39375	2.44478	2.49882
18	2.62254	2.38615	2.35174	2.38210	2.43292	2.48671
19	2.61106	2.37613	2.34158	2.37176	2.42240	2.47597
20	2.60079	2.36716	2.33251	2.36251	2.41301	2.46635
21	2.59155	2.35912	2.32432	2.35421	2.40454	2.45773
22	2.58320	2.35184	2.31693	2.34669	2.39689	2.44993
23	2.57560	2.34522	2.31022	2.33987	2.38995	2.44284
24	2.56868	2.33917	2.30409	2.33364	2.38362	2.43638
25	2.56233	2.33364	2.29848	2.32793	2.37782	2.43046
26	2.55649	2.32855	2.29331	2.32268	2.37248	2.42502
27	2.55110	2.32385	2.28855	2.31784	2.36757	2.42000
28	2.54612	2.31950	2.28414	2.31336	2.36301	2.41535
29	2.54149	2.31547	2.28005	2.30920	2.35879	2.41104
30	2.53718	2.31171	2.27624	2.30534	2.35486	2.40703
35	2.51945	2.29625	2.26057	2.28942	2.33869	2.39054
40	2.50629	2.28475	2.24893	2.27758	2.32672	2.37829
45	2.49611	2.27591	2.23994	2.26848	2.31745	2.36887
50	2.48802	2.26886	2.23279	2.26123	2.31009	2.36137
60	2.47596	2.25835	2.22214	2.25043	2.29914	2.35020
80	2.46101	2.24533	2.20894	2.23705	2.28557	2.33637
100	2.45212	2.23757	2.20109	2.22908	2.27750	2.32815
120	2.44621	2.23243	2.19588	2.22380	2.27215	2.32269
200	2.43447	2.22220	2.18551	2.21330	2.26152	2.31186
∞	2.41704	2.20702	2.17009	2.19771	2.24576	2.29577

TABLE D.3

One-sided Non-equicoordinate Percentage Point (g₂) for
Block Correlations (Case II with $\rho_1=\rho_2=0.5$, $\rho_3=0.0$)

$1-\alpha = 0.95$

ν ↓	(2,9)	(3,3)	(3,4)	(3,5)	(3,6)
5	3.31484	3.23119	3.17806	3.18162	3.21123
6	3.13238	3.05207	2.99923	3.00175	3.02964
7	3.01114	2.93277	2.88016	2.88206	2.90892
8	2.92486	2.84771	2.79529	2.79679	2.82297
9	2.86037	2.78404	2.73178	2.73301	2.75871
10	2.81038	2.73463	2.68247	2.68350	2.70890
11	2.77048	2.69515	2.64313	2.64402	2.66913
12	2.73792	2.66291	2.61098	2.61177	2.63669
13	2.71084	2.63608	2.58423	2.58494	2.60971
14	2.68798	2.61341	2.56163	2.56227	2.58692
15	2.66841	2.59400	2.54228	2.54287	2.56742
16	2.65148	2.57720	2.52553	2.52608	2.55055
17	2.63668	2.56251	2.51089	2.51140	2.53581
18	2.62364	2.54956	2.49798	2.49847	2.52282
19	2.61206	2.53806	2.48652	2.48698	2.51129
20	2.60171	2.52779	2.47624	2.47669	2.50098
21	2.59240	2.51852	2.46705	2.46747	2.49171
22	2.58399	2.51016	2.45871	2.45913	2.48334
23	2.57634	2.50256	2.45114	2.45154	2.47573
24	2.56937	2.49563	2.44423	2.44462	2.46879
25	2.56298	2.48927	2.43789	2.43828	2.46243
26	2.55710	2.48343	2.43207	2.43245	2.45658
27	2.55168	2.47804	2.42670	2.42707	2.45119
28	2.54667	2.47305	2.42173	2.42209	2.44620
29	2.54201	2.46841	2.41711	2.41747	2.44156
30	2.53768	2.46410	2.41281	2.41317	2.43725
35	2.51986	2.44637	2.39514	2.39548	2.41953
40	2.50664	2.43321	2.38199	2.38232	2.40639
45	2.49641	2.42302	2.37188	2.37221	2.39623
50	2.48829	2.41493	2.36382	2.36415	2.38816
60	2.47619	2.40288	2.35181	2.35214	2.37614
80	2.46120	2.38794	2.33693	2.33726	2.36126
100	2.45227	2.37905	2.32808	2.32840	2.35241
120	2.44635	2.37315	2.32220	2.32253	2.34654
200	2.43458	2.36142	2.31051	2.31085	2.33488
∞	2.41712	2.34400	2.29314	2.29350	2.31759

TABLE D.3

One-sided Non-equicoordinate Percentage Point (g_2) for
Block Correlations (*Case II with* $\rho_1=\rho_2=0.5$, $\rho_3=0.0$)

$1-\alpha = 0.95$

$\nu \downarrow$	(p_1,p_2) (3,9)	(4,4)	(4,5)	(4,6)	(4,9)
5	3.34289	3.40668	3.34530	3.32938	3.39013
6	3.15417	3.21069	3.15091	3.13531	3.19349
7	3.02887	3.08021	3.02152	3.00619	3.06287
8	2.93977	2.98722	2.92931	2.91421	2.96997
9	2.87323	2.91762	2.86031	2.84540	2.90056
10	2.82169	2.86360	2.80678	2.79200	2.84678
11	2.78057	2.82048	2.76402	2.74940	2.80389
12	2.74704	2.78525	2.72910	2.71460	2.76890
13	2.71918	2.75594	2.70005	2.68565	2.73982
14	2.69566	2.73117	2.67550	2.66119	2.71527
15	2.67555	2.70997	2.65448	2.64025	2.69428
16	2.65815	2.69161	2.63629	2.62213	2.67611
17	2.64295	2.67556	2.62038	2.60629	2.66025
18	2.62956	2.66141	2.60636	2.59233	2.64627
19	2.61768	2.64885	2.59391	2.57993	2.63386
20	2.60707	2.63763	2.58276	2.56882	2.62280
21	2.59752	2.62751	2.57277	2.55887	2.61282
22	2.58889	2.61838	2.56371	2.54986	2.60381
23	2.58106	2.61007	2.55549	2.54167	2.59563
24	2.57391	2.60250	2.54798	2.53420	2.58817
25	2.56737	2.59556	2.54110	2.52735	2.58134
26	2.56135	2.58918	2.53478	2.52106	2.57506
27	2.55580	2.58329	2.52894	2.51525	2.56927
28	2.55067	2.57784	2.52354	2.50988	2.56391
29	2.54590	2.57278	2.51853	2.50489	2.55894
30	2.54147	2.56807	2.51386	2.50024	2.55431
35	2.52325	2.54869	2.49467	2.48114	2.53529
40	2.50974	2.53432	2.48038	2.46694	2.52120
45	2.49930	2.52320	2.46941	2.45601	2.51030
50	2.49101	2.51436	2.46065	2.44731	2.50165
60	2.47867	2.50120	2.44761	2.43433	2.48877
80	2.46339	2.48488	2.43145	2.41826	2.47283
100	2.45430	2.47517	2.42182	2.40870	2.46336
120	2.44827	2.46873	2.41544	2.40235	2.45707
200	2.43630	2.45592	2.40275	2.38974	2.44459
∞	2.41856	2.43690	2.38387	2.37098	2.42611